PRINCIPLES
of
METEOROLOGICAL
ANALYSIS

PRINCIPLES
of
METEOROLOGICAL
ANALYSIS

by

WALTER J. SAUCIER

Professor Emeritus of Meteorology
North Carolina State University, Raleigh

DOVER PUBLICATIONS, INC., *New York*

Published in Canada by General Publishing Company, Ltd., 30 Lesmill
Road, Don Mills, Toronto, Ontario.
Published in the United Kingdom by Constable and Company, Ltd., 10
Orange Street, London WC2H 7EG.

This Dover edition, first published in 1989, is an unabridged, slightly corrected republication of the work first published by The University of Chicago
Press, Chicago, 1955.

Manufactured in the United States of America
Dover Publications, Inc., 31 East 2nd Street, Mineola, N.Y. 11501

Library of Congress Cataloging-in-Publication Data

Saucier, Walter J.
    Principles of meteorological analysis / by Walter J. Saucier.
        p.   cm.
    Includes bibliographies and index.
    ISBN 0-486-65979-8
    1. Meteorology—Methodology. 2. Numerical analysis. I. Title.
QC874.S28   1989
551.5—dc19   88-37536
                                                                    CIP

Preface

The teaching of weather analysis has long been handled quite apart from other basic courses in meteorology. Weather analysis, though in this respect short of analytical, has maintained a dominant character of empiricism and art in its evolution from synoptic weather observing and geographic weather representation. The more academic subjects—statics, thermodynamics, mechanics, physical processes and phenomena, and others—contributed the logic and theory in meteorological education but, much like the pure physics from which they derived, in a more reserved approach with little application. Too great a distinction between these two broad phases of training has led students to overemphasize one at the expense and criticism of the other. It is believed that both science and profession of meteorology have been retarded by the lack of integration. While the solution to such a problem lies beyond the scope of a textbook, and the proper proportions of each discipline is the subject of debate, an attempt is made here to narrow that gap.

Still another drawback in the approach to analysis stems from its historical development. There has been a continued separation of surface analysis, which dates to the earliest networks of simultaneous weather observation, from systematic co-ordinated analysis with the free atmosphere above which has become possible only in the last few decades. Of course, the atmosphere obeys no such division. It is a fluid system in motion which is greatly affected by the earth's surface and, in turn, which produces its own pronounced effects on the surface and in weather observed at the surface. If we must divide the atmosphere for purposes of daily analysis, then division appears more appropriate between the region of plentiful data and the high levels with scant daily observations.

Both ideas for improvement have been part of the Chicago philosophy in training. The viewpoint of analyzing local conditions and events in their broad three-dimensional settings was part of the Chicago school of meteorology since its beginning by Professor C.-G. Rossby. Bringing practice and scientific subjects together in the training program was a slow process made complex by the usual problems of curriculum plus the need for reorganizing courses, creating new ones, and trying to adapt existing textbooks. With the preliminary change in curriculum in 1948 came the origin of this volume (another change occurred while the book was in press). The new approach included the placing of work in weather analysis and introductory forecasting after basic courses in climatology, observations, statics, thermodynamics, and motions. Then analysis could proceed more rapidly and more thoroughly as the mode of extending and integrating basic subject matter in terms of the living atmosphere.

This book developed from the new course outline in analysis. Early in the preparation of the material for publication it became evident that existing references were short on fundamentals of method, procedure, and principle in weather analysis and that they were far behind the times in describing analysis except perhaps at the surface. For those reasons the text was lengthened for more detail on underlying principles.

In organizing the material into logical written form, it was necessary to modify the sequence radically from the order which seemed most suitable in the classroom. Most of the material in Chapters 6 and 7, for example, occurred earlier in the course work than might appear from the order in this text. While cross-section and constant-pressure analysis were being covered, so were certain elements of scalar analysis, graphical analysis, hydrostatics, and stability. The text can be followed in the order given, but it is flexible, and some of the material will appear less abstract if the order and emphasis are adjusted somewhat. In fact, much of the elementary kinematic analysis, appearing in Chapter 10, could be placed near the beginning.

A glance at the distribution of emphasis in this book might give the impression that it stresses certain apparently obscure points while briefly surveying topics described at length in the common meteorology textbooks. That peculiar distribution of emphasis was intentional, particularly in later chapters (7, 8, 9, 11, 12, and 13); but teaching should not follow such emphasis blindly. Parallel study is necessary. Because standard textbooks in the bordering or overlapping phases of meteorology go into as much detail on certain subjects of analysis as would be appropriate here, it was felt that much repetition could be sacrificed for inclusion of more detailed discussions on the useful topics they lacked. Mostly in Chapter 3, where the basic material on hydrostatics is given in detail, is there prominent repetition of classical textbook material. On the other hand, many pertinent discussions in physical meteorology were either shortened or omitted, assuming they would be covered thoroughly in a parallel course.

The presentation is intended for centering at intermediate levels of meteorology training as given at the leading institutions in this country. It is based on at least an elementary knowledge of climatology, observations, statics and thermodynamics, physical processes (involving radiation, change of water phase, turbulence, convection, etc.), and atmospheric motions. Of these, some of the last-named can be studied concurrently. It is also based on knowledge of differential calculus, in particular partial differentials, and on certain rudiments of vector algebra. Although the subject matter is based primarily on the courses in weather analysis as then given at the University of Chicago, material has been added and reoriented, mostly at a lower level but some more advanced, to make the text more comprehensive in its subject.

A list of reading references appears after each chapter to supplement the text, and many references are found in footnotes. Problems and exercises are included with the earlier chapters, where such can be used to advantage, but less are given thereafter when all or most of the work involves daily weather charts. But in no case should this book be considered a substitute for the most important feature of training in weather analysis—that of laboratory practice.

Without the contribution, assistance, and encouragement of a large number of associates this book never would have materialized. Most of the credit goes to the students whose scientific curiosity was the force and whose patience gave life to experiment. Professor H. R. Byers, through his interest and ideas in meteorological education, led to the approach embodied here. The book was his suggestion, and its progress was aided by his constant encouragement. Work on the later portions and editings of the manuscript were done at Texas A. and M. College with equal support by Professor D. F. Leipper and his staff of the Department of Oceanography.

Much of the mathematical treatment was developed in response to numerous stimulating discussions with Professor George W. Platzman of the University of Chicago. His many hours devoted to that, to reviewing most of the preliminary draft, to giving valuable aid in editing, and to fostering thought throughout were a great personal sacrifice of time and effort, motivated by his intense interest in the project and rewarded solely by his satisfaction with results. Other members of the Chicago staff, past and present, contributed ideas and direct help in the writing of this book. Included are Professors C.-G. Rossby, Erik Palmén, Herbert Riehl, and Dave Fultz. Technical assistance

was given by Dorothy Bradbury, Arnold Finkelstein, and Della Friedlander.

Credit is due the Geophysical Research Directorate, Air Force Cambridge Research Center, under whose sponsorship much of the research on kinematics was done. Finally, thanks are given to those authors whose illustrations are included in the text, and to the Chief, U.S. Weather Bureau, for permission to reproduce a number of Weather Bureau diagrams and for the assistance given by his staff.

WALTER J. SAUCIER

COLLEGE STATION, TEXAS
October 1, 1953

Academic affiliations of the author, Walter J. Saucier:

The University of Chicago,
Instructor in Meteorology, 1948–1952.

The A. and M. College of Texas,
Assistant Professor, then Professor, of Meteorology,
1952–1960.

The University of Oklahoma,
Professor of Meteorology, 1960–1969.

North Carolina State University,
Professor of Meteorology, 1969–1985, Professor Emeritus,
1986–

Symbols, Notations, and Abbreviations

GREEK SYMBOLS

α Specific volume (cm^3 gm^{-1}); wind direction (north wind 360°, east wind 90°, etc.).

β Slope angle of a surface or line relative to the horizontal; meridional variation of the Coriolis parameter, $\beta = (2\omega/a)$ cos ϕ; angle between axis of dilatation and scalar lines of an air property.

Γ Adiabatic vertical temperature lapse rate: dry adiabatic, Γ_d; moist adiabatic, Γ_s.

γ Vertical temperature lapse rate; also, motion of a pattern along the wind direction.

Δ A difference, usually over large distances or time intervals.

δ A difference over a small but finite distance or time.

ζ Vorticity (sec^{-1}) of the wind relative to the earth; ζ_g is geostrophic vorticity.

θ Potential temperature (° K); equivalent potential temperature, θ_E; wet-bulb potential temperature, θ_w.

κ R_d/c_p.

λ Longitude.

μ Dynamic viscosity.

ν Kinematic viscosity.

π 3.14159. . . .

ρ Density (gm cm^{-3}).

Σ A sum or summation.

σ Image scale of a projection (see p. 30).

T, τ Wave period.

Φ Geopotential.

ϕ Latitude.

ψ Colatitude; isentropic acceleration potential; in some cases, a certain angle.

ω Angular velocity (sec^{-1}), including earth's angular velocity.

ROMAN SYMBOLS[1]

A Area; cross-section area of a unit ap solenoid tube; arctic air mass.

a Radius of earth (mean, 6370 km, approx.).

c Wind speed (cm sec^{-1}, m sec^{-1}, knots, mi hr^{-1}); wave speed; continental air mass; specific heat (c_p, specific heat at constant pressure).

D Altimeter correction (see p. 54).

d Derivative or differential, from calculus; as subscript, usually denotes "dry."

E Stability of an air parcel or air mass: for vertical displacements, E_v; for horizontal displacements, E_h; for isentropic displacements, E_θ.

e Base of natural logarithms; water-vapor pressure (e_s is e at saturation).

1. Some of these symbols are used also as point or quantity designators.

F Viscous or frictional force.

f Coriolis parameter, 2ω sin ϕ.

f_w Correction factor to water-vapor pressure due to departure from ideal gas laws.

G Gravitational attraction; gradient.

g Acceleration of gravity, 980 cm sec^{-2} approx.

h As subscript, denotes horizontal.

I Rate of entrainment in convection (see p. 73).

K Hypsometric or barometric constant; coefficient of vertical transfer; curvature.

K Kelvin temperature.

k Arbitrary constant; air mass being warmed from below.

L Wave length; latent heat of vaporization

of water; distance between meridians in map projections.

M Mass.

m Map reduction scale (see p. 31); maritime air mass.

N Baroclinity, or solenoid density.

n Distance in the direction of an ascendant; distance to the right of the wind direction.

P Atmospheric pressure at a certain level or of a certain value; polar air mass.

p Atmospheric pressure (mb); pressure of the *dry* air, p_d.

Q An arbitrary atmospheric property.

q Specific humidity ($\%_0$); saturation specific humidity, q_s.

R Distance from the earth's axis of rotation ($R = a \cos \phi$); radius of curvature; gas constant (R_d is gas constant for dry air, 2.8704×10^6 ergs gm^{-1} deg^{-1}).

r Radius of latitude arc in map projections; mixing ratio ($\%_0$); saturation mixing ratio, r_s.

S Map scale: $S = m\sigma$.

s Distance in arbitrary direction; distance downwind; distance of latitude arc from equator in map projections; as subscript, denotes "saturated."

T Temperature (usually $^\circ$ K); T_E, equivalent temperature; T_w, wet-bulb temperature; T_s or T_d, dewpoint temperature; T^*, virtual temperature; tropical air mass.

t Time; also, centigrade temperature.

U Relative humidity; speed of a pattern; zonal (westerly) wind velocity.

u Velocity component along x-axis.

v Velocity component along y-axis; as subscript, denotes "vertical."

W Liquid equivalent of water vapor in a column of air.

w Velocity component along z-axis; air mass being cooled from below.

X An unknown.

x, y, z Cartesian coordinates.

Z Height in geopotential units; Greenwich Civil Time (GCT).

VECTOR NOTATIONS

i, j, k   Unit vectors along x, y, z-axes, respectively.

n   Unit vector along the n direction.

∇ The vector operator: $\nabla = \mathbf{i}\, \dfrac{\partial}{\partial x} + \mathbf{j}\, \dfrac{\partial}{\partial y} + \mathbf{k}\, \dfrac{\partial}{\partial z}$.

$\mathbb{C}$ The vector wind, defined as horizontal.

$f*$ The "Coriolis operator" (see p. 170).

ABBREVIATIONS

div   Divergence.

def   Deformation (applied in horizontal only).

Table of Contents

CHAPTER 1

A Review of the Atmospheric Variables

1.01. *Introduction.*—Atmospheric analysis is concerned with space and time distributions of several variables. We may speak of *physical variables*, which include those expressing physical state of the air (pressure, temperature, water-vapor content, and all other quantities dependent on one or more of these), and *kinematic variables*, which involve the motion of air particles or of systems of air particles. The most familiar example of the second group is the wind, which is the horizontal velocity of air at a point.

In the early chapters emphasis is placed on physical analysis of the atmosphere. This involves the concepts of statics, thermodynamics, and, to some extent, dynamics. Although it will be necessary at times to consider the wind and vertical motion, particularly as the wind can be determined from pressure distribution, a thorough treatment of distinctly kinematic analysis is delayed until Chapter 10.

Only a few of the variables are measured directly. These are pressure, temperature, surface wind, and one of the humidity quantities—either relative humidity, wet-bulb temperature (along with dry-bulb temperature), or dewpoint temperature, as the case may be. The wind in the free atmosphere is determined by trigonometry from the drift of ascending balloons, of aircraft in flight, or even of clouds. Each of the above four measured quantities can be determined independently of the others; they are the primary atmospheric variables. A much larger number of "secondary variables" are required in comprehensive analysis; these are evaluated by use of certain equations of physics from given values for each of two or more of the primary variables. In this group are potential temperature, density, humidity (i.e., water-vapor content), momentum, and others, of which several are really basic quantities even though they are evaluated indirectly.

1.02. *Temperature.*—There are three temperature scales in use in meteorology. In English-speaking countries surface temperatures are given in the Fahrenheit scale, ° F; in most other countries the Celsius (centigrade) scale, ° C, is used for that purpose. In upper-level temperature reports ° C is standard except in the British Isles, where ° F is still employed. The absolute, or Kelvin, scale, ° K, is basic in meteorological thermodynamics. Unless stated otherwise, the temperature in any physical equation is understood to be in this scale.

The relation between Kelvin (T) and Celsius (t) temperature is

$$T = t + 273.16° . \qquad (1)$$

From this it is evident that a given increment of temperature is the same in both scales: $\delta T(° K) = \delta t(° C)$. The formulas for conversion of ° C and ° F temperatures are

$$°F = (9/5)t + 32 ;$$
$$t = (5/9)(°F - 32) . \qquad (2)$$

One should be familiar not only with these equations but also with the following cor-

1

responding values of temperature (273.16 rounded to 273):

° K	233	243	253	263	273	283	293	303	313	255
° C	− 40	− 30	− 20	− 10	0	10	20	30	40	− 18
° F	− 40	− 22	− 4	14	32	50	68	86	104	0

Figures 1.021 give the normal distributions of mean temperature for the lower 10,000 feet of the atmosphere in January and July. (The lines are mean virtual isotherms for the layer drawn for intervals about 3° C; the line 960 corresponds to about 7° C.) The difference in pattern between these charts and the more familiar normal surface-temperature charts is due mainly to greater control by the underlying surface on the surface-air temperatures and to varying patterns of motion with height. Significant features illustrated are the general poleward decrease in temperature

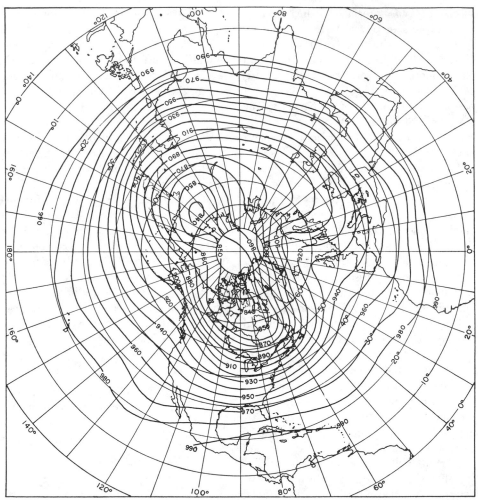

Fig. 1.021a.—Thickness (10's of feet) of the layer between 1000 mb and 700 mb in January

throughout the year but greatest in winter, the reversal of ocean-continent temperature difference from winter to summer, and the greater annual temperature ranges and extreme values of temperature over continents. Although not clearly shown, in the tropics horizontal and seasonal variations of temperature are very small.

The vertical variation of temperature is usually given as the vertical temperature *lapse rate*, $\gamma = -\partial T/\partial z$. A layer in which temperature *decreases* with height is said to have *positive* lapse rate. If temperature increases with height ($\gamma < 0$), the layer is a temperature *inversion*. Where temperature does not change with height, the vertical temperature distribution is said to be *isothermal*. More broadly, however, isothermal and isothermalcy imply a region of space with constant temperature or a physical process conservative in temperature.

Figures 1.022 give average north-south

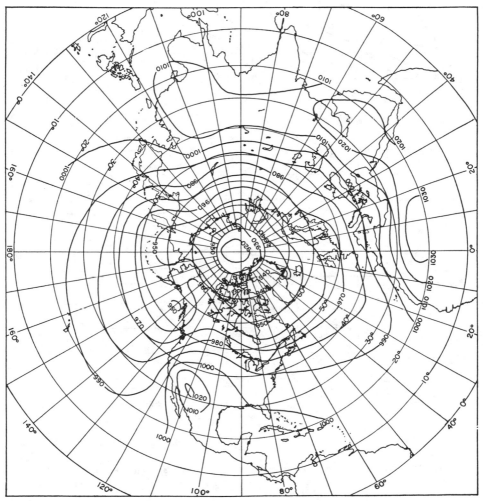

U.S. Weather Bureau

FIG. 1.021b.—Thickness (10's of feet) of the layer between 1000 mb and 700 mb in July

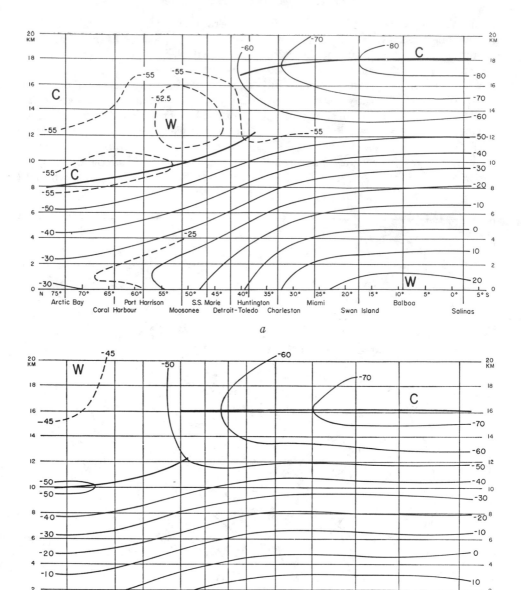

FIG. 1.022.—Distribution of virtual temperature in ° C (*thin lines*) and the tropopause (*heavy lines*) in (*a*) January–February, (*b*) July–August. (After S. L. Hess, "Some New Mean Meridional Cross Sections through the Atmosphere," *Journal of Meteorology*, Vol. V, No. 6 [1948].)

vertical cross sections of temperature[1] in the northern hemisphere.[2] These were prepared for a single meridian, and they hold strictly for only a small range of longitude. However, a principal difference to be found between meridians is some meridional shift of the patterns, which can be determined from Figures 1.021.

The name "tropopause" is given to the boundary between the *troposphere* below and the *stratosphere* above.[3] The troposphere is characterized normally by lapse of temperature in the vertical. In the average cross sections a significant positive lapse rate of temperature is found everywhere in the troposphere except near the ground in high latitudes in winter. Temperature variation upward in the stratosphere is distinctly different from the troposphere. In the lower stratosphere the temperature is constant with height (isothermal), increases with height (inversion), or decreases rather slowly with height. Thus, *the tropopause marks a change in the vertical temperature lapse rate from rather large values in the upper troposphere to relatively small or even negative values in the lower stratosphere.*[4] The vertical

1. The difference between true temperature and virtual temperature is negligible over most of the area of these cross sections. At the highest temperatures the difference would amount to about 3° C, and already at temperature 0° C it does not exceed 1°. The *patterns* of actual and virtual temperature are even more alike, as can be seen in Figure 6.10.

2. For the southern hemisphere see F. Loewe and U. Radok, "A Meridional Aerological Cross Section in the Southwest Pacific," *Journal of Meteorology,* VII (1950), 58–65.

3. For a more thorough discussion of nomenclature applied to various layers of the atmosphere refer to H. Flohn and R. Penndorf, "The Stratification of the Atmosphere," *Bulletin of the American Meteorological Society,* XXXI (1952), 71–77 and 126–30.

4. Refer to U.S. Weather Bureau Circular P, *Manual of Radiosonde Observations,* for regulations governing determination of the tropopause based on numerical values of the change in vertical temperature lapse rate.

temperature variation at still higher levels in the stratosphere might differ appreciably from the region just above the tropopause.

There are many interesting features shown by these patterns which deserve careful study. Among these are seasonal and meridional variations in vertical temperature lapse rate, height and nature of the tropopause, and horizontal temperature gradients.

1.03. *Pressure.*—Hydrostatic pressure is the force exerted by the free air on unit area due to the weight of the overlying atmosphere, and it is the same in all directions about a point. Defined as a force per unit area, pressure has c.g.s. unit dyne cm^{-2}. Because that unit is too small for our needs, the standard meteorological unit of pressure is 1 millibar (mb), which is 1000 dynes cm^{-2}. Mean sea-level pressure is slightly more than 1000 mb.

Since pressure is determined by the weight of the overlying atmosphere, it follows that *pressure decreases with height* in logarithmic manner. Pressure at a given elevation varies somewhat with time and geographic location. The *horizontal* variation of pressure is of prime importance in weather analysis because of its close relation to air motions. To determine the horizontal variation of pressure near the ground, it is necessary to reduce station pressures to a common reference level, that is, mean sea level, and the pressure so corrected is known as the "sea-level pressure." It is this corrected value of pressure that is found plotted in the common surface-weather chart.

Figures 1.031*ab* give normal sea-level pressure in January and July, and Figures 1.032*ab* the normal height of the 500-mb pressure surface.[5] The 500-mb contours

5. These were abstracted from *Normal Weather Charts for the Northern Hemisphere* (U.S. Weather Bureau Technical Paper No. 21 [Washington, D.C., October, 1952]).

drawn for intervals 200 feet may be interpreted as isobars at the 18,000-foot level spaced roughly 4 mb apart. In all seasons at upper levels, pressure increases from pole to equator in the normal patterns, with two or three centers of low pressure in the polar regions and almost flat pressure distribution over the tropics. The nearly circular pattern of isobars about the pole in middle latitudes is a significant feature of the upper-level charts. Important changes from winter to summer are decrease by about half in the

equator-to-pole pressure difference, poleward shift in location of maximum pressure gradient, and emergence of a more distinct high-pressure belt in subtropical latitudes.

At sea level we find (i) in winter pronounced oceanic low-pressure areas centered in the Aleutian and Icelandic regions, high pressure over continents and the pole, and the subtropical belt of high pressure centered near 25° or 30° latitude, and (ii) in summer large well-developed oceanic high-pressure areas centered in the east-central

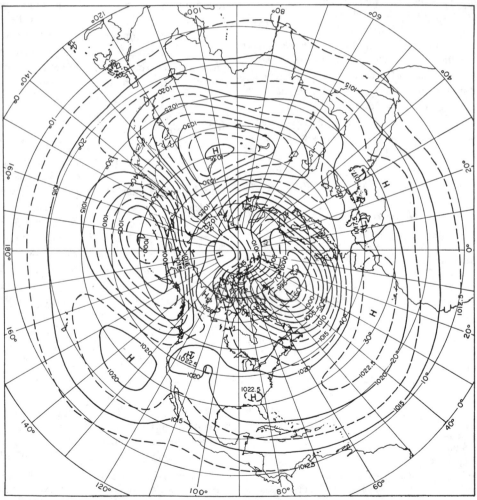

U.S. Weather Bureau

Fig. 1.031*a*.—Sea-level pressure (mb) in January

parts of the oceans at 35°–40° latitude, lower pressure over continents, and a weak pattern in the high latitudes, with slightly higher pressure at the pole. Except for the radical difference in continent and ocean distribution, pressure patterns in the southern hemisphere have similar but less marked seasonal changes.

In meridional profiles of average sea-level pressure for each latitude circle, the following major features in north-south order are found: (i) in January—pressure decrease from the north pole to the subpolar trough near 60° N, subtropical ridge near 30° N, tropical ("doldrum") trough near 5° S, subtropical ridge near 35° S, and circumpolar trough skirting Antarctica; (ii) in July—circumpolar trough near 65° N, ridge near 40° N, tropical trough from 10° to 15° N, ridge near 25° S, and trough around Antarctica.

The horizontal pressure patterns de-

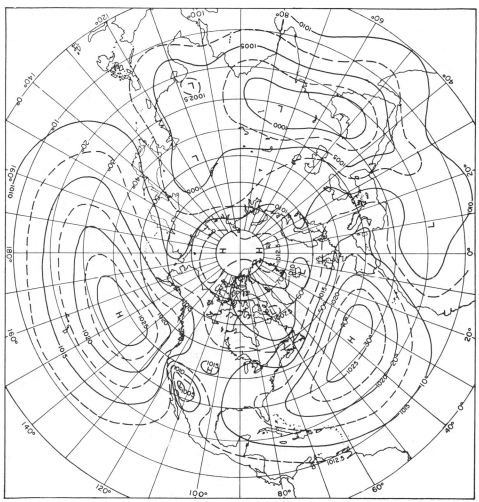

U.S. Weather Bureau

Fig. 1.031*b*.—Sea-level pressure (mb) in July

scribed above are long-term averages. At any given time the pattern of pressure, as also the temperature pattern, is moving and changing. It is likely to have some of the general features of the normal charts, especially in summer, but many details will differ. The synoptic charts in Figures 7.02 afford a comparison of a daily situation in winter with the January normals. Some important departures in the daily charts are the more cellular character, the greater amplitudes and shorter wave lengths aloft,

the different locations of cells, and locally larger pressure gradients. Pronounced centers in normal charts indicate preference in those areas for daily centers of the same type.

1.04. *Density and specific volume.*—Air density ρ and specific volume a are related by $\rho a = 1$. Specific volume is related to pressure and temperature through the equation of state.

$$a = RT/mp,\qquad(1)$$

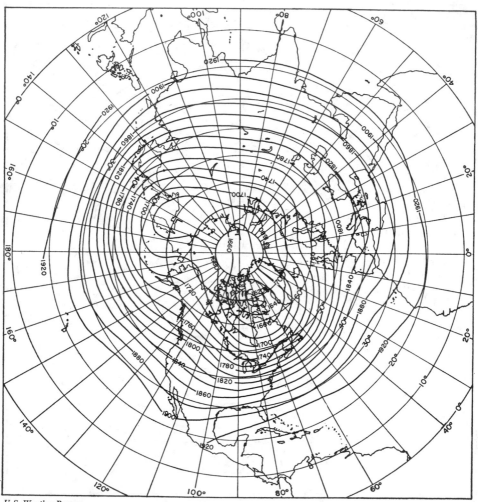

U.S. Weather Bureau

FIG. 1.032a.—500-mb height (10's of feet) in January

where R is the universal gas constant, and m the effective molecular weight for air which varies slightly with water vapor content.

1.05. *Humidity.*—The partial pressure due to presence of water vapor in the air is the *vapor pressure*, e. For an air sample there is a maximum value for the water-vapor pressure known as the *saturation vapor pressure*, e_s, which is a function only of temperature. The relation between saturation vapor pressure and temperature is shown most simply (though not quite exactly) by the empirical formula of Tetens:[6]

$$e_s = 6.11mb \times 10^{at/(t+b)}, \qquad (1)$$

where a and b are constants, different for ice and water,[7] and t is ° C. Below freezing, e_s may have either of two values at a given temperature, according as it is computed

6. O. Tetens, *Zeitschrift für Geophysik*, Vol. VI (1930).

7. Over water, $a = 7.5$, $b = 237.3°$; over ice, $a = 9.5$, $b = 265.5°$.

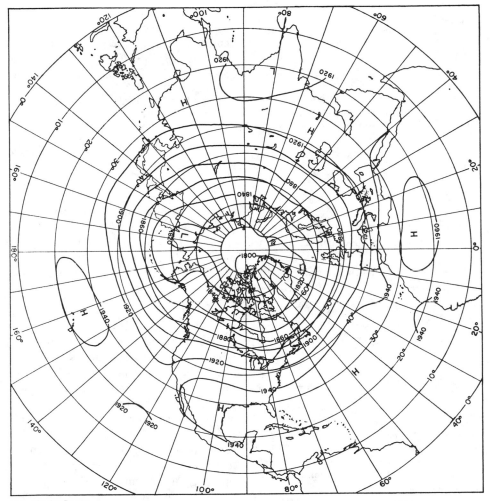

U.S. Weather Bureau

FIG. 1.032b.—500-mb height (10's of feet) in July

with respect to ice or water surface. At such temperatures e_s is greater over liquid water (Table 1.05a).

Any temperature above freezing has a single value for saturation vapor pressure. The variation of e_s is logarithmic with linear variation of temperature. At 0° C, e_s equals 6.11 mb; at 10° C, 12.3 mb; while at 20° C, 23.4 mb. This particular exponential de-

TABLE 1.05a

SATURATION VAPOR PRESSURES OVER
WATER (e_{sw}) AND OVER ICE (e_{si})*

Temperature (° C)	e_{sw} (Mb)	e_{si} (Mb)	e_{sw}/e_{si}	e_{si}/e_{sw}
0......	6.108	6.108	1.000	1.000
− 5.....	4.215	4.105	1.050	0.953
−10.....	2.863	2.597	1.102	0.907
−15.....	1.912	1.652	1.157	0.864
−20.....	1.254	1.032	1.215	0.823
−25.....	0.8070	0.6323	1.276	0.784
−30.....	0.5088	0.3798	1.340	0.746
−35.....	0.3139	0.2233	1.406	0.711
−40.....	0.1891	0.1283	1.474	0.678
−45.....	0.1111	0.0720	1.543	0.648
−50.....	0.0636	0.0394	1.615	0.619

* Extracted from *Smithsonian Meteorological Tables* (6th rev. ed.; Washington, D.C., 1951). Tables 94, 96, 100, and 101.

pendence of e_s on temperature is useful to remember in all thermodynamic relations involving water vapor.[8]

There are several other humidity variables employed. One is the *absolute humidity*, ρ_w, which is defined in the equation of state for water vapor,

$$\rho_w = m_w f_w e / RT , \qquad (2)$$

where m_w is the molecular weight of water and f_w is a correction factor for the de-

8. In Resolution 164 of the Twelfth Conference of Directors of the International Meteorological Organization, Washington, D.C., 1947, the Goff-Gratch formulas for e_s were adopted (J. A. Goff and S. Gratch, *Transactions of the American Society of Heating and Ventilating Engineers*, Vol. LII [1946]). Proper values for e_s as a function of temperature are found in *Smithsonian Meteorological Tables* (6th rev. ed.; Washington, D.C., 1951).

parture of the mixture of air and water vapor from ideal gas laws.

The (total) density of moist air is defined by the sum of dry-air density and absolute humidity. Thus,

$$\rho = \rho_d + \rho_w = m_d(p - f_w e)/RT$$
$$+ m_w f_w e / RT .$$

As $m_d = 28.966$, and $m_w/m_d = 0.62197$,

$$\rho = m_d(p - 0.37803 f_w e)/RT . \qquad (3)$$

This shows that *dry air is slightly more dense than moist air at the same pressure and temperature*. The influence of water vapor on air density is less than 2 per cent of the density.

The *specific humidity* is the mass of water vapor per unit mass of the moist air mixture:

$$q = \rho_w/\rho = 0.62197 f_w e /$$
$$(p - 0.37803 f_w e) . \qquad (4)$$

Specific humidity is often given by the approximate formula

$$q \simeq 0.622(e/p) , \qquad (5)$$

since $0.378e$ is so small in comparison with p, and $f_w e \simeq e$.

The *mixing ratio* is similar to specific humidity; it is the ratio of the mass of water vapor to the mass of the dry air with which the water vapor is mixed.

$$r = \rho_w/\rho_d = 0.62197 f_w e/(p - f_w e) . \qquad (6)$$

Although the mixing ratio is always slightly larger than the specific humidity, the difference between them is negligible for most purposes. For example, with air pressure 1000 mb and vapor pressure 35 mb, a rough maximum value for e, the specific humidity differs by only 2 per cent from the mixing ratio. As ordinarily this difference would be less, and in view of present inaccuracies in humidity measurement, this difference is negligible, and we assume $q \simeq r$.

Both specific humidity and mixing ratio

are dimensionless, and they are usually expressed in units parts per thousand ($^o/_{oo}$), viz., grams per kilogram. A mixing ratio of 5 represents 5 gm of moisture per 1000 gm of dry air. But, when substituting the mixing ratio into basic thermodynamic expressions, the exact numerical value (0.005) is the one usually implied.

The maximum values of q and r for given temperature and pressure are the *saturation specific humidity* and *saturation mixing ratio*, q_s and r_s, respectively:

$$q_s = 0.62197 f_w e_s / (p - 0.37803 f_w e_s) ,$$

$$r_s = 0.62197 f_w e_s / (p - f_w e_s) . \quad (7)$$

From this we observe that saturation specific humidity (or mixing ratio) is a

and is usually expressed as a percentage:

$$U = (r/r_s) \times 100 \text{ per cent} . \quad (8)$$

We therefore find $U \simeq q/q_s$ and $U \simeq e/e_s$. When $r = r_s$, the relative humidity is 100 per cent, and the air is said to be *saturated*. The relative humidity may exceed 100 per cent (i.e., supersaturation) in real conditions in the atmosphere; but, since present methods of observation do not measure relative humidities above saturation, we are forced to assume a maximum value 100 per cent for U as a working rule in analysis. Minimum values of relative humidity approach zero, but instrumental difficulties preclude accurate measurement of very low relative humidities. For tem-

TABLE 1.05b

DEWPOINT-FROSTPOINT CONVERSION*

Dewpoint ° C....	0	−5	−10	−15	−20	−25	−30	−35	−40	−45	−50
Frostpoint ° C...	0	−4.4	− 8.9	−13.4	−17.9	−22.5	−27.2	−31.8	−36.5	−41.3	−46.0
Difference ° C....	0	0.6	1.1	1.6	2.1	2.5	2.8	3.2	3.5	3.7	4.0

* Based on Tables 94 and 96 of *Smithsonian Meteorological Tables* (6th rev. ed., 1951).

function of both pressure and temperature. In the range of p and T with which we are most concerned, pressure varies by only one order of magnitude, from about 100 to 1000 mb, while e_s varies by several orders of magnitude. There is therefore only slight dependence by q_s on pressure but large exponential variation of q_s with temperature. In turn, this means that large values of specific humidity are possible only at high temperatures and that the distribution of specific humidity (mixing ratio) in the atmosphere is controlled largely by the distribution of temperature.

The *relative humidity* U is the ratio of actual mixing ratio to saturation mixing ratio at the same pressure and temperature[9]

9. Definition adopted by the Twelfth Conference of Directors of the International Meteorological Organization, Washington, D.C., 1947.

peratures above freezing reliable values of U can be measured to below 20 per cent, while the lowest limit of measurement is as much as 50 per cent at lower tropospheric temperatures. In fact, relative humidities are not reported at present for the free atmosphere at temperatures below −40° C.

Relative humidity has no preferred distribution in the atmosphere, such as snown by pressure, temperature, and even specific humidity. It is not an apparent function of latitude, elevation, or season. An idea of the great variability of relative humidity can be obtained by reference to the variability of cloudiness in space and time; yet the difference between clouds and no clouds represents only a small interval in the total range of relative humidity.

The element of humidity usually measured in the atmosphere is relative humidity, and other quantities—mixing ratio, dew-point temperature, etc.—are determined from it currently with reference to satura-tion over liquid water even at temperatures below freezing. This can lead to considerable error in vapor pressure, mixing ratio, and dewpoint whenever the values for ice are more valid than those for water. Table 1.05b shows the difference between dew-point over water and dewpoint over ice (frostpoint).

1.06. *Virtual temperature.*—The two gases, water vapor and dry air, have differ-ent equations of state. In practice, the two gases are treated as one, and the effect of moisture content is taken into account by adding a fictitious correction to the tempera-ture.

The equation of state for moist air is given by Eq 1.05(3). Upon substituting the approximate equation for specific humidity (Eq 1.05[5]) into the equation of state, we have

$$\rho = p/R_d T(1 - 0.61q) ,$$

if beginning here we substitute the symbol R_d for the ratio R/m_d, which is 2.8704×10^6 ergs gm^{-1} deg^{-1}. The reciprocal of $(1 - 0.61q)$ is to a close approximation $(1 + 0.61q)$. Then $\rho = (p/R_d T)(1 + 0.61q)$. If we set $T(1 + 0.61q) = T^*$, the equation of state for moist air becomes $\rho = p/R_d T^*$ or $pa = R_d T^*$. The quantity T^* is the *virtual temperature* and is evaluated from observed T and q. Its advantage arises from the fact that the equation of state, having the single gas constant R_d for dry air, can be applied to the moist atmosphere by adding a simple correction to the tempera-ture. *The virtual temperature is the tempera-ture of dry air having the pressure and density of the given sample of moist air.*

Virtual temperature is always greater than actual temperature. The difference be-tween them is

$$T^* - T = 0.61qT = 0.61rT . \quad (1)$$

In the atmosphere this difference varies more with humidity than with temperature, as seen by Table D in the Appendix. In the normal range of atmospheric temperatures the moisture necessary to increase the virtual temperature difference $T^* - T$ by 1° C varies between about 5.5 and 6.5 ‰. Since the moisture is usually between zero and 20‰, the following approximate equation is sufficient for many purposes in compu-tation of virtual temperature:

$$(T^* - T) \simeq q/6 \simeq r/6 .$$
$$(q, r \text{ in } \%_0) \quad (2)$$

1.07. *Potential temperature.*—The poten-tial temperature θ is that temperature the air would have if reduced dry adiabatically (i.e., at constant entropy for dry air) to a standard pressure 1000 mb. In terms of p and T, θ is defined by

$$\theta = T(1000/p)^{R_d/c_p} , \quad (1)$$

in which c_p is the specific heat of dry air at constant pressure. The adopted value of the ratio $\kappa = R_d/c_p$ for dry air is 2/7.

For pressure 1000 mb, $\theta°$ K = $T°$ K. Po-tential temperature is directly proportional to temperature and varies inversely with pressure. A region of the atmosphere in which potential temperature is constant is said to be dry adiabatic. A condition char-acterized by constant potential temperature along the vertical describes a dry adiabatic temperature lapse rate, which amounts to about 10° C km^{-1}.

Average cross sections of potential tem-perature are shown in Figure 1.07. It is seen that, on the average, potential temperature increases upward through both troposphere and lower stratosphere. Horizontally, po-tential temperature varies in the same sense as temperature. In the troposphere poten-tial temperature normally varies in value from about 250° K near the ground in cold arctic air to around 400° K near the tropical tropopause.

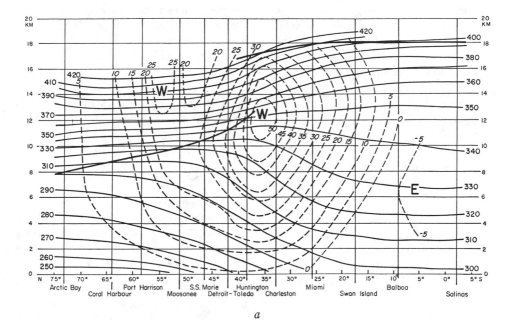

a

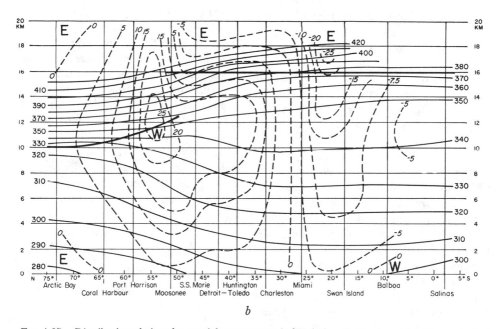

b

FIG. 1.07.—Distribution of virtual potential temperature in ° K (*thin solid lines*), geostrophic west wind in m sec⁻¹ (*dashed lines*), and the tropopause (*heavy lines*) in (*a*) January–February, (*b*) July–August. (After Hess.)

Occasionally need might arise for the *partial potential temperature*, θ_d, as defined by Rossby.[10] This is the potential temperature for the temperature and partial pressure (p_d) of the dry air, and is

$$\theta_d = T(1000/p_d)^\kappa = \theta(1 + rm_d/m_w)^\kappa . \quad (2)$$

Thus, θ_d is slightly larger than θ, but the difference is usually less than 3° K.

The *virtual potential temperature*, θ^*, is that potential temperature corresponding to the virtual temperature at the given pressure. If we multiply both θ and T in Eq 1.07(1) by $(1 + 0.61q)$, then

$$\theta^* = T^*(1000/p)^\kappa = \theta(1 + 0.61q) . \quad (3)$$

Therefore θ^* is slightly larger than θ, and the maximum difference amounts to about 5° K.

1.08. *Equivalent potential temperature.*— In weather analysis another form of potential temperature is necessary, one which is conservative not only for dry adiabatic processes but also for processes in which the possible gain (loss) of sensible heat in condensation (evaporation) is taken into account. The theory of saturation processes is very involved, and for that reason we make no attempt to derive the required formulas. Saturation processes are explained in most textbooks in physical meteorology.

We first should explain some pertinent nomenclature used in connection with thermodynamic processes. The adjective "potential," when used with temperature, implies that air-parcel temperature resulting from adiabatic reduction to a standard pressure 1000 mb. "Equivalent" implies the sum total of latent and sensible heat contained in an air parcel. Thus, the equivalent temperature represents essentially the sum of the actual air temperature and the

temperature increment corresponding to the heat latent in the water vapor. If equivalent temperature is T_E, then, according to the Robitzsch definition,[11]

$$T_E = T + Lr/c_p . \quad (1)$$

Since the latent heat of vaporization (condensation) varies only slightly with temperature and $c_p \simeq 0.24$ cal gm⁻¹ deg⁻¹, an approximate equation for T_E is

$$T_E \simeq T + 2.5r . \quad (r\%_{00})$$

It is valid for ordinary ranges of temperature and moisture.

Another definition for T_E was given by Rossby:[12]

$$T_E = Te^{Lr/c_pT} . \quad (2)$$

(The symbol e is here the base of natural logarithms.) This holds for saturated and unsaturated conditions; in the latter case L and T refer to values at adiabatic saturation. Rossby's definition of T_E, called the "pseudo-equivalent temperature" by Petterssen,[13] is that temperature resulting if the air pressure were lowered sufficiently in a pseudo-adiabatic process, such that all latent heat is realized and then returned dry adiabatically to its original partial pressure p_d.

The *equivalent potential temperature*, θ_E, is defined by dry adiabatic reduction of T_E to 1000 mb and is given by the proportionality $T/\theta_d = T_E/\theta_E$. By substitution from Eq 1.08(2),

$$\theta_E = \theta_d e^{Lr/c_pT} . \quad (3)$$

This will be our standard definition for θ_E. It is the "Rossby θ_E," which Petterssen calls the "pseudo-equivalent potential temperature."

10. C.-G. Rossby, *Thermodynamics Applied to Air Mass Analysis* ("Meteorological Papers, M.I.T.," Vol. I, No. 3 [Cambridge, Mass., 1932]).

11. M. Robitzsch, *Meteorologische Zeitschrift*, 1928.

12. *Op. cit.*

13. Sverre Petterssen, *Weather Analysis and Forecasting* (New York: McGraw-Hill Book Co., 1940).

For practical uses a rough approximation to θ_E as a linear function of θ and r might suffice:

$$\theta_E \approx \theta + 3r . \qquad (r\%_{00}) \qquad (4)$$

By this approximation, θ_E is evaluated conveniently from values of θ and r, both of which are known or easily computed from pressure, temperature, and relative humidity. The error resulting from the approximation can be as much as a few degrees at high values of θ and r.

We have listed the basic physical variables in atmospheric analysis. There are several additional thermodynamic quantities in use involving moisture, but discussion of these is reserved for the next chapter, where they are more easily described graphically.

PROBLEMS AND EXERCISES

1. Determine the height of the mercury column in a barometer for pressure 1000 mb, gravity 980 cm sec^{-2}, and temperature 10° C.
2. Convert the following pressures (mb) to their corresponding values in centimeters and inches of mercury.

$$1050 \qquad 1013.2 \qquad 1000 \qquad 850$$

3. On a sheet of ordinary graph paper lay out a temperature scale along the abscissa from $-80°$ C on the left to 20° C on the right and label the ordinate in height from 0 at the bottom upward to 20 km. Then from data in Figure 1.022a plot temperature-height curves for Arctic Bay, Moosonee, Huntington, and Swan Island in distinguishing colors.
4. Using Table H for the U.S. Standard Atmosphere, locate in the right margin of the above graph each hundredth isobar beginning at 1000 mb.
5. From Eq 1.05(1) compute e_s (over ice and water below freezing) for each 10° from $-20°$ C to 30° C. Plot the resulting curves on a sheet of ordinary graph paper with abscissa labeled in temperature and ordinate in pressure.
6. From data obtained in (5), determine the saturation mixing ratio for each of the following sets of values for temperature and pressure (assuming $f_w = 1$):

$$\begin{array}{ll} 0° \text{ C and } 1000 \text{ mb} & -10° \text{ C and } 1000 \text{ mb} \\ 0° \text{ C and } \;\;500 \text{ mb} & \;\;10° \text{ C and } 1000 \text{ mb} \end{array}$$

How does the variation of r with temperature compare with its variation with pressure?
7. Compute the relative humidity for each of the following sets of values:

$$\begin{array}{llll} (a) & p = 1000 \text{ mb}, & t = 0° \text{ C}, & \text{and } e = 5 \text{ mb} \\ (b) & p = 1000 \text{ mb}, & t = 0° \text{ C}, & \text{and } r = 1\%_{00} \\ (c) & p = \;\;500 \text{ mb}, & t = 0° \text{ C}, & \text{and } r = 1\%_{00} \end{array}$$

8. Find the virtual temperature ° C for each of the conditions below (use Table D):

$$\begin{array}{lll} (a) & t = 20° \text{ C}, & p = 1000 \text{ mb}, & \text{and } r = 10 \\ (b) & t = 10° \text{ C}, & p = \;\;900 \text{ mb}, & \text{and } U = 80 \text{ per cent} \end{array}$$

9. Determine the density and specific volume for each condition in (8) above.
10. Determine the potential temperature for the following sets of values, using Eq 1.07(1):

$$\begin{array}{lll} (a) & p = 1020 \text{ mb}, & t = 27° \text{ C} \\ (b) & p = \;\;980 \text{ mb}, & t = 27° \text{ C} \\ (c) & p = \;\;100 \text{ mb}, & t = -73° \text{ C} \end{array}$$

11. Given $p = 1000$ mb, $t = 10°$ C, and $U = 60$ per cent, find the potential temperature, virtual potential temperature, and partial potential temperature.
12. Determine the Rossby equivalent temperature and equivalent potential temperature for an air sample at $p = 800$ mb, $t = 13°$ C, and $U = 100$ per cent.
13. A given air parcel has pressure 700 mb and temperature 7° C. Find its temperature ° C if it followed a dry adiabatic process to 900 mb.
14. Derive the proper form of Eq 1.08(4) by expanding Eq 1.08(3). Explain the validity of Eq 1.08(4).

READING REFERENCES

BERRY, F. A., BOLLAY, E., and BEERS, N. R. (eds.). *Handbook of Meteorology*, pp. 83–101, 326–61. New York: McGraw-Hill Book Co., 1945.

BRUNT, D. *Physical and Dynamical Meteorology*, pp. 1–20. 2d ed. Cambridge: Cambridge University Press, 1939.

BYERS, H. R. *General Meteorology*, pp. 39–76, 129–42, 150–67. New York: McGraw-Hill Book Co., 1944.

HAURWITZ, B. *Dynamic Meteorology*, pp. 3–21, 36–38. New York: McGraw-Hill Book Co., 1941.

HEWSON, E. W., and LONGLEY, R. W. *Meteorology: Theoretical and Applied*, pp. 1–16, 20–24, 28–31, 40–52, 59–65. New York: John Wiley & Sons, 1944.

WILLETT, H. C. *Descriptive Meteorology*, pp. 1–4, 19–28. New York: Academic Press, 1944.

CHAPTER 2

Meteorological Charts and Diagrams

2.01. *Introduction.*—For a complete analysis of the physical and kinematic structure of the atmosphere, many observations in space and time are required. Observations at the surface and aloft are made once, twice, or more times daily at scheduled intervals from land stations and ships at sea. Through international agreement observation schedules are for the most part synchronous around the globe. Scheduled synoptic surface observations include measurements of pressure, temperature, humidity, and wind, plus descriptions of visible atmospheric phenomena such as clouds, hydrometeors, and lithometeors. Such observations are made four times daily, at 0030, 0630, 1230, and 1830 GCT (Greenwich Civil Time). In addition, each region or country maintains a network of hourly reporting stations to satisfy aeronautical and other local requirements.

In the free atmosphere the observations are of pressure, temperature, humidity, and wind, which are obtained from radiosonde and balloon wind soundings, and are supplemented by weather reconnaissance aircraft flights. The frequency and time of upper-air soundings vary somewhat from one region to another. Over most of North America pilot-balloon soundings are made at 6-hour intervals beginning at 0300 GCT, and radiosonde observations twice daily— 0300 and 1500 GCT. The upper-air observations in this country are therefore less frequent than surface observations and not synoptic with them, a fact which adds difficulty in trying to co-ordinate analysis of the free atmosphere with surface analysis and which is partly responsible for the tendency to consider events at the surface and aloft in different lights.

The distribution of surface and upper-air stations is irregular. In Europe and over a large part of North America the surface synoptic stations are sufficiently dense for most purposes in analysis, but in polar regions and in other remote land and ocean areas the number of reporting stations is still far from adequate. The paucity of upper-air data is even more serious, the number of sounding stations (Fig. 2.20) being only a small percentage of the surface stations.

The first objectives of conventional analysis are to represent by observed data, to find, and to interpret the state of the atmosphere as a function of space at fixed intervals in time. Accordingly, the meteorological charts and diagrams employed must best portray the space variations of the atmospheric variables with consideration for convenience in plotting and analysis, for accuracy in representation, and for interpretation. A three-dimensional representation drawn to scale would be the most accurate and natural form, but it would present numerous difficulties in plotting and analysis. That part of the atmosphere directly concerned is a thin spherical shell with lateral dimensions thousands of times its vertical depth. To simplify the plotting and analysis, the charts are flat surfaces which represent spherical level surfaces and cross-section planes in the atmosphere, and the scale of the vertical coordinate is exaggerated many times in comparison with hori-

17

zontal coordinates. Accuracy of space representation is thus sacrificed. The convenient two-dimensional forms of representation all have the undesirable feature of distortion, which may not interfere with mechanics in analysis, but it should be considered when interpreting the analysis.

2.02. *Classification of the charts.*—For most analytical work the Cartesian coordinate system is used. The z-axis is upward along the plumb line, and the xy-plane is tangent to level surfaces at the z-axis through the origin. The principal space charts may be classified as follows:

(i) z variable, x and y constant: the sounding charts
(ii) z constant, x and y variable: the meteorological maps
(iii) x (or y) constant, z and y (or x) variable: the cross-section charts

The first group is for illustrating and evaluating variations in the vertical. All necessary data are derived from a single radiosonde, aircraft, or wind sounding. Included in this group are simple graphs with elevation as ordinate and an atmospheric variable as abscissa. Somewhat different as a sounding chart is the *hodograph* diagram, which is merely polar coordinate paper and is used to show the variation of wind with height.

The most complete forms of sounding charts are the thermodynamic diagrams, which illustrate not only the variations of the primary variables with height but also certain hydrostatic and stability properties in the sounding. The ordinate of thermodynamic charts is usually some simple function of pressure but is nearly a true height scale. The abscissa is usually linear in temperature.

The meteorological synoptic maps are the most extensively used and the most essential of the charts. Their importance in analysis follows from the manner in which the

kinematics and dynamics of the atmosphere depend so much on horizontal distributions of certain variables, in particular, pressure and temperature. Equally important is the fact that in most cases the vertical structure of the atmosphere can be deduced or derived from a series of horizontal charts at different levels. Indeed, a large part of our work is aimed at developing ability to deduce three-dimensional structure from analysis of a few level charts.

Vertical cross sections are indispensable for illustrating and studying the atmosphere in three dimensions as a supplement to the horizontal charts. There is little acceptance of the synoptic cross section as a routine chart in daily weather analysis, partly because its preparation is so time-consuming and also because some of its advantages can be derived after sufficient training and practice by comparing series of charts for successive levels in the atmosphere. The cross section has the essentials of the sounding charts plus the added quality of showing atmospheric structure along an additional space coordinate.

2.03. *Thermodynamic charts.*—The most familiar and yet most complicated sounding charts are a type known as thermodynamic charts—often loosely called "adiabatic" charts. There are different thermodynamic charts in use by the various meteorological services,[1] but they are all projections of one another, and the difference between any two is mostly in appearance. Each complete chart contains five sets of lines: isobars, isotherms, dry adiabats, pseudo-adiabats, and saturation moisture lines.

Meteorological thermodynamic charts are transformations of the familiar α, $-p$ or

1. A summary of many existing thermodynamic charts was given by L. Weickman, *Über aerologische Diagrammpapiere* (Berlin, 1938), and extended by P. Defrise, *Contribution à une étude comparative des diagrammes aérologiques* ("Belgium, Institut Royal Météorologique, Miscellanées," Fasc. 33 [1948]).

Clapeyron diagram (Fig. 2.031). Specific volume is not measured directly in the atmosphere, and it is preferable to have temperature as abscissa. As the isotherms of a perfect gas are linearly spaced along any isobar, we need only straighten these isotherms from rectangular hyperbolae to vertical lines. In this transformation it is desirable to maintain pressure, or a simple function of pressure, as ordinate with pres-

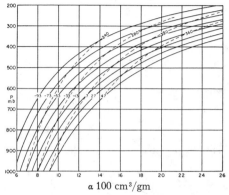

Fig. 2.031.—a, $-p$ diagram with isotherms (*full lines*) and dry adiabats (*dashed lines*).

sure decreasing upward. Another desirable feature of the projection is that of area preservation; that is, the area outlined by any two isotherms and any two isobars should be the same on the projected diagram as on the a, $-p$ diagram, except for a scale factor constant for the entire projection. The need for conservation of areas arises from the ratio between area and energy on the a, $-p$ diagram. Although exact area conservation is not necessary for many phases of analysis, it is desirable for thermodynamic computations such as pressure-height evaluation.

The *emagram* (Fig. 2.032a) is formed by rotating the isotherms of the a, $-p$ diagram into vertical positions. To make the new projection area preserving, the ordinate must be $-\log p$. This diagram is often called the T, $-\log p$ diagram. The dry adiabats

(potential temperature lines) are slightly curved concave upward and converge toward lower pressure and temperature. On most printed versions of the chart, dry adiabats make an angle about 45° with the isobars, and isopleths of saturation mixing ratio are almost straight and vertical. As in other thermodynamic charts the pseudo-adiabats, or isopleths of θ_E and θ_w, are curved (convex toward low pressure and high temperature) and diverge as they become asymptotic to the dry adiabats at low temperatures.

The *Stüve* or *pseudo-adiabatic* diagram[2] (Fig. 2.033b) is similar to the emagram, and one is often mistaken for the other. The Stüve chart can be derived from the emagram by making the dry adiabats straight and by keeping the isotherms vertical and linearly spaced and the isobars horizontal. This requires for the ordinate $-p^{\kappa}$ ($\kappa = R_d/c_p$), and all dry adiabats radiate from the point $p = 0$, $T = 0°$. The straight dry adiabats are the only advantage of this chart over the emagram; but, in achieving this improvement, the property of area preservation is sacrificed. The pseudo-adiabatic chart has been the most widely used chart in the United States, possibly because of the simplicity introduced by the three primary sets of straight lines. The departure from equal areas on the chart is small and, in most practical work, can be neglected.

A third thermodynamic chart, the Refsdal *aerogram*,[3] has coordinates $\log T$, $-T \log p$. On this chart only one isobar is straight; all others are curved and diverge toward higher temperature and therefore are not perpendicular to the isotherms,

2. G. Stüve, "Potentielle und pseudopotentielle Temperatur," *Beiträge zur Physik der freien Atmosphäre*, XIII (1927), 218–33.

3. A. Refsdal, "Das Aerogramm: Ein neues Diagrammpapier für aerologiske Berechnungen," *Meteorologische Zeitschrift*, Vol. LII (1935). Cf. Sverre Pettersen, *Weather Analysis and Forecasting* (New York: McGraw-Hill Book Co., 1940).

which are vertical. The aerogram has a unique property, because of the particular choice of coordinates, which simplifies the process of pressure-height evaluation; but it lacks simplicity in the configuration of the principal isopleths.

If the isotherms of the emagram are held fixed while the dry adiabats are rotated into straight horizontal positions, another thermodynamic chart is derived, the *tephigram*[4]

(Fig. 2.032b). Its coordinates are temperature and entropy (ϕ, $= c_p \log \theta$); thus, the ordinate of the tephigram is potential temperature in logarithmic scale. The tephigram is widely used in other countries, especially Great Britain and Canada. Along each isotherm the pressure varies logarithmically,

4. Sir Napier Shaw, *Manual of Meteorology* (4 vols.; Cambridge: Cambridge University Press, 1926–32), Vol. III, Chap. 7.

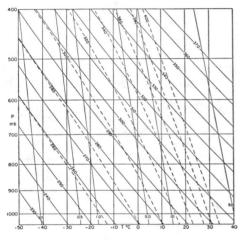

FIG. 2.032a.—Emagram

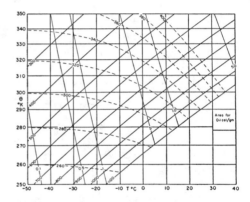

FIG. 2.032b.—Tephigram

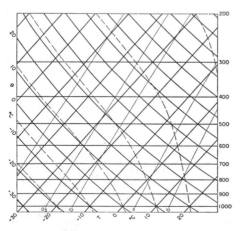

FIG. 2.032c.—Herlofson's Skew T, $-\log p$ Diagram

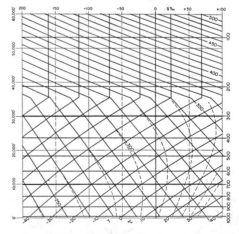

FIG. 2.032d.—Bellamy's Z_p, S pastagram (the U.S. Standard Atmosphere is represented by the vertical line for $S = 0$).

as in the emagram, but the pressure scale varies from one isotherm to the next. The isobars are curved and intersect dry adiabats and isotherms at angles which, on most versions of the chart, are roughly 45°.

Another chart, proposed by Herlofson,[5] is the "Skew T, $-\log p$ Diagram" (Fig. 2.032c). This chart incorporates the straight horizontal isobars of the emagram and the desirable feature of a large angle between are horizontal. Since this chart has so much in common with the tephigram, whatever is said about one usually holds for the other.

Bellamy's *pastagram*[6] (Fig. 2.032d) is similar in appearance to the Skew T, $-\log p$ Diagram, with exception of the upper portion of the chart, where it is actually an emagram. The pastagram is a true thermodynamic chart based on the U.S. Standard Atmosphere. The ordinate is standard-

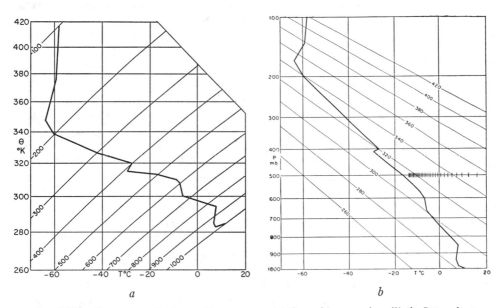

Fig. 2.033.—Comparison of pT sounding curves on (a) the tephigram and on (b) the Stüve chart

isotherms and dry adiabats found in the tephigram. The ordinate of the diagram is $-\log p$. Along any isobar the temperature varies linearly, but the isotherms slope upward to the right and form an angle approximately 45° with the isobars. Dry adiabats are slightly curved (concave to the upper right) and intersect the isotherms at almost right angles. It can be transformed from the tephigram by leaving isotherms fixed while straightening the isobars and finally rotating the chart clockwise so that the isobars atmosphere height above mean sea level; but, since in a standard atmosphere each height has a particular pressure, the ordinate scale is labeled also in pressure. The abscissa of the pastagram is specific temperature anomaly, or roughly temperature departure from the standard atmosphere. Thus, the standard-atmosphere, pressure-temperature sounding is a vertical line in the chart. A sharp discontinuity in the slopes of isotherms and dry adiabats occurs at the level of the tropopause in the stand-

5. N. Herlofson, "The T, $-\log p$ Diagram with Skew Coordinate Axes," *Meteorologiske Annaler* (Norway), Met. Inst., Band II, No. 10 (Oslo, 1947).

6. J. C. Bellamy, "The Use of Pressure Altitude and Altimeter Corrections in Meteorology," *Journal of Meteorology*, II (1945), 1–79.

ard atmosphere; this is the primary source of difficulty in using the chart. Below the discontinuity, isotherms are slightly curved and slope upward to the right; above, isotherms are vertical.

Selection of any individual chart for use in analysis is a matter of (1) its accuracy and adaptability for thermodynamic computations, particularly pressure-height and stability evaluations; (2) the geometric simplicity of the diagram, as related to the ease with which atmospheric soundings can be plotted and visually interpreted and to its adaptation as a reference model for mentally picturing various atmospheric processes; and (3) the degree with which it differentiates humidity and thermal stratification in a sounding.

Interpretation of a sounding usually involves reading absolute values along the sounding, which can be determined from any of the charts, and comparing the slopes of sounding curves. In the troposphere the slopes ("lapse rates") of pressure-temperature soundings usually vary between isothermal and dry adiabatic. Thus, the larger the angle between adiabats and isotherms on the chart, the easier it is to detect significant variations in temperature lapse rate. A comparison of the same sounding plotted on the tephigram and on the Stüve chart is given by Figure 2.033. (A portion of the same sounding appears on a pastagram in Fig. 3.09.) Observe how the tephigram magnifies variations in slope of the sounding when lapse rates are between dry adiabatic and isothermal. Because of this particular advantage in thermodynamic analysis, and no real disadvantage, the tephigram is adopted as the standard chart in this text.

2.04. *Details of thermodynamic charts.*—The thermodynamic diagrams are made for the ranges of pressure and temperature normally measured in atmospheric soundings. Most diagrams are printed in two sections to give sufficiently large scale on con-

venient size paper, the lower section extending upward to pressures near 400 mb and the upper section from there to 50 mb or less to accommodate the uppermost portion of the ascent. The complete diagram can be printed on a single sheet of paper if a reduced scale is used, and it is then possible to see the entire plotted sounding at once.

A number of special scales or additional lines can be printed on the thermodynamic chart for special computations. A scale usually found in a margin of the chart translates pressure to height in a standard atmosphere. This scale is useful for *approximate* evaluations of height. The standard atmosphere sounding may be printed within the grid of the chart as a reference sounding. Along the abscissa the scale might be given in both temperature and height, the latter of which is used in plotting pressure-height curves for the soundings. Particularly on the charts used for exact pressure-height evaluations, a series of short vertical marks are printed on each hundredth isobar (see Fig. 2.033*b*). The temperature difference between successive vertical marks on the same isobar gives the virtual temperature correction at saturation; the true virtual temperature difference for any level in the sounding is found from the product of the relative humidity and the temperature increment shown by this scale. Finally, on the true thermodynamic charts the ratio of area to energy can be indicated (Fig. 2.032*b*) for use in some thermodynamic computations.

2.05. *Plotting of soundings on thermodynamic charts.*—The radiosonde observation (RAOB) gives pressure, temperature, and a humidity quantity for the surface and for various points above the station, including the terminal point in the sounding. Heights are transmitted for specific pressures (*mandatory pressures*) along the sounding. In current specifications by the U.S.

Weather Bureau the mandatory pressures are 1000, 850, 700, 500, 400, 300, 200, 150, 100, and 50 mb. To give a more detailed outline of the observed sounding, data for several *significant points* are included. These points are locations of significant changes in the vertical variations of temperature or relative humidity.

In plotting a sounding on a thermodynamic chart, each level in the report is located as a point in the pressure-temperature grid for reported values of pressure and temperature. The pressure-temperature (pT) curve is drawn by connecting successive plotted points with straight-line segments. Because significant levels are included in the coded report, this curve will be a close facsimile of the true trace of temperature with pressure. Each plotted point in the pT curve represents observed values of pressure and temperature. Between two plotted points the straight-line segment gives for any pressure a temperature which differs in the troposphere by no more than 1° C from the observed temperature and in the stratosphere by no more than 2° C from the observed temperature, according to present specifications.

The humidity distribution in a sounding can be represented by several methods. If dewpoint is transmitted for each level, a pressure-dewpoint (pT_s) curve can be plotted in the pressure-temperature grid (Fig. 6.02a). If mixing ratio is the transmitted humidity quantity, then a pressure-mixing ratio (pr) curve is plotted in the grid of pressure and saturation mixing ratio. The pT_s and the pr curves are identical. Besides showing the variation of humidity with pressure, the humidity curve serves an additional purpose. Its proximity to the pT curve along an isobar gives indication of relative humidity; moist and dry layers in the sounding thus can be distinguished by visual comparison of those curves.

For some purposes it may be necessary

to plot the virtual temperature curve (pT^*) for the sounding. This can be done quickly if the pT and the pr (or pT_s) curves have been entered on the chart. The principle is indicated by Eq 1.06(2). To obtain the pT^* curve, the pT curve is displaced isobarically by about 1° C for each 6‰ of moisture.

2.06. *Graphical operations on the thermodynamic chart.*—All the commonly required thermodynamic quantities can be evaluated from the pT and pressure-humidity points on the chart. Pressure and temperature are observed, and they locate the pT point (point A in Fig. 2.06). Assume the air sample is at 850 mb and 10° C. The potential temperature is found from the value of the dry adiabat through point A; thus $\theta = 296.5°$ K, to the nearest half-degree.[7] Also, by following this dry adiabat to its intersection with the 1000-mb isobar, the intersection B occurs at 23.5° C or 296.5° K, which is by definition the potential temperature for the air parcel at A. The saturation mixing ratio r_s for the air sample is the value of the r_s line through its pT point on the chart. For this example r_s is 9.15‰.

Since all remaining quantities depend on moisture content, we shall assume that a moisture measurement for the parcel is also given; say, the dewpoint $T_s = 4°$ C (point D). The dewpoint temperature is the temperature at which saturation is reached by cooling at constant pressure without addition or removal of moisture. In this process r is constant, but r_s decreases as temperature decreases, until saturation is reached, when $r_s = r$. Therefore, the mixing ratio is the value of the saturation-mixing-ratio line through the dewpoint D, 6.0‰ in this ex-

7. The graphical operations for the example given here were performed on a thermodynamic chart constructed with the old value $R_d/c_p = 0.288$ and values of r_s (as function of p and T) which are outdated. The values given here for θ, r_s, r, θ_E, θ_w, etc., are thus subject to slight discrepancies.

ample. The ratio r/r_s = 66 per cent is the relative humidity.

The partial pressure of the dry air, $p_d = p - e$, for this air parcel is 842 mb. The partial potential temperature θ_d is the potential temperature for this pressure and the air temperature 10° C; thus, θ_d = 297° K.

For mixing ratio 6‰ and temperature 10° C, Eq 1.06(2) shows the difference

pressure decreases, it expands, and its temperature decreases accordingly. During this process the mixing ratio is constant, and the parcel reaches saturation when its saturation mixing ratio reaches the value of its initial mixing ratio, 6‰. The point of adiabatic saturation is located at C; it is the intersection of the dry adiabat through the pT point A with the saturation-mixing-

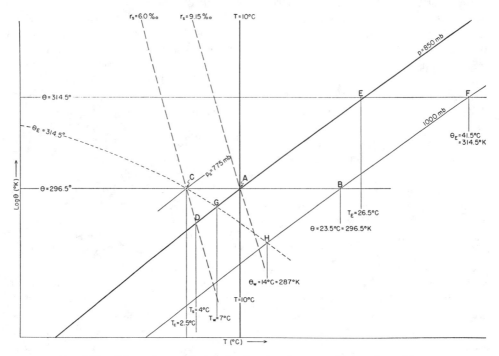

FIG. 2.06.—Scheme for graphical evaluation of various quantities on the tephigram

$T^* - T$ is very nearly 1.0° C, and thus T^* is 11° C or 284° K. By using the equation of state, $\alpha = R_d T^*/p$, we find $\alpha = 0.96 \times 10^3$ cm³ gm⁻¹. The reciprocal of α is the density; ρ is therefore 1.04×10^{-3} gm cm⁻³. Evaluation of α or ρ is done by substituting into the equation of state or by reference to special tables, as their isopleths do not appear on common thermodynamic charts.

Now suppose this parcel is lifted dry adiabatically, that is, in a constant potential temperature or *isentropic* process. As its

ratio line having the value of the mixing ratio for the air sample. This intersection has a number of names: the adiabatic saturation (condensation) point, the "characteristic" point,[8] the lifting condensation level, the isentropic condensation level, and others, all of which have the same meaning. The pressure at this point is the adiabatic saturation pressure, P_c; in this case P_c is

8. The characteristic point is defined more rigorously by the dry adiabat for θ_d instead of θ, but for simplicity we are avoiding this distinction.

about 775 mb. The temperature there ($T_c =$ 2.5° C) is the adiabatic saturation temperature. As seen presently, this adiabatic saturation level is the focal point for many graphical evaluations from the chart.

After saturation we suppose that further lifting of the parcel describes a pseudo-adiabatic process. In this lifting process the relative humidity is assumed to remain constant at 100 per cent; all excess moisture is immediately condensed and released, and the corresponding latent heat of condensation is realized as sensible heat. Release of latent heat contributes to warming, while expansion due to lifting contributes to cooling at the dry adiabatic rate. Thus, the pseudo-adiabatic rate of cooling is less rapid than the dry adiabatic rate. The difference between these two lapse rates is proportional to the rate of moisture condensation, which, in turn, is roughly proportional to the mixing ratio,[9] as can be seen by inspection of a thermodynamic chart. Hence, as a crude though practical approximation, we can say that the difference between dry adiabatic and pseudo-adiabatic lapse rates is proportional to the mixing ratio.

The pseudo-adiabat through the adiabatic saturation point, C, determines the future state of the parcel during continued lifting. The value attached to this particular pseudo-adiabat in our example is 314.5° K, which is the value of the dry adiabat to which this pseudo-adiabat becomes asymptotic.[10] The equivalent potential temperature is given by this value; $\theta_E = 314.5°$ K. This is also the absolute temperature obtained after lifting the parcel until it has lost all its moisture and is then returned dry adiabatically to the pressure 1000 mb (point F). The equivalent temperature is found by reading the value of temperature at the intersection E of the dry adiabat

9. It is almost a function only of $r(L/T)$, where L is the latent heat of vaporization of water.

$\theta = 314.5$ with the partial pressure 842 mb; $T_E = 26.5°$ C.

If the pseudo-adiabat through the adiabatic condensation point is followed in the reverse direction down to 1000 mb, the *wet-bulb potential temperature* is the temperature at that point (H in the diagram); $\theta_w = 287°$ K. The intersection of this pseudo-adiabat with the isobar of original pressure 850 mb gives the *wet-bulb temperature* ($T_w = 7°$ C).

2.07. *Variability of the different air properties.*—It is important in weather analysis to recognize the variability of each of the air properties in various atmospheric processes. Those properties which tend to be conserved for periods of hours or a few days provide a means for identifying air parcels and large masses of air from one time to the next and also for analyzing modifications during their recent history.

10. The pseudo-adiabatic equation (for the *water* stage) recommended by the Aerological Commission of the International Meteorological Organization (IMO) (1947) is

$$(c_p + cr)dT/T - R_d(dp_d/p_d) + d(Lr/T) = 0 ,$$

where c is the specific heat of water, L the latent heat of vaporization, and L, r, and T refer to saturation. Pseudo-adiabats are obtained by step-wise integration of this equation for selected initial values of T at 1000 mb; these pseudo-adiabats are therefore isopleths for known values of θ_w (values of θ_E not known exactly), but their construction involves a series of approximations.

Rossby's approximate definition of θ_E, Eq 1.08(3), is based on the assumption $c_p + cr \simeq c_p$. Isopleths of Rossby θ_E are thus *approximate* pseudo-adiabats, according to the differential equation above. However, the difference in slopes is negligible, and the discrepancy in θ_w (or θ_E) between the two methods is ordinarily less than 0.5° K, as well as can be determined from existing results based on the recommended equation. The Rossby method provides a more accurate procedure of integration which yields pseudo-adiabats each with constant known values for both θ_E and θ_w (of consistent definition). The discussion in the text implies this method.

The individual change in any property is the change in that property experienced by an air parcel during a given interval in its life-history, irrespective of the direction or character of motion. In other words, it is the change observed while following the particular air parcel. This time rate of change is denoted by the derivative d/dt. If a certain property Q is conservative for a given process, then $dQ/dt = 0$ for that process.

We are concerned with three fundamental processes in the atmosphere, viz., adiabatic, radiation, and change of water phase. These processes may occur singly or in combination. Air motions and modifications are dry adiabatic as long as the air remains unsaturated and is not influenced by radiation or evaporation.

Figure 2.06 can serve for examining the individual change of an air parcel (A) as it is taken through various processes. By definition, in dry adiabatic displacement the potential temperature θ of the air is conserved, and in either dry adiabatic or pseudo-adiabatic displacements θ_E and θ_w do not change. Unless the parcel remains at the same pressure, its pressure and temperature vary in both dry adiabatic or pseudo-adiabatic motions. Therefore, the properties defined by temperature and pressure—density, specific volume, virtual temperature, saturation humidity, absolute humidity, equivalent temperature, dewpoint temperature, and wet-bulb temperature—are not conserved during dry adiabatic and pseudo-adiabatic processes. Relative humidity varies widely in dry adiabatic motions, but it is assumed constant at 100 per cent in the pseudo-adiabatic *lifting* process. The mixing ratio undergoes no individual change in dry adiabatic displacement; after saturation the air parcel is assumed to follow the pseudo-adiabatic process, in which the mixing ratio decreases as potential temperature increases (Eq 1.08[4]).

Through radiation cooling (warming) the potential temperature of the air parcel

decreases (increases). Its mixing ratio is conserved unless the cooling process takes the temperature below saturation. The effect of radiation on all air properties can be studied by moving point A along its isobar p while holding point D fixed. With radiation, only mixing ratio, dewpoint temperature, and vapor pressure are conserved, until saturation is reached. For constant radiative loss of heat, cooling is lessened after condensation sets in, but all three quantities decrease.

During evaporation the humidity of the air parcel increases, while its temperature is lowered by an amount proportional to the increase of humidity, if the heat of vaporization is acquired from the air parcel. The relation is such that dewpoint temperature and air temperature (points D and A in Fig. 2.06) converge toward a fixed wet-bulb temperature (point G) during the evaporation process. Evidently, such a process is conservative[11] in T_w, θ_w, θ_E, and T_E; but, since the temperature and moisture are changing, all other properties also change.

In summary, θ_E and θ_w are conservative except for radiation. Since they are the most conservative of the air properties, θ_E and θ_w are the best identifiers. Next come mixing ratio and potential temperature, both conserved during dry adiabatic processes. Since motions in the free atmosphere are predominantly dry adiabatic, at least for periods between successive synoptic charts, θ and r are useful parameters for air-mass identification and for studying modifications in both thermal and moisture properties of the atmosphere.

2.08. *The Rossby diagram.*—The Rossby or equivalent potential temperature diagram[12] was devised as a tool for air-mass

11. In the manner we have defined them, the properties T_w, θ_w, θ_E, and T_E are only *approximately* conserved during evaporation. For a detailed discussion refer to Petterssen, *op. cit.*, pp. 23–25.

12. C.-G. Rossby, *Thermodynamics Applied to Air Mass Analysis* ("Meteorological Papers, M.I.T.," Vol. I, No. 3 [Cambridge, Mass., 1932]).

analysis. Although the chart has had little use in recent years, its principle is still a useful one which we can incorporate in the tephigram. The Rossby diagram has coordinates log θ_d as ordinate and r as abscissa. Isopleths of θ_E are also found in the chart. To plot the "sounding" curve on this chart, the values of θ_d and r first are evaluated for each level of the radiosonde observation by reference to special tables or directly from the thermodynamic chart. These determine the *characteristic curve*, whose significance lies in its slope relative to the θ_E lines. This relation is an indicator of the *potential stability* of the air column.

The tephigram is a convenient substitute for the Rossby diagram. On the tephigram the ordinate is log θ, and the mixing-ratio lines are almost vertical. Except for the slight inclination of moisture lines and their nonlinear spacing, the θr grid of the tephigram has a layout similar to the $\theta_d r$ grid on the Rossby diagram. The characteristic curve can be plotted on the tephigram in lieu of the dewpoint curve, or in addition to it, and a single chart then suffices.

The approximate characteristic curve is obtained on the thermodynamic chart by locating the adiabatic condensation point for each level in the sounding and then connecting successive points. This curve is not a true characteristic curve for the sounding, since θ is used in place of θ_d, but the difference between the two curves is small. This curve provides the same information as the characteristic curve on the Rossby diagram. In addition, its proximity to the pT curve (with respect to displacement along dry adiabats) provides an indication of relative humidity and thus serves a purpose of the dewpoint curve.

2.09. *Wind-sounding charts.*—In daily weather analysis little emphasis is placed on representing the *variation* of wind along the vertical. The vector wind at any level is usually considered more significant than the vertical wind shear through that level, and, whenever it is required, the vertical variation can be estimated by comparing charts of wind at different levels. The use of wind-sounding charts, however, is basic for understanding not only motions but also the physical structure of the atmosphere.

One method of plotting a wind sounding is that often employed on a vertical cross section. Winds are entered at proper heights along the vertical axis in shaft-and-barb form similar to that used on surface charts. Or it may be represented by an arrow through the point, with speed indicated either vectorially or by number placed near the terminus of the arrow. In any case, wind directions on a sounding are plotted such that north wind is from top to bottom, west wind from left to right, and the other directions accordingly.

On the hodograph chart (Fig. 2.091) wind direction is the azimuth from the origin of the coordinate system; wind speed, the radial distance from the origin. The plotted point is the terminus of the wind vector, and the origin of the chart is the common initial point for all vectors. Each wind of the sounding is plotted the appropriate distance along the direction azimuth, and the point is labeled in height (thousands of feet). Successive points along the sounding are connected by a curve, the hodograph curve or *hodogram*, which is not to be confused with the horizontal projection of the sounding-balloon trajectory. The hodogram between any two plotted points gives the "shear" or vector change of wind between the two levels.

Reference is often made to the westerly (u) and southerly (v) components of the vector wind $\mathbb{C}$ at a given point in space. These components are evaluated graphically from the hodograph by projecting the vector wind (magnitude c, direction a) upon the coordinate axes of the chart (Fig. 2.092). The west-wind component u is the projection of the vector wind upon the x or west-east axis, and the south component is the projection upon the y-axis. Either or both of the components u and v may be positive or negative, according as the wind has a com-

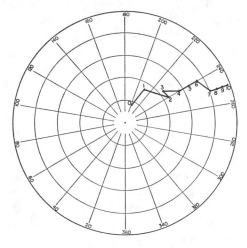

Fig. 2.091.—Hodograph chart

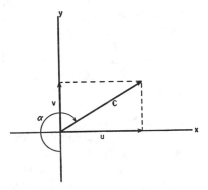

Fig. 2.092.—Components of the wind

ponent along or opposite to the direction by which u and v are defined.

As u and v are trigonometric functions of vector wind,

$$u = -c \sin \alpha , \qquad v = -c \cos \alpha .$$

By applying these equations or using tables made from them, the components of the wind can be evaluated directly from the wind reports.

2.10. *Meteorological maps.*—The most familiar tool of the weather analyst is the synoptic map, a chart with simultaneous observations for a reference surface in the atmosphere. The chart may be for a fixed level, for example, mean sea level, 10,000 feet above sea level, etc.; or it may be for some inclined constant property surface. In all cases, however, the same base map is employed.

The surface chart, commonly called the sea-level chart or simply "the synoptic map," is really not a constant-level chart. Pressure is the only element reduced to sea level. Temperature, dewpoint temperature, and wind refer only to the ground level; other elements of surface observations are visible phenomena in the free atmosphere. The complete surface synoptic observations nevertheless are plotted and analyzed on a single chart to give a composite picture.

Constant-level charts for the free atmosphere were standard in the United States until 1945; at that time they were replaced by constant-pressure charts. The constant-level charts contained synoptic observations of pressure, temperature, humidity, and wind. At present, constant-level charts are used systematically only for the horizontal wind fields at various levels in the atmosphere.

The most common of the constant property charts is that of constant pressure. Charts of mandatory pressure surfaces, with observations of temperature, humidity, wind, and also computed height above sea level, are prepared from data in the radiosonde reports. Isobaric charts serve the same general purposes as did the constant-level charts. Constant-potential-temperature (isentropic) charts have less general value in weather analysis.

From the mass of plotted data on a synoptic chart the analyst attempts to delineate with required accuracy the distributions of the variables—pressure, temperature, humidity, wind, etc. One soon learns that the dynamic factors in the weather, and therefore the basis for forecasting, are the patterns which these variables describe in space, considered either individually or together.

Since analysis is aimed at such geometric representation, the charts we use must be selected with that thought in mind. In the case of thermodynamic charts there is no restriction on the amount of distortion they create in the profile of a sounding curve; the thermodynamic chart contains implicitly only one space dimension, the vertical, and the sounding on the chart is not a curve in space. But other charts contain two space dimensions. Here the element of geometric similarity between patterns in the atmosphere and those on the charts is important. For that reason, and also because textbooks have omitted this subject, we shall examine the various maps and cross sections in some detail.

2.11. *Map projections.*—The process of map-making may be considered in elementary view as consisting of two steps. First, the surface of the earth is projected upon some fictitious geometric surface, the *image* surface, which is then developed by flattening into a plane surface. All properties of this image surface are prescribed by the type of map projection desired. Finally, the developed image surface is reduced in scale to a size suitable from the standpoint of economy, convenience, and required detail in the final map.

For our purposes the image is the surface of a right circular cone (Fig. 2.111a), the surface of a right circular cylinder, or a plane surface. The cylinder and the plane can be considered limiting cases of the cone, the former a cone of vertex angle zero and the latter a cone of vertex angle 180°. Although image surfaces can be placed over the globe at any angle with respect to the

earth's axis of rotation, we are concerned only with the case in which the earth's axis coincides with the axes of the three cones.

During the process of projecting the earth upon the image surface, the surface of the earth undergoes varying degrees of expansion and contraction. On the earth each geo-

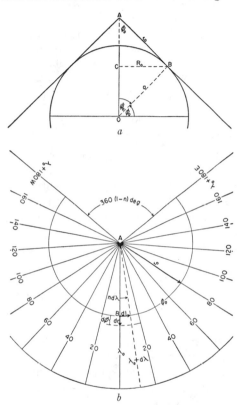

Fig. 2.111.—Scheme of the conic polar projection (tangent cone).

graphic element has correct shape and area, the scale is everywhere the same, directions or azimuths are true at any point, and the shortest distance between two points at the same level is an arc of a great circle. One or more of these properties must be sacrificed in projection; the properties to be retained in the final map determine the type of projection employed. *There is not a single projection satisfying all purposes for which maps are usually required.*

Map projections can be distinguished by the following principal properties:

1. Areas are exactly comparable anywhere on the map; these are *equal-area* or *authalic* map projections.
2. Correspondence between angles on the map and angles on the earth: *conformal*, isogonal, or orthomorphic maps. Conformality is also defined by equality of scale in all directions about a point, thus preserving the shapes of small geographical features at the expense of varying the area and scale over the map.
3. Conservation of azimuths or bearings from the center of the map to any other point, as in *azimuthal* projections.
4. Straight rhumb lines (loxodromes).
5. Straight great circles.

The Mercator and gnomonic projections are valuable for navigation. A Mercator map is unique in that any straight line is also a rhumb line; on a gnomonic map any straight line is a great circle. For climatic charts conservation of areas is often required, but that property is of secondary importance for meteorological charts. We are primarily concerned with representation of atmospheric patterns, necessitating preservation of true angles and shapes as far as possible, and, consequently, conformal projections are recommended. However, a compromise is made with the desirable feature of area preservation by using the conformal maps mostly for those latitudes of the map where the scale variation, and therefore the variation of areas, is relatively small.

The most obvious feature of conformal charts is the *right angle formed by the intersection of meridians and parallels* everywhere on the map. This feature is required by the definition of conformality; all angles are preserved in conformal projection except possibly at the pole. Consequently, the shapes of small geographic elements are preserved in projecting them conformally upon a flat map. Since the scale varies from one part of the map to another, distortion

must occur in the shapes of larger areas. The conformal maps have *straight meridians* because each line of projection is along a meridional plane. The latitude circles are orthogonal trajectories of the meridians.

The latitude at which the image surface is tangent to the earth is the *standard parallel* of the projection. For "secant" projections, in which the image surface inter-

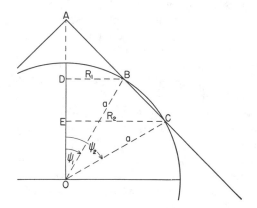

FIG. 2.112.—Scheme for conic polar projection (secant cone).

sects the surface of the globe, there may be two standard parallels, each at a latitude of intersection. Along any standard parallel, distance on the image surface is equal exactly to distance on the earth. The *image scale* σ is thus unity along a standard parallel, and at any other latitude it is greater or less than unity according as the image surface is above or beneath the surface of the earth. The scale is constant along any parallel for all projections with which we are concerned, but the scale varies outward from the pole in a manner determined by the form of the image surface. However, on a conformal map *the scale must be the same in all directions through any point*. This one principle is the basis for deriving all conformal projections mathematically and, in turn, for constructing the latitude and longitude grid of the maps upon which the outlines of land masses and all other geographic details are drawn.

The image scale σ is always the dimensionless ratio of image distance to earth distance. The reduction scale, m, is the final variation of S over the finished map is due entirely to the variation of σ. The ratio of area element dA on the map to that dA_E

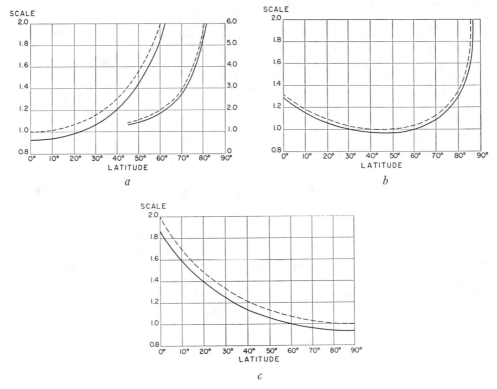

a

b

c

Fig. 2.113.—Scale variation with latitude on the (*a*) Mercator, (*b*) Lambert Conformal, and (*c*) polar stereographic projections. Dashed curves are for tangent projections, solid curves for secant projections. In (*a*) the scale on the right refers to the right set of curves.

map reduction at the standard latitude and is more familiarly known as just the "map scale." This reduction scale is constant for the entire map and is usually found printed on the map (e.g., 1:10,000,000). It is defined by the ratio:

$$m = \frac{\text{Distance on map at standard latitude(s)}}{\text{Corresponding distance on the earth}}.$$

The actual scale S at any point on the map is the product of m and σ; that is, $S = m\sigma$. Only at a standard latitude is the total map scale S equal to the reduction scale, and the on the earth for any projection is the square of S:

$$dA/dA_E = S^2 = (\sigma m)^2 = \sigma^2 \times \text{a constant for the map.} \qquad (1)$$

Area is conserved when the ratio, and thus where σ, is constant.

The three conformal maps required for synoptic charts are the *Mercator*, a cylindric projection; the *Lambert (or Gauss) Conformal Conic;* and the *polar stereographic*, a plane conformal-azimuthal projection. These are illustrated in Figures 13.03, 9.02*a*, and 1.021, respectively, and each is preferred for specific ranges of latitude. Since they are all special cases of a conformal conic projection, there is a general theory governing all three.

2.12. *Theory of conformal conic projections.*—In Figure 2.111a is the cross section of a tangent cone enveloping the northern hemisphere; this same image surface after development into a plane (by cutting along one meridian) is represented by the sector of a disk in diagram *b*. We may denote the earth's radius[13] by *a*, the standard latitude of the projection by ϕ_0, arbitrary latitude by ϕ and its colatitude by ψ, the radius of a latitude circle on the earth by $R(R = a \cos \phi = a \sin \psi)$, and the radius of an arc of latitude on the image surface by *r*.

The circumference of a given latitude circle on the earth is $L_E = 2\pi R$. The arc length of the same latitude line on the image is $L = 2\pi nr$, $(0 \leq n \leq 1)$, where $2\pi n$ is the angle subtended by the arc *L* on the developed image surface and *n* is a fraction which is a geometric property of the cone and is called the constant of the cone. As the scale is constant along a parallel, the scale along a circle of latitude is the ratio of arc length on the image surface to arc length on the earth:

$$\sigma_\phi = L/L_E = nr/R = (nr \csc \psi)/a \; . \; . \; (1)$$

At the standard latitude ϕ_0, $\psi = \psi_0$, $\sigma = 1$, and thus $n = \cos \psi_0$.

Along a meridian an increment of arc on the earth is given by $dr_E = ad\psi$. On the image surface the corresponding increment of radius for the same increment of latitude is $dr = \sigma_\lambda dr_E$, or $dr = \sigma_\lambda ad\psi$, in which the subscript λ denotes constant longitude. With conformality, $\sigma_\phi = \sigma_\lambda$, and substitution of σ_ϕ from Eq (1) for σ_λ in the last

13. The earth is an oblate spheroid with equatorial radius 6378.4 km and polar radius 6359.9 km. For most meteorological purposes the earth is assumed a sphere of mean radius 6370 km, but, where more accurate maps are required, the earth's ellipticity is considered.

equation gives

$$dr/r = n(\csc \psi)d\psi =$$
$$(\cos \psi_0)(\csc \psi)d\psi \; .$$

This leads to

$$r = (a \tan \psi_0)[(\tan \psi/2)/$$
$$(\tan \psi_0/2)]^{\cos \psi_0} , \quad (2)$$
$$\sigma = (\sin \psi_0)(\csc \psi)[(\tan \psi/2)/$$
$$(\tan \psi_0/2)]^{\cos \psi_0} . \quad (3)$$

Eq (2) gives the image radius of any latitude circle; multiplying *r* by the selected reduction scale *m* gives the radius of the latitude arc on the map. The angular spacing of unit meridians on the map is *n* compass degrees about the pole. If the standard latitude is 45°, the spacing of unit meridians is $n = \cos 45° = 0.707$ compass degrees, and the hemisphere is mapped within a 255° sector on the developed image surface. The resulting map is the Lambert conformal conic projection with one standard parallel. The scale is indicated by the dashed line in Figure 2.113b.

If the standard parallel is at the equator, the projection is then a Mercator. In this limiting case $\psi_0 = 90°$ and $n = 0$. Therefore, from Eqs (2) and (3),

$$r = \infty , \quad (4)$$

and

$$\sigma = \csc \psi . \quad (5)$$

The latitude circles are thus parallel straight lines, and meridians also are parallel straight lines. Meridians are equidistant, and their spacing on the map equals their spacing on the earth at the equator, multiplied by the reduction scale. The spacing of parallels on the map is a function of latitude.

If the cone is reduced to the opposite extreme, it becomes the polar stereographic projection. In this case, ψ_0 approaches zero, $n = 1$,

$$r = 2a \tan \psi/2 , \quad (6)$$

and

$$\sigma = 2/(1 + \cos \psi) = 2/(1 + \sin \phi) . \quad (7)$$

The parallels are closed concentric circles about the pole, and the meridians are spaced with the same angle on the map as on the earth. The scale at the equator is twice the scale at the pole (Fig. 2.113c); thus, the spacing of parallels is twice as large near the equator as near the pole.

The International Meteorological Organization recommends[14] use of the so-called "secant" projections instead of tangent projections. In secant projections the image surface intersects the surface of the earth, and there are two standard parallels in the Lambert and Mercator projections and only one in the polar stereographic. Secant projections are no substantial improvement over tangent projections; scale variations and distortions of shapes remain about the same. (In the common secant projections, the distortion is about 9/10 as large as in the corresponding tangent projections.) A main advantage served by secant projections is the smaller average absolute departures from the standard latitude scale over a large working area of the map.

The derivation of the secant conformal conic projection is similar to the previous one. In this case there are two standard parallels, ϕ_1 and ϕ_2, as shown by B and C in Figure 2.112. Then $dr/r = n (\csc \psi) d\psi$, and the radius of any latitude arc on the image surface is

$$r = (a/n)(\sin \psi_1)[(\tan \psi/2)/(\tan \psi_1/2)]^n$$
$$= (a/n)(\sin \psi_2)[(\tan \psi/2)/$$
$$(\tan \psi_2/2)]^n . \quad (8)$$

The constant of the cone, n, is determined by equating the scales at the two standard parallels:

$$n = (\log \sin \psi_1 - \log \sin \psi_2)/$$
$$(\log \tan \psi_1/2 - \log \tan \psi_2/2) . \quad (9)$$

In a projection with standard parallels 30° and 60°, $n = 0.716$.

14. See the "Reading References" at the end of this chapter.

This projection is the well-known Lambert Conformal Conic projection with two standard parallels. Its scale is

$$\sigma = (\sin \psi_1/\sin \psi)[(\tan \psi/2)/$$
$$(\tan \psi_1/2)]^n$$
$$= (\sin \psi_2/\sin \psi)[(\tan \psi/2)/$$
$$(\tan \psi_2/2)]^n . \quad (10)$$

The scale variation on a Lambert projection with standard parallels at 30° and 60° is shown by the solid curve in Figure 2.113b.

For a Mercator projection with two standard parallels equidistant from the equator, $n = 0$, and r must be infinite for all latitudes (thus again straight and parallel latitude lines). The scale is

$$\sigma = (\sin \psi_1)(\csc \psi) . \quad (11)$$

It is unity at the standard parallels, less than 1 between standard parallels, greater than 1 where $\phi > \phi_1$, and infinite at the poles (Fig. 2.113a). The spacing of meridians on the image surface is their spacing on the earth at the standard latitudes.

The other limiting case of the Lambert, as n approaches unity, is the polar stereographic with standard parallel ϕ_0. The radius of any latitude circle on the image plane is

$$r = a(1 + \cos \psi_0)(\tan \psi/2) , \quad (12)$$

and the scale is

$$\sigma = (1 + \cos \psi_0)/(1 + \cos \psi)$$
$$= (1 + \sin \phi_0)/(1 + \sin \phi) . \quad (13)$$

The variation of scale with latitude is shown by the continuous curve in Figure 2.113c.

2.13. *Lambert conformal projection.*—The Lambert projection, Eqs (2), (3), (8), and (10) above, is the ideal conformal projection for middle latitudes where the scale variation is a minimum, as deduced from the curves in Figure 2.113b. The recommended standard parallels for the secant projection are 30° and 60° in the northern hemisphere (10° and 40° in the southern hemisphere), giving a maximum scale variation between 25° and 65° latitude about 7 per cent.

Distortion of shapes in projection is a result of scale variation over the map. The degree of distortion is indicated by the slopes of the curves in Figure 2.113. Distortion is small between standard parallels, and it increases outward particularly in the direction of the pole. Since distortion occurs only in south-north direction, geographic features oriented predominantly west-east (e.g., Cuba) show little distortion in projection; but large land masses with longest diameter south-north (North America, Greenland) are thrown considerably more out of shape.

Since scale variation is smallest in the vicinity of minimum scale, the projection is most area-conserving in those latitudes (Eq 2.11[1]). In the Lambert with two standard parallels this occurs between standard latitudes, and the projection becomes more and more nonauthalic as the image scale increases toward equator and pole. For this reason the Lambert is the preferred conformal projection for middle latitudes—for meteorological maps of North America and the United States.

2.14. *The Mercator projection.*—The Mercator is no doubt the most misunderstood projection of the three. Because of the great distortions it creates in high latitudes, it does not readily appear as a conformal projection.

Since the location of parallels on the map could not be shown from the general equations for conformal projections, we give the necessary formula here. Instead of employing radial distances r outward from the pole, it is necessary here to work from the equator. From Figure 2.111a observe that geodetic distance ds_E along a meridian at sea level is the product of the earth's radius and the increment of latitude: $ds_E = ad\phi$. In the image the corresponding distance in that interval of latitude is $ds = \sigma_\lambda ds_E$, or $\sigma_\lambda = ds/(ad\phi)$. Along a parallel on the earth the relation between arc distance and the angle of longitude is $dL_E = a(\cos \phi)d\lambda$. On the image, however, the distance between meridians is constant over the entire Mercator projection and is given by $dL = a(\cos \phi_1)d\lambda$, where ϕ_1 is the standard parallel of the projection. By equating the two scales σ_ϕ and σ_λ, $ds = a(\cos \phi_1)(\sec \phi)d\phi$. Integration then yields

$$s = a(\cos \phi_1) \log_e \tan (\pi/4 + \phi/2) , \quad (1)$$

where s is image distance along meridians, measured from the equator. This gives the distance of any latitude line from the equator on the image surface. It is applicable to both tangent and secant projections.

The scale variation on the Mercator is shown for one hemisphere in Figure 2.113a for the projection with standard parallel at the equator (*dashed*) and for standard parallels at $22\frac{1}{2}°$ north and south latitude (*full line*). In the region within 20° of latitude on either side of the equator the scale varies by less than 10 per cent, thus indicating the Mercator has little distortion and is largely equal-area in that region. In high latitudes the variation of scale becomes very large, as evidenced by the slopes of the curves in Figure 2.113a. Nevertheless, just as the Lambert serves for middle latitudes, the Mercator is the best suited for the tropics.

2.15. *The polar stereographic projection.*—This is the simplest and most convenient of the three projections; it is a perspective projection for which the point of projection is the opposite pole. The polar stereographic is also the only one of the three projections giving a continuous map of a hemisphere. Consequently, it is the recommended projection not only for maps of the polar regions but also for hemispheric charts. For this it is preferable to the Mercator because of smaller scale variations. The recommended standard parallel for stereographic charts is 60° latitude.

On the polar stereographic projection the scale is smallest at the pole (Fig. 2.113c) and increases quite uniformly outward. The scale at the equator is precisely twice the scale at the central pole, irrespective of location of the standard parallel, and the scale at the opposite pole would be infinite. As in other projections, this chart is most area-conserving in the latitude of minimum scale, that is, near the pole.

2.16. *Recommended scales and details for surface maps.*—The scales required for working charts vary with the amount of detail necessary in the analysis and with the area of the earth desired within the borders of a map of convenient paper size. Working charts for areas the size of the United States have scales 1:5,000,000, 1:7,500,000, 1:10,-000,000, or 1:12,500,000. Recommended scales become progressively smaller as larger areas of the earth are included in the map; for hemisphere charts the scales 1:20,000,-000 or 1:30,000,000 are usually preferred.

On the printed chart there is usually a conversion scale relating distances on the chart to distances on the earth for every latitude. One should refer to this scale for distances instead of using the standard latitude scale for the map.

In addition to the latitude-longitude grid and the outlines of land masses, there are several other desirable features which should be entered in the base map. The larger lakes, wide river estuaries, and mountain ranges have an important bearing on the analysis of at least the lower atmosphere, and they should be part of the map. Surface topo-graphic contour lines and topographic layer shading or relief are desirable for reference purposes in analysis.

Station circles about 2 or 3 mm in diameter are provided on the map for each synoptic station; the size of the circle will vary somewhat with the scale of the map. The index number for the station may be given in small print. The numerical index system for North America, together with names, exact locations, and elevations of all stations, is given in the U.S. Weather Bureau publication *International Station Numbers for North and Central America*.

2.17. *Measurement of geodetic distances from the map.*—In analysis and forecasting we frequently need to convert map distances to (great-circle) distances on the earth. The conversion scale found printed on most maps provides one method of measurement, but this operation is somewhat involved for routine work. To avoid this difficulty, we use a system of length units based on the latitude-longitude grid of the map. We speak of distances in "degrees of latitude" or in "degrees of longitude" instead of so many inches or centimeters on the map or so many miles or kilometers on the earth.

Along a meridian the arc distance δs_E on the earth corresponding to latitude displacement $\delta \phi$ in degrees of latitude is

$$\delta s_E = (2\pi a/360)\delta\phi° = k\delta\phi°,$$

in which the constant k is the arc distance of 1° of latitude. It is independent of map scale and also independent of latitude on a spherical earth. The radius of the earth (a) has average value 6370 km. The above formula then becomes

$$\delta s_E \simeq 111 \, \delta\phi° \quad \text{in kilometers},$$

$$\delta s_E = 60 \, \delta\phi° \quad \text{in nautical miles},$$

and

$$\delta s_E \simeq 69 \, \delta\phi° \quad \text{in statute miles}.$$

From these conversion formulas, true distance along a meridian can be obtained readily from the latitude difference between two points on the map without referring to the map scale.

In a west-east direction the distance can be found from the angular longitude difference $\delta\lambda$ between two points, but it is necessary to specify latitude. Arc distance δL_E along any latitude circle on the earth is

$$\delta L_E = (2\pi a/360)(\cos\phi)\delta\lambda°$$

$$= k\,(\cos\phi)\delta\lambda° \,.$$

The constant k is the same as in the previous equation.

of large-scale features in the weather pattern.

2.18. *Measurement of geodetic directions from the map.*—In measuring directions of lines on maps as true directions on the earth, we have the difficulty that most straight lines on maps do not have constant direction. The Mercator projection is unique in that any straight line is a rhumb line—a line of constant direction on the earth.

On the Mercator, meridians and parallels are straight and mutually perpendicular; either set of grid lines can serve as the reference for angular measurement. The direc-

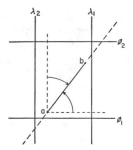
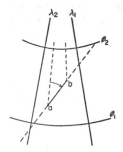

FIG. 2.18.—Measurement of direction on map projections

For most purposes it is desirable to have a length unit the same for all latitudes and directions. For this unit we usually adopt the arc distance 1° of latitude. The procedure for measuring short distances in any direction is to measure the map distance with a straight edge or dividers, lay off this distance on a meridian at the same latitudes, and read the distance in units degrees of latitude. It is best to apply this method to short successive segments of the total distance to provide for scale variation along the meridian.

Use of the degree-of-latitude unit is not restricted to distance. Speed can be expressed in degrees of latitude per day: 1° latitude per day is $2\frac{1}{2}$ nautical miles per hour ($2\frac{1}{2}$ knots). The degree-latitude-per-day unit is usually applied to the movement

tion of straight-line segment ab in the Mercator projection in Figure 2.18a can be measured with the ordinary protractor by setting the origin at a and the reference line of the protractor along either the meridian or the parallel through point a, and finally by reading the value of the angle at the intersection of ab, or its extension, with the angular scale. Since the straight line ab on the map has invariant direction on the earth, the same result is obtained by placing the origin at any point along ab or its extension. To measure the direction at any point of a *curved* line on a Mercator projection, draw the tangent to the curve at that point and measure the direction of the tangent in the manner described above.

On other conformal projections the only straight rhumb lines are meridians; all others are curved concave toward the nearer pole. Consider the straight-line segment through points *a* and *b* of Figure 2.18*b*. We desire the direction of this line at point *a*. Since meridians are the only straight grid lines, the meridian through *a* serves as the reference line in the measurement. The direction of *ab* at *a* is the angle between this particular meridian and segment *ab*. To measure the direction of the segment at any other point (e.g., point *b*), the reference meridian is the meridian through *b*. Obviously, the direction of the line at *b* is different from *a*.

2.19. *Great-circle arcs.*—The arc of a great circle on the earth or in the atmosphere is the shortest spherical distance between two points the same elevation from mean sea level. A great-circle arc between two designated points of equal elevation is the line of intersection of that plane and that spherical surface defined by the center of the (spherical) earth and the two points. There is an obvious reason for selecting the great-circle arc in nautical and aeronautical navigation, especially between distant points on the earth. Certain practicing meteorologists therefore should have at least a working knowledge of the concept of great circles.

The great-circle arc between the pole and any other point on the earth is along the meridian of that point. *All meridians are great circles through the poles.* Thus, on either of the three meteorological projections, the great-circle arc toward or across the pole is a straight line on the chart. The equator is the only latitude circle which is also a great circle.

The gnomonic projection is the only one for which all arcs of great circles are straight lines on the map. Figure 2.19 shows part of the coordinate grid in a polar gnomonic projection. The surface of the earth is projected from its center upon a plane tangent to the earth (in this particular case tangent at the pole). As the equator is an infinite distance from the pole in the polar gnomonic, it is the only great circle that cannot be defined in that chart. In the more general case the point of tangency with the earth is not the pole but some other desired central point on the earth's surface.

In the image surface of a polar gnomonic projection, the distance *r* from the pole to any point not beyond the equator is $r = a \cot \phi$, where *a* is the radius of the sphere and ϕ the latitude of the point in question. From this it can be shown that through any point in the projection the scale along the meridian is $\sigma_\lambda = \sec^2 \psi = \csc^2 \phi$, and the scale σ_ϕ along the parallel is $\sigma_\phi = \csc \phi$. Since $\sigma_\lambda \neq \sigma_\phi$ except at the pole, the projection is nonconformal. At all places other than the point of tangency in a gnomonic projection, the scale is greatest in a radial direction from the point of tangency and least in the orthogonal directions.

Because it is often necessary in analyzing motions and in preparing aeronautical forecasts to determine, at least qualitatively, great-circle arcs on common meteorological charts, we outline a procedure here. First, the initial and terminal points of the route are located in a gnomonic grid, and a straight line is drawn between them. Then a sufficient number of points along this line are located by their same latitude-longitude coordinates in the working chart, and a smooth curve is drawn connecting them. The result is the desired great-circle arc on the working chart. (A similar method applies for drawing rhumb lines using the Mercator.)

The above procedure is a basis for visualizing the general shapes of great-circle arcs on the adopted meteorological charts. For instance, since the scale increases more rapidly equatorward on the polar gnomonic than on the polar stereographic projection, a straight line other than a meridian on the gnomonic must appear as a line curved concave toward the pole on the stereographic (consider the equator, for example). On the Mercator all great circles, other than equator and meridians, are curved convex toward the nearest pole, and the curvature increases with latitude. The curvature of great-circle arcs on the Lambert conformal projection is left for deduction by the reader.

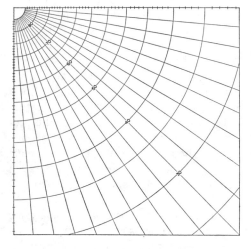

Fig. 2.19.—Section of latitude-longitude grid of a polar gnomonic projection.

2.20. *Plotting the synoptic map.*—Regulations and customs governing procedure in plotting synoptic charts would cover several pages. We will not discuss current plotting procedures here, as methods are constantly changing, and details depend on the individual weather services. Detailed instructions for plotting procedure can be found in manuals devoted specifically to that purpose.

Although isentropic charts are not standard at present and the necessary data are not transmitted in the required form in the radiosonde report, we describe the essential plotting procedure at this point. An isentropic chart represents a surface of constant potential temperature. Data are obtained for each station from the intersection of the plotted sounding curve with the particular potential temperature line on the thermodynamic chart. The pressure and temperature of the isentropic surface are obtained at the intersection of the pT curve with this "isentrope." The mixing ratio for the isentropic surface is the value on the pr curve *at this same pressure*. The condensation pressure P_c is a more useful humidity measure for isentropic analysis; it is the pressure at which the isentrope intersects the characteristic curve on the thermodynamic chart. The wind is taken from the appropriate level in the wind report.

2.21. *Cross sections.*—The third general class of meteorological diagrams is designed for illustrating and studying the structure of the atmosphere along a vertical plane. The coordinates of the simple cross section are height above sea level and horizontal distance along the line of the cross section. The abscissa scale usually is taken the same as the standard latitude scale of the base map or some convenient multiple or fraction thereof. In order to bring out necessary vertical detail in the analysis while maintaining satisfactory dimensions of the paper, it is necessary to have the vertical scale a hundred to five hundred times as large as the horizontal scale. The exact ratio of scales will vary with the range of horizontal distances to which the base cross section is applied.

Even though the surface of the earth and all level surfaces are curved, the abscissa of the chart is made perfectly straight. This is analogous to using the same base map for all levels in the atmosphere without bothering about vertical variation in scale. In fact, between two geographic points the horizontal distance at 10 km is only 100.2 per cent of the distance at sea level. The discrepancy introduced is no greater than in locating station circles in the map and is smaller than the error of assuming that a

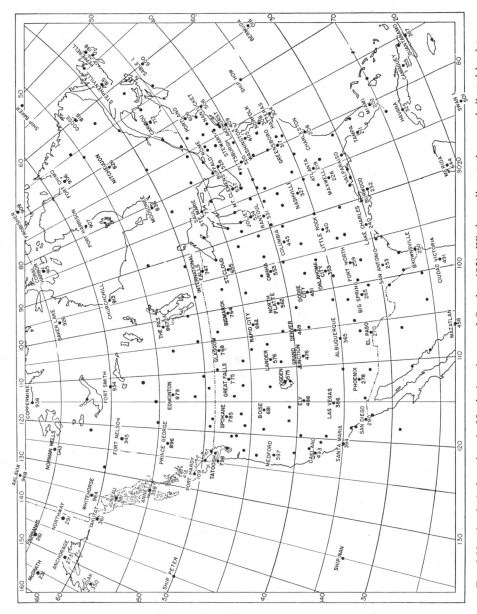

Fig. 2.20.—Aerological stations in North America, as of October 1951. Wind-sounding stations are indicated by dots; radiosonde stations by name and numerical index. (Note.—Some of the military stations were omitted to avoid overcrowding in the drawing.)

sounding-balloon trajectory is vertical. Thus, for practical and theoretical reasons, the base cross section is drawn with straight and parallel lines for constant elevation above sea level.

Cross sections may be classified in two types on the basis of purposes they serve. The *synoptic cross section* is the more important for analysis. On it, simultaneous sounding (and surface) observations are plotted for each reporting station along the section. It permits a detailed analysis and study of the atmosphere at a given time. In contrast, the *route cross section* is not synoptic and usually is a forecast rather than analysis. It is used primarily for flight briefings. On it the forecaster enters expected locations, in both dimensions, of phenomena of interest to flying personnel. These include clouds, weather, critical temperatures, turbulence, and so on. All elements are drawn schematically in a route section in a conventional system which permits a quick picture of the forecast. The chart is often used in reverse by the flight crew to enter descriptions of actually observed phenomena en route. It is a valuable and practical aid in flight briefing and in weather observation by reconnaissance.

2.211. *The synoptic cross section.*—We stated earlier that the coordinates of the cross section are height and horizontal distance. However, in radiosonde reports the vertical reference for each point is pressure and not height. There are two methods by which the height scale is converted to pressure. The first involves use of the pressure-height relation in a standard atmosphere. Certain values of pressure are located on the linear height scale of the ordinate, and straight isobars are then extended across the chart. The vertical scale of the new chart is still linear in height and is approximately logarithmic in pressure. The other method in preparing the base chart makes the ordi-

nate the logarithm of pressure decreasing upward (Fig. 6.02*b*). With this modification the vertical scale is not strictly linear in height.[15]

The cross section is an effective link between the thermodynamic sounding charts and the various synoptic charts. The ordinate of the cross section is similar to that of the thermodynamic chart, and the horizontal scale of the cross section can be made exactly the same as the map scale. The plotted sounding on a thermodynamic chart illustrates atmospheric structure only along the vertical coordinate. On the cross section a plotted sounding shows the same structure in the vertical, but, in addition, when compared with adjacent soundings, the structure in a horizontal dimension is obtained also.

Whenever attention is to be placed on details of the surface layer of the atmosphere, it is desirable to enter the earth's topographic outline within the coordinates of the cross section if the surface is very irregular. Such a procedure is recommended on cross sections across western North America,

15. In a cross section made with linear height scale in a standard atmosphere, the vertical scale of a plotted sounding varies inversely with the (virtual) temperature in the following manner:

$$\frac{\text{Vertical scale in sounding}}{\text{Vertical scale of cross section}}$$

$$= \frac{\text{Temperature in the standard atmosphere}}{\text{Actual temperature}}.$$

In a $-\log p$ cross section the vertical scale in any sounding is inversely proportional to the temperature:

Vertical scale in sounding

$$= \frac{\text{Constant for the cross section}}{T}.$$

As the temperature normally decreases upward in the troposphere, the vertical scale increases upward in this region. With isothermal conditions along a sounding, the scale does not vary with height.

In both cross sections the vertical scale varies between adjacent soundings according as the temperature varies horizontally.

where the various mountain ranges have to be given special attention in a careful analysis. The outline can be made by referring to a physiographic map.

2.22. *Time sections.*—A variety of diagrams, in some respects related to the cross sections described above, find some limited use as supplementary aids especially in local analysis. Most of them can be classified as *time sections.* One coordinate is time; the other is either a space dimension (x, y, or z) or some other quantity—pressure, temperature, a wind component, etc. The latter type of chart might include such elementary forms as the barograph chart and the thermograph chart—time profiles of pressure and temperature at a fixed point.

Of the time sections with a space dimension as the second coordinate, the vertical time section (t, z diagram) is the most extensively used (Figs. 12.06). Along the ordinate is height or some function of pressure almost linear in height, as in the synoptic cross section. Sounding and surface data are entered in the chart at the proper location above the station along the vertical line for the time of the observation. This section is analyzed in the same manner as the synoptic cross section. The result is a motion-picture scheme giving the sequence of events at a fixed station. This chart is particularly useful for analysis in regions of sparse data where a great deal of emphasis must be placed on the weather history at one station to determine the synoptic analysis in the vicinity. A simple modification with surface observations only is also useful (Fig. 9.02*d*).

The t, x (or t, y) section is an aid in following the progression of weather along a line of closely spaced observing stations (Fig. 12.04*a*). Observations from each station are plotted as a function of time and distance from an arbitrary origin. From the analysis of this chart the velocity, acceleration, lateral spreading, change in shape, or change of intensity of such things as cloud areas and rain areas, for example, can be studied in greater detail. These could be developed into useful tools for local analysis and forecasting.

PROBLEMS AND EXERCISES

I. *Construction of emagram.*

1. On two-cycle semilogarithmic paper lay out a pressure scale from 1000 mb to 10 mb on the logarithmic ordinate, with pressure decreasing upward. Label the abscissa in a linear temperature scale from $-100°$ C on the left to $40°$ C on the right (scale should be about $20°$ C per inch, or larger). Within this T, $-\log p$ coordinate grid, draw the following lines:

 (*a*) Potential temperature $240°$, $260°$, $280°$, $300°$, $320°$, $340°$, and $360°$ K. Use Eq 1.07(1) for potential temperature and the following values for the constants involved: $R_d = 2.8704 \times 10^6$ and $c_p = 1.004 \times 10^7$ ergs per gram-degree.

 (*b*) Saturation mixing ratio 0.1, 1.0, and 10.0‰, using a table for saturation vapor pressure as a function of temperature. These lines need not be drawn for pressures less than 100 mb.

 (*c*) Pseudo-adiabats for $\theta_E = 300°$ K and $360°$ K.

2. From the first law of thermodynamics, the relation of energy (specific energy) to pressure and temperature can be expressed by

$$\oint dq = -\oint \alpha dp = -R_d \oint T d(\ln p) \, .$$

Within the upper right portion of your diagram, outline a rectangular area $10°$ C in width and corresponding to 0.1 calorie per gram of energy. The conversion from mechanical to heat units is 1 calorie = 4.185 joules = 4.185×10^7 ergs per degree.

3. Summarize the characteristic features of the emagram.

II. *Construction of tephigram.*

    4. On a sheet of graph paper lay out a log θ scale along the ordinate (θ increasing upward) and draw horizontal lines for 240°, 260°, 280°, 300°, 320°, 340°, and 360° K. Along the abscissa enter a linear temperature scale from $-100°$ C to 40° C, and draw the isotherms for each multiple of 10°. Enter the isobars for each 100 mb and the additional lines listed in (I) above. Observe that the equation for an isobar is log $T =$ log $\theta +$ constant.

    5. Expressed in terms of T and θ, the work done in a cyclic process on a tephigram is given by

$$\oint dq = c_p \oint T d(\ln \theta) \ .$$

Within the lower right portion of the diagram, draw a rectangle of width 10° C whose area corresponds to 0.1 calorie per gram of specific energy.

    6. Summarize the characteristic features of the tephigram.

III. *Evaluations from a sounding.*

    7. Plot the pT, pT_s, and characteristic curves for the following sounding on a thermodynamic chart.

ppp	TT	T_sT_s	hhh	ppp	TT	T_sT_s	hhh
1000.....	....	....	290	536......	-13.2	-29.8	....
978.....	8.8	-0.8	....	514......	-15.2	-37.5	....
910.....	4.2	-8.2	....	500......	-17.7	MB	18,250
850.....	2.2	-17.5	4650	400......	-27.8	MB	23,630
763.....	-0.5	-4.2	....	366......	-33.5	-45.8	....
732.....	-1.8	-4.2	....	315......	-39.2	-47.5	....
700.....	-3.5	-6.2	9740	300......	-41.2	....	30,210
677.....	-4.8	-8.2	....	200......	-56.3	....	38,920
663.....	-4.8	-15.2	....	154......	-52.4	....	....
567.....	-13.2	-18.5	....	104......	-59.8	....	....
....	....	....	....	100......	....	....	53,400

    8. Evaluate and tabulate the following quantities for each reported level in the sounding: p, T, T_s, r_s, r, r/r_s, θ, θ_E, θ_w, T^*, T_E, T_w, T_c, and P_c. Compute α and ρ at 700 and 300 mb·

    9. Determine T, T^*, θ, θ_E, and r/r_s at 600 mb.

    10. In the lower margin of the chart label the abscissa as a linear height scale with a range of at least 50,000 feet on the left to zero on the right. Then, using the reported heights in the sounding, draw a pressure-height curve from 1000 to 100 mb. From this curve determine the approximate heights of the significant points in the sounding.

    11. Compute the average vertical lapse rates of temperature and potential temperature between 500 and 400 mb.

    12. Compute the vertical lapse rate of temperature in the vicinity of 800 mb.

    13. Compute the vertical change in temperature lapse rate at 300, 200, and 154 mb. At which pressure is the tropopause located in this sounding?

    14. Determine the change in T, T_s, T_w, θ, θ_E, θ_w, r, r_s, r/r_s, α, and P_c of an air parcel initially at 850 mb if subjected to

      (a) Adiabatic lifting (lowering) by 100 mb,

      (b) Radiation heating (cooling) by 10° C,

      (c) Increase of 2‰ of moisture through evaporation.

Repeat steps (a) and (b) for a parcel at 700 mb. Summarize in detail your conclusions concerning the conservativeness of these various thermodynamic quantities.

IV. *Wind sounding.*

00228	0328	20333	0433	40519	2705	62703	2808
82812	2915	02820	22726	42731	62733	82742	02847
52757	02663	52658	(1949 U.S. Wind Code; wind speed in knots.)				

15. Plot the above wind sounding on a hodograph chart. Draw the hodogram.
16. Draw the vector difference in wind between 2000 and 12,000 feet.
    (a) Determine its direction and magnitude.
    (b) Determine its westerly and southerly components.
17. Evaluate the westerly and southerly components for each reported level.
    (a) On graph paper draw the vertical profiles of west and of south wind.
    (b) Check your result in 16(b), using differences in components from 2000 to 12,000 feet.

V. *Map projections.*
   18. Lay out the grid lines for each 10° latitude and each 10° longitude covering an area 60° longitude in width and extending from 30° N to 60° N latitude with standard latitude scale 1:50,000,000 for the following conformal projections:
       (a) Mercator, standard at the equator,
       (b) Lambert, standard at 30° and 60° latitude,
       (c) Polar stereographic, standard at 60° latitude.
       Show all computations.
   19. A circular curve on the surface of the (spherical) earth is centered at 45° N; the surface diameter of this curve is 222 km. Describe the shape and compute the approximate area inclosed by this curve when projected by
       (a) Mercator, standard at $22\frac{1}{2}°$ latitude,
       (b) Lambert, standard at 30° and 60° latitude,
       (c) Polar stereographic, standard at the north pole,
       assuming standard latitude scale 1:10,000,000. Compute the net change in projected area on each of these projections as the curve moves along the surface of the earth from 45° N to 70° N without change of shape or area.
   20. Repeat the procedure of (19) for a square whose sides are 222 km on the earth and centered at 60° N.
   21. Two concentric circles centered at 45° N on the surface of the earth have radii 9° and 11° of latitude, respectively. Determine the distance between these two curves at the four extremities of latitude and longitude if projected upon a Mercator map of scale 1:10,-000,000 standard at the equator.
   22. Compare the shapes of the six great-circle arcs between New York and London and between Los Angeles and New York on the three recommended meteorological projections. Explain the differences.
   23. Distortion (or deformation) of shapes by projection may be expressed by the difference $\partial\sigma_\lambda/\partial y - \partial\sigma_\phi/\partial x$, in which σ_λ and σ_ϕ are the scales along meridians and parallels, respectively; y is the sea-level distance along meridians (positive poleward); and x is this distance along parallels (positive eastward in the northern hemisphere). Using this formula and the equations for scale in the text, tabulate the distortion in units $(100 \text{ km})^{-1}$ at 0°, 15°, 30°, 45°, 60°, 75°, and 90° latitude for the Mercator standard at $22\frac{1}{2}°$, Lambert conformal standard at 30° and 60°, polar stereographic standard at 60°, and polar gnomonic standard at the pole. How does distortion in a given conformal map projection correlate with variation of scale?
   24. Compare numerically the distortion in the tangent projection to the distortion in the corresponding recommended secant projection for the Mercator, Lambert, and polar stereographic. Make the same type of comparison for area instead of distortion.

VI. *Cross sections.*
   25. Using footnote 15, determine the percentage variation of vertical scale in a $-\log p$ cross section corresponding to a (virtual) temperature change from $-60°$ C to 25° C.
   26. In a certain $-\log p$ cross section the distance between the horizontal lines for 700 mb and

300 mb is 20 cm. For an individual plotted sounding on this cross section the actual 700-mb height is 9900 feet and the 300-mb height is 30,500 feet.

(a) What is the average vertical scale of this cross section between the two levels for this particular sounding?

(b) If the mean virtual temperature between these two pressures is $-20°$ C, what would be the average vertical scale between these same pressures when the mean virtual temperature is $-40°$ C?

27. On a given vertical cross-section chart the distance from 1000 mb to 700 mb is 10 cm. If the horizontal scale in use is 1 cm = $1°$ ϕ, for a situation in which the average heights of 1000 mb and 700 mb are 200 feet and 10,200 feet, respectively,

(a) find the actual slope of a line in the atmosphere corresponding to a slope of 1 on the chart;

(b) determine the true shape of a curve represented by a circle on the chart.

READING REFERENCES

BERRY, F. A., BOLLAY, E., and BEERS, N. R. (eds.). *Handbook of Meteorology*, pp. 361–71, 594–600, Sec. IX. New York: McGraw-Hill Book Co., 1945.

BRUNT, D. *Physical and Dynamical Meteorology*, pp. 76–82. Cambridge: Cambridge University Press, 1939.

BYERS, H. R. *General Meteorology*, pp. 165–67, 175–79. New York: McGraw-Hill Book Co., 1944.

DEETZ, C. H., and ADAMS, O. S. *Elements of Map Projection*. 5th ed. Coast and Geodetic Survey, U.S. Department of Commerce, Special Publication No. 68. Washington, D.C., 1944.

HAURWITZ, B. *Dynamic Meteorology*, pp. 78–84. New York: McGraw-Hill Book Co., 1941.

HEWSON, E. W., and LONGLEY, R. W. *Meteorology: Theoretical and Applied*, pp. 54–65. New York: John Wiley & Sons, 1944.

HOLMBOE, J., FORSYTHE, G. E., and GUSTIN, W. *Dynamic Meteorology*, pp. 34–39. New York: John Wiley & Sons, 1945.

INTERNATIONAL METEOROLOGICAL ORGANIZATION (ORGANIZATION MÉTÉOROLOGIQUE INTERNATIONALE), Publication No. 71, 1949. Resolutions 26–33, pp. 55–61.

PETTERSSEN, SVERRE. *Upper Air Charts and Analyses*. NAVAER, 50-IR-148. Washington, D.C., 1944.

———. *Weather Analysis and Forecasting*, pp. 16–22, 50–56. New York: McGraw-Hill Book Co., 1940.

STEERS, J. A. *An Introduction to the Study of Map Projections*. 7th ed. London: University of London Press, Ltd., 1949.

UNITED STATES DEPARTMENT OF THE AIR FORCE. *Preparation of Weather Maps and Charts*. Washington, D.C.: Hdqrs. A.W.S., 1950.

ZÖPPRITZ, K., and BLUDAU, A. *Leitfaden der Kartenwurfslehre*, Vol. I: *Die Projectionslehre*. Leipzig and Berlin: B. G. Teubner, 1912.

CHAPTER 3

Hydrostatics and Static Stability

I. HYDROSTATICS

3.01. *Gravity.*—Gravity is the force exerted by the earth on unit mass at rest relative to the earth and is approximately 980 dynes at sea level.

According to Newton's law of universal gravitation, the absolute gravitational force imparted by the earth to unit mass in the

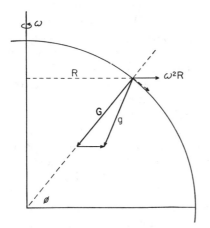

Fig. 3.01.—Relation between gravitational attraction (G) and gravity (g).

atmosphere is of magnitude $G = C/r^2$, where C is a constant and r the distance from the center of the earth to the unit mass. The centrifugal force acting on unit mass, at the surface and at rest relative to the earth, due to the earth's rotation is represented by the vector of magnitude $\omega^2 R$ in Figure 3.01, where ω is the angular velocity of the earth, 7.29×10^{-5} sec^{-1}, and R the distance from the earth's axis. If we assume the earth is spherical, the gravitational vector G is along the earth's radius, and apparent gravity g is the vector resultant of G and

$\omega^2 R$. The component of centrifugal acceleration along the spherical surface is $\omega^2 R \sin \phi$ and is responsible for the elliptical shape of the earth. The mass and shape of the earth are so adjusted as a consequence that apparent gravity is perpendicular to sea level, and neither this nor G is directed precisely toward the center of the earth. From empirical determinations the sea-level value of g at 45° latitude is[1] 980.616 cm sec^{-2}. At any latitude ϕ,

$$g_\phi = 980.616(1 - 0.0026373 \cos 2\phi$$
$$+ 0.0000059 \cos^2 2\phi) . \quad (1)$$

As gravitational attraction is inversely proportional to the square of the distance from the center of the earth, gravity varies also with elevation in the atmosphere. If elevation above sea level is z and the appropriate value of the earth's radius is a, then gravity g_z at any point is related to sea-level gravity g_0 at the same latitude by $g_z = g_0/(1 + z/a)^2$. The local acceleration of gravity anywhere in the atmosphere is

$$g = g_\phi/(1 + z/a)^2 . \quad (2)$$

At sea level, gravity is approximately 978 cm sec^{-2} at the equator and 983 cm sec^{-2} at the pole—a total variation of about 0.5 per cent. Along the vertical gravity decreases by about 3 cm sec^{-2} in the first 10 km above sea level. This small variation in any direction suggests that, for many approximate computations involving gravity, a constant value 980 cm sec^{-2} may be used.

1. *Smithsonian Meteorological Tables* (6th rev. ed.; Washington, D.C., 1951).

Eq (2) is a general formula accounting for gravity variations with latitude and elevation only. These are the only space variations of gravity with which we need be concerned in large-scale statics. Besides these, there are more localized gravity variations controlled by the topography of the earth's solid surface and the distribution of mass beneath. Local variations can be large in mountainous territory and near steeply inclined ocean bottoms.

3.02. *Geopotential.*—Displacement of mass upward or downward in the permanent gravitational force field of the atmosphere requires work done against or by gravity. This change in potential energy per unit mass, $\delta\Phi$, is the product of the force (per unit mass), g, and the displacement in the direction of the force, δz.

$$\delta\Phi = g\delta z, \qquad (1)$$

where Φ is the specific potential energy of the mass (ergs gm^{-1}). The potential at any level z_1 is

$$\Phi_1 - \Phi_0 = \int_{z_0}^{z_1} g\delta z.$$

If mean sea level is the level of reference, at which $z_0 = \Phi_0 = 0$, the potential at any point is

$$\Phi = \int_0^z g\delta z, \qquad (2)$$

where Φ is *geopotential*. Since it is a function only of position, fixed surfaces of constant geopotential can be defined in the earth's atmosphere. The surface $\Phi = 0$ is at mean sea level, and Φ increases upward. Except at sea level, geopotential surfaces are not surfaces of equal height, that is, not parallel to sea level; nor are they surfaces of constant gravity. Since at any height above sea level gravity decreases from pole to equator, a layer of geopotential depth $\delta\Phi$ has greatest geometric depth at the equator. Geopotential surfaces diverge from pole to equator

and intersect equal-height surfaces except at sea level.

Even though its geometric height above sea level might change in the process, displacement of an air parcel along a geopotential surface does not change its potential energy. Hence, a geopotential surface is a level surface—a surface everywhere perpendicular to gravity. Level surfaces (and charts) are preferred to equal-height surfaces for atmospheric analysis; a level surface has nowhere a component of gravity along it.

The concept of geopotential is used in measurement of heights in the atmosphere; it combines the two variables g and z in a single variable Φ. Geopotential, when interpreted as height, gives dynamic distances between level surfaces. From Eq (2), the geopotential Φ in c.g.s. units is numerically about 980 times the geometric height. To make the geopotential and geometric coordinates nearly equal numerically, the right side of Eq (2) is divided by the dimensionless constant 980 if g is in c.g.s. units. The new unit of height we shall call the *geopotential "length" unit,*[2] and height in this unit will be denoted Z. If z is in meters, then Z is in geopotential meters (gpm); if z is in feet, Z is in geopotential feet (gpft). Formulas for conversion of geometric length to geopotential length, or to geopotential, are

$$\delta\Phi = 980\delta Z = g\delta z,$$

$$\Phi = 980Z = gz. \qquad (3)$$

2. Most textbooks in meteorology describe the *dynamic meter* as the standard unit of height in the atmosphere. However, the geopotential unit was and still is being used under a different name (0.98 dynamic meter) by the U.S. Weather Bureau for all height computations in the atmosphere. The geopotential unit has now become the international standard (IMO Resolution No. 78, 1947).

The relation between dynamic and geopotential length units is

1 geopotential meter = 0.98 dynamic meter.

The dimensions of Z are specific energy. However, here we are concerned only with the numerical value. Wherever $g = 980$ cm sec^{-2}, 1 Z-meter (or foot) equals 1 geometric meter (or foot). Where $g > 980$ cm sec^{-2}, a Z-unit is smaller than the geometric unit. At sea level the variation in length of 1 gpm from pole to equator is 0.5 cm.

3.03. The hydrostatic relation.

—In a vertical column of unit cross-sectional area, the pressure p_2 at level z_2 is the total downward force per unit area due to the weight of the overlying air. At a lower level z_1 the pressure p_1 is, in hydrostatic equilibrium,[3] the sum of the downward force per unit area at z_2 plus the force contributed by the weight of the column segment between z_1 and z_2. This segment has volume $\delta z = z_2 - z_1$, mass $\rho \delta z$, and weight $g\rho \delta z$. Therefore,

$$p_1 - p_2 = -\delta p = g\rho \delta z . \qquad (1)$$

Thus,

$$dp/dz = -\rho g , \qquad (2)$$

and, with reference to Eq 3.02(3),

$$dp/dZ = -980\rho . \qquad (3)$$

Eqs (2) and (3) give common forms of the hydrostatic equation, the basic statical formula in aerological observation and analysis. In Eq (3) p is in dynes cm^{-2}, Z in geopotential cm, and ρ in gm cm^{-3}. By substituting for density from the equation of state, we get

$$dp/dZ = -980(p/R_d T^*) . \qquad (4)$$

The drop of pressure in the vertical is directly proportional to the pressure and inversely proportional to virtual temperature.

3.04. The barometric or hypsometric formula.

—The preceding equation is simplified

3. Hydrostatic equilibrium is a balance between all forces along the vertical. The actual acceleration of a parcel along the vertical is ordinarily so small in comparison with gravity that the hydrostatic assumption is sufficiently accurate for all practical purposes.

by introducing the value 2.8704×10^6 ergs gm^{-1} deg^{-1} for R_d.

$$dZ = -KT^* dp/p . \qquad (1)$$

The factor K, the barometric or *hypsometric* constant, is 29.28980 gpm per degree or 96.09494 gpft per degree.

From Eq (1) we find that near 1000 mb (near sea level) and temperature 280° K, an increment 1 mb corresponds to vertical depth 8.2 meters, about 27 feet. The same temperature at 100 mb would give ten times

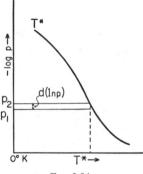

FIG. 3.04

that depth per millibar. At 220° K unit isobars are 21 feet apart near 1000 mb, and about 210 feet apart near 100 mb. Hence, the vertical change of pressure varies considerably with pressure and temperature in the atmosphere, but within usual ranges in the troposphere the rate varies much more with pressure than with temperature. In the portion of the atmosphere where we are most concerned, pressure varies by a whole order of magnitude, from 1000 mb to 100 mb, while temperature varies about one-tenth as much, from roughly 200° K to 300° K. Thus we may use pressure as an *approximate* though crude coordinate of height.

From observed values of pressure and temperature in a sounding it is possible to determine the height of any pressure by integrating Eq (1), which we rewrite

$$dZ = -KT^* d(\ln p) . \qquad (2)$$

The depth of a layer is a linear function of its virtual temperature and of the difference in logarithms of the bounding pressures. Now consider the virtual temperature sounding on the T, $-\ln p$ diagram in Figure 3.04. A thin stratum of air is bounded by pressures p_1 and p_2, and its abscissa extends from $0°$ K to T^* °K. Its area on the chart is therefore

$$dA = -T^* d(\ln p) .$$

This area (multiplied by a scale factor) gives the depth of that layer in the atmosphere.

3.05. *Pressure-height evaluation.*—The procedure followed in height evaluation is in principle merely integration of such areas on the thermodynamic chart. Height evaluation is done in steps by integrating thicknesses between isobars bounding *standard pressure layers*. The standard pressures presently in use by the U.S. Weather Bureau are 1000, 900, 850, 800, 700, 600, 500, 400, 350, 300, 250, 200, 175, 150, 125, 100 mb, and so on (Circular P, 1950 ed.). The thickness of each such layer above the station, the thickness of the layer between station pressure and the first standard isobar, and the elevation of the station above sea level are all the data necessary to complete pressure-height determination in a sounding.

The depth of a layer represented between two standard isobars on the chart is the integral of Eq 3.04(2):

$$Z_2 - Z_1 = \Delta Z$$
$$= -K \int_1^2 T^* d(\ln p) . \qquad (1)$$

If mean virtual temperature between levels Z_2 and Z_1 is denoted by $\bar{T}^*$, then[4]

$$\Delta Z = K\bar{T}^* \ln p_1/p_2 . \qquad (2)$$

4. If common logarithms ($\log_{10}$) are used, the value of the constant is

$$K = 221.267 \text{ gpft per degree, or}$$

$$K = 67.442 \text{ gpm per degree .}$$

Values of $K \ln p_1/p_2$ are given in Table 3.05 for convenient pressure layers. Multiplying the constant by $\bar{T}^*$ for the layer gives its depth. The values in the table are increments of thickness corresponding to each $1°$ C. The columns for $100/(K \ln p_1/p_2)$

TABLE 3.05

NUMERICAL VALUE OF THE FACTOR $K \ln p_1/p_2$ AND ITS RECIPROCAL FOR CERTAIN PRESSURE LAYERS

PRESSURE LAYER (MB)	$K \ln p_1/p_2$		$100/(K \ln p_1/p_2)$	
	For gpft	For gpm	For gpft	For gpm
1000–850....	15.617	4.760	6.403	21.007
850–700....	18.658	5.687	5.360	17.584
700–600....	14.813	4.515	6.751	22.148
600–500....	17.520	5.340	5.708	18.726
500–400....	21.443	6.536	4.664	15.300
400–300....	27.645	8.426	3.617	11.868
300–250....	17.520	5.340	5.708	18.726
250–200....	21.443	6.536	4.664	15.300
200–150....	27.645	8.426	3.617	11.868
150–100....	38.963	11.876	2.567	8.420
200–100....	66.608	20.302	1.501	4.926
300–200....	38.963	11.876	2.567	8.420
500–300....	49.088	14.962	2.037	7.569
700–500....	32.333	9.855	3.093	10.147
1000–700....	34.275	10.447	2.918	9.572

indicate the increment of temperature to change the thickness by 100 gpft or 100 gpm.

The mean virtual temperature of a layer is determined graphically on the thermodynamic chart (Fig. 3.051). It is the isotherm found by equalizing area *a*, the area added by the mean virtual isotherm, with area *b*, the area excluded by the mean virtual isotherm. When the virtual temperature sounding through the layer is a straight line, the mean virtual temperature is the average of the virtual temperatures at the bounding pressures, or the virtual temperature at the midpoint of distance in the layer.

In current Weather Bureau practice evaluating mean virtual temperature in a pressure layer is done as follows. First, the mean temperature of the layer is determined graphically from the pT curve in the manner shown by Figure 3.051. Then the mean relative humidity in the layer is determined. Multiplication of the mean relative humidity by the virtual temperature increment at saturation, which is indicated by the distance between successive vertical marks along particular isobars on the working thermodynamic chart (see Fig. 2.033b), gives the mean virtual temperature correction. This mean correction for the layer is added to

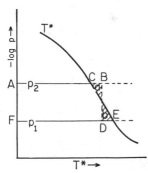

Fig. 3.051

the previously obtained mean temperature to get the desired mean virtual temperature of the layer.

From Eq (2) a table can be prepared for each standard pressure layer, giving ΔZ as a linear function of $\bar{T}^*$. The first such tables, prepared by V. Bjerknes, gave thickness in units dynamic meters (1 dynamic meter equals 1.02 gpm).[5] The U.S. Weather Bureau (Circular P) uses a modified version of the original Bjerknes tables; the thicknesses are in units 0.98 dynamic meter, or geopotential meter according to our definition. Those tables are standard for pressure-height evaluation in the United States. A few similar tables are found in the Appendix.

Finding the height of the first standard isobar above the station is more difficult, since for that layer there are three variables: depth ΔZ, virtual temperature, and surface pressure. These are provided for by two

separate tables. The principle can be shown by rewriting the barometric formula

$$\Delta Z = K(273 + \bar{t}^*)(\ln p_1 - \ln p_2) , \quad (3)$$

where $\bar{t}^*$ is now mean virtual temperature ° C, p_1 is surface pressure, and p_2 is the constant pressure of the first standard isobar. Expansion of Eq (3) gives

$$\Delta Z = K(273)(\ln p_1 - \ln p_2)$$
$$+ K\bar{t}^*(\ln p_1 - \ln p_2) . \quad (4)$$

The first term on the right is the thickness $(\Delta Z)_0$ of the layer for $\bar{t}^* = 0°$ C. The other term is the correction $(\Delta Z)'$ for departure of the mean virtual temperature from $0°$ C. Thus $\Delta Z = (\Delta Z)_0 + (\Delta Z)'$. From Eq (4), $(\Delta Z)' = (\Delta Z)_0(\bar{t}^*/273)$. Therefore,

$$\Delta Z = (\Delta Z)_0 + (\Delta Z)_0(\bar{t}^*/273) . \quad (5)$$

Thus for each $1°$ C in $\bar{t}^*$, the thickness is corrected by $1/273$ of its value at $0°$ C. The first of the two tables gives $(\Delta Z)_0$ as a function of surface pressure for a given first standard isobar. In a second table is the correction $(\Delta Z)'$ in terms of $(\Delta Z)_0$ and $\bar{t}^*$. The correction is added to or subtracted from $(\Delta Z)_0$ according as $\bar{t}^*$ is positive or negative; the result is the true thickness of the layer. Although one table would suffice, the advantage of having the $(\Delta Z)'$ table separate is its independence of pressure, and it can be used for any pressure layer. Addition of the depth of the first layer to the station elevation[6] gives the height of the first standard isobar above sea level.

In pressure analysis of the free atmosphere we frequently must recompute heights of mandatory pressures in transmitted soundings, or check the consistency of reported heights, because of errors in coding or transmission of the report or even in the original height evaluation. The analyst therefore should be familiar with techniques

5. See the "Reading References" at the end of this chapter.

6. More exactly, the floor of the instrument shelter.

for checking pressure-height computations.

For more rapid recomputing it is convenient to have sets of thickness scales for each standard pressure layer printed on the thermodynamic chart or on a transparent overlay for the chart (Fig. 3.052). These section with the thickness scale the layer thickness is read directly, without needing the numerical value of temperature. Thus at least one step in the computation is omitted, and the inaccuracy from rounding off temperature to the nearest whole or

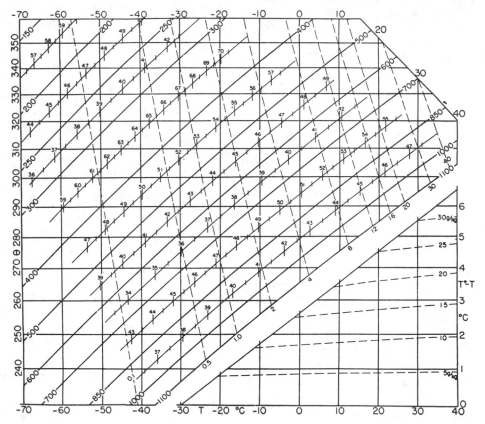

FIG. 3.052.—Tephigram with thickness scales (printed values of thickness are in hundreds of geopotential feet).

scales are entered *equidistant* from the bounding isobars of the layer. The location of each 100-foot interval (or smaller intervals) of thickness along the scale can be found as a function of temperature from Eq (2); Table E in the Appendix gives some of these. The advantage of the scales is readily apparent. The mean virtual isotherm for the layer can be located on the chart in the manner described previously. At its inter-

half-degree does not enter. Such a diagram with complete sets of scales was suggested by Väisälä for ordinary pressure-height computations;[7] experience shows that it is also a valuable auxiliary in pressure analysis of the upper atmosphere. Practical techniques for deriving the height of the first

7. See J. Holmboe, G. E. Forsythe, and W. Gustin, *Dynamic Meteorology* (New York: John Wiley & Sons, 1945), pp. 115–19.

standard isobar, to accompany the thickness diagram, are described in the following section.

3.06. *Approximate calculations of depth.*—

There are several empirical methods for determining thicknesses and heights on plotted soundings when time is a factor and close approximations can suffice. A crude measure of thickness is obtained by using the standard atmosphere scale printed on most charts. That scale is useful for estimating depths of shallow strata in the sounding and also, more crudely, the heights of

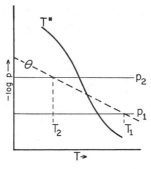

Fig. 3.06

points up to a few thousand feet above sea level. The accuracy of this method will be evident in Section 3.09.

Greater accuracy in approximation of the height of an arbitrary pressure is obtained from the pressure-height curve made from the transmitted heights of mandatory pressures in a sounding. The same technique can be used to determine the depth of any layer in the sounding. The thermodynamic chart also can be provided with a number of pressure-height scales or nomograms for certain values of temperature which give the thickness of layers at different temperatures.[8]

The depth of a thin stratum in the sounding also can be computed independently on

8. Cf. N. Herlofson, "The T, $-\log p$ Diagram with Skew Coordinate Axes," *Meteorologiske Annaler* (Norway), Met. Inst., Band II, No. 10 (Oslo, 1947).

any chart by the "adiabatic" method. It is quick and gives results of sufficient accuracy for many purposes. Consider the layer bounded by pressures p_1 and p_2 in Figure 3.06; the virtual temperature is indicated by curve T^*. The mean dry adiabat θ determines temperatures T_1 and T_2 at the bottom and top of this layer, and the difference is $\delta T = T_2 - T_1$. Now, the dry adiabatic lapse rate is $-(\delta T/\delta Z)_\theta = 980/c_p \simeq 9.8°$ C/ gpkm $= 0.98°$ C/100 gpm. Then along a dry adiabat on the chart,

$$\delta Z \simeq -102\ \delta T \text{ gpm },$$

or

$$\delta Z \simeq -335\ \delta T \text{ gpft }.$$

Thus, each $1°$ C difference in temperature along the adiabat corresponds to about 100 meters or 330 feet in depth. After little practice, one finds this a rapid method for approximating depths of arbitrary pressure layers.

The differential barometric formula (Eq 3.04[1]) provides a simple direct method for computing depths of shallow strata. Suppose we are interested in the depth of a 10-mb layer in the vicinity of 480 mb at temperature $250°$ K. Substitution into the formula gives

$$dZ = 96.1 \times 250 \times (10/480)$$
$$= 500 \text{ gpft }.$$

In general, the value of pressure to be used in the denominator of this formula is a logarithmic mean pressure for the layer. That pressure $\bar{p}$ must satisfy the relation $dp/p = \ln(p_1/p_2)$, where p_1 and p_2 are the bounding pressures and dp the difference between them. Actually, if the layer is thin, as in the above example, that pressure is sufficiently close to $(p_1 + p_2)/2$ and not significantly different from either boundary pressure. The differential barometric formula can be applied also to deeper layers, in which case the pressure to be substituted is the logarithmic mean, and the virtual temperature is the mean for the layer. For example, for the layer 700/1000 mb the proper mean pressure is 841 mb. The value 850 mb leads to an error only 1 per cent in thickness of the layer. The differential formula can be used for evaluating the height of a first standard or mandatory pressure surface above a station.

Bellamy's *Pressure-Height Slide Rule* is another method for obtaining the depth of thick layers in a sounding.[9] The slide rule is designed so that any one of the variables T^*, ΔZ, p_1, or p_2, can be evaluated quickly given the three others.

By use of Table F in the Appendix, or the nomogram prepared from it, the height of the 1000-mb surface above ground is obtained from the surface pressure and the mean virtual temperature between. If the surface pressure is less than 1000 mb, the distance down to the 1000-mb surface is obtained as follows: Assume a virtual temperature lapse rate between the surface and 1000 mb by standard procedure (as outlined in Sec. 3.11), compute $\bar{T}^*$ for the layer, and obtain the thickness from the table by interpolation. The height of the first standard isobar (e.g., 850 mb) above the surface can be obtained indirectly by use of this table also. Take an arbitrary lapse rate of T^* from the surface down to 1000 mb, evaluate the thickness of this layer from the table, evaluate the thickness of the layer from 1000 to 850 mb by methods described before, and subtract the two values.

3.07. *Extrapolation of upper-level pressures and heights from indications at the surface.*—Aerological networks over most of North America and large parts of other continents are sufficiently dense that pressure analysis can be made with confidence at various levels from the surface to 200 or

9. J. C. Bellamy, *Pressure-Height Slide Rule* (Chicago: University of Chicago Press, 1943).

100 mb. However, owing to the significance we attach to broad-scale pressure patterns, pressure analysis over the oceans is just as important. Since few aerological observations are made at sea, especially over the Pacific Ocean (where they are most needed for subsequent events over North America), we must devise methods for improvising or extrapolating upper-level pressure-height data over oceans. Surface reports over the sea are sufficiently numerous to determine a reliable surface pressure pattern in many areas. From surface pressures, 1000-mb heights can be derived with necessary accuracy. For this purpose there are tables and nomograms presently in use by the various services and at analysis centers (Table F in Appendix).

Given the height of the 1000-mb surface over any geographic point, all that is needed to determine the height of an upper pressure surface p_2 is an estimate of the mean virtual temperature in the layer between 1000 mb and p_2. There are two methods for obtaining this temperature. First, it is possible to estimate the mean virtual temperature (and therefore the thickness) of the layer by considering present indications and the past history of the air mass. This so-called "layer method" of extrapolation makes use of thickness, or departures from normal values of thickness, as an air-mass property. Adding the thickness thus obtained to the height of the 1000-mb surface gives the height of the upper pressure surface.

The second method of extrapolation derives thickness from observed surface (or 1000-mb) temperature and an appropriate lapse rate of virtual temperature, $\bar{\gamma}$, which can be defined by

$$\bar{\gamma} = (T_2^* - T_1^*)/\Delta Z$$

$$= (T_2^* - T_1^*)/K\bar{T}^* \ln p_1/p_2 .$$

Since $\bar{T}^* = (T_2^* + T_1^*)/2$, the barometric

equation is

$$\Delta Z = (K \ln p_1/p_2)T_1^*/$$
$$(1 - \tfrac{1}{2}K\bar{\gamma} \ln p_1/p_2) . \quad (1)$$

Thus, for a particular layer defined by p_1 (usually 1000 mb) and upper pressure p_2, the depth is given uniquely by T_1^* at the base pressure and the mean lapse rate in the layer. Since T_1^* is known, the only estimation is a suitable value of $\bar{\gamma}$. This is estimated from local indications in surface observations—from cloud and weather types—and also from the previous history of the air mass. Tables and nomograms have been suggested for this purpose and for solving Eq (1) in terms of T_1^* and $\bar{\gamma}$.

3.08. *Standard atmospheres.*—The concept of a standard atmosphere arose through certain aeronautical requirements, in particular, evaluation of aircraft performance and calibration of altimeters. In weather analysis the use of a standard atmosphere is limited at present, but it serves as a reference for the relation between pressure, height, and also temperature in the vertical.

The NACA[10] (U.S. Standard) Atmosphere, the ICAN[11] Atmosphere, and the new ICAO[12] Atmosphere are similar in many respects. In all cases the temperature is virtual temperature. The U.S. Standard Atmosphere is defined as follows:

Pressure at mean sea
  level: 1013.25 mb
Temperature at mean
  sea level: 15° C

10. U.S. National Advisory Committee for Aeronautics (see "Technical Reports," Nos. 147, 218, and 538).

11. International Commission for Air Navigation.

12. Resolution 78 of the Twelfth Conference of Directors, International Meteorological Organization, Aerological Commission, Washington, D.C., 1947.

Lapse rate of temperature in troposphere: 6.5° C per kilometer
Tropopause: −55° C (35,332 feet; 234 mb)
Lower stratosphere: −55° C (isothermal)
Acceleration of gravity: 980.665 cm sec⁻² (at any altitude)

The ICAN Atmosphere differs only in the following respects: gravity = 980.62 cm sec⁻², sea-level pressure = 1013.2 mb, change of lapse rate occurs at 11 km (36,080 feet), and the temperature above 11 km is −56.5° C. The ICAO Atmosphere assumes a standard gravity $G_s = 980.665$ cm sec⁻², and the height H is given in a geopotential unit according to $G_s dH = 980dZ = gdz$. At $H = 0$ the pressure is 1013.25 mb and the temperature 15° C = 288.16° K. At the tropopause the temperature is −56.5° C; the lapse rate above is isothermal to at least 20 km, and the lapse rate below is 6.5° C per kilometer (all height units ICAO geopotential).

The pressure, temperature, and height are uniquely defined for any point in either standard atmosphere. These data for intervals of pressure and height in the U.S. Standard Atmosphere are given in the Appendix. One should be familiar with the standard heights (nearest 100 feet) of pressures 1000, 850, 700, 500, 400, 300, 200, and 100 mb. These are helpful for approximate positions in the vertical and also for comparative purposes in pressure analysis. For instance, 300 mb is instantly thought of in terms of 30,000 feet. As a more detailed example, consider the following: In a radiosonde report the 1000-mb height is 560 feet and the 700-mb height 10,460 feet, giving thickness 9900 feet. Since the depth of the same layer in the U.S. Standard Atmosphere is 9520 feet, or 380 feet less than the reported sounding, reference to Table 3.05 shows that the layer must be 380/34, or about 11° C warmer than in the standard atmosphere. Similar deductions can be made concerning lapse rates of temperature by comparing vertical temperature differences in the sounding with the known lapse rate in the standard atmosphere.

3.09. *The* D *method of pressure-height evaluation.*—With the U.S. Standard Atmosphere as reference, Bellamy devised a convenient technique for pressure-height analysis which involves departures from the standard atmosphere.[13] For a complete discussion of the idea, the reader is referred to the original paper. The principle can be extended to space analysis of pressure, temperature, and wind, and the reference also can be any sounding besides the U.S. Standard Atmosphere.

The "altimeter correction" D is the difference between the *actual* height Z of a pressure surface and the height Z_p of the same pressure surface in the standard atmosphere. Thus,

$$D = Z - Z_p. \qquad (1)$$

For instance, if the actual height of 700 mb is 10,000 feet,[14] $D = 10,000 - 9880 = 120$ feet. Or, if the actual height of 700 mb is 8800 feet, $D = 8800 - 9880 = -1080$ feet. Thus in the first example, if the altimeter is set with sea-level pressure in the standard atmosphere, an aircraft flying at 700 mb (9880 feet indicated altitude by the pressure altimeter) would be actually 120 feet above the indicated altitude, and in the second example it would be 1080 feet below the indicated altitude, as the altimeter was calibrated according to the standard atmosphere.

The actual temperature is indicated through the specific temperature anomaly S at the given pressure

$$S = (T - T_p)/T_p, \qquad (2)$$

13. J. C. Bellamy, "The Use of Pressure Altitude and Altimeter Corrections in Meteorology," *Journal of Meteorology*, II (1945), 1–79.

14. Henceforth, in all references to height in the atmosphere, we will use simply "feet" or "meters," on the understanding that geopotential units are implied.

where T is actual temperature and T_p the standard atmosphere temperature at the same pressure. Thus, for a temperature $10°$ C ($= 283°$ K) at 700 mb, where $T_p = -4.6°$C $= 268.4°$ K, $S = +0.055$. Usually S varies between -0.2 and 0.1. The virtual specific temperature anomaly is, in analogous form,

$$S^* = (T^* - T_p)/T_p. \qquad (3)$$

It can be shown that

$$dD/dZ_p = S^*. \qquad (4)$$

Eq (4) is the hydrostatic equation in the D and S system of parameters. In integral form it is

$$D_2 - D_1 = \int_1^2 S^* dZ_p. \qquad (5)$$

This is the analogue of Eq 3.05(1). With the mean value of S^* for the layer, Eq (5) is

$$D_2 - D_1 = \bar{S}^*(Z_{p2} - Z_{p1}), \qquad (6)$$

which is the formula used for pressure-height computations.

The left side of Eq (6) is the difference in departures of height of each boundary pressure from their respective heights in the standard atmosphere. In other words, $(D_2 - D_1)$ is the departure of actual thickness from the known thickness of the pressure layer in the standard atmosphere. Once $(D_2 - D_1)$ has been found, the actual thickness is given by the algebraic sum of $(D_2 - D_1)$ and the standard atmosphere thickness. For example, if D at 500 mb is $+360$ feet and D at 700 mb is -40 feet, $(D_{500} - D_{700}) = +400$ feet, and the actual thickness is $8400 + 400 = 8800$ feet (standard atmosphere thickness 8400 feet). Accordingly,

$$\Delta Z = (1 + \bar{S}^*)(Z_{p2} - Z_{p1}), \qquad (7)$$

obtained by adding $(Z_{p2} - Z_{p1})$ to both sides of Eq (6). Eq (7) states that *the thickness of any pressure layer is the product of* $(1 + S^*)$ *for the layer and its known standard thickness.*

Eq (7) can be rewritten in the form

$$\Delta Z = (\Delta Z)_p + (\Delta Z)_p(T^* - \bar{T}_p)/\bar{T}_p.$$

We may rewrite Eq 3.05(5) as

$$\Delta Z = (\Delta Z)_0 + (\Delta Z)_0(\bar{T}^* - 273)/273,$$

in which the zero subscripts refer to mean virtual temperature 0° C. The analogy between the two equations is striking. In the first the reference thickness and specific temperature anomaly are based on the standard atmosphere, while in the second they are based on the isothermal atmosphere of temperature 0° C.

tion is the actual height Z of the 998-mb surface, D is the difference between station elevation and Z_p for surface pressure.

If 1000-mb height is desired, we may use (for this example) the surface virtual temperature as representative of the fictitious air column extending downward from 998 mb to 1000 mb. For this layer $\bar{S}^* = -0.008$. Since the standard atmosphere thickness of the 2-mb layer is 50 feet, to the nearest 10 feet, Eq (6) gives about -0.5 foot for the difference in altimeter corrections between

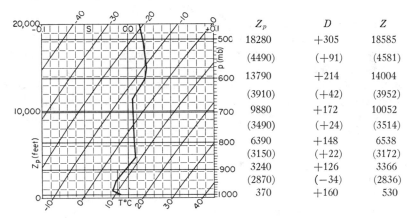

	Z_p	D	Z
500	18280	+305	18585
	(4490)	(+91)	(4581)
600	13790	+214	14004
	(3910)	(+42)	(3952)
700	9880	+172	10052
	(3490)	(+24)	(3514)
800	6390	+148	6538
	(3150)	(+22)	(3172)
900	3240	+126	3366
	(2870)	(−34)	(2836)
1000	370	+160	530

FIG. 3.09.—Illustration of the D method of pressure-height evaluation on the pastagram

The only variable to be evaluated on the thermodynamic chart is the mean value of S^* for the layer. This is accomplished quickly if isopleths of S are provided in the printed chart along with isobars and isotherms. On the pastagram (Fig. 2.032d) the S lines are vertical and linearly spaced. On ordinary thermodyanamic charts, all S lines roughly parallel the pT curve for the standard atmosphere, and they are nearly equally spaced.

In Figure 3.09 is a pT^* sounding with the following surface conditions: pressure = 998 mb, elevation = 580 feet, $Z_p = 420$ feet, and $D = +160$ feet. The surface pressure and station elevation are known. Surface Z_p is the height of the surface pressure (998 mb) in the standard atmosphere and is obtained from tables. Since station eleva-

998 and 1000 mb. At 1000 mb D is evidently $+160$ feet, and the height of the 1000-mb surface is $(370 + 160) = 530$ feet above sea level. These results are tabulated in the right margin of Figure 3.09. The thickness of the 2-mb layer could have been evaluated directly by substitution into Eq (7).

To find the height of 900 mb, first evaluate $\bar{S}^*$ in the layer from 998 mb to 900 mb; $\bar{S}^* = -0.012$. For this layer $(Z_{p2} - Z_{p1}) = (3240 - 420) = 2820$ feet. Eq (6) gives $(D_{900} - D_{998}) = -34$ feet. At 900 mb D is thus 126 feet, and the 900-mb surface is 3366 feet above sea level. The operation is continued for the remaining pressure layers until the D and the height of the uppermost standard pressure level are determined.

On the chart in Figure 3.09 observe that Z_p lines are drawn for each 1000 feet and S

lines for each 0.01, thus forming a series of squares each with area corresponding to 0.01 × 1000, or 10 feet. Since the right side of Eq (6) is merely the summation of such squares in the area bounded by $S = 0.0$, by the actual sounding curve T^*, by the upper pressure p_2 (or height Z_{p2}), and by the lower pressure p_1 (or height Z_{p1}), it is evident that the thickness of an arbitrary pressure layer can be determined by adding algebraically ten times the number of squares in this area to the standard atmosphere thickness of the layer.

A principal benefit introduced by the D method is the way in which it simplifies and co-ordinates pressure analysis. The pattern of D on a pressure surface is the same as the pattern of height. But D is a small numerical quantity representing the residual after the height in the standard atmosphere has been subtracted at each point. The D value also carries a climatic inference, since it represents the departure in height of a pressure surface from a reference condition. Furthermore, comparison of D on adjacent pressure surfaces gives an immediate indication of the temperature between them. For this method of pressure analysis, pressure-height computation would be simplified, since the work is complete once D for each mandatory pressure surface is found.

3.10. *Pressure reduction to reference levels below the surface.*—The vertical decrease of pressure near the earth's surface is about 1 mb per 27 feet; that ratio varies with temperature and considerably more with elevation. The horizontal variation of pressure is thousands of times smaller than the vertical variation. Yet, the small horizontal pressure variation is of great importance in determining motions in the atmosphere. Our interest in motions, and therefore in the pattern of pressure in *level surfaces*, demands that surface-pressure measurements be reduced to a reference level. This reference can be any level in the vicinity, but mean

sea level is the standard reference for pressure reductions from the surface.

Reduction of pressure to sea level uses the barometric formula. The procedure is similar in principle to height evaluation in a sounding, except now the temperature in the fictitious air column below the surface must be *assumed* by some manner or means. At stations in the United States reduction to sea level is done quickly with the aid of a specially prepared table unique for the station. This table gives sea-level pressure from station pressure and a representative value T_S for the surface temperature. The value T_S is the average of the surface temperatures at present and 12 hours earlier. This is done to minimize effects of diurnal temperature variation on the sea-level pressure.

When the differential barometric formula, Eq 3.04(1), is integrated from sea level $Z = 0$ to station level $Z = Z_s$, where pressures are P_0 and P_s, respectively,

$$\ln P_0 - \ln P_s = Z_s/K\bar{T}^* , \qquad (1)$$

or

$$P_0 = P_s e^{Z_s/K\bar{T}^*} . \qquad (2)$$

The units of Z_s are either gpm (for $K = 29.28980$, numerically) or gpft (for $K = 96.09494$); the pressures can be in any units, provided they are consistent; $\bar{T}^*$ is mean virtual temperature °K in the fictitious air column between sea level and the surface. Z_s is a known constant for the station, P_s is observed station pressure, and $\bar{T}^*$ must be defined as a function of the surface temperature T_S. The proper relation between T_S and $\bar{T}^*$ is therefore the *only* problem in sea-level pressure reduction.

In preparing tables for pressure reduction to sea level at individual stations in the United States, the Weather Bureau assumes that $\bar{T}^*$ equals T_S for any station of elevation less than 1000 feet. For stations of higher elevation there are several empirical corrections made to derive a suitable mean virtual temperature. They consist of an appropriate assumed lapse rate of temperature for the layer, a correction for moisture, and the Plateau Correction. Each is explained in the reading references and in Weather Bureau Form WB 1154A. Because of differences in reduction methods for stations above and those below 1000 feet, it is possible that discrepancies as large as 0.5 mb might be found in the sea-level pressures of two closely neighboring stations if one is below and the other above 1000 feet.

The mean virtual temperature $\bar{T}^*$ for the column, from sea level to station level, may be considered the sum of a mean temperature $\bar{T}$ and a moisture correction C_m for the column

$$\bar{T}^* = \bar{T} + C_m . \tag{3}$$

The mean temperature $\bar{T}$ is defined conveniently in terms of the surface temperature T_S and a mean lapse rate $\bar{\gamma}$ for the column

$$\bar{T} = T_S + \bar{\gamma} Z_s/2 . \tag{4}$$

In general, the lapse rate would be half the dry adiabatic, that is, about $5°$ C per kilometer or $1.5°$ C per 1000 feet. Thus, $\bar{T} = T_S + Z_s/400$ for Z_s gpm, or $\bar{T} = T_S + Z_s/1312$ for Z_s gpft. The mean temperature for the column thus would be the sum of T_S and a constant factor for the station.

Although the correction for lapse rate is basically as outlined above, the technique in use in the United States is considerably more involved. The following quotation is taken from WB Form 1154A, with exception of change in symbols:

In the United States, when Z_s is greater than 1000 feet, the lapse rate is taken as a function of T_S; that is, a different suitable value of the lapse rate is employed for each T_S. The dependency of the lapse rate on T_S may be represented by means of a diagram. The relationship involved in this may then be used to produce a second diagram showing $\bar{T}$ as a function of T_S on the basis of Eq (4), which is very convenient for expediting the computations. In elevated regions where marked extremes of temperature occur, better results are obtained by this procedure than by assuming the lapse rate is a

constant. The method which has been used in the United States for finding the relationship between T_S and the lapse rate, or between T_S and $(\bar{T} - T_S)$, is very involved, requiring extensive compilations of comparative data for regions including both high- and low-level stations. The principle used is that the synoptic reduced, sea level pressures, P_0, for all stations when plotted on a weather map should be self consistent, and yield reasonably smooth isobars in harmony with the prevailing distribution of cyclones and anti-cyclones, both on a monthly average basis and during extremes of low temperatures. The values of P_0 for low stations establish the reference basis to which the values of P_0 for high stations must conform since the former are reliable because $\bar{\gamma} Z_s/2$ must be small and discrepancies therein can produce little error in $\bar{T}$. Generally satisfactory results are obtained by adjusting $\bar{\gamma} Z_s/2$ as a function of T_S (or $\bar{T}$) until consistency between the values of P_0 for all stations is obtained, in accordance with the reference basis. For further details concerning this subject, the reader is referred to Report of the Chief of the Weather Bureau, 1900–1901, Volume II, entitled "Report on the Barometry of the United States, Canada, and the West Indies," by F. H. Bigelow.

This refinement on the assumed lapse rate of temperature makes the lapse rate vary with the surface temperature in a range from isothermal to nearly dry adiabatic at some stations and within smaller range at others. The variation of lapse rate with temperature is not the same for all stations but is a function of geographic location. In the report cited above, Bigelow defined three distinct curves relating T_S to the mean lapse rate in the column. These he ascribed to the Pacific Slope, the Middle Plateau, and the Eastern Slope regions of the United States and Canada. This represents an attempt to reduce the natural effects on the air temperature by the surface of the earth and by large orographic features which otherwise would be assessed to the entire column down to sea level.

This solution to the problem of suitable lapse rates is by no means complete and exact; discrepancies in sea-level pressure for neighboring high stations can still be of the order 10 mb in some situations. Under the widely differing topographic conditions in western North America, with the consequent local variations of temperature and daily extremes of temperature for such heights, it is next to impossible to derive statistically a set of lapse rates which give reasonable results in all conditions. A new study on the subject of sea-level reduction of pressure is being undertaken at the present time, and, although perfection cannot be anticipated, any improvement is welcomed by the one who uses these data.

The moisture correction (Eq [3]) to the temperature is of comparatively less importance than temperature lapse rate. For stations of high elevation the maximum effect on sea-level pressure by assumed moisture content in the column is of the order 1 mb. At low temperatures the moisture effect is negligible. Hence, little error can result from assuming the mean moisture content for the column is a statistical function of mean temperature. In computing tables for sea-level pressure reduction, a variable virtual temperature increment is added to $\bar{T}$ to get $\bar{T}^*$.

The third consideration is the Plateau Correction, first suggested by Ferrel.[15] It is an increment of pressure added algebraically to the sea-level pressure found by Eq (2). The Plateau Correction, P_c, is defined by

$$P_c = 0.064 \ (\bar{T} - \bar{T}_n)(Z_s/1000) \ , \qquad (5)$$

in mb, where Z_s is in gpft and $\bar{T}_n$ is the annual normal value of $\bar{T}$ for the station. Since $\bar{T}_n$ and Z_s are constants for the station,[16] P_c is determined uniquely by $\bar{T}$ and therefore by T_S. The Plateau Correction is added to the previously determined sea-

level pressure to give the final reduction table for the station.

The object of the Plateau Correction to sea-level pressure is to reduce the departures from mean annual sea-level pressure to about the same amount as found for stations nearer sea level. Without this correction the departures are about twice as large at plateau stations as at low-level stations. The Plateau Correction may be large in extremes of temperature. For example, if $\bar{T}$ differs by 20° C from the annual mean, the magnitude of the correction for a station of 5000-foot elevation is 6.4 mb. Under warm conditions the Plateau Correction increases the sea-level pressure, while under cold conditions it decreases the sea-level pressure. The Plateau Correction thus offsets the prevailing surface temperature inversions of winter and the large lapse rates near the ground in summer, which otherwise would affect too greatly the assumed mean temperature in the column between the surface and sea level.

To summarize the entire procedure in sea-level pressure reduction in the United States, we may write the final equation

$$P_0 = P_s e^{Z_s/\kappa \bar{T}*}$$

$$+ 0.064 \ (\bar{T} - \bar{T}_n)(Z_s/1000) \ . \quad (6)$$

Mean temperature $\bar{T}$ is defined in terms of surface temperature T_S and a lapse rate which is a function of T_S. It follows that $(\bar{T} - \bar{T}_n)$ is given also by T_S. Since $\bar{T}^* = \bar{T}(1 + 0.61\bar{r}\%_0)$, and C_m is assumed a function of T_S, then $\bar{T}^*$ is also determined explicitly by T_S. Therefore, the sea-level pressure P_0 is given by two observed variables, station pressure P_s and 12-hour average surface temperature T_S, which are the sole arguments in the reduction table for each station.

15. William Ferrel, "On Reduction of Barometric Pressure to Sea-Level and Standard Gravity," in *Annual Report of the Chief Signal Officer of the Army, 1886* (Washington, D.C., 1887), Appendix 23.

16. In the Plateau Correction the elevation Z_s is not in all cases the true elevation of the station. For a station whose elevation differs greatly from the surrounding starions, Z_s in the average elevation of the neighboring stations.

3.11. *Height reductions to pressure references.*—In computing heights of mandatory pressure surfaces for a radiosonde report at a station where surface pressure is less than 1000 mb, it is necessary to extrapolate downward to the 1000-mb surface. Even stations at sea level occasionally must do this. In mountainous areas some stations have surface pressures less than 850 mb, and the height of the 850-mb surface is determined also by such extrapolation.

The barometric formula is again the basis, only the dependent variable is now the depth of the layer between the pressure surface in question and the station pressure. In Figure 3.11 surface pressure is P_s, the

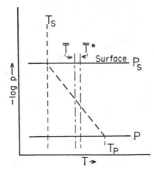

FIG. 3.11

appropriate surface temperature T_S, and the mandatory pressure P. The temperature at the desired pressure surface, T_P, is defined by T_S and a suitable lapse rate γ. Of these quantities, only P is fixed, and both P_s and T_S vary with surface conditions. The thickness of the layer between P and P_s is

$$\Delta Z = K\bar{T}^*(\ln P - \ln P_s) . \quad (1)$$

The first problem in evaluation of ΔZ is to specify $\bar{T}^*$ in terms of the prescribed surface temperature T_S and an appropriate virtual temperature correction.

The theory described is the one presently employed by the U.S. Weather Bureau. Surface temperature T_S is not the same as used in sea-level pressure reduction; it is the average of twice the present temperature and

the temperature 6 hours earlier. This T_S is therefore subject to greater diurnal variation than the one in sea-level reduction.

Temperature T_P can be defined by $T_P = T_S + \gamma\Delta Z$, in which γ is the assumed lapse rate of temperature from P to the surface. It follows that the mean temperature for the column is $\bar{T} = T_S + \gamma\Delta Z/2$. The standard value of γ adopted by the Weather Bureau is half the dry adiabatic.

As the mean virtual temperature is given by $\bar{T}^* = \bar{T}(1 + 0.61\bar{r}\%_{00})$, then $\bar{T}^* = (T_S + \gamma\Delta Z/2)(1 + 0.61\bar{r}\%_{00})$, and, by substitution for ΔZ from Eq (1), we have

$$\bar{T}^* = T_S\{[1 + 0.61\bar{r}]/$$
$$[1 - K\gamma(1+0.61\bar{r})(\ln P$$
$$- \ln P_s)/2]\} . \quad (2)$$

We may denote $(\bar{T}^* - T_S)$ by C, and C can be determined by using Eq (2). Then, from the barometric equation,

$$\Delta Z = K(T_S + C)(\ln P - \ln P_s) . \quad (3)$$

According to instructions in U.S. Weather Bureau Circular P, evaluation of ΔZ for either 1000 mb or 850 mb is handled in two steps. In a first set of tables, C is found as a function of P_s and T_S. These tables, each valid for a certain value of P, were computed from assumed values of $\bar{r}$ as a function of temperature. Next, ΔZ is found with $\bar{T}^*$ and P_s as arguments. The elevation of P above sea level is then station elevation minus ΔZ.

3.12. *Pressure tendencies.*—The pressure tendency is the time rate of change of pressure at a fixed point, denoted $\partial p/\partial t$. Similarly, the height tendency of a constant-pressure surface is $(\partial Z/\partial t)_p$. Exclusive of station barograph traces, the only measurements of pressure tendencies we have are made from integrated pressure changes over intervals of 3, 6, or 12 hours. The surface barometric tendency reported in the synoptic observation is integrated over 3 hours, and computations of upper-level pressure or height tendencies are over the time intervals between radiosonde observations. Although none of the measured tendencies is strictly instantaneous, we use the same mathematical notations as though they were.

The pressure at level Z in the atmosphere is obtained by integrating Eq 3.03(3), from Z to the top of the atmosphere, where $p = 0$ and $Z = \infty$.

$$p_Z = 980 \int_Z^\infty \rho \, dZ . \qquad (1)$$

By partial differentiation with time, we obtain the local pressure tendency at Z

$$(\partial p / \partial t)_z = 980 \int_Z^\infty (\partial \rho / \partial t) dZ . \qquad (2)$$

This shows that the pressure tendency at any point depends only on the integrated mass change above the point. Net increase in mass above Z results in pressure rise at Z, which follows directly from definition of pressure.

The difference in pressure tendency between two levels is a function of both pressure and temperature. For two levels Z_1 (lower) and Z_2 (upper), with pressures p_1 and p_2, respectively, $\ln p_1 - \ln p_2 = (Z_2 - Z_1)/K\bar{T}^*$. Then

$$\partial(\ln p_1)/\partial t - \partial(\ln p_2)/\partial t$$
$$= -[(Z_2 - Z_1)/K\bar{T}^{*2}](\partial \bar{T}^*/\partial t) . \qquad (3)$$

During a short interval of time $1/\bar{T}^{*2}$ may be treated approximately constant. It follows that the difference in tendencies of the logarithm of pressure at two levels is approximately a linear function of the tendency of mean virtual temperature. To get the difference in tendencies of pressure, the last equation can be transformed by the identity $\partial(\ln p)/\partial t = (1/p)(\partial p/\partial t)$ to give

$$(\partial p/\partial t)_1 = (p_1/p_2)(\partial p/\partial t)_2$$
$$- [p_1(Z_2 - Z_1)/K\bar{T}^{*2}](\partial \bar{T}^*/\partial t) . \qquad (4)$$

The factor p_1/p_2 is greater than unity. If

mean virtual temperature is unchanged, the pressure tendency at the lower level is of the same sign but of greater magnitude than the tendency above. For example, consider two levels where the pressures are 1000 mb and 700 mb. Then $(p_1/p_2) = 10/7$, and the pressure tendency at the bottom is 10/7 the tendency above, if temperature remains constant.

It is often observed that pressure tendencies at two levels are radically different, perhaps of opposite sign, particularly when the levels are a large distance apart. This is the effect of time variation of mean temperature between levels. That term in Eq (4) has its greatest effect on the pressure tendency at the lower level when $Z_2 - Z_1$ is large, when $\bar{T}^*$ is small (not very significant), and when the tendency of temperature is large in magnitude. In the example above, and with $\bar{T}^* = 273°$ K, the factor $p_1(Z_2 - Z_1)/K\bar{T}^{*2}$ is approximately 10/7, which is also the ratio of p_1 to p_2. Therefore,

$$(\partial p/\partial t)_1 \simeq (10/7)[(\partial p/\partial t)_2 - \partial \bar{T}^*/\partial t] .$$

Each 1° C change in temperature for that layer has the same absolute effect on the pressure tendency at the lower level as does 1-mb change in pressure at the upper level. If pressure is rising aloft, surface pressure may be rising, falling, or steady, depending on the tendency of temperature in the column. Temperature change increases the lower-level pressure tendency in comparison with the upper one when temperature falls and decreases the lower-level tendency in comparison with the upper one when temperature rises.

The tendency equation gives interesting results when applied to sea-level pressure in elevated regions. From Eq 3.10(6)

$$(\partial p/\partial t)_0 = (p_0/p_s)(\partial p/\partial t)_s$$
$$- Z_s[(p_0/K\bar{T}^{*2})(\partial \bar{T}^*/\partial t)$$
$$- 0.000064 \, (\partial \bar{T}/\partial t)] , \qquad (5)$$

with Z_s in gpft. If we neglect the small difference between $\partial \bar{T}^*/\partial t$ and $\partial \bar{T}/\partial t$ and evaluate numerically $p_0/K\bar{T}^{*2}$ for usual ranges of p_0 and $\bar{T}^*$, the equation reduces to

$$(\partial p/\partial t)_0 \simeq (p_0/p_s)(\partial p/\partial t)_s$$
$$- (0.5 \text{ to } 1)(Z_s/10{,}000)(\partial \bar{T}^*/\partial t) . \quad (6)$$

Both terms on the right are of comparable magnitude for high-elevation stations.

When we recall the method by which $\bar{T}^*$ in the fictitious air column was obtained, it is evident that $\partial \bar{T}^*/\partial t$ is given nearly by the tendency of surface temperature, $\partial T_s/\partial t$. Then by Eq (6) we find that sea-level pressure tendency in elevated regions is greatly affected by surface-temperature change and that this effect is greater the higher the station elevation. For a station at 10,000 feet, its sea-level pressure changes up to 1 mb for each 1° C change in surface temperature. This is ample reason for attempting to remove the effect of diurnal temperature variation, which might cause diurnal variation of 10 mb or more in sea-level pressure.

Diurnal variation is not the only surface-temperature change of importance in the sea-level pressure tendencies for elevated stations. Other changes are the large- and small-scale variations by motions of air masses, differences in radiation due particularly to the distribution of cloudiness, and air drainage, sheltering, and other orographic effects. Any of these may produce fictitious sea-level pressure variations overshadowing real variations in the atmosphere. Therefore, caution should be exercised when interpreting sea-level tendencies as indications of real pressure changes in the atmosphere. *The reported 3-hour pressure tendency in the surface synoptic observation is the only reliable measure of surface-pressure tendency in mountain and plateau regions.* In the vast areas where the earth's surface departs little from sea level, sea-level pressure change at any station is a good estimate of atmospheric pressure change.

3.13. Height tendencies at constant pressure.—In the case of two constant pressure surfaces p_1 and p_2, $(p_1 > p_2)$, the respective heights Z_1 and Z_2 are the variables. Differentiation of Eq 3.05(2) gives

$$(\partial Z/\partial t)_2 - (\partial Z/\partial t)_1$$
$$= K(\ln p_1/p_2)(\partial \bar{T}^*/\partial t) . \quad (1)$$

Thus, the difference in vertical displacements of two pressure surfaces varies only with the change in mean virtual temperature. If temperature remains constant, height tendencies of the two isobars are identical.

The factor $K(\ln p_1/p_2)$ is constant for any pair of isobaric surfaces. For the (700/1000)-mb layer it is 34.3 gpft/° C, and the tendency equation for the layer is

$$(\partial Z/\partial t)_{1000} = (\partial Z/\partial t)_{700} - 34.3(\partial \bar{T}^*/\partial t)$$

gpft per unit of time. Interpretation of this formula is simplified by recalling that the depth of a pressure layer varies directly with its virtual temperature. Where the 700-mb surface remains fixed while the mean temperature below rises by 1° C, the 1000-mb surface falls by 34 gpft.

When considering height tendencies of fictitious pressure surfaces below ground, we find essentially the same difficulties as with pressure tendencies at the subterranean reference levels. From Eq 3.11(3)

$$(\partial Z/\partial t)_P$$
$$= - K(\ln P/P_s)(\partial T_s/\partial t + \partial C/\partial t)$$
$$+ K(T_s + C)(1/P_s)(\partial P_s/\partial t) . \quad (2)$$

The pressure P is the mandatory pressure in question, and the subscript s refers to the surface. Since a fixed lapse rate of temperature is used for this reduction, $\partial C/\partial t$ is the variation of moisture correction and is thus negligible compared to $\partial T_s/$

∂t. The tendency of a pressure surface below ground is composed of two terms, one controlled by surface-temperature change and the other by surface barometric tendency. The second term contributes in the manner such that rising pressure at the surface results in lifting of the pressure surface—visualize isobars lifted through the ground when surface pressure rises and forced below ground when surface pressure falls. The first term on the right of Eq (2) shows that, as temperature rises at the station, pressure layers expand vertically, and the particular isobar is forced downward relative to the surface isobar.

To show the relative importance of the two terms, let us assume certain values for the pressures and temperatures in an example. Suppose the station pressure P_s is 800 mb, corresponding to elevation approximately 6200 feet, $T_S = 0°$ C $= 273°$ K, and $C = 5°$ C, while the pressure surface considered is 1000 mb. The tendency of this surface in gpft is then, numerically,

$$(\partial Z/\partial t)_{1000} = -K(0.223\ \partial T_S/\partial t$$
$$- 0.348\ \partial P_s/\partial t)\ .$$

In this example the 1000-mb height tendency is affected just as much by 1.5° C change in surface temperature as by 1-mb change in surface pressure. It is evident that the change in temperature can on occasion make the 1000-mb tendency opposite to tendencies of pressure surfaces above ground. For stations nearer sea level, the role of surface temperature variations in the tendency of 1000 mb is obviously less.

3.14. *Space variations of pressure.*—Consider an s, Z cross section (Fig. 3.14a) of two isobaric surfaces p_1 and p_2, where s is horizontal distance in arbitrary direction. In the drawing the slopes are greatly exaggerated. The slope of p_2 at any point along it is $(\partial Z/\partial s)_2$; similarly, the slope of p_1 is

$(\partial Z/\partial s)_1$. Now, along any ordinate of the cross section the thickness of layer $p_1 p_2$ is $Z_2 - Z_1 = \Delta Z$. Hence, $(\partial Z/\partial s)_2 - (\partial Z/\partial s)_1 = \partial \Delta Z/\partial s$. The difference in slopes of the pressure surfaces equals the lateral variation of the layer thickness, a fact evident from simple geometry.

Differentiation of the barometric formula along s gives $\partial \Delta Z/\partial s = K(\ln p_1/p_2)(\partial \bar{T}^*/\partial s)$. Thus,

$$(\partial Z/\partial s)_2 - (\partial Z/\partial s)_1$$
$$= K(\ln p_1/p_2)(\partial \bar{T}^*/\partial s)\ .\quad (1)$$

This is the basis for an important corollary to the hydrostatic equation. *The difference between slopes of adjacent pressure surfaces is a linear function of the lateral variation of virtual temperature.* Another interpretation is: *Isobaric surfaces diverge toward warmer air.* Here we have physical explanation for different patterns of pressure at sea level and aloft.[17] The troposphere is colder in the higher latitudes, and isobaric surfaces slope more strongly downward toward the north aloft than near the surface. If pressure surfaces near sea level are nearly horizontal, as in fact they tend to be, then those aloft slope downward to the north with amount determined by the south-north virtual temperature gradient.

In the pressure tendency equation we found the change of pressure at a fixed point consisted of the integrated mass change in the column above that point. In a similar way the horizontal variation of pressure in any direction s is the integrated horizontal variation of mass in that direction from this level to the top of the atmosphere. That is,

$$(\partial p/\partial s)_z = 980 \int_z^\infty (\partial \rho/\partial s)dZ\ .$$

Again, it is convenient to compare pres-

17. Examine the pressure and temperature charts in Chap. 1.

sure variations at two levels. From Eq 3.05(2),

$$\partial(\ln p_1)/\partial s - \partial(\ln p_2)/\partial s$$
$$= -[(Z_2 - Z_1)/K\bar{T}^{*2}](\partial \bar{T}^*/\partial s) \text{ , (2a)}$$

and

$$(\partial p/\partial s)_1 = (p_1/p_2)(\partial p/\partial s)_2$$
$$- [p_1(Z_2 - Z_1)/K\bar{T}^{*2}](\partial \bar{T}^*/\partial s) \text{ .   (2b)}$$

simply: Isobars diverge vertically in the direction of increasing temperature, as shown in Figure 3.14c. Thus, if $(\partial p/\partial s)_2$ and $(\partial p/\partial s)_1$ are both positive, as indicated, the divergence of isobars causes $(\partial p/\partial s)_1$ to be lessened relative to $(\partial p/\partial s)_2$. If $Z_2 - Z_1$ and $\partial \bar{T}^*/\partial s$ are sufficiently large, it is entirely possible for $(\partial p/\partial s)_1$ and $(\partial p/\partial s)_2$ to be

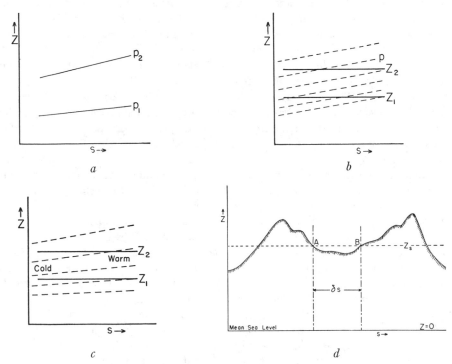

Fig. 3.14

Here Z_1 is the lower level and Z_2 the upper, and p_1 and p_2 are their variable pressures. It follows that, in absence of horizontal temperature differences, horizontal pressure variation is larger below than above, but both are the same sign. This is due to logarithmic distribution of pressure with height and is demonstrated schematically in Figure 3.14b.

The effect of temperature on the difference between horizontal variations of pressure at two levels can be visualized quite

reversed in sign, a condition frequently observed.

Later on we will be confronted with some of the perplexities in analysis of sea-level pressure in mountainous regions. Since most of the difficulties can be explained from the above principles, we should examine that subject now. To simplify the problem, assume that two stations A and B (Fig. 3.14d), a short horizontal distance δs apart, have the same elevation so that their sea-level reduction tables are approximately the same. The prescribed surface temperature for reduction is T_S, station elevation is Z_s, surface pressure is P_s, and sea-level pressure is P_0. From Eq (2b)

$$(\delta p/\delta s)_0 = (P_0/P_s)(\delta p/\delta s)_s$$
$$- (P_0 Z_s/K\bar{T}^{*2})(\delta \bar{T}^*/\delta s) . \quad (3)$$

When temperatures at A and B are the same, $(\delta p/\delta s)_0 = (P_0/P_s)(\delta p/\delta s)_s$, and the average variation in sea-level pressures between A and B is the same sign but greater magnitude than the horizontal pressure variation at the surface. This deduction is the same as found for the free atmosphere with reference to Figure 3.14*b*. Therefore, radical inconsistencies between the sea-level pressure pattern and that in the free atmosphere must be due to the second term on the right of Eq (3).

The reason for the large differences which do exist must be found in the departure of the horizontal temperature pattern near the surface from its *representative* horizontal pattern in the free atmosphere. In Figure 3.14*d* we ordinarily should find a rather simple and uniform distribution of temperature at levels beyond immediate control by the ground. Near the surface, however, the pattern of air temperature is controlled largely by the very irregular character of the earth's surface, including all details of orography, exposure, soil cover, and so on. The main trouble is inherent in the lack of correlation between surface and free-air temperatures, that is, the great variations in stability of the surface layer. The most suitable temperature for reduction of pressure to sea level is not surface temperature but temperature at some height beyond the immediate masking effects by the surface, for, if it is desired that sea-level pressure pattern be an indication of atmospheric pressure patterns, then the temperatures (and temperature patterns) which go into sea-level pressure reduction also should represent the atmosphere. We are told that

surface temperatures in elevated regions are not reduced to sea level partly because of the misleading and weird patterns resulting. Yet, it is a paradox that the basis for pressure reduction is just that. Thus, in elevated regions, sea-level pressure distribution consists partly of horizontal pressure distribution in the free air and partly of the irregular surface-temperature pattern.

To make this one point more concrete, we take an example based on Figure 3.14*d*. Assume the stations A and B are 5000 feet above sea level and that their station pressures are 840 mb. Therefore, $(\delta p/\delta s)_s = 0$, and the average sea-level pressure variation between them is $(\delta p/\delta s)_0 = -(P_0 Z_s/K\bar{T}^{*2})$ $(\delta \bar{T}^*/\delta s)$. The 12-hour average surface temperature is 50° F (283° K) at A and 32° F (273° K) at B, the difference between the stations due perhaps only to orography. As $\delta \bar{T}^*/\delta s < 0$, $(\delta p/\delta s)_0 > 0$. With a lapse rate 5° C per kilometer and appropriate moisture corrections, one finds the sea-level pressure is 1009 mb at A and 1016 mb at B. Thus, excluding the Plateau Correction, the reported sea-level pressure is 7 mb larger at B than at A, owing only to different surface temperatures.

Now, let us examine the temperature distribution in the free air above the surface layer. Suppose that beginning just above the surface the air is at each level 2° C warmer at B than at A, a condition not at all unlikely. If we consider this free-air temperature distribution in the pressure reduction, sea-level pressure at A should be larger than B, contrary to that found by using surface temperatures. Such inconsistency between sea-level and atmospheric pressure patterns will be verified many times in pressure analysis, and the problem will be respected after one is familiar with the complexities in mountainous areas.

In like manner it could be shown that the 1000-mb surface has similar unnatural and often misleading configurations below the earth's surface as a result of unrepresentative temperatures used in reduction. It should be evident also that less error results in reduction downward to 5000 feet or 850 mb than to sea level or 1000 mb, thus making the 5000-foot (850-mb) chart preferable to the sea-level (1000-mb) chart in elevated regions such as the western United States.

II. STATIC STABILITY

3.15. *The concept of hydrostatic stability.* —Stability in the atmosphere governs many weather processes and phenomena. Indirectly, it plays a major part in determining vertical and horizontal distributions of many of the variables, since vertical transport and vertical mixing are controlled largely by stability in air masses.

Stability may be examined from the state of equilibrium of a sample air parcel displaced a short distance from its initial position. If after displacement the parcel tends to return to its original point, the air of which this parcel is a sample is said to be in stable equilibrium, or "stable." If the perturbed parcel continues to move farther from its reference point, the air is "unstable." In the limiting condition, when the displaced parcel develops no acceleration as a result, the equilibrium is neutral or indifferent. Stability describes a state of the atmosphere. It does not express any particular motions of an air parcel, nor does it imply existence or nature of any forces which might disturb it. However, it does suggest the probable effect on an air parcel by an arbitrary perturbing force.

The conditions for stability can be analyzed with respect to both vertical and horizontal displacements, but our present discussion is restricted to the vertical, namely, *static stability.* For statically stable equilibrium, the atmosphere must be such that, if the air parcel is displaced either upward or downward, a buoyant force acts to restore it to the initial position; the buoyant force acts opposite to the original displacement. In static instability the buoyant force acts in the same direction as the displacement.

Consider a parcel of air displaced a short distance upward. In its new position it has density ρ', and the density of its new and undisturbed environment is ρ. From Archimedes' principle, the buoyant vertical acceleration, dw/dt, of the parcel in its new position is

$$dw/dt = g(\rho - \rho')/\rho' . \qquad (1)$$

Acceleration of the parcel is given in terms of temperature by substituting for ρ and ρ' from the equation of state and letting $p = p'$, that is, the parcel has at all times the pressure of its surroundings. If T^* and $T^{*\prime}$ are the virtual temperatures of the environment and the displaced parcel, respectively, then

$$dw/dt = g(T^{*\prime} - T^*)/T^* . \qquad (2)$$

Stability will be visualized more easily in terms of temperature lapse rates on thermodynamic charts. The virtual temperature $T^{*\prime}$ of the displaced parcel is a function of its initial or reference level temperature, T_0^*, and an appropriate lapse rate which it follows in displacement. The latter is assumed to be either the dry or the saturated adiabatic, according as the parcel is unsaturated or saturated. If the process lapse rate is γ' and if the upward displacement is δz, then $T^{*\prime} = T_0^* - \gamma' \delta z$.

The environment or sounding temperature T^* at the second position can be defined in terms of the reference level temperature T_0^* and the lapse rate γ of virtual temperature in the air column: $T^* = T_0^* - \gamma \delta z$. Substitution into Eq (2) gives

$$dw/dt = (g/T^*)(\gamma - \gamma')\delta z$$
$$= (980/T^*)(\gamma - \gamma')\delta Z . \qquad (3)$$

In all further considerations of stability, we substitute T for T^* and $\partial T/\partial z$ for $\partial T^*/$

∂z in Eq (3). The difference between T and T^* is seldom more than 1 per cent of the temperature. However, $\partial T/\partial z$ may differ from $\partial T^*/\partial z$ by as much as 2° C km⁻¹ with large vertical variations of moisture, as can be seen from differentiating Eq 1.06(1) with respect to height.

3.16. (a) *Stability of unsaturated air.*—As long as the parcel is not saturated and does not reach saturation in displacement, it follows the dry adiabatic process. In Figure 3.161a a segment of a temperature sounding is shown on a tephigram. Let parcel A be

accelerated farther downward. Notice the condition $\partial\theta/\partial z < 0$.

In each of the diagrams above there is a different state of stability, and in each there is a characteristic slope of the sounding curve on the thermodynamic chart. Since the slope of the curve indicates the vertical lapse rate of temperature in the atmosphere and since the slope of dry adiabats on the chart defines the dry adiabatic lapse rate, it follows from Eq 3.15(3) that stability in an unsaturated air column is found by comparing the slope of the pT (strictly pT^*) curve with the dry adiabats.

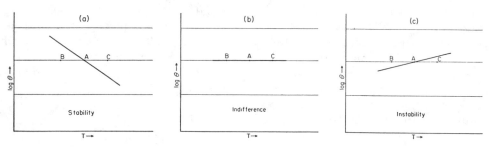

Fig. 3.161.—Type of stability in relation to the slope of a sounding curve on the tephigram (unsaturated conditions).

any point along this segment. If it is displaced upward dry adiabatically to B, it is colder than its new environment and therefore is forced back to its initial point A. Similarly, if the parcel is set in motion to point C, it is warmer than the environment and accelerated back toward A. Observe that in stability the air column is characterized by $\partial\theta/\partial z > 0$.

A different condition exists in Figure 3.161b. Here the parcel comes to rest at any point to which it might be displaced. Thus, Figure 3.161b, showing $\partial\theta/\partial z = 0$, represents neutral or indifferent stability. Diagram (c) shows the state of instability. If parcel A is pushed adiabatically toward B, it becomes warmer than its environment and is accelerated farther away from its initial level; and, if set in motion downward, it is colder than surroundings and therefore

The value of the dry adiabatic lapse rate Γ_d can be derived from Eq 1.07(1). That equation is differentiated with respect to height at constant potential temperature to give $(1/T)(\partial T/\partial z)_\theta = (\kappa/p)(\partial p/\partial z)$. By substitution from the hydrostatic equation and further simplifying the results,

$$\Gamma_d = -(\partial T/\partial z)_\theta = g/c_p .$$

For $g = 980$ cm sec⁻² and $c_p = 1.004 \times 10^7$ ergs (gm deg)⁻¹, $\Gamma_d = 9.76$ deg km⁻¹.

(b) *Stability of saturated air.*—If the test sample of air is saturated, the critical conditions for stability and instability differ somewhat from the unsaturated case. If the saturated parcel contains no liquid water, in lifting the critical lapse rate is the saturated pseudo-adiabatic, Γ_s, while in sinking the critical lapse rate is still the dry adiabatic, Γ_d. If the parcel contains liquid water, descent takes place moist adiabatically (nearly along the pseudo-adiabat) if the water in the parcel is evaporated during the descent. Stability criteria for pseudo-adiabatic lifting are represented in Figure 3.162.

When criteria for both saturated and unsaturated conditions are combined, five different types of lapse rates are required to de-

3.17. Degree of stability.—It is evident that static stability tends to suppress random vertical motions; it implies a certain degree of damping or suppression by the atmosphere on displacements of small masses of air. On the other hand, instability in an air column is associated with a tendency for turbulence and "bumpiness" re-

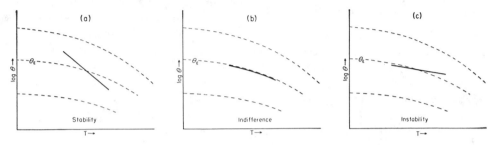

FIG. 3.162.—Type of stability in relation to the slope of a sounding curve on the tephigram (saturated conditions).

scribe adequately all possible conditions. They are:

$\gamma < \Gamma_s$, stable for both processes ;

$\gamma = \Gamma_s$, indifferent for saturation; stable for nonsaturation;

$\Gamma_s < \gamma < \Gamma_d$, unstable for saturation, stable for nonsaturation;

$\gamma = \Gamma_d$, unstable for saturation, indifferent for nonsaturation;

$\gamma > \Gamma_d$, unstable for both processes.

The third condition, $\Gamma_s < \gamma < \Gamma_d$, describes a pT curve with slope between dry and pseudo-adiabatic. It is a stable state if the parcel is unsaturated but unstable if saturated; this is *conditional instability*. The first state listed above, $\gamma < \Gamma_s$, is called "absolute stability." Lapse rates for which $\gamma > \Gamma_d$ are *superadiabatic* lapse rates, and this may be called *absolute instability*, since the air column is unstable for any adiabatic process. However, "absolute instability" is a term sometimes reserved for autoconvective or mechanical instability, the condition with increase of density with height.

sulting from extension of the vertical displacements by the buoyant accelerations. The degree of suppression (stability) and magnification (instability) of the ever present tendency for vertical turbulence within the atmosphere is a direct function of the lapse rate of temperature and varies in smaller degree with the temperature itself (Eq 3.15[3]).

Stability can be viewed as the opposition by the atmosphere to unstable vertical accelerations. We define static (vertical) stability E_v as the negative acceleration per unit height δz, so that, with reference to Eq 3.15(3),

$$E_v = -(1/\delta z)(dw/dt)$$
$$= (g/T)(\gamma' - \gamma) . \quad (1)$$

For unsaturated conditions $\gamma' = \Gamma_d$. The lapse rate γ is, by definition, $-\partial T/\partial z$. Therefore,

$$E_v = (g/T)(\partial T/\partial z + \Gamma_d) . \quad (2)$$

The usual order of magnitude of E_v is 10^{-4} sec^{-2}. In extremely stable conditions, such as develop near the surface over arctic snow covers in winter, stability can be as large as 1×10^{-3} sec^{-2}. Large negative values approaching -1×10^{-3} sec^{-2} may also occur, but these are restricted to a very shallow layer over hot ground.

Stability is defined also in terms of potential temperature. By Eq 4.13(11),

$$E_v = (g/\theta)(\partial\theta/\partial z) = g\ \partial(\ln\theta)/\partial z . \quad (3)$$

Observe that E_v has dimension (time^{-2}) and that stability is actually the percentage rate of increase of potential temperature with height, multiplied by gravity.

For the more restricted case of saturated air, stability is

$$E_{vs} = (g/T)(\partial T/\partial z + \Gamma_s) . \quad (2a)$$

This is more complicated than for unsaturated states, because Γ_s varies with temperature and moisture.

3.18. Conditional instability and the parcel method of convection.

—An air column whose temperature lapse rate is between dry and saturated adiabatic is unstable for saturation and stable for nonsaturation.[18] But with such a lapse rate in unsaturated air of high relative humidity it is possible for air to become unstable if lifted beyond its adiabatic saturation point, beyond which the air parcel is assumed to follow the saturated pseudo-adiabatic lapse rate, which may make the parcel warmer than its environment and therefore unstable.

Point A in Figure 3.181 is characterized by a conditionally unstable state. It is initially unsaturated; for minor vertical impulses it is stable. But assume the upward impulse carries it to saturation while not affecting the environment materially. In the course of vertical displacement, the parcel

18. Technically, this is true even with lapse rates less than saturated adiabatic if in a short distance above the lapse rate is larger than saturated adiabatic.

first describes a dry adiabatic process to its saturation level B. After this it follows a pseudo-adiabatic process and, as the parcel was initially not too far from saturation, may reach a level C, where its temperature is the same as the environment. Here the pseudo-adiabat describing the process of the parcel intersects the sounding curve. Beyond C the parcel continues along the same pseudo-adiabat; there the parcel is warmer than its environment and is ac-

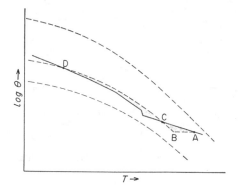

Fig. 3.181.—Conditional instability

celerated upward. Finally, the parcel trajectory on the chart again intersects the sounding curve at a point D, and thereafter the parcel is colder than its environment.

Throughout its course from A through B to C the parcel is denser than surroundings; stabilizing accelerations are opposite to the motion. In order for the parcel to arrive at C, work must be done on it against these stabilizing forces (and also against friction). At point C the stabilizing acceleration is zero. If the initial impulse is sufficient to carry the parcel beyond C, the parcel is freely accelerated upward. Since above C the parcel develops an upward velocity component separate from the initial impulse, level C is known as the *level of free convection* (LFC).

The change in kinetic energy of unit mass displaced a distance dz with acceleration dw/dt is $d(w^2/2) = w\ dw = (dw/dt)dz$. Substitution from Eq 3.15(2), transformations by the equation of state and the hydrostatic equation, and then integration give

$$\tfrac{1}{2}w_2^2 - \tfrac{1}{2}w_1^2$$

$$= R_d \left[\int_1^2 T^{*\prime} d(-\ln p) \right.$$

$$\left. - \int_1^2 T^* d(-\ln p) \right]. \quad (1)$$

The integral of $T^* d(-\ln p)$ between two levels is proportional to the area on a thermodynamic chart bounded by the isobars p_1 and p_2, the isotherm $T^* = 0°$ K, and the curve of T^*. In a similar way, the first integral in Eq (1) is proportional to the area lying to the left of the adiabatic process curve (ABC in Fig. 3.181). The difference between integrals is therefore given by the hatched area below C, and, since the difference is negative in this region, kinetic energy is *lost* by the parcel between levels A and C. Applied to layer CD, Eq 3.18(1) shows that the energy *gained* is given by the hatched area between the sounding curve and the particular pseudo-adiabat through C and D. The lower hatched area is called the *negative area* (kinetic energy lost by the parcel) and the upper hatched area the *positive area* (parcel gains kinetic energy). The size of the negative area and the ratio of the negative to the positive area are of significance in vertical convection. If the negative area is small, a comparatively small impulse can send the parcel through its LFC and result in an unstable condition. When the negative area is large, a larger initial upward push is required to reach instability. Hence, the smallness of the negative area may be interpreted as probability for setting off instability, other things the same.

The ratio of negative and positive areas establishes the types of conditional instability:[19]

(i) *real latent unstable type:* positive area larger than negative area;

19. C. W. B. Normand, "On Instability from Water Vapour," *QJRMS*, LXIV (1938), 47–69.

(ii) *pseudo-latent unstable type:* positive area smaller than negative area;

(iii) *stable type:* positive area absent.

These are discussed in the "Reading References." To summarize here, with real latent conditional instability the energy which can be released by latent heat during convection is greater than the energy expended in lifting the parcel, while with stable conditional instability the lifted parcel never reaches the LFC, and therefore work has to be done on it everywhere in its ascent.

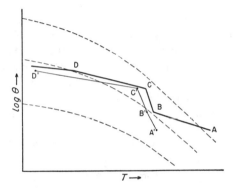

FIG. 3.182.—pT curve (*heavy line*) and characteristic curve (*fine line*).

The type of conditional instability of any point in the sounding can be estimated at a glance from the characteristic curve on the thermodynamic chart (Fig. 3.182). This curve connects the adiabatic lifting condensation levels (LCL's) of successive points along the sounding. The negative area for any point on the characteristic curve is bounded by the pseudo-adiabat extending upward from the LCL, by the dry adiabat extending to the right of this point, and by the pT sounding curve. Its positive area is bounded by the same pseudo-adiabat and by the pT curve above the LFC.

From the foregoing it is now apparent that two conditions usually must be satisfied to obtain actual instability from conditional instability of an air mass. The negative area must be small enough that a small upward displacement is sufficient to send the parcel into regions where it accelerates upward. Small negative area is favored by high humidity and nearly dry adiabatic lapse rate. The other condition to be met is the existence of an upward force to give the parcel the required impulse. This may be supplied mechanically, as by air motion over rough terrain, or supplied through localized convection currents over a warm surface, as exemplified by most cumulus clouds.

The parcel method of convection has been used here as a test for possible instability of a mass of air. Its validity in terms of what might actually occur depends on a number of assumptions made in the theory. The main assumptions are that the kinetic energy of the environment can be neglected and that the individual parcel of air maintains its identity and does not mix with surroundings. Such a physical process is difficult to justify. The "slice-method" approach[20] to the explanation of convection considers the compensating downward motions in the environment and the net effects over a large region, but it does not account for mixing processes in the individual ascending and descending currents. The effect of mixing, which must occur in some degree, is a restriction on the validity of the parcel and slice methods, particularly in the light of recent evidence on entrainment during vertical convection.

3.19. *Objective use of the parcel method of convection.*—Even with all the drawbacks

20. J. Bjerknes, "Saturated Ascent of Air through a Dry-adiabatically Descending Environment," *QJRMS*, LXIV (1938), 325–30; N. R. Beers, "Atmospheric Stability and Instability," in F. A. Berry, E. Bollay, and N. R. Beers (eds.), *Handbook of Meteorology* (New York: McGraw-Hill Book Co., 1945), pp. 693–711; G. P. Cressman, "The Influence of the Field of Horizontal Divergence on Convective Cloudiness," *Journal of Meteorology*, III (1946), 85–88; Sverre Petterssen, *Contribution to the Theory of Convection* ("Geofysiske Publikationer," Vol XII, No. 9 [Oslo, 1939]).

the parcel method might have in theory, it has been used successfully with some discretion in estimating the probability of cumulus and thunderstorm development through daytime heating, as well as the time and height of cloud formation. This approach has had its greatest value in stagnant summer conditions over continental interiors.

The technique may be described as follows with reference to Figure 3.19, which we suppose typifies the pT and pT_s curves for an early-morning sounding having condi-

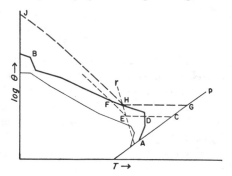

Fig. 3.19.—Changes in pT sounding due to diurnal heating from below; initial pT curve (*heavy*), initial pT_s curve (*fine*).

tional instability above the surface layer. By addition of heat below, the nighttime inversion is removed, first at the ground and then progressively higher and higher as heating continues. The sounding so modified is approximately adiabatic in the bottom layer. The area on the chart between the initial pT curve, the later pT curve, and the isobar of surface pressure is proportional to the energy added by insolation. As the accumulated energy received at the ground is a function of time of day, geographic position, season, and cloudiness, it is possible to predict for a given station either the time at which the pT curve in the mixed layer occupies a certain position on the chart or the position of the curve at a given future time, if the sounding is not changed materially by horizontal air motion. Incidentally, since a maximum possible area can be defined for a given station and time of year, it is possible to predict by this means a maximum surface temperature during the day corresponding to a given initial sounding.[21]

To examine the conditional instability of the heated air, a representative value must be selected for mixing ratio in the layer. This is difficult, since the choice depends on the amount of evaporation from the surface during heating, the vertical distribution of moisture, and also the vertical extent and thoroughness of mixing. There are different rules used in practice. Let us say, for simplicity, that the representative figure is the average moisture in the lowest 1500 feet. Thus, the adiabatic saturation point lies along mixing-ratio line r.

At some later time the pT curve resembles $CDHFB$, and the adiabatic saturation level for parcels heated near the ground is E. As distance DE is still large, the downward acceleration developed by a mass or stream of air from near the surface would not permit any appreciable amount of this air to reach condensation and would prohibit even more this surface air from penetrating the LFC at F. With additional heating the sounding curve for the lower atmosphere approaches GH, and the chances for condensation and free convection are improved. The critical condition for cumulus development is generally taken as GH, which is determined by the dry adiabat through the intersection H of moisture line r and the pT curve. Point H defines the *convective condensation level* (CCL). It is unique in being both LCL and LFC for the buoyant surface air. Once condition GH is obtained from surface heating, convection starting near the ground is free to reach condensation at H and then accelerate on higher in the form

21. M. Neiburger, "Insolation and the Prediction of Maximum Temperatures," *Bulletin of the American Meteorological Society*, XXII (1941), 95–102.

of cloud. The surface temperature (and time) at which this first occurs is obtained from G, the height of the cloud base from H, and the cloud depth from the depth of the positive area.

Several factors were ignored in this presentation. Superadiabatic conditions near the ground are the rule during surface heating. Allowing for this, the critical state is not reached until a higher surface temperature is attained. Also, convection could penetrate a small negative area and perhaps destroy it, thereby permitting cloud formation somewhat earlier. Mixing with drier air aloft might limit the depth of the resulting clouds and perhaps prohibit their formation altogether. Therefore, although this technique serves well as a first step in estimating convection, other considerations must be made in proportion to the degree of refinement desired in the results. Included among these are the relative humidity above the CCL, diurnal variations in temperature and moisture, effects of cloud cover, and local changes in the air column due to horizontal advection.

Slightly modified versions of the above method are used to delineate geographic areas of conditional instability where showers and thunderstorms are most likely to develop due to strong surface heating. The "instability index" is determined on the thermodynamic chart from the temperature difference at some upper level (say, 600 mb) between the pT sounding curve and the pseudo-adiabat for the θ_E of air at some representative lower level (say, 900 mb). With reference to Figure 3.181, if we assume that A is the lower level in question, then the temperature difference obtained is merely the isobaric width in temperature of the indicated positive area at the particular upper level. Since the positive area is directly proportional to this temperature difference, the manner by which the difference indicates instability is evident. There is room for questioning the reliability of this technique, but simplicity is a valid point in its favor.

3.20. *Entrainment in the parcel method of convection.*

—The parcel theory of convection has served as a useful and practical tool in weather analysis and forecasting in spite of its serious theoretical defects. A more realistic approach to the problem recently has considered qualitatively the effect of entrainment,[22] whereby the moving parcel or

Figure 3.20 serves to illustrate this process. Only upward displacements are shown, but similar results are found for downward motions, and the idea is applicable also to horizontal motions. In both examples some exaggeration was made for clarity in drawing. Bearing this in mind, one will take the corresponding results to indicate the qualitative effect of entrainment rather than the numerical differences it might usually produce.

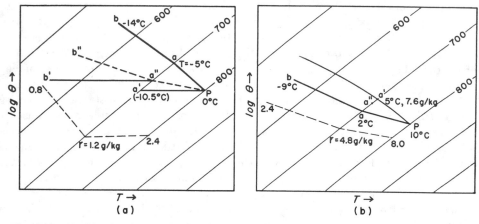

FIG. 3.20.—Modifications in soundings due to convection with entrainment; (*a*) unsaturated, (*b*) saturated.

stream of air drags or draws in some of the surrounding air which then becomes mixed with it. Through mixing, the original current of air may lose its identity, as the properties of the mixture will be intermediate between the properties of the air following the adiabatic process and of the environment air mixed in with it. The discrepancy resulting from this refinement in theory depends on the difference between the moving parcel and its environment and on the rate of entrainment.

22. H. Stommel, "Entraining of Air into a Cumulus Cloud," *Journal of Meteorology*, IV (1947), 91–94; J. M. Austin, "A Note on Cumulus Growth in a Non-saturated Environment," *Journal of Meteorology*, V (1948), 103–7, and "Cumulus Convection and Entrainment," in American Meteorological Society, *Compendium of Meteorology* (Boston, Mass., 1951), pp. 694–701.

Example (*a*) is intended to show the effect of entrainment on the process curve of an air parcel P ascending dry adiabatically through a stable environment. The pT curve for the air column is given by the heavy line extending to the upper left of P; the pT_s curve is dashed in the lower left. If P ascended dry adiabatically without mixing, the process would describe a path on the chart toward a'. Now, if the parcel in its ascent is allowed to draw in and continually intermix with environment air, its path on the chart is between the pT curve (Pa) of the environment and the adiabatic process curve (Pa'). It would reach 700 mb at some point a'' between a' and the environment temperature for that level. It is evident that the position a'' relative to a' and a is proportional to the added mass of air from the surroundings relative to the mass of the initial parcel P.

Assuming a given rate of entrainment, it is possible to find the temperature and moisture of the mixed parcel after finite vertical displacement. If the mass of the original parcel is M and the mass of entrained air δM, then, with complete mixing, the temperature at a'' is

$$T_{a''} = (T_{a'}M + T_a\delta M)/(M + \delta M) , \quad (1)$$

and the mixing ratio at a'' is

$$r_{a''} = (r_{a'}M + r_a\delta M)/(M + \delta M) . \quad (2)$$

If the rate of mixing is invariant during vertical displacement, a curved line from P to a'' describes the process.

The *rate of entrainment*, I, can be expressed in terms of increase in mass δM of the ascending parcel through an interval of pressure δp.

$$I = (1/M)\delta M/\delta p ,$$

where I is thus the fractional change in mass of the vertically moving parcel per unit change in pressure. It is expressed as a percentage per millibar, a percentage per 100 mb, or the like. From meager evidence we have on entrainment rates it is not yet possible to give an average working value. Byers gives a computed rate 100 per cent per 500 mb (doubling of the mass in a vertical displacement 500 mb) for a thunderstorm,[23] while Stommel's results determined by a different method for a cumulus cloud show an appreciably greater rate.

If the rate of entrainment is multiplied by the pressure displacement, then $I\delta p = \delta M/M$. By substituting this into Eqs (1) and (2) and then simplifying results,

$$T_{a''} = (T_{a'} + T_a I\delta p)/(1 + I\delta p) \quad (3)$$
and
$$r_{a''} = (r_{a'} + r_a I\delta p)/(1 + I\delta p) . \quad (4)$$

23. H. R. Byers and R. R. Braham, *The Thunderstorm* (Washington, D.C.: Government Printing Office, 1949), pp. 32–37.

To apply this to the example in Figure 3.20a, let us assume a rate of entrainment 100 per cent per 300 mb (1/300 mb). Then, at 700 mb,

$$T_{a''} = \frac{-10.5° - 5° \,(1/300)\,(100)}{1 + (1/300)\,(100)}$$

$$= \frac{-10.5° - 1.7°}{4/3} \simeq -9°C ,$$

$$r_{a''} = \frac{2.4 + 1.2\,(1/300)\,(100)}{1 + (1/300)\,(100)}$$

$$= \frac{2.4 + 0.4}{4/3} = 2.1‰ .$$

Under those conditions the mass of air reaching 700 mb from P would be warmer and less moist than without mixing.

Computations of the relative effect of entrainment on convection were meant for the saturated portion of the parcel ascent above the level of free convection, usually as a means of explaining observed temperatures in cumulus and cumulonimbus clouds lower than suggested by the original parcel method. In the discussion of saturated ascent we follow the outline by Austin.[24]

Assume P in Figure 3.20b represents the CCL for the parcel method. If mixing with the environment is ignored, a parcel ascending through P would continue upward with increasing speed along pseudo-adiabat Pa'. Now consider the alternative of mixing with the colder and less humid surroundings during ascent from P. The parcel arrives at the upper level following a process different from both Pa and Pa'. For an assumed rate of entrainment it is possible to determine by Eqs (3) and (4) and by additional considerations the state of the parcel when it arrives at 700 mb. For the same value of I, 1/300 mb, by mixing alone the resulting temperature at 700 mb is about 4°C (weighted average of 5° C and 2° C) and mixing ratio 6.9‰ (weighted average of 7.6 and 4.8).

24. "Cumulus Convection and Entrainment," *op. cit.*, pp. 694–701.

Now we must account for the water supposedly condensed between P and a'. Part or all of the excess moisture can be said to evaporate into the mixture at 700 mb to bring the mixture to or toward saturation. The parcel had no condensed water and a mixing ratio 9.7‰ at P. Thus the condensed water between P and a' is 2.1‰, or 2.1 gm if the mass of the parcel was 1 kg at P. Evaporating water to saturate the mixture must lower the temperature and raise the mixing ratio from the preliminary values 4° C and 6.9‰. The adjusted values for the two variables are determined by the wet-bulb temperature for the pressure at a corresponding to the preliminary values 4° C and 6.9‰. This gives temperature 3.8° C and mixing ratio 7.1‰ for the mixed parcel at a''. The increase of 0.2‰ is for the entire $1\frac{1}{3}$ kg of mixed air (1 kg starting from P plus $\frac{1}{3}$ kg entrained). Therefore, to bring the mixture to saturation, $1\frac{1}{3}$ kg $\times$ 0.2 gm/kg $\simeq$ 0.3 gm of condensed water is used. This leaves the mixture with 2.1 − 0.3 = 1.8 gm of water at a'' for each 1 kg of air starting from P. The important result is the lowered temperature and the less condensed water of the ascending parcel compared to a process with no mixing at all. Furthermore, with mixing, the ascending air would have a lapse rate of temperature intermediate between the pseudo-adiabatic and the environment lapse rates.

In the technique above we (1) lifted the saturated parcel (pseudo-)adiabatically from P to the next significant point on the sounding; (2) mixed the parcel with the entrained air accumulated from the environment, not accounting for presence of water; and (3) allowed some of the excess water to evaporate into the mixed air by the wet-bulb process, obtaining the final point a'' on the chart. Following these steps simplifies the procedure, but, as mixing should occur continuously during ascent, the amount of water we allowed to evapo-

rate in step 3 would not have condensed in ascent and, hence, does not evaporate at the upper level. One can see that for a very dry atmosphere above P the initially saturated parcel could follow a dry adiabatic process if entrainment is sufficient.

3.21. *Layer stability: Unsaturated case.*— Up to now we considered only displacements of individual small masses of air through resting media. Vertical displacement of entire layers or air columns should be investigated also, for there are many conditions in which air layers can be lifted or lowered bodily in the atmosphere. One obvious example occurs in strong air motion up or down the slopes of the surface topography. Air layers also may be forced upward through the effects of strong horizontal convergence of air near the surface or of organized upper-level divergence, and similarly they may be brought downward with horizontal divergence of motion in the lower atmosphere. In each case we must consider vertical displacements of large parts of the sounding. This produces significant changes in stability of the layer which may lead to different behavior of the atmosphere.

First, we consider displacements when saturation is not reached in the process; this may also include downward displacements of initially saturated layers of air if all downward motion is assumed to follow the dry adiabatic process. In Figure 3.21a is given a stable layer with pT curve ab. Let ab refer to a vertical column of unit cross-sectional area. If the column is lifted dry adiabatically to $a'b'$ without change of mass and without lateral expansion or contraction, such that the pressure difference from top to bottom remains the same, the column expands vertically, and the air becomes less stable by vertical expansion. If column ab is lowered dry adiabatically to position $a''b''$, it is more stable as a result of vertical compression.

Now consider a different air column *cd* with lapse rate initially superadiabatic. As this column is lifted under similar provisions to position *c'd'*, the lapse rate tends toward dry adiabatic. If *cd* is lowered to position *c''d''*, the lapse rate departs even more from dry adiabatic. By combining the results above, we obtain the useful rule: *If an air*

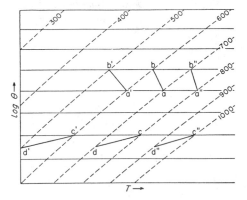

Fig. 3.21*a*.—Effects of ascent and descent on stability.

column or layer is displaced dry adiabatically without lateral expansion or contraction, ascent makes the lapse rate approach the dry adiabatic, and descent makes the lapse rate depart more from the dry adiabatic, irrespective of the initial stability. Evidently, the larger the vertical displacement, the greater is the change in lapse rate. For a layer having a dry adiabatic lapse rate initially, the lapse rate is unchanged in the process.

From the fact that layers or columns do undergo lateral expansion and contraction during ascent or descent, these factors must be considered, in addition to conservation of mass, when considering change of stability. To incorporate all these factors, we may use Eq 3.17(3). For a thin stratum of air of vertical thickness $(\delta z)_1$, bounded on the bottom by potential temperature θ and on the top by $\theta + \delta\theta$ (Fig. 3.21*b*), the stability is

$$E_1 = g(\delta \ln \theta)/(\delta z)_1 .$$

After the stratum is lifted or lowered dry

adiabatically, so that its final thickness is $(\delta z)_2$, its stability is

$$E_2 = g(\delta \ln \theta)/(\delta z)_2 .$$

By combining the above two equations,

$$E_2 = E_1(\delta z)_1/(\delta z)_2 . \qquad (1)$$

Therefore, *the change of stability during dry adiabatic displacement is a function only of the change in depth of the layer.* If an initially stable layer undergoes vertical stretching, the layer becomes less stable. Eq (1) should be analyzed for initial stability or instability and vertical stretching or shrinking.

When we introduce conservation of mass into Eq (1), we obtain

$$E_2 = E_1(A_2/A_1)(p_2/p_1)^{1-\kappa} , \qquad (2)$$

if p_1 is the mean pressure of the layer before displacement and p_2 after displacement. This is the complete expression for modification of stability by dry adiabatic motions. *The stability after displacement is of the same*

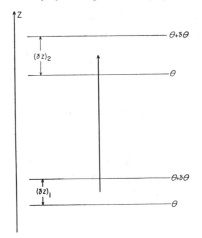

Fig. 3.21*b*

sign as the initial stability, but its magnitude is modified by compressibility in the atmosphere and by lateral divergence (and convergence) within the layer.

The effect of p_2/p_1 on stability was shown in Figure 3.21*a*. When we consider the effect of A_2/A_1, it is evident that increase in horizontal area (lateral divergence) of individual air columns, $(A_2 > A_1)$, increases the magnitude of stability whether it is positive or negative initially. Lateral convergence within the layer decreases the magnitude of stability. As the layer is expanded horizontally, it contracts vertically and thereby carries adjacent potential temperature surfaces closer together in the vertical. When the layer is given no net vertical displacement, $(p_2/p_1) = 1$, and the change of stability is governed only by the change in horizontal area of the column.

3.22. Layer stability: Saturated case.— We now consider the more restricted case of an initially saturated layer *lifted* in a pseudo-adiabatic process. The effects of lifting and lateral divergence are of the same sense as in the unsaturated process, only the reference lapse rate is now the saturated pseudo-adiabatic. A saturated layer whose lapse rate of temperature is pseudo-adiabatic, that is, $\partial \theta_E/\partial z = 0$, remains indifferent for lifting and lateral divergence. If the saturated layer has lapse rate greater than pseudo-adiabatic, in which case $\partial \theta_E/\partial z < 0$, lateral divergence promotes greater instability; lateral convergence decreases the lapse rate. With sufficient lifting and lateral divergence, or simply with enough lifting alone, the lapse rate in the layer can become even larger than dry adiabatic. In event the saturated layer is characterized initially by a lapse rate less than pseudo-adiabatic, lifting and horizontal convergence both favor increase of lapse rate.

3.23. Layer stability: Air becomes saturated during lifting.—An air layer not saturated initially may become partially or totally saturated if given reasonable upward displacement. In this case none of our previous deductions is entirely valid. The problem is complicated also by the fact that

various parts of an air column may require different amounts of lifting to reach saturation.

For an initially unsaturated stable layer to become unstable after being lifted to saturation, Rossby[25] has shown that the necessary condition is $\partial \theta_E/\partial z < 0$. A stable layer in which θ_E decreases with height is said to be *convectively or potentially unstable;* with lifting, it becomes unstable after satu-

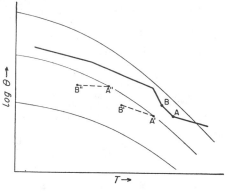

FIG. 3.23*a*

ration. In the opposite condition, the layer remains stable no matter how much it is lifted; this layer is *potentially stable.*

Consider layer AB of the sounding in Figure 3.23*a*; it is unsaturated and also stable. The characteristic curve for the layer is $A'B'$. As the entire layer is lifted bodily, all parts of layer AB cool first at the dry adiabatic rate until the part with highest relative humidity reaches saturation. With further lifting the saturated portion of the layer cools pseudo-adiabatically, and the remainder of the layer dry adiabatically, until the entire layer becomes saturated. Thereafter, all parcels follow the pseudo-adiabatic process, and each parcel takes a path on the thermodynamic chart defined by the pseudo-adiabatic curve through its

25. C.-G. Rossby, *Thermodynamics Applied to Air Mass Analysis* ("Meteorological Papers, M.I.T.," Vol. I, No. 3 [Cambridge, Mass., 1932]).

adiabatic saturation point. Since $\partial\theta_E/\partial z <$ 0, and since θ_E of each parcel is conserved, after saturation the sounding through the layer, now $A''B''$, has lapse rate greater than pseudo-adiabatic. Hence, this layer is now unstable in saturation. If the layer is lifted still further, it can become superadiabatic.

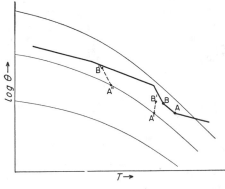

F<small>IG</small>. 3.23*b*

A layer characterized by $\partial\theta_E/\partial z > 0$ undergoes different behavior as it is lifted. In Figure 3.23*b* consider layer AB with characteristic curve $A'B'$. As the layer is displaced upward beyond saturation, the top follows a pseudo-adiabatic curve of higher value than does the bottom, and after saturation the lapse rate is less than pseudo-adiabatic at all times. Thus, a layer with potential stability ($\partial\theta_E/\partial z > 0$) becomes absolutely stable when lifted until the entire layer is saturated.

Since θ_E at any point can be *approximated* roughly by a linear function of θ and r, we may also analyze potential stability in terms of those two variables. From Eq 1.08(4),

$$\partial\theta_E/\partial z \approx \partial\theta/\partial z + 3\partial r/\partial z. \qquad (1)$$

$\partial\theta/\partial z$ is normally positive in the atmosphere, while $\partial r/\partial z$ is most often negative. Therefore, $\partial\theta_E/\partial z$ is usually the difference between opposite variations of θ and r, and $\partial\theta_E/\partial z$ is less than $\partial\theta/\partial z$ for those condi-

tions. For a layer with dry adiabatic lapse rate ($\partial\theta/\partial z = 0$), the sign and degree of potential stability are determined very nearly from the distribution of humidity alone. This equation shows that, with strong static stability, potential instability is possible if humidity falls off rapidly with height. For example, a temperature inversion in which $\partial\theta/\partial z = +10°/km$ is potentially unstable if $\partial r/\partial z = -4\%_{00}$ per kilometer. Such distributions of θ and r are observed occasionally, but little significance can be attached to the potential instability, since the required amount of lifting is large.

Eq (1) provides a basis for comparing the potential stability of arctic and tropical air masses. At low temperatures r_s is small, r is limited to a small range of values, and $\partial\theta_E/\partial z$ is nearly the same as $\partial\theta/\partial z$. Since arctic air masses are usually very stable, at least in the lower troposphere, they are also

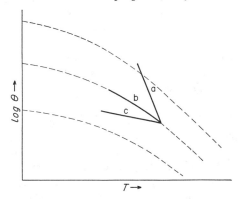

F<small>IG</small>. 3.23*c*.—Relation of potential (convective) stability to the slope of the characteristic curve relative to the pseudo-adiabats: (*a*) stability, (*b*) indifference, (*c*) instability.

potentially stable. In contrast, in tropical temperatures r_s is large, large vertical variations of r can occur, and potential stability can easily be opposite in sign to static stability. We observe the most frequent and extreme cases of potential instability occurring in tropical air masses.

Although evident by now, potential

stability can be analyzed visually from the characteristic curve on a thermodynamic chart. In Figure 3.23c characteristic curve *a* is potentially stable $(\partial\theta_E/\partial z > 0)$; curve *b* is potentially indifferent $(\partial\theta_E/\partial z = 0)$, and curve *c* represents potential instability $(\partial\theta_E/\partial z < 0)$. It follows that *a layer is potentially stable or potentially unstable according as its characteristic curve slopes upward to the right or to the left of the pseudo-adiabats on the thermodynamic diagram.*

3.24. *Effect of mixing on vertical lapse rates.*—Turbulent or eddy transfer of heat in the atmosphere has pronounced effects on lapse rates. This modification is evident es-

external source or sink of heat, is the dry adiabatic curve $A'B'$, which has the same total heat content existing in the state AB; $A'B'$ is therefore at the average potential temperature for the layer. Mixing of a stable layer results in cooling at the top and warming at the bottom, while the mean potential temperature of the layer is unchanged.

If an initially unstable layer is subjected to a similar mixing process (diagram *b*), heat is transported upward, so that the top of the layer is warmed while the bottom is cooled. The final lapse rate is dry adiabatic, and the potential temperature in the mixed layer is the average of initial potential temperatures.

Figs. 3.24*ab*.—Effect of vertical mixing on the *pT* sounding curve

pecially near the ground, where mechanical stirring is the general rule and where the rapid horizontal temperature variations at the surface tend to perpetuate turbulent motions. Vertical mixing of air through turbulence is also effective in the free atmosphere, and in some extreme cases an air column may become completely stirred from the surface up to high levels in the troposphere.

Mixing within a column of air gives an approach to homogeneity in certain properties throughout the affected portion of the column. Heat and humidity are redistributed in the column so that in the idealized final state $\partial\theta/\partial z$, $\partial r/\partial z$, and $\partial\theta_E/\partial z$ are all zero if the layer remains unsaturated, while only $\partial\theta_E/\partial z = 0$ if saturation occurs in the process. The change of lapse rate in a stable layer subjected to thorough mixing without saturation is illustrated in Figure 3.24*a*. The initial *pT* curve is AB. The theoretical end result of mixing, if there is no

This gives some insight into why the atmosphere is normally in a stable state. Because of rapid reaction to turbulent impulses, a superadiabatic lapse rate is a temporary phenomenon found only in relatively shallow regions of the atmosphere and is infrequently observed except in a thin layer in contact with a hot surface.[26] In contrast, for a stable layer to be mixed, work must be performed against stabilizing forces. Unstable conditions tend to become more stable, while stable conditions tend to remain stable.

Redistribution of humidity by vertical mixing without saturation, and without addition or loss of moisture, is similar to the effect on potential temperature. The end

26. Because superadiabatic lapse rates are so seldom found except in this thin layer, use of the word "unstable" is usually broadened to include dry adiabatic lapse rates and also lapse rates slightly less than dry adiabatic.

result of thorough mixing is $\partial r/\partial z = 0$ and relative humidity increasing upward. Since vertical mixing without saturation tends toward homogeneity of potential temperature and mixing ratio through the layer, after complete mixing all air in the layer has the same adiabatic saturation level. Therefore, thorough mixing reduces the characteristic curve to a point on the thermodynamic chart.

Vertical mixing of an initially unsaturated layer of air often leads to saturation in at least a portion of the layer. Saturation

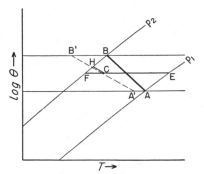

FIG. 3.24*c*.—Mixing of a moist layer

occurs first in the upper part of the mixed layer, and the depth of the saturation depends on the mean relative humidity of the layer and total depth over which mixing occurs.[27] After saturation, further mixing produces sensible warming of the mixed layer by release of latent heat in condensation, and the mean potential temperature of the layer is increased while the mean mixing ratio is decreased.

In Figure 3.24*c* is given the initial state of a layer bounded by pressures p_1 and p_2; its pT curve is AB and characteristic curve $A'B'$. As the layer is mixed, the sounding curve tilts toward the dry adiabats and finally approaches EF. In the meantime $A'B'$ reduces to CH, a curve located at the

27. This relation is shown graphically by Sverre Petterssen, *Weather Analysis and Forecasting* (New York: McGraw-Hill Book Co., 1940), p. 95.

mean equivalent potential temperature for the layer. The *mixing condensation level* (MCL) is located near intersection C of the mean potential temperature curve EF with the "mean pseudo-adiabat" CH for the layer. The final state of the mixed layer is roughly the new pT curve ECH and the new characteristic curve CH. In this state $\partial \theta_E/\partial z = 0$, and θ_E for every parcel is the initial mean value of θ_E for the layer. Area CFH is proportional to the difference in mean mixing ratio between curves $A'B'$ and CH.

Locating the MCL of a layer serves a useful and practical purpose in analyzing certain weather conditions. The MCL approximates the base of a stratus or stratocumulus cloud deck maintained by low-level mixing and the bases of convective cumulus cloud in daytime heating. After thorough mixing, the temperature lapse rate is dry adiabatic and mixing ratio is constant from the ground to the cloud base (the subcloud region). Thus, surface temperature and dewpoint can be used to approximate the height of the adiabatic saturation level for all air samples in the subcloud region and thus the elevation of the cloud base. If T_0 and T_{s_0} are the surface temperature and dewpoint, respectively, and if T_c and T_{s_c} are the temperature and dewpoint at the adiabatic saturation level of the mixed layer, then

$$T_c = T_0 + z_c(\partial T/\partial z)_\theta \quad \text{and}$$
$$T_{s_c} = T_{s_0} + z_c(\partial T_s/\partial z)_r ,$$

where z_c is the height of the saturation level above ground. At that level $T_c = T_{s_c}$, and therefore

$$z_c = -(T_0 - T_{s_0})/[(\partial T/\partial z)_\theta$$
$$- (\partial T_s/\partial z)_r] . \quad (1)$$

In the denominator $(\partial T/\partial z)_\theta$ is $-9.8°$ C per kilometer, and the second quantity varies from about $-1.4°$ C per kilometer to

−1.9° C per kilometer in normal range of temperatures. An average value for the denominator is −8.2° C per kilometer, or −2.5° C per 1000 feet. The numerical formula relating surface dewpoint depression to height of the MCL is then

$$z_c = (T_0 - T_{s_0})/8.2 \quad (\text{° C and km}) ; \quad (1a)$$

$$z_c = (T_0 - T_{s_0})/2.5 \quad (\text{° C and 1000's}$$
$$\text{feet}) ; \quad (1b)$$

$$z_c = (T_0 - T_{s_0})/4.5 \quad (\text{° F and 1000's}$$
$$\text{feet}) . \quad (1c)$$

If the surface temperature is 60° F and dewpoint 51° F, the MCL in a completely mixed layer in contact with the ground is 2000 feet.

3.25. *Additional processes affecting stability.*—There are several other processes acting independently or in combination which may alter stability. We must consider also the modifications by simple radiation, by exchange of heat through contact with the earth's surface, and by differing horizontal motions in the vertical. In each case the effect may be viewed as variation of warming or cooling with respect to height. Obviously, any process which gives increased warming (or decreased cooling) with height serves to stabilize the air layer, and the reverse is destabilizing.

The radiation-insolation process has the greatest control on the stability of the air layer at the ground. The atmosphere is comparatively transparent to solar radiation, while land surfaces absorb strongly. Thus, the ground and the thin layer of air in contact with the ground are capable of rapid warming during the day, while upper levels are relatively unaffected. At night the surface air cools more rapidly than the free air above. Thus, surface-air temperatures react quickly to diurnal and seasonal changes, and the temperature of the free air is affected more slowly. Nighttime cooling is

stabilizing, and daytime heating destabilizing, in the lower atmosphere. There is large diurnal variation of stability in the surface layer over land, largest under clear skies. Diurnal variation of stability over a water surface is small in comparison.

Another process changing stability is the movement of air over colder or warmer surfaces. The temperature of the air at the boundary is changed by radiation, convection, and conduction, and this change may

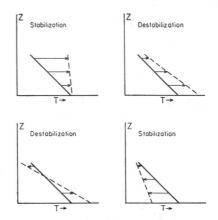

Fig. 3.25.—Some of the possible changes of vertical temperature lapse rate due to differing horizontal temperature advection with height.

be distributed over finite depth through turbulent mixing. Air moving over a colder surface experiences stabilization, and the reverse is true for air motion over a warmer surface.

Finally, as illustrated by Figure 3.25, differential horizontal advection of temperature in the vertical modifies stability locally. Such modification is a frequent occurrence in the friction layer, where horizontal motion usually varies radically from level to level.

3.26. *The role of stability in the modification of air columns.*—Heat, moisture, and momentum may be transported upward or downward through turbulent mixing. The direction of transport by mixing depends on the distribution of each property, and the rate of transfer is a function both of the distribution of the properties and of the intensity of mixing.

a) Heat.—Upward flux of heat across a unit horizontal plane (isobaric surface) in unit time is given by [28]

$$F_1 = -K_1 \rho c_p (\partial T/\partial z + \Gamma_d) \qquad (1)$$

for unsaturated air. (In saturated conditions Γ_d is replaced by Γ_s.) The coefficient of eddy diffusion of heat is K_1; it is a posi-

deep layers, whereas with stable conditions redistribution of heat is restricted to a more shallow extent of the atmosphere. Figure 3.261 illustrates the different modifications of an air column by heating and cooling from below. The former leads to a temperature ·curve dry adiabatic from the ground to the top of the mixed region. In the right diagram of Figure 3.261, cooling from below increases stability and suppresses the turbulence which otherwise would mix the cooling effect through a deep layer. Cooling therefore remains concentrated in a shallow skin layer

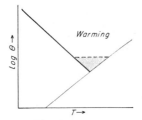

FIG. 3.261.—Different types of modification by warming and cooling from below

tive quantity which is zero at the earth's surface and increases upward in the friction layer. It depends on stability, and it is thousands of times larger than the coefficient of molecular conductivity. Eq (1) may be given also in terms of stability

$$F_1 = -(K_1 p E_v)/\kappa g . \qquad (\kappa = R_d/c_p) \qquad (1a)$$

Vertical flux of heat is controlled by the sign and numerical value of stability in the atmosphere. When $\partial\theta/\partial z > 0$, heat transfer is downward; when $\partial\theta/\partial z < 0$, it is upward. It was previously suggested that the intensity of turbulent motions, and therefore the coefficient K_1, is small with stability and large with instability. If this is true, upward transfer of heat is more rapid and thorough than downward transfer. In the same light, unstable conditions are associated with penetration of turbulent mixing through

in contact with the cold surface, unless mechanical turbulence is maintained by fresh winds. At higher levels, perhaps only a few hundred feet above the surface, temperatures remain much less affected by the cooling below.

Figure 3.261 also serves to illustrate the fact that, with equal amounts of solar energy added to the early-morning sounding, the more stable the surface-air layer, the more rapid is the daytime heating and the larger is the diurnal temperature range. It is also evident that the highest maximum surface temperatures are found with high temperatures at upper levels.

b) Moisture.—Upward flux of water vapor[29] through a unit plane may be given by

$$F_2 = -K_2 \partial r/\partial z , \qquad (2)$$

where K_2 is an appropriate coefficient of eddy diffusion of water vapor and r is

28. See, e.g., D. Brunt, *Physical and Dynamical Meteorology* (2d ed.; Cambridge: Cambridge University Press, 1939), p. 225.

29. Vertical flux of atmospheric impurities would be similar, but with different coefficients.

mixing ratio. Flux of moisture is upward when $\partial r/\partial z < 0$. On the average, there is upward flux of water vapor to compensate for the loss of moisture from the atmosphere by precipitation.

The coefficient K_2 is again related to instability; for given $\partial r/\partial z$ the transfer of water vapor is greater in unstable conditions. With extreme stability little vertical exchange of moisture is possible in the atmosphere no matter how rapidly humidity may vary with height. Therefore, the existence of large temperature lapse rates is of primary importance for transport of moisture from the source at the surface upward into the atmosphere.

The control stability maintains on turbulent transfer, and therefore on vertical redistribution of moisture, extends itself to the rate at which moisture is introduced into the atmosphere by evaporation. Evaporation from a water surface proceeds at a rate proportional to the vapor-pressure difference $(e_s - e)$, where e is the vapor pressure of the air in contact with the water surface and e_s is the saturation vapor pressure for the temperature of the water. In absence of mixing, only a thin boundary layer of air is affected by evaporation, but this skin layer soon becomes saturated, and evaporation practically ceases. Any of the factors favoring suppression of turbulence—strong stability, smooth surfaces, and very light winds—restricts evaporation and transport of water vapor from the earth's surface into the free air. On the other hand, if humidity decreases with height, turbulence maintains upward transfer of water vapor from the bounding layer, thus maintaining positive $e_s - e$ at the surface, and evaporation can proceed until a deep mixed layer is saturated.

c) Momentum.—Turbulent mixing is important also in vertical distribution of horizontal momentum, especially in the surface layer, where friction with the earth

contributes large vertical shear of the wind. In Figure 3.262 the diagram on the left is a profile of wind speed in stable conditions; the frictional effect is concentrated near the surface, the surface wind is subnormal, and motion is unusually smooth. The second diagram is a wind profile for less stable conditions. Here turbulent mixing distributes frictional retardation over a deeper layer, carries down momentum from above, and

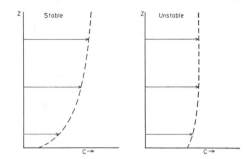

Fig. 3.262.—Vertical profiles of wind speed in the friction layer for stability and instability.

gives above-normal surface-wind speeds with greater gustiness.

The effect of turbulent mixing on the distribution of wind *directions* is similar. In unstable conditions mixing maintains small differences of wind throughout the mixed layer. When the surface layer is stable, wind directions can vary rapidly in the vertical, as the drag is restricted to a thin layer. The angle between the wind direction above the friction layer and the surface wind direction, as well as the difference in speed, is a direct function of stability. Diurnal variation of stability gives diurnal variations in these quantities and also in the depth of the affected layer.

It is important to examine the relation between vertical mixing and large-scale modification of the atmosphere with unstable conditions and addition of moisture from below. The increase of humidity at upper levels resulting from upward flux of moisture can lead to condensation as a part of the process. Release of latent heat may produce not only such convective weather phenomena as cumulonimbus and showers but also rapid warming of the atmosphere. This is shown schematically by drawing *a* in Figure 3.263, in which the initial sounding is *ADB*. As the air column is heated from below without saturation, the sounding may assume the form *EDB*, such as typified by daytime heating or by air movement over a warm dry surface. However, if attending this heating there is also addi-

sufficiently dry, so that the difference $e_s - e$ is positive, then evaporation can occur from a moist surface, but humidity is increased only in a shallow layer. Fog may develop, or, if sufficient turbulence is maintained by fresh winds, a low stratus or stratocumulus cloud deck may form. The important fact is that surface cooling shields the free atmosphere from modification by the earth's surface. In summary, *air moving over a warm moist surface is subject to rapid and radical modification to great depths; air moving over a colder surface, either moist or dry, is modified only in a thin surface layer.*

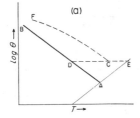

FIG. 3.263

tion of moisture by evaporation, the instability present fosters upward transport of moisture to the extent that the MCL might be lowered appreciably. Let us assume the resulting MCL is located at *C*. Those buoyant surface particles reaching saturation are accelerated upward along *CF*, and thorough "overturning" of that part of the atmosphere will ensue. An equilibrium condition for this MCL is curve *ECF*, which is warmer and moister than *ADB*. Modification of extensive masses of air is therefore more thorough and violent if cold air moves over a warm ocean surface than if this cold air moves over a warm land surface.

To contrast the effect of air motion over cold moist surfaces on the modification of the atmosphere, consider drawing (*b*). In the initial state the air column is characterized by *pT* curve *ADB*. As it moves over a colder surface, whether moist or dry, the sounding assumes form *EDB*. If the air is

3.27. Formation of temperature inversions. —The processes favoring stabilization, which were described previously, are the explanations for temperature inversions and other layers of strong stability in the atmosphere.[30]

A. Lower stratosphere
B. Free tropospheric inversions
  1. Differential horizontal advection
  2. Differential vertical motions
  3. Differential radiation
  4. Differential mixing (turbulence)
C. Ground inversions
  5. Radiation cooling
  6. "Contact" cooling

1. *Differential horizontal advection.*— Overrunning by warm air and underrunning by cold air, occurring singly or in combination, form and maintain stable

30. Another process worth considering is differential heating or cooling through change in water phase, that is, by *differential evaporation* or *condensation* in an air layer.

layers. For this process to be most effective, large-scale shearing motion and large horizontal temperature gradients are required. The commonly observed "frontal" inversions of the free atmosphere are usually attributed to this cause.

2. *Differential vertical motions*, accompanied by horizontal divergence of mass in a stable atmosphere, leads to the subsidence and convective inversions. To clarify this process, we may refer to Figures 3.21*ab* and 3.27*a*. In the latter, the drawing at the left shows an initial potential temperature distribution with equally spaced lines θ_1, θ_2, . . . , θ_5, and potential temperature increasing with height. If subsidence (sinking) occurs uniformly, each isentropic surface is brought down the same, and no layer of concentrated stability results. Therefore, subsidence alone does not create the observed "subsidence inversions" in the free atmosphere. But suppose that during subsidence the air in a particular layer, for example, between θ_4 and θ_3, is permitted to spread out laterally. Then from continuity the upper boundary of the layer sinks more than the lower boundary, and intermediate isentropic surfaces converge vertically. In the atmosphere above and below the layer of maximum lateral divergence the increase of stability is less. The necessary criterion for producing a subsidence-type inversion in an initially stable atmosphere is vertical shrinking of that layer, with attendant horizontal mass divergence as required by continuity. A final state in the process is indicated by the dashed potential temperature lines. The greatest stabilization coincides not with the largest downward displacements but where vertical shrinking is largest. At some distance above the subsidence inversion, nearly dry adiabatic lapse rates result from vertical stretching required for mass continuity.

On the right side of Figure 3.27*a* is shown stabilization with ascending motions and horizontal divergence. If the air remains unsaturated during the lifting, the effect of ascent is similar to the effect of subsidence, except that vertical stretching now occurs below the region of stabilization. Because ascent of air is likely to result in saturation and subsequent pseudo-adiabatic processes, the scheme for convection in Figure 3.27*a* is not so simple in real conditions as for subsidence. There is evidence, however, that formation of stable layers by differential up-

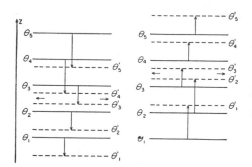

FIG. 3.27*a*.—Formation of stable layers through differential vertical motions and lateral divergence of mass in an initially stable atmosphere.

ward displacements does occur frequently in the atmosphere. A cumulonimbus cloud is a manifestation of intense upward motion; lateral spreading of the stratiform anvil cirrus gives visual evidence not only of lateral divergence in this region but also of a stable layer overlying the cloud mass.

A characteristic feature of true subsidence inversions is extreme dryness and upward decrease of relative humidity (and usually moisture) through the inversion.[31] This feature, usually taken as evidence that an existing inversion has resulted from subsidence, may not occur if in the initial state the moisture increased with height or if during subsidence there was also strong

31. A discourse on subsidence and subsidence inversions was given by J. Namias, *Subsidence within the Atmosphere* ("Harvard Meteorological Studies," No. 2 [Cambridge, 1934]).

advection of moisture in the upper regions of the layer.

3. *Differential radiation.*—A suggested explanation for formation and strengthening of inversions in the free atmosphere considers the varying radiative power of the atmosphere in the vicinity of cloud tops, moisture discontinuities, and haze layers. Mal,[32] Namias,[33] and Neiburger[34] have dis-

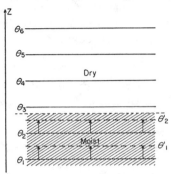

FIG. 3.27b.—Illustrating a stabilizing process by radiation at a sharp moisture discontinuity in the free atmosphere.

cussed this problem in some detail. For an air stratum which is moist below, dry above, and with sharp moisture discontinuity between, it is reasoned that radiation cooling of the moist air and indifference to radiation by the dry air can produce a temperature inversion at the moisture discontinuity. Figure 3.27b illustrates schematically this possible effect. Radiative cooling in the moist lower portion of the diagram displaces isentropic surfaces θ_1 and θ_2 to positions θ_1' and θ_2'. Since there is no change in temperature in the dry part of the layer, θ_3 and all higher isentropic surfaces remain fixed.

There is less divergence of opinion on the effect of radiation from a *cloud sheet* on producing temperature inversions in the free atmosphere. Ordinary cloud decks radiate

32. S. Mal *et al.*, *Beiträge zur Physik der freien Atmosphäre*, XX (1932), 56–77.
33. J. Namias, *Monthly Weather Review*, LXIV (1936), 351–58.
34. M. Neiburger, *Journal of Meteorology*, I (1944), 29–41.

as black bodies, and, if the air above the clouds is relatively dry, upward radiation from the clouds escapes. Cooling at the cloud tops, while the air above remains relatively unaffected, results in stabilization above the clouds and possible formation of an inversion.

It appears more reasonable to assume as a whole that differential radiation is a process which accentuates vertical variations in stability rather than initiates them. In order to create and maintain a sharp moisture discontinuity or well-defined upper boundary of a cloud layer, there should exist a stable layer above the boundary and a mixed layer below, about where they would be attributed to differential radiation.

4. *Turbulence.*—The effect of turbulence in stabilizing air strata is most common just

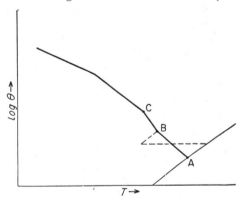

FIG. 3.27c.—Schematic illustration of a turbulence inversion formed above the mixing layer.

above the surface friction layer, but turbulence may be effective also at higher levels. A first condition for the formation of temperature inversions by turbulence is $\partial\theta/\partial z > 0$, as shown by ABC in Figure 3.27c. Suppose the air column is subjected to strong turbulent mixing near the ground. Turbulent exchange of heat in layer AB produces cooling in the upper portion[35] and smaller change in levels above B. The in-

35. Similar cooling by mixing should create an inversion just above the heated layer indicated in the left of Figure 3.26l.

tensity of the resulting temperature inversion, as well as the depth of the mixed layer, depends on the initial stability, the shape of the original sounding curve, and the strength of the mixing process.

5. *Radiation cooling.*—The two types of ground inversions given in the outline are basically the same; the difference is only in the way we view their formation. In the first there is net loss of heat from the earth's surface; in the second there is exchange of heat between the colder surface and the air moving over it.

The true radiation inversion is a feature of early-morning soundings at land stations under clear and stagnant conditions. During the course of morning heating, this inversion is then rapidly destroyed.

The nocturnal temperature inversions are most pronounced in winter, when nights are long and the lower atmosphere over continents is generally most stable. In the snow-covered regions of the arctic, the long winter nights produce the strongest inversions observed; through some the temperature may increase as much as 40° C in the first kilometer above ground. (Examine the arctic soundings in Chap. 6.) These develop most effectively over snow fields when air is relatively stagnant, when skies are clear and humidity low, and when the air is sufficiently far removed or protected from warm bodies of water. A detailed discussion of their development by radiative cooling alone has been given by Wexler.[36]

6. *"Contact" cooling.*—The rate of contact cooling of surface air, or, more properly, the rate of cooling due to exchange of heat with the surface, depends on the temperature gradient of the underlying surface in the direction of air motion and on the speed of motion. But since moderate to strong winds are conducive to turbulent mixing of the surface layer, and thus tend to

retard a *surface* inversion, such an inversion by contact cooling alone must be restricted only to conditions of rather light winds and large horizontal contrasts in temperature of the underlying surface. The more common type of inversion so produced is located not at the surface but at the top of the mixed layer; it is a turbulence inversion.

It might be suggested here that a primary cause in the formation of temperature inversions by motion of air over colder surfaces is differential advection of temperature by vertical wind shear in the friction layer, as distinguished from loss of heat in the lowest extremity of individual air columns by contact with a cold surface. The air in the upper part of the friction layer moves farther in unit time than the air at the surface, and the air trajectories can differ widely. Stability results as warmer and warmer air is brought over a relatively stagnant cold film of air at the surface. Finally, an inversion of temperature may result which has not required loss of heat from individual air parcels.

We have outlined the various processes which can be responsible for zones of great stability in the atmosphere. More often than not, an inversion results from a combination of two or more of the processes. It is common to find an inversion above the friction layer resulting from both surface cooling and turbulent mixing with moderate to fresh winds. Many of the lowest subsidence inversions are strengthened below by mixing and also by cooling at the surface. There are many other possible and applicable combinations.

3.28. *Local temperature changes due to adiabatic vertical motions.*—Potential temperature is a function of time (t) and the three space dimensions (x, y, z), so that its total differential can be expressed

$$d\theta = (\partial\theta/\partial t)dt + (\partial\theta/\partial x)dx$$
$$+ (\partial\theta/\partial y)dy + (\partial\theta/\partial z)dz . \quad (1)$$

36. H. Wexler, "Cooling in the Lower Atmosphere and the Structure of Polar Continental Air," *Monthly Weather Review,* LXIV (1936), 122–36.

Division of both sides by dt and rearrangement gives

$$\partial\theta/\partial t = d\theta/dt - u(\partial\theta/\partial x)$$
$$- v(\partial\theta/\partial y) - w(\partial\theta/\partial z) , \quad (2)$$

where u, v, and w are the velocity components in the x, y, and z directions. The term on the left is the rate of local change in θ. The first term on the right is the rate of change in θ experienced by an individual air parcel, and $-w(\partial\theta/\partial z)$ is the local rate of change due to vertical motion. In absence of all motions, the change observed at a fixed point $(\partial\theta/\partial t)$ is identical with the individual change $d\theta/dt$.

If we restrict attention to dry adiabatic vertical motions only,

$$\partial\theta/\partial t = -w(\partial\theta/\partial z) . \quad (3)$$

In moving upward or downward, unsaturated air parcels carry their θ surfaces with them, thereby changing potential temperature locally at a rate proportional to the vertical velocity and stability.

The analogous expression for $\partial T/\partial t$ is obtained from Eqs (3) and 1.07(1):

$$\partial T/\partial t = -w(\Gamma_d + \partial T/\partial z) . \quad (4)$$

With unsaturated stable conditions, upward motion results in cooling at fixed levels, and downward motion in warming. To show the magnitude of local temperature variations due to vertical motions, consider an atmosphere with lapse rate 6° C km⁻¹; then $(\Gamma_d + \partial T/\partial z) = 4°$ C km⁻¹. With vertical velocity 1 km per day, local change is 4° C per day. It appears that local temperature change in the free atmosphere by vertical motion can be of magnitude comparable to changes by horizontal advection. In very stable regions with rapid vertical motion, temperature variations aloft due to subsidence or convection can at times completely overshadow the effects of horizontal advection.

For upward motions in saturated conditions, Γ_d should be replaced by Γ_s in Eq (4). In a saturated adiabatic process the effect of ascent on local temperature change is usually somewhat smaller than in dry adiabatic, and it even may be of opposite sign for the same lapse rate and upward motion.

3.29. *Relation of stability and vertical motion to weather.*—It is now evident that the control by static stability on vertical transport of heat and water vapor, and therefore on modification of the atmosphere by the earth's surface, must also manifest itself in various weather phenomena. This introduces a lengthy subject in air-mass meteorology which can be summarized only briefly.

Most types of visible weather can be classified according as the air is saturated or unsaturated and is stable or unstable and also according to the nature of the vertical motions producing the weather. The same mechanism affecting moist and dry air masses results in different visible phenomena and weather sequences. Weather observed with stable conditions differs from that with unstable conditions. It is necessary to consider also the nature of the vertical motions; turbulent mixing in a shallow layer, local convection penetrating to great depths in the atmosphere, and large-scale ascent or descent result in quite different forms of weather. All this obviously depends on more extensive considerations than just stability.

The predominant process in the formation of clouds and precipitation in the atmosphere is ascending motion; radiative processes aid in cloud formation and are effective in cloud dissipation. At the surface the condensation phenomena are mostly a result of isobaric cooling and, indirectly, also the result of addition of moisture by evaporation from the surface or by falling precipitation. Owing to decrease of relative humidity during adiabatic descent, downward motions are on the whole free from condensation. But absence of cloudiness should not be taken as evidence of broad-scale descent; there can be clear skies without vertical motion or even with strong ascent.

a) Unsaturated and stable conditions.— The visible atmospheric phenomena attending stable conditions in unsaturated air are layers of haze and smoke particles (lithometeors) suspended in the air. Over continents in autumn and winter, when the friction layer is most stable and when great quantities of smoke are injected into the air, horizontal visibility at the surface can be greatly impaired, particularly in the lee of industrial areas.

Through vertical mixing provided by moderate winds or daytime heating, the smoke or haze can be lifted above ground, or at least diluted through a deeper layer. If a stable layer tops the mixed region, the impurities will be trapped beneath, and the upper boundary of haze or smoke is well defined with smooth appearance. If no such stable layer exists, impurities can be distributed over greater depths in the atmosphere. Although haze and smoke layers can be mixed by turbulence or lifted bodily in organized convection, it is also possible for them to intensify by vertical shrinking.

The presence of smoke and haze layers, on the ground or in the free atmosphere, usually is evidence of stability just above; but the converse does not hold, since presence of such impurities is conditioned by a source. Continental arctic air masses in their source regions are usually very clear, yet they are the most stable of all air masses. Thus, there is no way of distinguishing between stability and instability in weather reports when condensation phenomena or atmospheric pollution is absent, except perhaps from strength and gustiness of surface wind.

b) Unsaturated and unstable conditions.— With the more unstable conditions, and in absence of condensation, the air is normally

clearer than in the previous case. But there are conditions in which instability causes lowered visibility. Unstable air motion over a powdery snow cover favors blowing or drifting snow. Unstable air masses acquire considerable amounts of dust in moving over dry bare surfaces, and, if motion is very turbulent, large particles of sand can be lifted and carried over considerable distances. Duststorms and sandstorms originating over the western Great Plains result from cold air moving rapidly over a dry, warm, and unprotected soil surface. Over water, visibility can be lowered by vertical and horizontal transport of ocean spray with choppy seas and turbulent conditions.

c) Saturated and stable conditions.—The principal condensation phenomena associated with stability near the surface include frost, dew, and shallow fogs. All these are the result of cooling by the surface. Fog can result from radiation cooling of stagnant moist air, by motion of moist air over a colder surface, by evaporation of rain into a cold surface layer, or by adiabatic cooling of moist surface air as it moves upward along sloping terrain. Explanations of these processes and discussions of favorable conditions for fog formation are covered in the "Reading References."

Clouds associated with stable conditions have generally stratified appearance. They may be formed and maintained in the upper portions of a mixed layer, as *Sc* and most *St*, or they may be formed as a result of adiabatic cooling in general upglide motion in the atmosphere with stable conditions. In ideal cases the surfaces of layer clouds are quite smooth and distinct; with rain falling from the clouds, the lower surface may become soft and indistinct. Particularly with the middle clouds, if the air was convectively unstable initially, isolated cumuliform masses might coexist with clouds of distinctly stratiform character.

Stratocumulus and the layer types of

altocumulus clouds are closely related in genesis and structure to the stratiform clouds, and often no distinction can be made. Altocumulus is frequently an attenuated form of altostratus, and also frequently the two types are coexistent. However, the principal distinguishing feature of cumuliform layer clouds is the presence of rolls or individual cells which might indicate mixing by wind shear or by shallow convection of the nature of Bénard cells.

Precipitation from stratiform clouds is more of the gentle continuous type. Drizzle is associated with low clouds of *St* and *Sc* genera, although drizzle is observed occasionally with low clouds of more convective character. It is also possible for light drizzle to fall initially as rain from clouds at higher levels.

d) Saturated and unstable conditions.—A unique condensation phenomenon forming at the surface with unstable conditions is steam fog, which results from cold air passage over a warm water surface (or warm wet-land surface).

Although we associate cumuliform clouds in general with instability, the low-cloud types cumulus humilis, fractocumulus, stratocumulus, and most varieties of altocumulus can be signs of convective or turbulent mixing in a layer of restricted depth. Cumulus congestus, cumulonimbus, altocumulus castellatus, and other chaotic forms of altocumulus signify convection of greater vertical penetration, and such clouds are better indicators of static, conditional, or potential instability. However, very often the distinct outlines of these clouds are obscured by their flattened and extensive bases or by a layer of lower clouds, so that the cloud reports frequently do not give a true picture of existing cloud conditions.

Precipitation occurring with unstable conditions is usually in showers, and again no clear distinction can be made. Some unstable forms of precipitation can be prolonged or of quite steady character, and also there are many occasions in which some convective activity can occur along with generally stable precipitation. In particular, it is extremely difficult on occasion to distinguish between showers and steady precipitation at the onset of precipitation.

Tornadoes, waterspouts, mammatiform clouds, thunderstorms, lightning, and hail are all indications of strong convection and instability.

3.30. Remarks on the nature of the vertical motions.—Precipitation resulting from turbulent mixing or from slight topographic uplift in the friction layer usually is light and consists of small drops; it may be intermittent or quite persistent; and, if the surface air is very moist, this precipitation can occur with fog. Precipitation usually is not observed at the ground with the analogous process occurring at higher levels in the atmosphere, perhaps only because small hydrometeors can easily evaporate before reaching the earth.

Bodily lifting of large masses of moist air in the free atmosphere covering areas hundreds of miles in diameter, such as observed with the usual storm areas and on windward mountain slopes, result in precipitation forms dependent on the potential stability of the air initially. With slight lifting the clouds may be totally stratiform in character and the precipitation of a steady type. If the air has potential instability, further lifting leads to local instability clouds and precipitation.

3.31. Brief survey of space and time variations of stability.—To simplify our discussion of stability variations, we may consider the atmosphere in three more or less distinct layers—the lower stratosphere (or base of the stratosphere), the free troposphere (or middle and upper troposphere), and the lower troposphere. From synoptic view, the tropopause is usually a rather sharp discontinuity in stability; but, because the height of the tropopause varies geographically and also interdiurnally, it is often better to speak of a tropopause *region* in a climatological sense. There is no distinct boundary between that we call the free troposphere and the lower troposphere.

In Figure 3.31*a* are mean January and July temperature soundings for three North American stations: Barrow, Alaska (71°20′ N, 156°24′ W); Omaha, Nebraska (41°18′ N, 95°54′ W); and Swan Island, West Indies (17°24′ N, 83°56′ W). These are based on averages of from 3 to 5 years of data[37] at each station. The most obvious

feature shown is the extremely small annual variation of temperature in the troposphere in the tropics and the great seasonal change in higher latitudes. For example, at 5 km the change in temperature from January to July is +17° C at Barrow, +19° C at Omaha, and +0° C at Swan Island.

Several peculiarities of the soundings result from the averaging process and are mis-

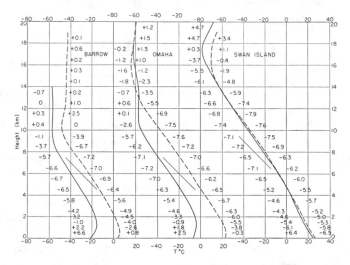

Fɪɢ. 3.31*a*.—Mean January (*full lines*) and July (*dashed lines*) soundings for three North American stations. Tabulated figures are vertical temperature differences per kilometer; short sloping lines represent the dry adiabatic lapse rate.

37. *Upper Air Average Values of Temperature, Pressure, and Relative Humidity over the United States and Alaska* (United States Weather Bureau Technical Paper No. 6 [Washington, D.C., 1949]). Only the nighttime ascents were used in compiling the average data for each station, thus making the temperatures in the surface layer unrepresentative of average conditions throughout the day. The discrepancy is particularly evident in the soundings for Omaha in Figure 3.31*a*.

leading when interpreted in terms of typical daily soundings, First, since the height of the tropopause oscillates locally with time, particularly in middle latitudes, the tropopause region is shown by rounded change in slope of the mean soundings. Another result of averaging is that only the more permanent features of the curve are retained—the stable stratosphere above the tropopause region, the rather large lapse rates in the middle troposphere, and the strong low-level stability in the arctic winter. Features of more temporary nature, such as the free tropospheric stable layers to which we attach significance in daily analysis, are removed through averaging.

In high latitudes the two most prominent changes from winter to summer in the mean curves are (1) displacement of the entire sounding (troposphere and lower stratosphere) toward higher temperatures and (2) large change in stability in the lowest 1 or 2 km above the surface. Over midlatitude continental regions similar changes in the temperature differences for each, it is also evident that in the free troposphere there is general equatorward increase of temperature lapse rate. Another point of interest is the zone of greater stability in the Swan Island soundings between 2 and 3 km. This is the effect of the western portion of the Atlantic trade-wind inversion, which is less

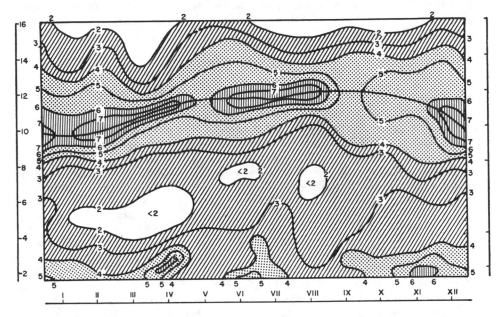

FIG. 3.31*b*.—Amplitude of the monthly mean interdiurnal variations of vertical temperature lapse rate, in units 0.1° C per kilometer per 24 hours, as function of height (km) and month. (After F. Herath, *Studie über interdiurne Temperaturänderungen in der freien Atmosphäre über Mitteleuropa nach Registrierballonaufstiegen in Lindenburg und Munich* ["Berichte des deutschen Wetterdienstes in der US-Zone," No. 25 (1951)].)

troposphere occur in the transition from winter to summer—the troposphere warms, and there is decrease of stability in the lower troposphere corresponding to reversal in the ratio of outgoing to incoming radiation. However, the changes in temperature and lapse rate at and above the tropopause are quite different from the changes observed in the polar regions.

From the curvature of each sounding, stability normally decreases with height in the free troposphere. From the slopes of the mean soundings and the tabulated vertical marked and also less persistent than in regions to the east.

Figure 3.31*b* shows the amplitude (arithmetic sum of increase and decrease) of mean monthly interdiurnal variations of vertical temperature lapse rate for two German stations, Lindenberg and Munich. The atmosphere below 16 km can be broken down into four rather distinct zones on this basis. Beginning at the top, observe that the stratosphere at some distance above the tropopause region undergoes very small average variations in stability from day to

day. The maximum occurs in the tropopause region. The free troposphere is another minimum, and, finally, a secondary
maximum variation occurs near the surface.

With exception of local peculiarities in
the surface layer, most indications given by
Figure 3.31*b* can be considered typical of
middle-latitude stations. In the tropics one
should expect the tropopause region to be
higher and show smaller mean interdiurnal
variations in lapse rate by virtue of smaller

variations in height of the tropopause. In
addition, the surface zone of secondary
maximum variation in lapse rates would be
absent; in fact, within the trade-wind region
maximum variation would appear in a zone
near 2 km. In polar regions the interdiurnal
variations of stability would be similar to
the pattern shown, the principal differences
being lower tropopause, different pattern
near the surface, and smaller maximum
variation in the tropopause region.

PROBLEMS AND EXERCISES

(Most of the following refer to the sounding listed in Section III of the "Problems and Exercises" in Chapter 2.)

1. *a*) Draw the pT and pU curves on the thermodynamic chart prescribed for pressure-height
   evaluation in the Weather Bureau. Mark the value of mean temperature and mean virtual
   temperature on the chart in each *standard pressure layer*. Using Weather Bureau pressure-
   height tables, determine the thickness (gpm) of each layer. Determine the height of the first
   standard isobar above ground if station elevation is 270 meters (886 feet). Find the height
   of each standard isobar above sea level, including 850 and 100 mb. Draw a pressure-height
   curve for this sounding, using a height scale along the abscissa of the chart.

   b) Convert the height of each mandatory pressure into gpft units. Compare your results with
   the reported heights.

2. *a*) If a thermodynamic chart is available with imprinted thickness scales, or if an overlay with
   thickness scales has been prepared for a certain chart, draw the pT and pT_s curves for the
   above sounding on that chart. Enter the pT^* curve; extend it downward to 1000 mb (for
   this problem only). Mark the mean virtual temperature for each layer and then tabulate the
   thickness of each layer to the nearest 10 gpft from the scales. Determine the height of 1000
   mb above sea level. Find the height of each mandatory pressure, as in Problem 1.

   b) Compare your results with the previous values.

3. Follow through the similar procedure of pressure-height evaluation from the sounding plotted
   on a pastagram, if available.

4. Determine the height of 1000 mb above sea level, using the Weather Bureau method and assuming the surface temperature is the same as 6 hours earlier.

5. Compute the height of 1000 mb by applying the formula $\Delta Z = 96.1 \, T^* \, (\delta p/p)$ gpft to the fictitious air layer between the surface and 1000 mb, using the value of T^* found in Problem 2.

6. Calculate the errors in height of the pressures 1000, 700, 500, 300, 200, and 100 mb corresponding to a consistent error of 1° C throughout the sounding. Explain the upward variation of
   these errors. How does the sign of error in height agree with the sign of error in temperature?

7. Calculate the errors in height of the same mandatory pressures resulting from 1-mb error in
   surface pressure. Explain.

8. What changes in surface pressure must occur (i) as the 300-mb surface rises (falls), while the
   temperature sounding is constant, and (ii) as the surface pressure remains fixed, while the
   mean temperature in the column rises (falls)?

9. What changes must occur in 300-mb height (i) as the surface pressure rises, while the mean
   temperature in the column rises (falls), and (ii) as the surface pressure falls, while the mean
   temperature in the column rises (falls)?

10. Determine the depth of the layer between 763 and 700 mb using the following methods:
    a) Weather Bureau pressure-height tables

  b) The D method (given the standard depth of this layer 2240 gpft and standard mean temperature 271° K)
  c) The differential barometric formula
  d) The adiabatic method
  Give your critical opinion of the relative accuracy and effort involved in each of these procedures.
11. Determine the depth of the layer between 700 and 600 mb by each of the above methods and also by the use of thickness scales (Fig. 3.052 may be used). Again, comment on the efficiency of each.
12. Consider a plateau station in the western United States of elevation 4800 feet. The mean annual temperature at the station is 50° F. Following are the 6-hourly observations of station pressure and temperature:

	0030Z	0630Z	1230Z	1830Z	0030Z	0630Z
Pressure..........	847.4	847.0	846.1	843.0	843.0	845.3
Temperature.......	50° F	46° F	41° F	57° F	41° F	33° F

Since the specially prepared sea-level reduction tables for this station are unavailable, we shall assume that the temperature lapse rate used in the fictitious air column below the surface is 5° C per kilometer and that the mean relative humidity in the column is 60 per cent. (WB Form 1154A could be used for this example.)
  a) Determine the appropriate surface temperatures T_S for sea-level reduction for the last four observations.
  b) Find the relation between mean temperature in the fictitious column as a function of T_S (Eq 3.10[4]). Compute $\bar{T}$ for each desired reduction.
  c) By referring to a thermodynamic chart, get the approximate mean moisture content for the column and convert to virtual temperature increment. Add this to $\bar{T}$ to obtain the required $\bar{T}^*$ for each observation.
  d) From Eq 3.10(1) or (2), determine each sea-level pressure P_0.
  e) Compute the Plateau Correction for each observation and add to the preliminary sea-level pressure determined in step (d). This result corresponds to the reported value of sea-level pressure.
  f) Draw profiles for surface and sea-level pressure as a function of time. Explain the departures in shape of the two curves. What is the effect of the Plateau Correction on the relation between these two curves? (NOTE.—Since the Weather Bureau uses temperature lapse rates which vary somewhat with the value of T_S, use of the tables unique for the station would give slightly different results.)
13. Calculate the 5000-foot pressure using the same T_S and the same assumptions for lapse rate and moisture. For this purpose, use Eq 3.10(1) in reverse. Enter the profile of 5000-foot pressure in the graph prepared previously. Comment on the relation between this 5000-foot pressure profile and each of the others. Which of the two reference-level pressures would give the more representative pressure variations for the atmosphere?
14. Using appropriate tables in U.S. Weather Bureau Circular P, determine the 1000-mb and 850-mb heights for the last three observations.
15. Find by Eq 3.12(4) the change in pressure at 5000 feet for each of the following combinations of 10,000-foot pressure change and change of mean virtual temperature in the bounded layer. For values of p_1, p_2, and $\bar{T}^*$, the values in the U.S. Standard Atmosphere may be used.

10,000-foot δp :	0	0	+1 mb	+1 mb	+1 mb	−1 mb	−1 mb	−1 mb
$\delta\bar{T}^*$:	0°	+1°	0°	+1°	−1°	0°	+1°	−1°

16. From Eq 3.13(1) and Table 3.05, supply the missing data.

700 mb	δZ :			0		+10′		−10′		−10′
	$\delta\bar{T}^*$:	0°	+1°	+1°	0°	+1°	−1°	0°	+1°	−1°
850 mb	δZ :	0	0		+10′		+10′		−10′	

17. The vertical acceleration developed by an air parcel displaced in its environment is the product of gravity and the specific (virtual) temperature anomaly with respect to the surroundings (Eq 3.15[2]).

   a) Compute the vertical acceleration developed by displacing a parcel in the sounding from 850 mb to 800 mb in a dry adiabatic process. By displacing the parcel from 850 mb to 900 mb.

   b) Suppose the parcel was saturated at 850 mb. Find its resulting acceleration if displaced to 800 mb.

   c) Take a surface parcel in the sounding and lift it to 900 mb (i) dry adiabatically and (ii) moist adiabatically and find the resulting accelerations.

18. Eq 3.17(3) is an expression for stability in terms of potential temperature distribution in the vertical.

   a) Compute the average stability in the layer between 500 mb and 400 mb. Between 200 mb and 150 mb.

   b) Determine the resulting stability if each of these layers shrank vertically to half its original depth. If they stretched to twice their original depth.

19. Locate in the sounding each layer having conditional instability and, using the midpoint of each, classify its conditional instability according to the list on page 69.

20. The stability and the vertical potential temperature distribution within a layer of air are sensitive to bodily vertical displacements and to the effect of lateral divergence of mass.

   a) Compute graphically the resulting stability and temperature lapse rate in the layer between 500 mb and 400 mb if it descended 200 mb without change of mass. If it ascended 200 mb without change of mass. Check results with Eq 3.21(2).

   b) Compute the resulting stability and temperature lapse rate if this same layer is lowered without change in depth so that its base is located at 700 mb.

   c) Compute the resulting stability if this same layer is lowered so that its base is located at 700 mb and its vertical depth is (i) halved and (ii) doubled. (Check, using Eq 3.21[2] with an approximate value for p_2.)

   d) Compute the resulting stability if this same layer is lifted so that its top is located at 200 mb and its vertical depth is (i) halved and (ii) doubled.

   e) Give a physical interpretation for the comparison of your results in (*c*) and (*d*) above.

21. Classify the layers between each reported level in the sounding as to their convective stability. Determine the minimum amount of lifting (in terms of pressure difference) required to realize instability.

22. *a)* Determine approximately the surface heating resulting from thoroughly mixing the layer bounded above by 910 mb. Repeat for the entire layer from the surface to 850 mb. Explain the difference.

   b) Determine the resulting relative humidities at the top and bottom of the surface–850-mb layer after ideal mixing.

23. Assume this air column moves over an ocean surface whose temperature is 65° F and that it absorbs moisture from below by evaporation.

   a) If after the air column reaches equilibrium with the surface the MCL is located 60 mb above the surface, find the net change in temperature and moisture at each standard pressure in the sounding for the equilibrium state of the air column.

   b) How would bodily descent of the air column by 30 mb before reaching the ocean surface affect your results?

24. From the original sounding compute the instantaneous tendency of temperature at 450 mb and at 550 mb due to upward (downward) velocity of 30 meters per hour.

25. The following data are from a sounding made at sunrise under clear skies and over level terrain:

	p (mb)	T (° C)	r (‰)
Surface..........	1000	14	9.5
	950	16	10.5
	900	15	10.0
	800	9	7.5

a) Determine the CCL and the associated surface temperature for convection, using the average value of r in the lowest 1500 feet.

b) Now suppose that, for the time of year and prevailing surface conditions, the fraction of solar energy received effective in heating the atmosphere from below is given by the area $\delta T = 10°$ C in width and $\delta \theta = 7°$ K in height on the thermodynamic chart. Find the representative maximum surface air temperature for that column of air.

c) Would convective cloud formation by surface heating be expected, and, if so, about what time of day?

d) Repeat (a) and (c), using the average r in the lowest 3000 feet.

e) What effect(s) would the following have on your answers to (a), (b), and (c)?
   (i) Evaporation from the ground increasing the average r by 1‰
   (ii) Subsidence warming by 1° C at 900 mb and 5° C at 800 mb
   (iii) Overcast of middle clouds beginning in mid-morning

READING REFERENCES

I. HYDROSTATICS

BIGELOW, F. H. "Report on the Barometry of the United States, Canada, and the West Indies," in *Report of the Chief of the Weather Bureau, 1900–1901*, Vol. II. Washington, D.C., 1902.

BERRY, F. A., BOLLAY, E., and BEERS, N. R. (eds.). *Handbook of Meteorology*, pp. 371–81. New York: McGraw-Hill Book Co., 1945.

BJERKNES, V., and SANDSTRÖM, J. W. *Dynamic Meteorology and Hydrography*, Part I: *Hydrostatics*, pp. 9–23, 41–48, 61–88, and Appendix B. Washington, D.C.: Carnegie Institution, 1910.

BRUNT, D. *Physical and Dynamical Meteorology*, pp. 28–36. 2d ed. Cambridge: Cambridge University Press, 1939.

BYERS, H. R. *General Meteorology*, pp. 167–71. New York: McGraw-Hill Book Co., 1945.

HAURWITZ, B. *Dynamic Meteorology*, pp. 11–17. New York: McGraw-Hill Book Co., 1941.

HEWSON, E. W., and LONGLEY, R. W. *Meteorology: Theoretical and Applied*, pp. 16–26. New York: John Wiley & Sons, 1944.

HOLMBOE, J., FORSYTHE, G. E., and GUSTIN, W. *Dynamic Meteorology*, Chap. 4. New York: John Wiley & Sons, 1945.

SPILHAUS, A. F., and MILLER, J. E. *Workbook in Meteorology*, pp. 79–87. New York: McGraw-Hill Book Co., 1942.

UNITED STATES WEATHER BUREAU. Form 1154 (series).

II. STATIC STABILITY

BERRY, F. A., BOLLAY, E., and BEERS, N. R. (eds.). *Handbook of Meteorology*, pp. 402–9.

BRUNT, D. *Physical and Dynamical Meteorology*, pp. 39–68, 136–59.

BYERS, H. R. *General Meteorology*, pp. 91–128, 146–48, 162, 179–82, 235–44, 247–49, 252–54.

HAURWITZ, B. *Dynamic Meteorology*, pp. 20–26, 44–57, 75–78.

HEWSON, E. W., and LONGLEY, R. W. *Meteorology: Theoretical and Applied*, pp. 31–40, 47–52, 150–59, and Chap. 14.

HOLMBOE, J., FORSYTHE, G. E., and GUSTIN, W. *Dynamic Meteorology*, Chap. 5, pp. 125–41.

PETTERSSEN, SVERRE. *Weather Analysis and Forecasting*, pp. 56–110. New York: McGraw-Hill Book Co., 1940.

SPILHAUS, A. F., and MILLER, J. E. *Workbook in Meteorology*, pp. 127–41.

CHAPTER 4

Theory and Practice of Scalar Analysis

4.01. *The coordinate system.*—The Cartesian (x, y, z) system is placed with the xy-plane tangent to a level surface, and the origin is located arbitrarily in the atmosphere but usually at sea level and at a convenient geographic point. The z-axis is the plumb line (though not everywhere straight). The y-axis is usually implied northward and the x-axis eastward, unless specified otherwise. For the present we assume the xy coordinates have arbitrary horizontal orientation. Our use of coordinates here is only for description of variations about a point in space. For that purpose, also, reference to rectangular coordinates is simpler than spherical coordinates.

4.02. *Fields of atmospheric properties.*—The quantities with which we shall deal consist of scalars and vectors. A scalar field for a given time is a region of space in which at each point there is a specific value of the scalar quantity. Pressure distribution in the atmosphere is such a scalar field, namely, the pressure field; and scalar fields of other quantities are defined similarly. If a vector is associated with each point, then the points and vectors constitute a vector field over the region. Both scalar and vector fields in the atmosphere may be given convenient forms of representation.

For a particular value of a scalar quantity to be identified with each point in the atmosphere it is necessary that the quantity be a continuous single-valued function of space. That is, in the atmosphere at a given time a value, and only one value, of the quantity can be specified at any point. We shall assume for analysis that the physical quantities p, T, ρ, r, θ, etc., are continuous in the scale with which we are dealing.

4.03. *Outline of scalar representation and analysis.*—Any of the usual atmospheric variables, which here we designate Q for generality, is a function of space and time; that is, $Q = Q(x, y, z, t)$. This quantity can be observed, represented, and analyzed in various combinations of the independent variables. Many possible forms are made evident below.

a) Value of Q *at a fixed point* (x, y, z *constant*).—If time is the only variable, a sequence of values of Q can be obtained for the fixed point of observation and used to represent the local variation of Q as function of time. The traces by stationary recording instruments (e.g., the barograph) are continuous forms of such representation. It is also possible to draw these Q, t profiles from observations of Q at finite intervals of time. The profile so derived will depart in some degree from a continuous recording. This departure affects the accuracy of interpolated values and of time rate of change (tendency), and it varies with the nature of the quantity and the length of time between observations. Evidently, the most suitable time interval between observations is determined by the nature of the quantity and the tolerance in approximation desired in reproducing the tendency form a finite frequency of observations.

If each observation contains measurements of several quantities, as the case with surface observations, a simple type of representation is shown by examples in Figure 9.02*d*—each observation is entered at its proper location along a time coordinate. The sets of profiles in Figure 12.02 bring out the time variations more clearly.

b) Value of Q *along a fixed line.*—Since time may or may not vary, there are two forms of representation to consider.

(i) *Time constant.*—All observations along the axis *s* are synoptic. This may be illustrated by Q, *s* graph or profile or by tabulated values of Q at the proper points on the line. From either form, the value of Q at any point on the line is obtained by interpolating. The variation of Q in the direction of the line about any point on the line is obtained by finite differences.

(ii) *Time variable.*—If variation with time also is considered, more involved forms of representation are necessary. Expansion of the Q, *s* diagram above into a three-dimensional form gives a Q, *s*, *t* system of coordinates in which the observed value of Q can be located at proper *s* and *t*. But, since three-dimensional schemes are impractical for plotting and analysis, a simplified form is adopted The *s*, *t* coordinates are maintained—to form the coordinate grid on a flat paper surface—but the values of Q are tabulated at respective observation points in the *s*, *t* grid. This is the basis of the vertical and horizontal time sections (*t*, *z* and *t*, *x*) described in Chapter 2 and illustrated by Figures 12.04*ab*. The scheme may be simplified further, but with less convenience for analysis, by tabulating the observed data in columns about their points along a line serving as the axis of either *t* or *s*.

c) Value of Q *in a plane.*—These forms of representation include coordinate planes (horizontal charts and cross sections), inclined planes, and even inclined surfaces of a constant property, but all may be considered two-dimensional space charts.

(i) *Time constant—the familiar synoptic*

charts.—If a single quantity Q is in question, the more practical form requires plotting the value of Q next to the reference point in the plane (see Fig. 4.04*a*). If other quantities are measured and desired for analysis, their values also are plotted at each point, as done in routine charts. To determine the variation of Q in the space dimension not incorporated in the chart, the value of Q in this chart is compared with the value on the adjacent concurrent chart. To determine the time variation of Q, one chart is compared with the later one. It is evident here that, with common synoptic charts, we cannot analyze the space and time variations of a quantity without requiring a large number of charts.

(ii) *Time variable.*—There are several conceivable methods which could be adapted to represent the time variation of a quantity in a plane chart. With two-dimensional schemes only, time variations can be shown by plotting time profiles (small Q, *t* diagrams) of the quantity at each observation point in the plane. A familiar example is indication of barometric tendency on the surface map by a symbol resembling the station barogram. Another form of representing time variations of a quantity in a plane is to plot its net change at each point over a selected time interval. A more cumbersome form might involve tabulating values of the quantity for successive times in a column adjacent to each observation point in the chart. Finally, there is the method of placing the analyzed patterns of the quantity for successive synoptic times on the same chart, using a different color for each.

d) Value of Q *in three space dimensions.*— We now encounter the most complicated forms of scalar representation. Because of the multiplicity of conceivable methods, and because so few are practical, we limit the discussion accordingly.

(i) *Time constant.*—One method of data

representation for three dimensions consists of small Q, z profiles at each observation point in a horizontal plane. A related method involves tabulation of Q for specified elevations or layers in a column adjacent to the point. A practical method, though indirect and incomplete, for representing and inferring space distributions of certain quantities is the use of topographic charts; these are discussed later in detail.

(ii) *Time variable.*—To show how the four variable dimensions can be provided for, one may expand the methods listed above. The variation of Q with time can be represented in Q, z profile system by separate profiles for each synoptic time or by additional graphic symbols. In the topographic method of analysis a simple manner of indicating time variations is to plot the net change of height of the surface during one or more time intervals or to tabulate successive heights of the surface at each observation point.

In view of the variety of ways in which the data may be plotted and analyzed, and considering that each method may have unique advantages, we cannot adopt any single method at present to provide for efficient utilization of data and time in analysis. Those charts selected are considered the most simple and practical methods for routine analysis with consideration for the system of observation, ease in preparation of the analysis, interpretation of the analysis, and the purpose of the entire procedure.

The primary system of observation and analysis established universally provides for the time dimension by synoptic observations made at sufficiently small time intervals that time differentials of the quantities between successive observations might be used roughly as tendencies. Thus, complete synoptic analyses can be made only for certain agreed times, and the individual analy-

sis is a static picture of space distribution. From practical necessity, the forms of representation must be two-dimensional. The following discussion on analysis is regulated by conventional ideas on observation and analysis and by the principal types of charts in use at the present time.[1]

4.04. *Scalar analysis in a plane.*—In Figure 4.04*a* are point values of Q in a plane. From the limitations specified for the variation of this quantity, we can outline its field by *scalar analysis*—by drawing isopleths of Q in the plane. Through each of these points we may draw a continuous curve assigned the numerical value of Q at that point. Each isopleth is a *scalar curve* or "isoline" of Q. In the case of pressure this curve is an isobar, for temperature an isotherm, for density an isopycnic, etc.

Since an infinite number of such lines can be drawn in the field of data, we first select a convenient number of *integer values* for which to draw isopleths. These differ by a certain *interval* (line interval) of Q which depends on the variation of Q within the region and on the detail required in the analysis. By custom, the integer values selected are multiples of the line interval. Thus, if the desired interval for analysis of Figure 4.04*a* is 20, then integer values are . . . 960, 940, 920, 900, ...

Figure 4.04*b* shows one possible analysis of this field of data. The most obvious feature is the invariance of Q along each line. No line breaks abruptly anywhere except at the limits of data. Scalar curves either are closed or reach the boundaries of the area. Lines with different values of Q *never intersect*, and a line *never branches or "forks."* It

1. Worth-while reading at this time is "Observations and Analysis," in American Meteorological Society, *Compendium of Meteorology* (Boston, Mass., 1951), pp. 705–27. (Do these two important problems of meteorology really deserve less than 2 per cent of the space in a critical survey of the science?)

is possible, however, for two lines of the same value to meet at a certain point.

Another interesting implication can be drawn from this analysis. Consider, for example, the line 920. From one end of the line to the other, without reversing direction, to one side Q is invariably less than 920, while on the other side all values are greater than 920. A scalar curve divides the plane space into two regions, one where the quantity is numerically less than the value of the curve, and another where it is greater.

$\partial Q/\partial z$. In the xz-plane $\partial Q/\partial x = 0$ where lines of Q parallel the x-axis and $\partial Q/\partial z = 0$ where they parallel the z-axis. The derivatives are both zero only where Q is homogeneous in the xz-plane.

4.05. *Singular points in the scalar field.*— As a rule, a distinct scalar line can be drawn through any point of the field. In Figure 4.04*b* only a few of an infinite number of possible lines were drawn, with interval $\delta Q = 20$. If more detailed analyses are de-

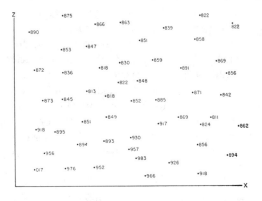

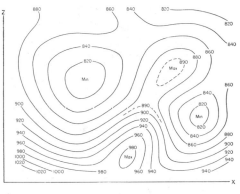

a   *b*

Fig. 4.04

The scalar analysis of any quantity aids in visual interpretation and measurement of its space variations. Where lines are dense, the quantity varies rapidly in value (perpendicular to the lines). Where lines are widely spaced, the quantity varies little, and the field is weak. When there are no lines at all, the field is strictly "flat," and the quantity is homogeneous in that region of the plane.

The average rate of variation of Q along any axis s is the change in Q over a certain distance of that axis divided by the distance. Hence, the average rate of variation of Q along s in finite distance δs is $\delta Q/\delta s$. As δs approaches zero, the local variation of Q along s is $\partial Q/\partial s$. Variations of Q along the three coordinate axes are $\partial Q/\partial x$, $\partial Q/\partial y$, and

sired, smaller intervals of Q are used. By taking the interval successively smaller and smaller, the number of lines is increased until, as $\delta Q \to 0$, the number becomes infinite. Then, each point in the field is traversed by a *single continuous line* with the possible exception of three *singular points*.

First, it is possible to find a value of the quantity only at a single point in a region. No line can be drawn through it, because nowhere in the vicinity is the same value repeated. This point must have either a maximum or a minimum value. *Maximum and minimum points*, usually called "centers," *are not traversed by a scalar line and are the only such points.*

In Figure 4.04*b* observe that the value of Q *decreases* outward in all directions from a maximum point and *increases* outward from a minimum point. This leads to another definition for centers in the scalar field. *A maximum or a minimum point is enveloped by at least one closed scalar curve in that region.* This closed line may be for one of the integer values used in the analysis, or it may be the line for an intermediate value of Q which is not drawn.

While the maximum and minimum points feature absence of scalar lines, the other

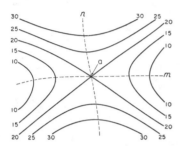

Fig. 4.05.—The col in a plane pattern

singular point is a point of contact between two lines of the same value. This is the hyperbolic singular point, which is called the *saddle-point* or *col* for reasons evident from topography. Point a in Figure 4.05 illustrates a saddle-point in a plane. *The saddle-point is the only possible place at which scalar lines can meet in the analysis, and the lines must have the same value.* However, it is not necessary that a saddle-point be shown by intersection of lines in the analysis. There can exist a saddle-point even though the choice of line interval does not allow this point to be represented.

A usual feature of all three singular points is the weak field in the vicinity; the gradient is zero across any of them. Such flatness is clearly verified in Figure 4.04b; lines are widely spaced near maxima, minima, and cols, and the spacing of lines gradually decreases outward from each of them.

4.06. *Scalar field in three dimensions.*— If we now consider the analysis of Q in Figure 4.04b also in the y direction, we find that the Q lines extend as *surfaces* in x, y, z space. In other words, the scalar curves in Figure 4.04b are nothing but lines of intersection of *scalar surfaces* with the xz-plane. Each surface has a single constant value of Q. In the atmosphere scalar surfaces for any quantity form closed volumes, form continuous sheets enveloping the globe, or intersect the surface of the earth.

In the hypothetical case of absolute homogeneity of the quantity in a region, its surfaces are not defined. Only in very small regions of the atmosphere is there homogeneity of any variable. With hydrostatic equilibrium a homogeneous state in pressure is not possible, although such a state of pressure can exist in a horizontal plane. The same is generally true of the density field. However, other physical variables can be homogeneous in limited regions.

4.07. *Gradients.*—The field of any atmospheric quantity can be described by sets of curves in two dimensions and by surfaces in three dimensions. The scalar field can be defined further by the space variation of the quantity. This leads to the concept of gradients.

Consider the arbitrary point P in Figure 4.07a. Through it we draw a line n normal to the Q lines and set the positive direction of n in the direction of increasing Q. It can be demonstrated that n must be the direction of most rapid horizontal variation of Q. This is the *horizontal ascendant direction* of Q. The *horizontal ascendant vector*, $\nabla_h Q$, is directed along the normal n:

$$\nabla_h Q = \mathbf{n}(\partial Q/\partial n) = \mathbf{i}(\partial Q/\partial x)$$
$$+ \mathbf{j}(\partial Q/\partial y) . \quad (1)$$

This vector has magnitude $\partial Q/\partial n$ and is directed toward increasing values of Q.

The horizontal vector operator, ∇_h, is defined by

$$\nabla_h = \mathbf{i}(\partial/\partial x) + \mathbf{j}(\partial/\partial y) , \quad (2)$$

and $\mathbf{i}$ and $\mathbf{j}$ are unit vectors along x and y, respectively. It is seen from Figure 4.07b,

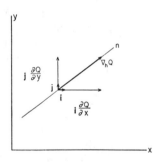

FIG. 4.07a

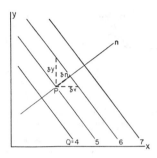

FIG. 4.07b

which corresponds to Figure 4.07a, that $\nabla_h Q$ is the vector resultant of the x and y ascendant vectors $\mathbf{i}(\partial Q/\partial x)$ and $\mathbf{j}(\partial Q/\partial y)$.

The ascendant at a given point in a plane is a simple and complete description of the plane scalar field in the vicinity. The vector direction gives the orientation of scalar lines. Its magnitude or length $\partial Q/\partial n$ expresses the density of lines—the *rate* Q varies in the plane.

Each of $\partial Q/\partial x$, $\partial Q/\partial y$, and $\partial Q/\partial n$ may be measured by two convenient means, as the average variation of Q per unit of distance:

$$\delta Q/\delta x = (Q_{x+1} - Q_x)/1 ; \quad (3a)$$

or as the reciprocal of the spacing of unit Q lines:

$$\delta Q/\delta x = 1/(x_{Q+1} - x_Q) . \quad (3b)$$

In dynamics we are concerned also with the vector equal and opposite to the ascendant. This we may call the *descendant vector*, whose component in a horizontal plane is $-\nabla_h Q$. Since the descendant has all the properties of the ascendant except difference in sign,

$$-\nabla_h Q = -\mathbf{i}(\partial Q/\partial x) - \mathbf{j}(\partial Q/\partial y) . \quad (4)$$

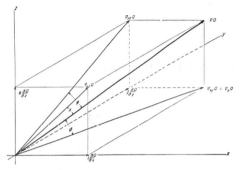

FIG. 4.07c.—Geometric relation among the total gradient, the plane gradients, and the coordinate components of the gradient.

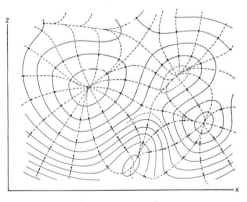

FIG. 4.07d.—Plane gradient vectors and gradient lines for the pattern in Figure 4.04b.

The horizontal descendant is normal to the scalar lines and toward lower values of the quantity. The descendant expresses the direction and rate of *decrease* of a quantity about a point.

In much of the work one is concerned directly with variations in certain planes, but it is often necessary to consider three-dimensional distributions. In three dimensions n is the local axis normal to the scalar surfaces, and the vector ascendant is

$$\nabla Q = \mathbf{i}(\partial Q/\partial x) + \mathbf{j}(\partial Q/\partial y)$$
$$+ \mathbf{k}(\partial Q/\partial z) . \quad (5)$$

Geometric interpretation of this vector is the same as before, except that now it refers to surfaces instead of lines. Its component magnitude along an arbitrary axis s or plane is $\partial Q/\partial s = |\nabla Q| \cos \theta$, if θ is the angle between ∇Q and s. To get the full implication of these relations, refer to the scheme in Figure 4.07c. The projection of ∇Q on the horizontal plane is $\nabla_h Q$. If ϕ is the angle between ∇Q and the xy-plane, the magnitude of the horizontal vector is

$$|\nabla_h Q| = |\nabla Q|\cos \phi . \quad (6)$$

In the case when ∇Q is vertical, its horizontal component is zero (e.g., geopotential). In the other extreme, when ∇Q is horizontal, $\nabla_h Q = \nabla Q$.

Component vectors in the other coordinate planes have similar interpretation from Figure 4.07c: $|\nabla_{xz}Q| = |\nabla Q| \cos \chi$; $|\nabla_{yz}Q| = |\nabla Q| \cos \psi$. The vector component along each coordinate axis now should be determined from the plane-vector components.

Figure 4.07d gives the plane descendant field corresponding to the scalar analysis in Figure 4 04b. On comparing those, one finds that several important features in the scalar field are clarified by the vector field. For example, scalar maxima show up as singular divergence points and minima as focal points of convergence. The saddle-point is the intersection of an axis of divergence and an axis of convergence. The divergence (convergence) of laterally adjacent vectors indicates the curvature of scalar lines and surfaces. Also, observe that along any scalar line all vectors cross the line at right angles and that *all vectors must cross the line without reversal of sense throughout the entire extent of the line.*

Although we refer constantly to vectors in space, it is usually unnecessary to represent fields of the physical variables vectorially as in Figure 4.07d. The vectors always can be inferred and measured from scalar analyses.

Use of the word "gradient" was suppressed above because of its latent confusion. The descendant is what has been called in meteorology the gradient. In mathematics the gradient is what we have called the ascendant. Since gradient is used also to infer only magnitude of variation, as we shall do here, it might be convenient to think of the ascendant as "up gradient" and the descendant as "down gradient." Thus, the gradient will imply the density of scalar lines or surfaces, measured along n, and the ascendant and descendant are vectors along that axis.

4.08. *Topography of property surfaces.*— Any physical variable can be analyzed by the topography of its individual surfaces. For example, in the case of pressure we analyze certain of its isobaric surfaces in space by drawing their elevation contour fields. The contour analysis on any surface can be interpreted in the same manner as an elevation contour chart of the earth's surface.

Contours of a scalar surface are defined by its intersections with consecutive level surfaces. For example, the 10,000-foot contour of the 700-mb pressure surface marks the intersection of the 700-mb surface with the level 10,000 feet. If geopotential height is used, all contours on a scalar surface are therefore level lines for that surface.

An advantage of topographic charts is that analysis can be done on a flat map and then interpreted in three dimensions. In Figure 4.08 is shown a portion $ABCD$ of a Q

surface. Lines z_9, z_{10}, and z_{11} are contours formed by the intersections of levels z_9, z_{10}, and z_{11}. The vertical projection of the surface upon the xy-plane is $A'B'C'D'$, and the projection of its contours z_9', z_{10}', and z_{11}'. It is obvious that the projected contours have precisely the same orientation, length, and horizontal spacing as on the Q surface itself. Since contours on the surface describe the shape and all details of the surface, the

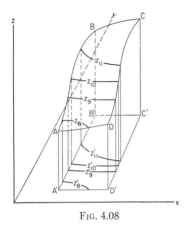

Fig. 4.08

pattern of contours drawn on a plane map contain everything necessary to reconstruct mentally the topography of that surface.

4.09. *Slope and curvature of surfaces.*—In general, the slope of any line (surface) is defined as the tangent of the angle between this line (surface) and a reference axis (plane). If we refer to the illustration in Figure 4.09a and define the slope angle of the Q surface by β_Q, the average slope of the surface in the x direction between geographic points k and m is

$$\tan{}_{xz}\beta_Q = (\delta z/\delta x)_Q . \qquad (1)$$

This leads to

$$\tan{}_{xz}\beta_Q = (\partial z/\partial x)_Q , \qquad (2)$$

for the slope of the surface along x at any point on the surface, provided the slope is

continuous at that point. Thus, *the slope of any surface along a given azimuth is equal numerically to its gradient of height along that azimuth.*

As the slope of the Q surface is inversely proportional to the horizontal spacing of its contours, it is apparent that the direction of maximum slope about any point on the surface lies along the ascendant of height on its topographic chart. If contours are oriented north-south with increasing heights to the east, we say that the surface has slope upward to the east. Both direction and magnitude of slope are given by the ascendant of height in the topographic map for the surface, and the general definition of the slope is

$$\tan \beta_Q = (\partial z/\partial n)_Q , \qquad (3)$$

in which n is normal to contours on the topographic chart. The relation between resultant slope and the component of slope along any horizontal azimuth s is

$$\tan{}_{sz} \beta_Q = (\tan \beta_Q) \cos \theta , \qquad (4)$$

where θ is the angle between this arbitrary direction and n.

The slope of the surface can be readily visualized and measured from the pattern of contours on its topographic map (Fig. 4.09b). First, it is seen whether the surface slopes upward or downward along a certain azimuth by observing if height increases or decreases in that direction. In Figure 4.09b the Q surface has a component of slope upward along x. The magnitude of slope can be obtained by the method of finite differences either from dividing the contour interval by the measured horizontal distance between successive contours or from dividing the height increment per unit of distance by the distance (Eq [3]). Where the slope is infinite, the surface is vertical, and all contours coincide. Where the slope is zero, the surface is level, and there are no contours in the vicinity.

The curvature of a surface profile in xz cross section is

$$K_{xz} = (\partial^2 z/\partial x^2)_Q/[1 + (\partial z/\partial x)_Q^2]^{3/2} . \quad (5)$$

If the cross-section plane is normal to the contours, then the general definition for curvature of a surface in vertical plane is

$$K_{nz} = (\partial^2 z/\partial n^2)_Q/[1 + (\partial z/\partial n)_Q^2]^{3/2} . \quad (6)$$

This equation defines what we shall call the *vertical curvature* of the Q surface. Because the slopes of scalar surfaces in the atmosphere are usually so small that the denominator of the equation is negligibly different from

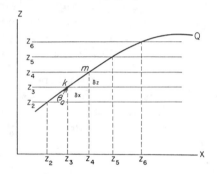

Fig. 4.09a.—*xz* cross section of a Q surface showing the vertical projection of its contours upon a level reference axis.

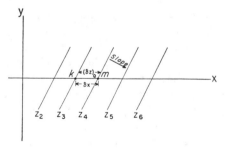

Fig. 4.09b.—Portion of the topographic pattern corresponding to Figure 4.09a.

unity, we consider for the time being only the approximate formula:

$$K_{nz} \simeq (\partial^2 z/\partial n^2)_Q . \quad (6a)$$

In finite differences this formula for curvature can be interpreted as the change in slope per unit horizontal distance along the normal to the contours. A *negative* value for K means the surface is curved *convex upward;* a *positive* value, *concave upward.*

To clarify the meaning of the above

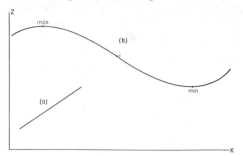

Fig. 4.09c

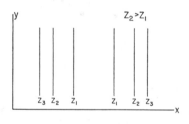

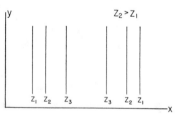

Fig. 4.09d

formula and the significance of topographic curvature in general, refer to Figure 4.09c. Line a has everywhere the same slope; its curvature is zero. Contours on such a surface would be equally spaced. However, a plane surface such as illustrated by line a is not a *natural* surface in space, and we seldom find equally spaced contours on topographic charts. A more natural surface is represented by curve b. We say this line is curved convex upward to the left of point $i;$ throughout this region the slope decreases along x. The portion of the curve to the right of i is curved concave upward. At i the curvature is zero; this is the point of transition from negative to positive curvature and is called the *inflection point* in the curve. It is apparent that contours for this surface are most closely spaced at the inflection point and most widely spaced at the maximum and minimum points.

viously defined. For example, a surface may have straight contours and yet itself be curved, as in Figure 4.09d. Thus the vertical curvature of the *surface* depends on the relation between *neighboring* contours, while the curvature of a contour is determined only by the shape of that individual contour.

4.10. *Special lines and points in the topographic field.*—Figures 4.101–4.105 give schematic drawings of a few important

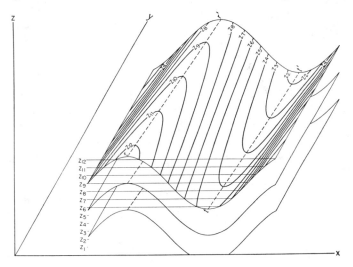

Fig. 4.101.—Idealized ridge and trough patterns in a property surface

From a pattern of contours it is possible to visualize both the slope and the vertical curvature of the property surface. In a region where contours are almost equally spaced the slope is nearly constant, and the vertical curvature of the surface is small. In a region where contours are distributed as in Figure 4.09d (upper) the surface is curved concave upward. From left to right, the contours first decrease to a minimum value and then increase; this indicates a dip in the property surface. Analysis of the lower drawing reveals that the surface is bulged upward (curved convex upward).

It is well to mention that the curvature of the contour lines themselves is in general not the same as the curvature of the *surface*, pre-

topographic features commonly found on scalar surfaces. These features in large scale can be followed from one synoptic chart to the next, and they constitute primary references for describing the contour patterns of the surfaces.

a) The ridge and trough.—In Figure 4.101 is a series of parallel wave-shaped surfaces in space for unit intervals of Q. On the uppermost surface, contours are entered as usual for equal height intervals; the dashed lines mark the maximum and minimum elevations of the surface. We call line rr' the ridge line or simply the *ridge* in the surface; this is analogous to a divide or "line of watershed" on the surface of the earth. It usually, though not necessarily, coincides with the axis of maximum convex curvature in the surface; it is always a line of maximum elevation of the surface. Line tt' is the trough line or *trough* and is similar to the "line of watercourse" on the earth. It marks a line of minimum elevation of the surface and is usually a maximum of concave curvature. If the y-axis is placed parallel to the

ridge lines are usually curved in both vertical and horizontal sense. Curvature of the ridge or trough line in the vertical is shown by varying contour gradients along that line. In horizontal perspective, curved ridge and trough lines are analogous to the curved courses of mountain ranges and valleys.

b) The line of inflection.—In Figure

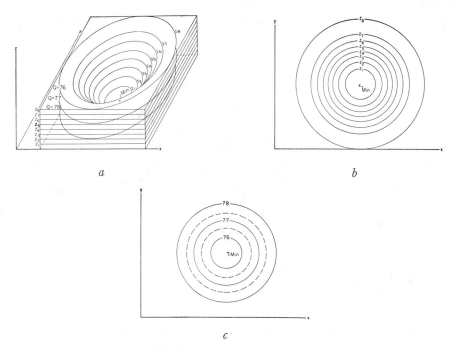

a   *b*

c

Fig. 4.102.—Idealized topographic pattern of a depression (*a*), with the contour pattern of the uppermost surface (*b*) and the horizontal pattern at level z_1 (*c*).

ridge or tough line, the general definitions for ridges and troughs are:

Ridge: $(\partial z/\partial x)_Q = 0$, $(\partial^2 z/\partial x^2)_Q < 0$.

Trough: $(\partial z/\partial x)_Q = 0$, $(\partial^2 z/\partial x^2)_Q > 0$.

In these $(\partial z/\partial y)_Q$ is or is not zero according as the point in question is or is not also a dome, depression, or col.

Figure 4.101 suffices for a general impression of ridges and troughs in atmospheric scalar surfaces, but there are additional features worth mentioning. Trough and

4.101 there are lines between trough and ridge of each surface which are axes of maximum slope of the surface. These are also lines of inflection in vertical curvature of the surface. On the corresponding topographic map the *inflection line* is located as the axis of maximum contour gradient. In Figure 4.101 the inflection lines connect the straight sections of the contours.

c) The dome and depression.—Figure 4.102 illustrates a *depression* in a series of surfaces; its opposite is the *dome*. The contours in this idealized drawing are concentric circles; actually, contours are seldom circular and seldom concentric.

The structure of the dome shows a certain change in curvature from its midsection up to its summit. In approaching the summit of the dome, the slope of the surface decreases, and the summit is characterized by

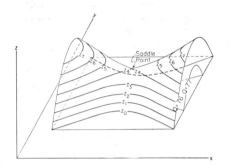

Fɪɢ. 4.103.—Topographic pattern of a saddle

zero slope and maximum convex curvature. The depression has all the analogous characteristics except it is curved concave upward. Drawings *b* and *c* in Figure 4.102 represent the contour pattern of the uppermost Q surface and the horizontal pattern of Q at level z_1.

A dome may be considered the intersection of two ridges and a depression the intersection of two troughs. Hence, the mathematical definitions in the topographic field are:

Dome (summit of): $(\partial z/\partial x)_Q = (\partial z/\partial y)_Q$

$$= 0; (\partial^2 z/\partial x^2)_Q < 0 > (\partial^2 z/\partial y^2)_Q.$$

Depression (pit of): $(\partial z/\partial x)_Q = (\partial z/\partial y)_Q$

$$= 0; (\partial^2 z/\partial x^2)_Q > 0 < (\partial^2 z/\partial y^2)_Q.$$

d) *The saddle or col.*—The saddle is a cross or intersection of a trough and a ridge, which becomes evident on examining the model in Figure 4.103. In this drawing the ridge is parallel to the *x*-axis, and the

central portion of the ridge is depressed by presence of the transverse trough. At the low point of the ridge line (the high point of the trough line) is the saddle-point. It is clear that two inclined ridges and two inclined troughs radiate from the simple saddle-point.

The drawing shows the saddle-point is the intersection of contours having the same value. By depressing this point the slightest amount, contour z_4 breaks into two parabolic curves similar to z_5, and neither z_4 nor z_3 touches at the saddle-point. Contour intervals used for drawing topographic charts are of the order of hundreds or thousands of feet, and, since the saddle-point in one surface exists at only one height, the probability of the saddle-point lying on one of the selected contours is extremely remote. Rather than attempt to draw lines with all the precision required otherwise, we follow the general rule of *not drawing intersecting lines in the analysis of cols*, and *no intersect-*

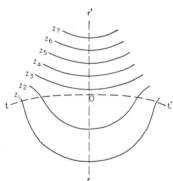

Fɪɢ. 4.104.—Illustration of the "semisaddle"

ing scalar lines are permissible in the analysis.

e) *The semisaddle.*—A feature frequently appearing in topographic charts for any of the variables, but for which no suitable name has been given, is illustrated by Figure 4.104. This pattern resembles the saddle, and also the ridge, but is usually referred to incompletely as just a ridge. The contours indicate downward slope of the surface from top to bottom of the drawing and downward slope laterally outward from rr', as characteristic of the ridge. However, in the vicinity of tt' the slope along rr' undergoes distinct change in magnitude. The form of the ridge line in space is curved similar to the ridge line of the saddle—concave upward. Also, tt' is similar to the trough line of the saddle. The only real difference between this pattern and the saddle is the sign of the slope along *or*. Because of its similarity and its relation in time and in space to the saddle, we might call it the "semisaddle."

stable layers particularly in temperature and humidity surfaces. It is a sharp rise (or fall) in the surface between two more level regions and gives the appearance of a steep bluff or *escarpment* in the surface. In the topographic pattern the contours are concentrated in the escarpment. Occasionally the rise in the surface is vertical, and in that case several contours coincide. At times also the surface is inverted in the escarpment, and the contour chart shows two reversals in the sequence of contours.

Figure 4.105*b* illustrates a gentle *fold*.

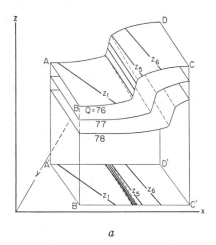
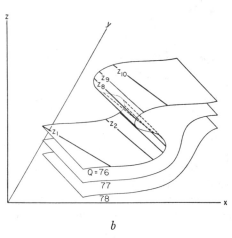

a	*b*

FIG. 4.105.—(*a*) Escarpment in property surfaces in zones of transition; (*b*) a surface fold

In Figure 4.104 the dominant feature in the pattern of the semisaddle is the ridge. The opposite variety also exists, that is, one in which the principal feature is a trough.

f) "Escarpments" and folds.—The other topographic features to be described are more limited in nature than the previous ones in that they occur in surfaces of some of the variables but not in others. None is found in large enough scale in pressure surfaces,[2] and only rarely would they be found in density surfaces.

Figure 4.105*a* shows a topographic feature frequently found across well-marked

Here the surface is inverted, but it has more rounded bends than the escarpment. Folds are often found in regions of weak gradient, and surfaces adjacent to a folded one can have radically different shapes. Surface $Q = 76$ has a pronounced fold, but $Q = 77$ never folds at all, and an upper surface $Q = 75$ may not bear any resemblance to surface $Q = 76$.

g) Space centers.—Occasionally there are isolated points in the atmosphere about which the numerical value of a variable increases in all directions. This point is a space minimum of the variable. A minimum point is surrounded by one or more closed surfaces. Space maxima are identical except

2. Except perhaps "pressure-jumps" of the type in Figure 4.105*a*.

in the direction of gradient. Collectively, maxima and minima can be called *space centers*.

4.11. Lines of first-order discontinuity.—

In introducing scalar analysis, we stated that scalar lines and surfaces are continuous in space. Without contradiction, it is possible for those lines and surfaces to have *dis-*

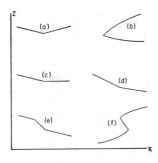

FIG. 4.111.—Various types of discontinuities

$\partial Q/\partial y$, or $\partial Q/\partial z$ is discontinuous. A point or limited region of abrupt change in the vertical temperature lapse rate is a discontinuity, as $-\partial T/\partial z$ is discontinuous at that point. A discontinuity in ∇Q is seen in the patterns of Q surfaces.

Discontinuities may be considered of two types—discontinuity in direction or discontinuity in magnitude of the gradient. The former case is illustrated by the center drawing in Figure 4.112; there is a finite angle between the discontinuity line and the lines of the pattern, which are kinked. Where the discontinuity line parallels the contours (scalar lines), it is a discontinuity of gradient magnitude only, and there are no kinks in the lines. Both types appear in Figure 4.104 if axis tt' is considered a discontinuity.

Figure 4.111 shows a variety of possible kinks in surfaces. In some (*c* and *e*) the

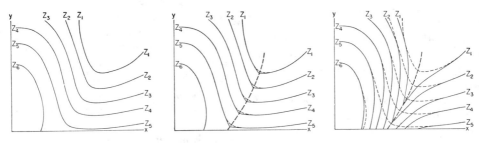

FIG. 4.112.—Illustration of the usual effect of a discontinuity on a contour pattern. *Left:* trough without discontinuity; *center:* same trough with discontinuity; *right:* same trough with exaggerated discontinuity.

continuous orientation. For example, an isobar may extend with nearly constant direction, or with gradual change of direction, and then at a certain point change its direction so abruptly that it appears to be refracted at that point. Similar features appear on scalar surfaces; an otherwise smooth surface will be sharply creased along a certain line or narrow zone. These crease lines in the surfaces are *discontinuity lines.*

A first-order space discontinuity[3] in Q occurs where any of its derivatives $\partial Q/\partial x$,

change in slope is small. In others (*b* and *f*) it may exceed 90°; these are large discontinuities or *sharp kinks*. Often the discontinuity is not associated with reversal in sign of the slope (*d* and *e*). Examples *e* and *f* (compare Fig. 4.105*a*), which show a double discontinuity, are associated with a *zone* of discontinuity or, more broadly, with a *zone of transition* for the variable.

The middle drawing in Figure 4.112

3. Since first-order discontinuities are the only ones to be considered, we shall henceforth refer to them simply as "discontinuities."

shows a contour pattern corresponding to example *c* in Figure 4.111. To the left is the contour pattern as it would exist in the absence of a discontinuity; it shows a rather smooth trough in the surface. When the discontinuity is introduced into the trough, the second drawing shows that the contours are made to kink at the discontinuity line, but the general shapes of the contours at some distance are relatively unaffected. The pattern of contours in the vicinity of a discontinuity line consists of (i) the general pattern as it would exist without introducing a discontinuity and (ii) the abrupt change in direction of the contours as required by the discontinuity. The last drawing in Figure 4.112 shows one possible result of exaggerating the discontinuity artificially. However, Figure 4.112 should not suggest that troughs are synonymous with discontinuity lines.

4.12. *Relations between topographic and horizontal scalar patterns.*—In much of the future work it is necessary to interpret the horizontal pattern of a variable from its topographic chart, and conversely. Conversion from one reference to the other is based on the relation of slope of a property surface to the horizontal gradient of that property. We may define the finite difference δQ by

$$\delta Q = (\partial Q/\partial x)\delta x + (\partial Q/\partial y)\delta y + (\partial Q/\partial z)\delta z , \quad (1)$$

applicable when derivatives are continuous. If we place *n* normal to scalar lines of Q in the horizontal plane and directed toward increasing values of Q, then

$$(\partial Q/\partial x)\delta x + (\partial Q/\partial y)\delta y = (\partial Q/\partial n)\delta n , \quad (2)$$

and

$$\delta Q = (\partial Q/\partial n)\delta n + (\partial Q/\partial z)\delta z . \quad (3)$$

If the two points in question lie along the same Q surface, then between the points

$\delta Q = 0$, and $(\partial Q/\partial n)\delta n + (\partial Q/\partial z)\delta z = 0$, or

$$(\delta z/\delta n)_Q = -(\partial Q/\partial n)/(\partial Q/\partial z) . \quad (4)$$

Comparison of this with Eq 4.09(3) shows that

$$\tan \beta_Q = (\delta z/\delta n)_Q$$
$$= -(\partial Q/\partial n)/(\partial Q/\partial z) . \quad (5)$$

The slope of a scalar surface is now defined by the horizontal and vertical gradients of the property. From Eq (5) it becomes evident why a discontinuity in either horizontal or vertical gradient gives discontinuity in slope of the surface.

The quantity $(\nabla_h z)_Q$, of magnitude $(\partial z/\partial n)_Q$, is the ascendant of the projected contours of the Q surface; it describes the spacing and orientation of contours. The quantity $\nabla_h Q$, of magnitude $\partial Q/\partial n$, describes the spacing and orientation of the Q lines in horizontal plane. Direction *n* on the horizontal chart is normal to the Q lines, and direction *n* on the topographic chart is normal to contours. Since each contour in the topography defines a line with the same value of the property in a horizontal plane, the two *n* directions must be parallel locally, and $(\nabla_h z)_Q$ lies in the same or opposite direction to $\nabla_h Q$. To determine the relative directions of the two ascendants, the sign of $\partial Q/\partial z$ is needed. If Q decreases upward, Eq (5) shows that the contour ascendant and horizontal scalar ascendant are the same geographic direction. Then domes in the topography correspond to maxima in the horizontal field, and depressions to minima in the horizontal field (Fig. 4.102). The sequence of contours is of the same sense as the sequence of Q lines in the horizontal. Pressure and density (and most often temperature in the troposphere) obey this relation. If $\partial Q/\partial z$ is positive, as usually the case with θ and θ_E, the orders of contours and horizontal lines are reversed. Domes in the surfaces correspond to horizontal minima and depressions to horizontal maxima.

An additional relation between the two types of patterns is the spacing of the respective sets of lines, that is, the relation between $(\partial z/\partial n)_Q$ and $\partial Q/\partial n$. If first we consider the hypothetical case of $\partial Q/\partial z$ everywhere the same, then $(\partial z/\partial n)_Q$ varies only with $\partial Q/\partial n$, and the spacing of contours on the topographic chart is a linear function of the spacing of Q lines in the horizontal chart. Where contours are crowded, so are the lines on the horizontal chart; where the contour field is flat, the horizontal field is also flat. This invariance of $\partial Q/\partial z$ was assumed in drawing the pattern of Figure 4.102; the Q surfaces are equidistant vertically. The resulting horizontal chart has great similarity to the pattern of the topographic chart, and the ratio of contour spacing to spacing of horizontal lines is invariant. Such great similarity between topographic and horizontal charts most nearly exemplifies the field of pressure.

If the vertical gradient undergoes large variations in the vicinity of a property surface, Eq (5) shows that the spacing of contours on the topographic chart may vary in a manner very different from the spacing of scalar lines in the horizontal. In extreme cases the pattern on one type of chart can give the wrong impression of the pattern on the other type of chart.

4.13. *Comparison of topographic and horizontal fields of the various quantities.*—As might be expected, there are important differences in the fields of pressure, temperature, humidity, potential temperature, equivalent potential temperature, and density. Some have smooth and regular topographies; others have numerous folds, wrinkles, and centers in their patterns.

A convenient basis to distinguish between the variables is the steadiness of slope of their surfaces, that is, the dominance by the vertical gradient of the quantity. A quantity whose topographic slopes vary in a narrow range is a *uniformly distributed quantity;* its surfaces have no folds, and its discontinuities in slope are relatively minor. Furthermore, since the range of slope for this surface is small, adjacent surfaces tend to be parallel, and topographic and horizontal patterns of the variable are a great deal similar. The less uniformly distributed of the variables have greater variety of topographic features, including folds and space centers. In accord with this classification, the atmospheric variables are described below in order of decreasing uniformity.

a) *Pressure.*—The most uniformly distributed of the atmospheric variables (other than geopotential) is pressure. Since pressure always decreases upward, such features as folds and space centers do not exist. Besides, the total absolute variation of isobaric slopes in the atmosphere is extremely small, and adjacent isobaric surfaces appear much alike.

The slope of a pressure surface is

$$\tan \beta_p = -(\partial p/\partial n)/(\partial p/\partial z)$$
$$= (980/g)[(\partial p/\partial n)/(\partial p/\partial Z)] . \quad (1)$$

From the hydrostatic equation, $\partial p/\partial z = -\rho g$. Air density varies from about 1.6×10^{-3} to 0.2×10^{-3} gm cm^{-3} in the troposphere for ordinary ranges of pressure and temperature. For an average value we may take $\partial p/\partial z = -1.0 \times 10^{-3}$ mb cm^{-1}.

The horizontal pressure gradient can assume any direction, but its magnitude is usually of the order 10^{-7} mb cm^{-1} (i.e., 1 mb/100 km). Under extreme conditions it may reach magnitude 10^{-6} mb cm^{-1}. Substitution of the average values above into Eq (1) gives $\tan \beta_p = 10^{-7}/10^{-3} = 10^{-4} = 1/10,000$. This slope corresponds to an angle only $0.006°$. Thus, the slope of a pressure surface is always extremely small in absolute value, and the vertical pressure gradient is a good approximation to the total pressure gradient.

The vertical spacing of isobaric surfaces ($\delta p = 1$ mb $= 10^3$ dynes cm^{-2}) is $\delta Z = 10^3/980\rho = 10^3\alpha/980$, which follows from Eq 3.03(3). For the tropospheric range of density given above, the spacing of unit millibar pressure surfaces varies from 6 gpm (19 gpft) in very cold air near sea level to about 50 gpm (165 gpft) in certain condi-

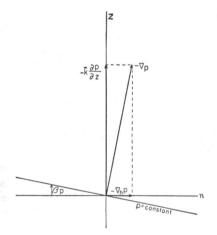

FIG. 4.13a.—Relation between the pressure gradient and the slope of an isobaric surface.

tions near the tropopause and to larger values above. The range of spacing is much smaller about any one pressure surface, however. For example, the spacing of 1-mb isobaric surfaces at 700 mb varies from about 10 gpm to 12 gpm in the observed range of temperature; thus, comparatively speaking, neighboring pressure surfaces are very nearly parallel and have nearly identical topographies.

In summary, in the neighborhood of any pressure, isobaric surfaces are to a large extent mutually parallel and equally spaced, and isobaric surfaces depart from level surfaces only slightly. Topographic features to be found on isobaric surfaces consist of ridges, troughs, domes, depressions, cols, and discontinuity lines, but in each case the slope of the surface is very small. Because of all this, horizontal pressure charts bear

close resemblance in pattern to topographic pressure charts, and the two charts can be used interchangeably.

b) Density.—Next in order of simplicity come density and specific volume. From the relation

$$\nabla\rho = \nabla_h\rho + \mathbf{k}(\partial\rho/\partial z),\qquad(2)$$

and substituting for pressure and temperature from the equation of state,

$$\nabla_h\rho = \rho[(1/p)\nabla_h p - (1/T)\nabla_h T].\quad(3)$$

When the horizontal temperature field is flat, $\nabla_h T = 0$, and the horizontal density ascendant is along that of pressure. Where $\nabla_h p = 0$, $\nabla_h\rho$ has direction opposite to $\nabla_h T$. When the quantity within brackets is zero, there are no isopycnics in the horizontal plane.

In form similar to Eq (3),

$$\partial\rho/\partial z = \rho[(1/p)(\partial p/\partial z) - (1/T)(\partial T/\partial z)].\quad(4)$$

Substitution for $\partial p/\partial z$ from the hydrostatic equation leads to

$$\partial\rho/\partial z = -(\rho/T)(34.1\times10^{-5}\,^{\circ}\mathrm{C\ cm}^{-1} + \partial T/\partial z).\quad(5)$$

The term $\partial\rho/\partial z$ is zero only when $-\partial T/\partial z = 34.1^{\circ}$ C km^{-1}, and, in order for density to increase with height (autoconvective instability), the temperature lapse rate must exceed that large value. This critical temperature lapse rate occurs only rarely in a surface microlayer. Thus, for our purposes, we can assume that density decreases upward.

The magnitude of the vertical density gradient can be found by substituting usual values for ρ, T, and $\partial T/\partial z$ into Eq (5), and similarly for the horizontal gradient in Eq (3). Then, from

$$\tan\beta_\rho = -(\partial\rho/\partial n)/(\partial\rho/\partial z),\quad(6)$$

it is found that the slope of an isopycnic surface is ordinarily larger than the slope of an isobaric surface by approximately one order of magnitude. Hence, isopycnic surfaces have somewhat greater undulations than isobaric surfaces, but in the free atmosphere they are not folded.[4]

c) Specific volume.—As one would expect from the definition of specific volume, its distribution is similar to that of density. As

$$\nabla \alpha = -(1/\rho^2)\nabla \rho = -\alpha^2 \nabla \rho , \quad (7)$$

the ascendant of α is always along the descendant of ρ, and their magnitudes differ by the factor α^2.

d) Potential temperature —From the discussion of stability, it is evident that potential temperature usually increases upward. In the specific case of dry adiabatic conditions in the vertical, the ascendant of θ is either horizontal or absent, according as $\partial\theta/\partial n$ is finite or zero; while, with superadiabatic lapse rates, the ascendant has a component downward. Since potential temperature increases upward or downward, or neither, we expect to find significant differences between the field of θ and the fields of p and ρ.

Eq 1.07(1) for potential temperature can be differentiated to give

$$\nabla\theta = -\theta[(\kappa/p)\nabla p - (1/T)\nabla T] , \quad (8)$$
$$\nabla_h\theta = -\theta[(\kappa/p)\nabla_h p - (1/T)\nabla_h T] , \quad (9)$$

and

$$\partial\theta/\partial z = -\theta[(\kappa/p)(\partial p/\partial z)$$
$$- (1/T)(\partial T/\partial z)] . \quad (10)$$

It is interesting to compare Eq (8) with the analogous one for density (of the form in Eq [3]). Within brackets,

for θ: $(\kappa/p)\nabla p - (1/T)\nabla T$;

for ρ: $(1/p)\nabla p - (1/T)\nabla T$.

The only difference between the two is the factor κ in the equation for θ. But this difference is significant—the effect of pressure distribution on the variation of θ is only 2/7 as large as its effect on density. This is proof that distribution of θ must be less uniform

4. Examine Figure 6.11.

than that of ρ. At the same time it is evident that θ is more uniform than T.

To show the relative effects of the horizontal distributions of p and T on the horizontal gradient of θ, one should analyze the magnitudes of the two terms within brackets in Eq (9). It is found that the horizontal pattern of θ is more a function of temperature distribution than of pressure distribution. Figure 4.13b represents schematically

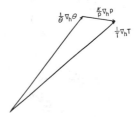

FIG. 4.13*b*

the relation of the three vectors in Eq (9). The angle between isopleths of θ and T is usually acute (Fig. 5.04). As the gradient of temperature becomes smaller, the relative effect of pressure distribution increases, and the angle between $\nabla_h\theta$ and $\nabla_h T$ becomes larger. In particular, when $\nabla_h T$ is zero, the horizontal ascendant of θ is opposite in direction to that of p.

The relation of gradients of T and θ in the vertical is a different matter. For a detailed analysis of $\partial\theta/\partial z$ in Eq (10), substitute the values $\kappa = R_d/c_p$ and $\partial p/\partial z = -\rho g$; then

$$\partial\theta/\partial z = (\theta/T)(\partial T/\partial z + g/c_p)$$
$$= (\theta/T)(\partial T/\partial z + \Gamma_d) , \quad (11)$$

where Γ_d is the dry adiabatic lapse rate. The factor θ/T is always positive and is greater than unity for $p < 1000$ mb. Deductions are made similar to those in Section 3.16(*a*).

The slope of a surface of constant θ (*isentropic surface*) is

$$\tan\beta_\theta = -(\partial\theta/\partial n)/(\partial\theta/\partial z) . \quad (12)$$

With dry adiabatic temperature lapse rates,

$\tan \beta_\theta = \infty$. If both $\partial\theta/\partial n$ and $\partial\theta/\partial z$ are zero in a region, then, isentropic surfaces are not defined, and there is homogeneity of θ. Superadiabatic conditions are usually confined to relatively shallow layers. If in such a region $\partial\theta/\partial n$ does not reverse direction, Eq (12) shows that a superadiabatic layer is associated with a fold in isentropic surfaces. An example is illustrated by Figure 4.13c.

When $\partial\theta/\partial z$ is large (stratosphere and tropospheric stable layers), slopes of isentropic surfaces are small and neighboring

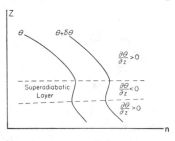

FIG. 4.13c

surfaces are similar. A large value for $\partial\theta/\partial z$ obviously gives smaller slopes for given values of $\partial\theta/\partial n$, but it also suppresses fluctuations of slope for given variations in $\partial\theta/\partial n$. In these stable regions the potential-temperature field is most uniform, and topographic charts have the greatest resemblance to horizontal charts. In contrast, in the less stable regions the slopes of isentropic surfaces are larger and neighboring isentropic surfaces can differ widely in shape for given horizontal gradient. However, conclusions drawn from topographic charts of θ surfaces are *in general* not greatly different from the interpretation of horizontal θ analysis, and, since air motions are isentropic to a large extent, analysis of isentropic charts is desirable.

e) Temperature.—The pattern of T in space bears similarities with, and significant differences from, the pattern of θ. The *horizontal* field of T is much like that of θ because of the relatively small effect of $\nabla_h p$

in most cases (see Fig. 5.04). Thus from Eq (9), $\nabla_h T \approx (T/\theta)\nabla_h\theta$. Besides being in the same general direction, the horizontal gradients of θ and T may differ in magnitude. At pressures greater than 1000 mb the horizontal spacing of isotherms is less than spacing of isentropes. On upper-level charts the relation is reversed.

The vertical gradient of T differs from that of θ vectorially and in variability. Isothermal surfaces are less regular than isentropic surfaces. Isothermal and inversion layers are more common than are adiabatic and superadiabatic layers, and the topographic features of isothermal surfaces are numerous and varied.

We thus have the general impression that the temperature field is less uniform than potential temperature, but there are exceptions. In the less stable regions of the atmosphere the isentropic surfaces tend to be irregular because of the smallness of $\partial\theta/\partial z$. But for this condition the magnitude of $\partial T/\partial z$ is relatively large, in turn giving smaller slopes and more regular patterns of isothermal surfaces, by the equation

$$\tan \beta_T = -(\partial T/\partial n)/(\partial T/\partial z) . \quad (13)$$

In regions of low stability isothermal surfaces frequently have smaller slope, greater parallelism, and fewer irregularities than isentropic surfaces. The upper part of the tropical troposphere often fits that description.

There is a useful relation between the slopes of isentropic and isothermal surfaces. In xz cross section

$$(\tan_{xz} \beta_T)/(\tan_{xz} \beta_\theta) \simeq 1$$
$$- (\partial T/\partial z)_d/(\partial T/\partial z) , \quad (14)$$

where $-(\partial T/\partial z)_d = \Gamma_d$, the dry adiabatic lapse rate. Eq (14) shows that θ and T surfaces slope in the same sense when the quantity on the right is positive, and the numerical ratio of the slope of T to θ surfaces is determined by $\partial T/\partial z$ only. It follows that, *with lapse rates between isothermal and dry adiabatic, isentropic and isothermal surfaces slope in opposite sense.* If temperature lapse rate is zero, isothermal surfaces are vertical; if dry adiabatic, isentropic surfaces are vertical; if superadiabatic, θ and T surfaces are oriented in the same sense. An inversion of temperature also gives θ and T surfaces in the same sense.[5]

The methods for observing and reporting aerological data create a problem with vertical gradients of temperature and potential temperature. Most reported significant points in a temperature sounding can be treated technically as discontinuities in lapse rate and in resultant gradients of temperature, potential temperature, and to lesser degree in gradients of density and pressure. But many of these possible discontinuities are created artificially by limiting the number of significant points in the report. Accordingly, we are forced to disregard all minor discontinuities in lapse rate and treat as real discontinuities only the larger ones which could not be removed easily by drawing a smooth curve through significant points.

f) Equivalent potential temperature.—The field of θ_E is in some regions similar to θ and more uniform than T, but in others it is considerably more irregular than both. From Eq 1.08(4)

$$\nabla \theta_E \approx \nabla \theta + 3\nabla r \, . \quad (r\%_0) \quad (15)$$

At low temperature r_s is small, and r varies within narrow limits. With such conditions ∇r is likely to be small, and $\nabla \theta_E \simeq \nabla \theta$. Thus in the high troposphere, in the stratosphere, and throughout the atmosphere in cold arctic air masses, the field of θ_E is almost identical with the field of θ. In contrast, in the warm tropical air masses of the lower troposphere the difference between $\nabla \theta_E$ and

5. Check these rules with the cross sections in Chap. 6.

$\nabla \theta$ can be large in direction and magnitude, and surfaces of θ and θ_E can be radically different in appearance.[6]

In summary, uniformity of θ_E and similarity between fields of θ and θ_E are roughly an inverse function of temperature. Charts of constant θ_E would be desirable in weather analysis, since θ_E is conservative for most processes, and therefore parcels tend to remain on their respective θ_E surfaces; but the extreme irregularities in θ_E surfaces where they would be of most value prohibits preparation and use of those charts on a routine basis.

g) Moisture content.—Last in order and by all means the most difficult of the variables listed is moisture content. Moisture has the most troublesome field to analyze because of its numerous irregularities. Another factor giving reason for doubt in analysis is the present inaccuracy of moisture observations in the atmosphere, especially at low temperatures. After giving some thought to these problems, one concludes that all we can expect to derive through moisture analysis is a rough picture in large scale.

From average distributions of moisture one is led to believe that humidity decreases with height from the surface upward. But, of all the variables listed, humidity can show the greatest differences in distribution between climatic and synoptic charts. In any given situation, expect to find at least one stratum in a sounding through which moisture increases with height, perhaps near the surface. Centers, folds, and the like in the patterns of moisture surfaces are the rule. Another significant manner in which the distribution of moisture differs from other variables is the great tendency for moisture to be stratified or layered, as suggested by the frequency of layer-type clouds.

6. Examine Figure 6.09.

4.14. *Gradient of a quantity on the surface of another quantity.*—It is often necessary to determine gradients of other variables along a constant property surface as a function of their horizontal and vertical gradients. The conversion from coordinate to surface references can be made through a general formula derived below.

In Figure 4.14 the slope of the Q surface in arbitrary horizontal direction s is $\tan_{sz} \beta_Q = (\delta z/\delta s)_Q$. Now consider the distribution of another quantity N whose scalar

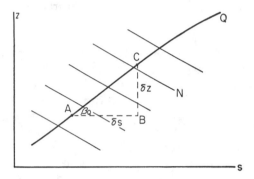

FIG. 4.14

lines are dashed. It is evident that $N_B - N_A = (\partial N/\partial s)\delta s$, $N_C - N_B = (\partial N/\partial z)\delta z$, and $N_C - N_A = (\partial N/\partial s)_Q\delta s$. The distance δs in the third equation is the same as in the first, since analysis at constant Q is performed on a flat map where the projection of AC is the same as AB.

From an algebraic consideration we see that $N_C - N_A = (N_B - N_A) + (N_C - N_B)$. Then, by substitution and dividing through by δs.

$$(\partial N/\partial s)_Q = \partial N/\partial s + (\partial N/\partial z)(\delta z/\delta s)_Q$$
$$= \partial N/\partial s + (\partial N/\partial z)\tan_{sz}\beta_Q.$$

Hence,

$$(\partial/\partial s)_Q = (\partial/\partial s)$$
$$+ (\tan_{sz}\beta_Q)(\partial/\partial z). \quad (1)$$

As one example, consider that N is temperature and that the Q surface in question

is an isobaric surface. Then,

$$(\partial T/\partial s)_p = \partial T/\partial s + (\partial T/\partial z)(\delta z/\delta s)_p.$$

The same principle holds for any other pair of variables, but the equation is not applicable when the surface is vertical. In case both quantities N and Q are the same, Eq (1) reduces to Eq 4.12(4).

4.15. *Approach to the problem of analysis.*—Our discussion of physical scalars in the atmosphere thus far was limited to the theory of lines and patterns; no attempt was made to discuss the *procedure* for obtaining those patterns from the data. We will now touch on general principles of procedure. For the present we can do little more than survey this comprehensive problem.

It is apparent that simplicity in the process of analysis is a function of the density of synoptic observations in space. If the data were sufficiently dense that each line could be drawn merely by connecting a series of data points, then there would be no problem at all, and the average person might do as good a job of drawing lines as the one with years of training and experience in analysis. But, owing to limited data, analysis is knowledge, attitude, and application above the mechanical art of line-drawing.

For evidence of the problem before us, consider the experiment illustrated by Figure 4.15. The first drawing (*a*) is an idealized pattern which we suppose represents the actual topography of an atmospheric surface. The rectangular grid of dots is the observation network, and we assume each station gives a report agreeing precisely with the actual pattern.[7] Observe that, in addition to the concentric circular pattern, a primary feature is the variation in spacing of contours.

Briefly stated, through analysis we attempt to reproduce patterns as they exist

7. This is not always a valid assumption.

in the atmosphere (to within a practical degree of accuracy determined by the use of the analysis). Here we observe how well this is accomplished with varying amounts of data. The complete field of data was analyzed first by linear interpolation be-

adequate. Second, in fitting smooth curves through interpolated points, one is guided by instinct, not necessarily by the data.

In diagram (*b*) alternate stations were omitted. In relation to the scale of the topographic pattern shown, this grid approaches

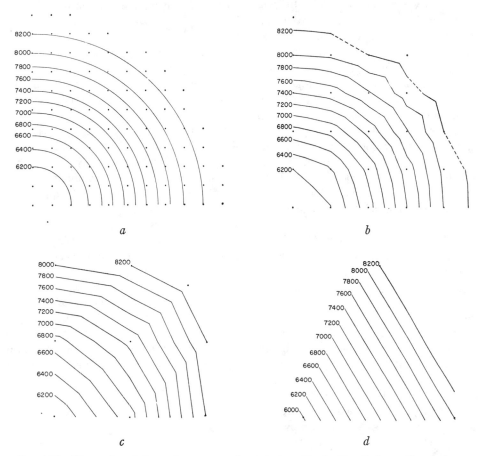

FIG. 4.15.—Illustration of decreasing accuracy in analysis as the density of observations is decreased

tween adjacent points and by connecting interpolated points with smooth curves. The result was so nearly identical with the actual pattern that it has not been reproduced. It was necessary nevertheless to introduce two assumptions in this elementary procedure. First, the slope of this surface was assumed so nearly constant between points that linear interpolation was

some of the better observation networks at the surface. The broken lines shown are the result of linear interpolation between points. But interpolation was selective; it was used along each ordinate and abscissa, and also along each diagonal from lower left to upper right, but not along opposite diagonals, where it would have given too erratic an appearance to contours. If each line is now rounded into a smooth curve (to suit better judgment), we soon conclude that more of the interpolations are wrong. In this further selection we again let reasoning guide us. Meantime, an inseparable feature of analysis is introduced, that of trial and error in curve-fitting. But, no matter how careful the interpolation and curve-fitting may be, the result disagrees with the pattern it intends to duplicate.

In (c) the network is again halved. Similar interpolation results in the broken lines shown, and again the interpolated points may be used to obtain nearly circular arcs bearing some resemblance to the pattern in (a); but the other significant feature has now almost disappeared. In (a) the gradient was concentrated about contour 7200. In (c) that gradient is diminished radically, and its maximum is displaced toward higher contours.

Another stage (d) in subtraction of data finds only three stations in the region, the least number of points from which a likely pattern can be drawn by interpolation alone; but the station density in relation to the scale of the pattern is still not less than one frequently has. The contours in (d) bear only rough resemblance to the actual pattern; both curvature of lines and variation of gradient have disappeared and cannot be determined from these raw data alone. Briefly, the "analysis" in (d) does not duplicate the existing pattern, and modifying the curvature and gradient is just as likely to produce the wrong effect as the right.

The pattern in Figure 4.15(a) is certainly as uniform and symmetric as can be expected. Actual patterns usually are more complicated and would undergo greater modification when subjected to such experiment. It also becomes evident that analysis is subjective. In either (c) or (d), for instance, introduction of smooth curves gives not a unique solution but one reflecting the

ability and judgment of the individual. Without further discourse, we conclude that technique of analysis must involve more than given data, more than ability to perform simple interpolation, and more than artistic talent for drawing smooth curves. The complete solution is unknown, but we will attempt in what follows to stimulate critical and constructive attitude toward procedures in analysis.

We see by the series in Figure 4.15 that, the more data, the more accurately the true state can be reproduced by scalar analysis. If it were possible to increase the amount of data, then surely efficiency of analysis should also improve. Besides defeating ourselves by recommending merely that weather services increase their observation networks, there are ways and means to supplement the reported data. In locating lines between adjacent reports, for example, the gap is filled not just by linear interpolation, which at times if used alone is in essence dilution of the data, but by considerations of physical controls in the atmosphere, which make it possible to infer information not directly observed and which the analyst soon applies with confidence and for the most part unawares. A basic tool is practical knowledge of the structure and behavior of the atmosphere.

The considerations employed by the analyst to improvise when confronted with limited data may be outlined briefly as follows:

1. Association (familiarity, reference to models or analogues, reference to normals and climatic controls, etc.)
2. Interpolation ⎱ ⎰ continuity in space
3. Extrapolation ⎰ ⎱ continuity in time
4. Coordination (of the patterns of related quantities in space and in time)

These four general concepts are closely related, and some overlapping is involved in meaning and application; none should be used singly.

4.16. *Association.*—A prime asset in analysis is ability to sense or "to feel" at once the logical distribution of a variable in space given only the barest amount of data. For the most part, this ability to picture the situation quickly is developed through familiarity and careful analysis of natural patterns where data are most dense. By training himself to form mental pictures of patterns, individually and in relation to each other, the analyst becomes capable of distinguishing between natural and artificial and between usual and unusual. By considering physical principles applicable to the given conditions, the analyst discovers that knowledge facilitates the choice of details in the pattern not adequately described by the observations and minimizes the chances of imagination guiding the analysis.

It is impossible to obtain the necessary familiarity from a textbook or the meteorological literature. Practical experience is indispensable. Moreover, we strongly suggest careful study of patterns and sequences from published daily map series, for both large and small areas, as a means of establishing ideas on geometry and development of atmospheric structure. The limited number of illustrations in this book were selected for other reasons, while assuming familiarity would be gained by practice.

While on the subject of association, we should not skip over the use of hydrodynamic models (and geometric similarities). We already mentioned a few elementary models in topographic analysis. Knowledge of these and the many others is a guide for deriving at least a preliminary idea of patterns from the plotted data. Where data are dense, our preconceived notions of the patterns should be modified to conform with the data, and the final analysis is thus obtained with less loss of time in trial and error. The value of models as a guide for supplementing the data is greatest in areas of sparse data, but unfortunately the final

analysis conforms more closely with the preconceived patterns, as there is more freedom for placing preferred models into the chart.[8]

For purposes of orientation, models are a definite aid in the process of analysis, but, like anything else, they also have their limitations. Subservience to and conformity with models can just as well be detrimental if used to the extent that analysis is more prejudiced than natural. One who calls himself practical and deprecates application of physical "theory" to his work might constantly be led by crude models which comprise perhaps the most raw form of theory. We should not lose sight of the fact that procedures in analysis are aimed at *finding* rather than *accommodating*.

4.17. *Interpolation in space.*—Scalar analysis involves positioning of lines with selected values within a field of data. Interpolation between points can be the most objective method for locating those lines. But interpolation can be equally misunderstood, and it often becomes a detriment instead of serving its best advantage. For that reason and because proper procedure in interpolation is basic in theory and practice, the concept is discussed at some length.

To begin, we take a simple example. Given points $A(x = 7, y = 2)$ and $B(x = 17, y = 7)$, find y at $x = 9$. The function $y(x)$ connecting the two points is unknown, so ordinarily we assume the function is linear between points, of the form $y = ax + b$. With this assumption interpolation can be done mentally. Since $x = 9$ is $2/10$ the abscissa distance from A to B, the desired value of y is $2/10$ the ordinate distance from A to B, or $2 + (2/10)(5) = 3$. Since we assumed the function is linear, A, B, and the interpolated point all lie on a straight line. The procedure is *linear interpolation;*

8. In less serious mind: It is hard to dispute theory without evidence.

for equal intervals in x, it gives equal intervals in y. This same technique applied to scalar analysis produces equally spaced lines between data points.

In all probability the most natural curve connecting A and B is not strictly a straight line; that is, y is not a linear function of x. In the atmosphere, property surfaces are not planes, and they are seldom equally spaced. *A straight line approximates this curve only if the points are close enough to-*

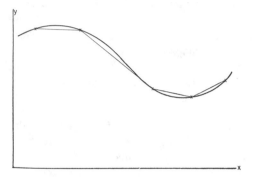

Fig. 4.17a

gether. The greater the distance between points, the more the straight-line approximation is subject to error. The two points can be connected by an infinite number of smooth curves. The shape of the proper curve is described by second- and higher-order derivatives of the function, which require knowledge of more than two points of data.

While the example above suggests the advantage of considering higher-order derivatives, it does not point out the harm that might result by ignoring them. In Figure 4.17a the five data points, giving height of a surface as a function of distance x, are intended for drawing an x, z profile of the surface. From these observations the best fit is some smooth curve, perhaps sinusoidal, and not by a series of straight-line segments connecting successive points. Then use of linear interpolation only is con-

tradictory. Linear interpolation also gives a discontinuity at each data point. Although discontinuities are possible, their occurrence on each data point is highly improbable.[9]

Any quantity Q may be represented synoptically as a function of distance s in a given direction about a point. Thus, $Q = Q(s)$. Here $Q(s)$ is the function and s the argument of the function. Knowing the character of the function permits evaluation of Q at any position s. Now suppose this function is known; at successive intervals of s we may tabulate the corresponding values of Q, as shown in the first two columns of Table 4.17a. Then in a third column are entered the first differences of the function, designated Δ^1. Each first difference when divided by the interval i is a finite-difference approximation to the first derivative of the function with respect to s in that interval of s. From the first differences we may compute second differences, indicated by Δ^2. When divided by the square of the interval i, the second differences correspond to second derivatives of the function. The process may be continued through higher differences. The number of orders of differences to be computed depends on the type of function in question, on the accuracy demanded by interpolation, and on the number of given points. On the right-hand side these considerations are shown symbolically with respect to the point $s = 2$. The difference table is a basis for interpolating between known values of a function.

A point of interest revealed by the table is the effect of external data on interpolation between two adjacent points. Assume it is desired to interpolate between s and

9. At this juncture we are touching on the problem of choosing between *accurate* procedure and *practical* procedure in analysis. Since we have not yet succeeded in achieving both, we are taking the viewpoint that the guiding principle should be accurate and its application simplified to suit the particular stringencies of time and effort.

$(s + i)$. The first difference $\Delta Q_{1/2}$ involves values only at these two points, but any additional refinement brings in higher derivatives, which are controlled not just by s and $(s + i)$ but also by data beyond. Although the effect of external data usually decreases with distance from the region interpolated, it is obvious nonetheless that best rules for interpolating *must account for the general character of the function over a larger region than between two data points.*

To find the value of Q at some point $(s + pi)$, where p is a fractional part of

and the second near the midpoint between data points.

In the above formulas the terms on the right are usually of descending order in magnitude; the terms of third and higher order might be neglected in most cases. With reference to topographic analysis and analysis in a plane, we are concerned primarily with (i) the value at data points (first term); (ii) the average gradient between data points (second term); and (iii) the variation of gradient (third term).

To see more clearly what effects the

TABLE 4.17*a*

s	Q	Δ^1	Δ^2	Δ^3	s	Q	Δ^1	Δ^2	Δ^3
0.......	65				$s-2i$.....	$Q(s-2i)$			
		5					$\Delta Q_{-3/2}$		
1.......	70		2		$s-i$......	$Q(s-i)$		$\Delta^2 Q_{-1}$	
		7		-4			$\Delta Q_{-1/2}$		$\Delta^3 Q_{-1/2}$
2.......	77		-2		s........	$Q(s)$		$\Delta^2 Q$	
		5		-1			$\Delta Q_{1/2}$		$\Delta^3 Q_{1/2}$
3.......	82		-3		$s+i$......	$Q(s+i)$		$\Delta^2 Q_1$	
		2					$\Delta Q_{3/2}$		
4.......	84				$s+2i$.....	$Q(s+2i)$			

interval i (so $1 > |p| > 0$), there are interpolation formulas which can be used with the difference table. Only the two most widely used central difference formulas are mentioned here. The *Stirling* formula is

$$Q(s + pi) = Q(s) + \tfrac{1}{2}p(\Delta Q_{1/2} + \Delta Q_{-1/2})$$
$$+ (p^2/2!)\Delta^2 Q + \tfrac{1}{2}[p(p^2-1)/3!](\Delta^3 Q_{1/2}$$
$$+ \Delta^3 Q_{-1/2}) + [p^2(p^2-1)/4!]\Delta^4 Q$$
$$+ \cdots; \tag{1}$$

and the *Bessel* formula is

$$Q(s + pi) = \tfrac{1}{2}[Q(s) + Q(s + i)]$$
$$+ (p - 1/2)\Delta Q_{1/2} + \tfrac{1}{2}[p(p - 1)/2!]$$
$$\times (\Delta^2 Q + \Delta^2 Q_1) + [p(p - 1)(p - \tfrac{1}{2})/3!]$$
$$\times \Delta^3 Q_{1/2} + \tfrac{1}{2}[p(p^2 - 1)(p - 2)/4!]$$
$$\times (\Delta^4 Q + \Delta^4 Q_1) + \cdots. \tag{2}$$

The first is more accurate near a data point

second differences have on interpolation as applied to immediate needs, consider the example in Figure 4.17*b*, which is a vertical cross section of an idealized 700-mb surface. The height of the surface along x is

$$Z = 10,000 + 800 \cos(\pi x/30) .$$
$$(Z \ gpft, \ x \ 100 \ km) .$$

A difference table for intervals $x = 500$ km is given below.

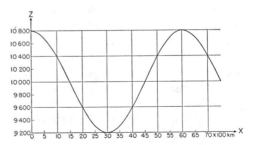

FIG. 4.17*b*

The points $x = 15$ and $x = 45$, as seen by Table 4.17*b*, are the inflection points of the curve. Here the first difference is a maximum (maximum slope), the second difference is zero (absence of curvature), and higher-order differences are small in comparison with the first difference; thus, linear interpolation gives little error in geographic location of contours in the im-

TABLE 4.17*b*

x (100 Km)	Z (Gpft)	Δ^1	Δ^2	Δ^3	Δ^4
0........	10,800				
		−107			
5........	10,693		−186		
		−293		79	
10........	10,400		−107		28
		−400		107	
15........	10,000		0		0
		−400		107	
20........	9,600		107		−28
		−293		79	
25........	9,307		186		−51
		−107		28	
30........	9,200		214		−56
		107		−28	
35........	9,307		186		−51
		293		−79	
40........	9,600		107		−28
		400		−107	
45........	10,000		0		0
		400		−107	
50........	10,400		−107		

mediate vicinity of those points. But, in both directions from inflection points, the first differences (slopes) decrease in magnitude, slowly at first and then more rapidly. At the maximum and minimum elevations of the curve the slope is least and spacing of contours greatest. Here, because of rapid variations in gradient, linear interpolation is subject to greatest error in geographic location of contours.

In practice we must depart in procedure from the outline of interpolation method given above for several reasons, among which is that time does not permit preparing tables or applying a formula. Interpolation is done mentally by applying a set of general rules and graphically in a process of approximations.

A method for interpolating between data is suggested by Figure 4.17*c*. Suppose the selected interval of Q is one unit and also that no discontinuities exist in the region. The problem is to locate the intersections of unit Q lines with axis AB. As a first approximation, we interpolate linearly between data points and obtain intersections shown by the marks on line (1). This gives a break at each data point, which we remove by varying the spacing more gradually across each point. By successively adjusting the intersections, the final interpolated points should be about as shown by the x-marks in line (2). Notice that the largest corrections by the second approximation occur where

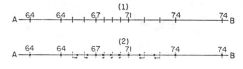

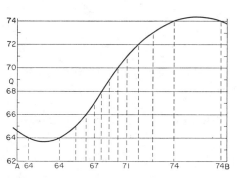

Fig. 4.17*c*—A method for interpolation along a line.

the second derivative of Q is relatively large and that the smallest corrections are where the second derivative is small in comparison with the first.

The principle in the above example can be used in routine analysis. First locate tentatively, perhaps by linear interpolation, the regions of maximum and minimum and the zone of maximum gradient and then adjust the points of intersection to give gradual sequence of spacings.

Up to this point we have considered

interpolation of a function continuous through at least its second derivative. There is no restriction on second- and higher-order discontinuities in atmospheric variables, but we shall not consider them here. Discontinuities of first order are assumed a frequent occurrence in any of the variables. There the first derivative of the function is discontinuous, and first- and higher-order differences of the function are indeterminate. Hence, *interpolation cannot be done across a discontinuity*. Assuming that the location of the discontinuity already is known from other considerations, or from tentative analysis in large scale, a practical method is to *treat the discontinuity as a boundary to interpolation*. Appropriate interpolation is used in the region to one side with least reference to data beyond it, and the sequence is ended abruptly at this boundary. A similar procedure is applied in the other region, and finally the scalar lines are connected at the boundary to give (zero-order) continuity to the function itself.

Effectiveness of interpolation in drawing an analysis is determined by the number of data per cycle or wave in the pattern. When the pattern in question is of small diameter or wave length, proper methods of interpolation even in best networks of data cannot be expected to yield satisfactory results; yet interpolation in this same network will do a reasonable job in finding patterns of larger scale. Where directly observed data are very sparse, as in upper levels over oceans, reliable results are impossible by interpolation alone and such should not be attempted, since the space between data might contain one or more significant cycles of the pattern. Whenever interpolation is used there, it is only to supplement other aids.

The principles are applicable to any type of analysis. In addition to analysis of an individual chart, the rules of interpolation can be applied less rigorously to coordinating analyses in a series of charts. For example, given the analyses of all other constant pressure charts (1000, 850, 500, 400 mb, etc.), one can visualize how a contour pattern for 700 mb might be derived by interpolation alone. The principle in this instance is continuity of patterns in space; any chart is a step between adjacent charts.

4.18. *Interpolation in time.*—Those atmospheric phenomena of primary concern at present have periods longer than the time intervals between synoptic charts. In fact, our concept of "synoptically important" features must be based on the synoptic time interval, and conversely. Most changes are so slow and gradual that synoptic charts even 12 or 24 hours apart show remarkable resemblance.[10]

Historical sequence—what we know as *time continuity* in the state of the atmosphere—is useful as expedient in procedure and as method for implementing the data. By studying the sequence of past charts, the analyst knows what to expect in the general pattern of the next. Even many minute details carry over. Reference to time continuity gives the basis on which to work with less waste of time and effort. However, like all other tools, continuity as an aid also has its limitations.

Mathematical methods for interpolating in time are given by Eqs 4.17(1) and (2). Only now reference is to a fixed point, the interval i is an increment of time, and the dependent variable is the value for a station at a certain time. This can be applied directly to many calculations in related problems, such as approximating missing observations and computing corrections for delayed radiosonde observations, but always with discretion.

4.19. *Extrapolation in space.*—Space interpolation requires data on both sides of the point in question, and the required

10. See Figure 9.02c and compare Figures 9.02ab

scalar lines are located by extending data inward. A problem is confronted at continental boundaries, beyond which no data may be found for large distances, and also in drawing the latest map, for that map is the limit of data in time. The problem can be approached by *extrapolation*, which extends the analysis outward from the end of a sequence of data.

A principle to consider is a Taylor series expansion:

$$Q(s + \delta s) = Q(s) + (\partial Q/\partial s)_s \delta s$$
$$+ (1/2)(\partial^2 Q/\partial s^2)_s (\delta s)^2 + \ldots \quad (1)$$

Point s designates the last point of data, and δs is an increment of distance beyond that point. In equal-interval form and notations used in the interpolation formulas, we may write, for example,

$$Q(s + pi) = Q(s) + p\Delta Q_{-1/2}$$
$$+ (p^2/2)\Delta^2 Q_{-1} + \ldots \quad (2)$$

It is possible to use the difference table for extrapolating by Eq (2).

In Figure 4.19 is the profile of Q along an axis through s, the last point of data. Suppose we wish to determine Q a distance δs beyond. Barring discontinuities near s, Eq (1) states that Q at $(s + \delta s)$ is a function of Q at s and all derivatives of Q at s. For locating scalar lines of Q in δs, this implies that their spacing should be the same as the spacing within the margin of data plus a correction for higher-order derivatives, with steps similar to those for interpolation. This possible procedure is thus a simple one, but it is also evident that limitations are many and that pure extrapolation is more precarious than interpolation.

Since any extension of the curve beyond s in Figure 4.19 depends on knowledge of the function at s (and in a short interval to its left), the validity of extrapolation depends upon density of data near the boundary, upon accuracy of the analysis also near

this boundary, upon scale of the pattern of the quantity, and upon amount of detail required and latitude of tolerance in approximation. We already know by reference to interpolation that analysis in the margin of data usually cannot be as accurate as elsewhere. Hence, extrapolation is based on a relatively unreliable area of analysis.

In Figure 4.19 the three dashed lines beyond s are a few of the many possible shapes the actual curve might assume in the

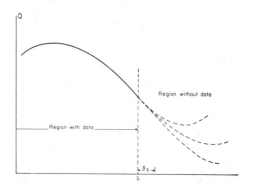

FIG. 4.19.—Difficulty with extrapolation

region of no data. Each is continuous with the distribution of Q to the left of s, both in slope and in curvature. In view of the many possible forms of the function in the unknown region, it is inconceivable how extrapolation can give any reasonably accurate analysis over a large distance. The error increases with distance, and soon analysis by extrapolation alone becomes random. Therefore, analysis by extrapolation only is not reliable beyond a small fraction of the wave length of the pattern. To proceed any further requires additional principles, which might include historical sequence, periodicities in the pattern, relation to other current patterns, use of normals, and indirect indications based on other elements.

4.20. *Extrapolation in time.*—The primary value of time continuity in weather analysis is its application to extrapolation as distinct from interpolation. One is mainly concerned with drawing the latest in a series of maps and charts and in preparing forecasts or *prognoses* from them. Use of extrapolation in sparse data to carry the analysis to the current time is in essence forecasting.

The Taylor series expansion of Q with time is

$$Q(t + \delta t) = Q(t) + (\partial Q/\partial t)_t \delta t$$
$$+ (1/2)(\partial^2 Q/\partial t^2)_t (\delta t)^2 + \cdots . \quad (1)$$

With reference to point values, Q at any point on the present map can be determined from its value on the previous map, from the local tendency of Q, from time variation of the tendency, etc., provided time intervals are small and discontinuities excluded. Now, considering the pattern of Q instead of values at fixed points, Eq (1) can be interpreted in terms of motion of patterns over the chart. A certain feature can be located by considering its last position, by computing its velocity from the most recent charts in the sequence, by computing its acceleration from those velocities, and so on. Similar reasoning holds for individual changes in the patterns, which include "deepening," "filling," and deformation. It is evident that most of the coming analysis can be visualized before consulting the latest data. But the truth of these deductions depends on the time interval between maps and on the accuracy of the last analysis, showing why the best analyst can err if preceded by an unreliable one!

The usefulness of extrapolation in time for extending the analysis points out the value of continuity and tendency charts[11] as auxiliary aids in analysis and forecasting, particularly where data are relatively sparse or unreliable. Continuity charts can be maintained by entering several previous positions and intensities of identifiable systems on the chart to be analyzed or on a

separate chart maintained solely for that purpose. Tendency charts may be prepared for the synoptic time interval (6 hours at the surface, 12 hours aloft) or convenient multiples thereof. It is usually recommended that intervals be used which minimize the semidiurnal and diurnal periodicities and preserve the ones desired. Tendency patterns, when extrapolated into the future, are forecasts of the changes in the particular variables at fixed points; if used with the last analyzed chart, the forecast changes provide point data and indications of internal changes in the patterns.

4.21. *Coordination of the various patterns.* —Since many of the refinements in analysis will be evident or developed in what follows, it is superfluous to go into any detail now. Here we enumerate a few examples of the physical concepts used in mutually adjusting the various patterns to build an analysis from limited data, but one will see that coherency is a most important approach to the problem of analysis, using our present system of observation and state of knowledge.

The horizontal patterns of pressure and wind are closely related,[12] to such an extent that a reasonably accurate pressure analysis can be deduced from wind data. Conversely, from the pressure analysis it is possible to derive the field of motion with the same accuracy. Another practical method for deducing upper-level wind and pressure systems is the use of detailed cloud observations, a practice which seems to have deteriorated with increased pilot-balloon wind measurements.

Equally important, patterns of pressure and temperature are uniquely related through hydrostatics. Given a horizontal pressure pattern and knowledge of the temperature distribution, it is possible to derive the pressure distribution at some lower or higher level.

11. Figures 9.02*c* and 9.08*c*.

12. As also the state of the sea, weather, visibility, etc., with either pressure pattern or wind.

It is also possible to derive upper-level pressure and wind patterns from the sea-level pressure pattern, surface temperatures, and indications of stability from clouds, visibility, and weather in the surface report and from knowledge of the history of the air mass. This relation is most useful in deriving upper-level flow patterns over oceans.

Since moisture distribution is so dependent on three-dimensional motions, knowledge of air trajectories in the horizontal and vertical is a clue to the pattern of humidity, and conversely. Clouds and precipitation can be used similarly. Patterns of temperature, potential temperature, and equivalent potential temperature are in varying degrees indicators of previous trajectories and modifications of air masses. Hence, their patterns give some indication of previous air motion and pressure fields. Correlations between the various patterns are not completely known and understood, but they are still important in many ways.

These are only a few of the methods by which one pattern is used to find others. Constructing and enlarging the analysis in this way is in a sense association, interpolation, and extrapolation from one analysis to another and checking all for physical coherency. In the process we are "diagnosing" the atmosphere and preparing for the forecast that follows.

4.22. *Remarks on direction of emphasis.*—
In the geometric patterns of lines in the analyzed chart lie the real significance for whatever uses might be made of the analysis. The patterns describe the gradient fields of the variables, which determine motions and dynamics of the atmosphere—briefly, the weather. In using a pressure analysis, little reference is made to the value of an isobar; any thorough study of the analysis concerns more the *pattern* of pressure.

Since the important criterion in judging and using the analysis is the pattern of the variable in space, *primary emphasis should be placed on drawing patterns correctly*. But that is not always the case in practice; too often major stress is placed on such matters of form as where and how to label a line or on artistic appearance of the chart.

Distinguishing between the important and the unimportant in analysis, particularly as concerns scale significance, is a serious problem to the student and to the forecaster in the field. The framework of our present-day system of observation and analysis is ages old, yet many new applications of analysis and forecasting have arisen since, one of the most prominent of which is short-period terminal forecasting for aviation. What may be important in analysis for local and short-period forecasting may be unimportant for forecasting the general weather pattern over large areas a week in advance. But, unfortunately, all must cope with a single set of standards; the forecaster at the airport whose emphasis is on small areas and short periods must use and even prepare the analysis on the same large scale as does the extended-period forecaster at a weather central. We cannot suggest a solution here, but there is room for thought.

4.23. *Limitations on the accuracy of analysis.*—
The accuracy with which actual patterns of atmospheric variables can be reproduced by analysis from a scatter of observations depends on a large number of additional factors. These can be classified as follows, but their discussion will be made very brief: (*a*) accuracy of the plotted data (instruments, observation, encoding, transmission, plotting); (*b*) density of the data; (*c*) nature of the particular variable; (*d*) local conditions; (*e*) scale of the phenomenon; (*f*) type of chart; and (*g*) skill of the analyst.

a) *Accuracy of the plotted data.*—Precision of meteorological instruments is a separate and lengthy subject in itself. No attempt is made here to discuss merits of various instruments and the many errors which have their source in faulty response or indication by instruments. For discussions of instrumental errors the reader is referred to textbooks, handbooks, and manuals treating the subject.

There is much in the literature about errors by surface meteorological instruments during certain weather conditions. To expand on this, radiosonde measurements of temperature and humidity, and therefore also of pressure, are made unreliable by icing, precipitation, and even passage of the instrument through a cloud. In wind soundings there are discrepancies in measurement of strong winds because of the large probable error at low elevation angles.

Errors in observation are a result of several different factors, including exposure of the instruments, time differences in observation, insufficient training, carelessness during both measurement and calculation, approximations, and assumptions in theory. Many errors resulting from improper observation are discussed at length in observers' manuals. The effect of exposure on indications by surface instruments is micrometeorological in nature. It is not possible for the analyst to familiarize himself with all peculiarities of instrument exposure at each station, but he should be cognizant of the fact that differences do exist and that occasionally some of the confusion in analysis can be traced to that source.

Exposure of the stations to the elements is as important as exposure of instruments at the individual stations (especially with reference to surface observations). Few stations have identical surroundings, and, although error in observation may not be involved, reports from adjacent stations are not always compatible. Natural variations between near-by stations are what we call *local effects* or local modifications on the air mass. They are confined mostly to the air layer in contact with the surface, except with strong vertical mixing when modifica-

tions may extend higher. Properly accounting for local effects is a vital part of surface-weather analysis and forecasting. The analyst should acquaint himself with all important details of physiography.

Time differences in scheduled synoptic observations amounting to several minutes usually are negligible. Some the order of an hour in upper-air observations are frequently unavoidable. In conditions of slow local change, time differences are inconsequential, but with rapid changes the error must be taken into account.

Approximations and rounding-off values for computing can lead to error in pressure-height evaluation, for example. Some error is introduced in upper-wind measurements by reporting for the wind at one level the *average* wind in a rather deep layer. The error is significant where the wind varies rapidly with height. Also, owing to the need for minimizing the number of significant points in a radiosonde ascent and because temperatures at mandatory pressure levels must be reported, there is an affinity for the tropopause to appear at 300, 200, or 150 mb when it really is ill defined.

The above discrepancies and errors have their source either in the instrument or in the process of observation. There is little one can do to remedy them directly, but he should be aware of possible errors in reports and take precautions in using the data. There are certain conditions in which a report is likely to be more unreliable than usual. Also, he might distinguish reliable from unreliable stations and thus place selective emphasis on individual reports. This is needed in analysis over oceans. Discrepancies among upper-air reports are noticeable over Europe, where different types of instruments are used, and even over the United States, where there has been no complete agreement among civil and military radiosonde reports.

Errors introduced during encoding the

report consist of two types: those due to approximation and those due to carelessness. Encoding temperature and wind, for instance, in most cases requires approximation to the nearest whole degree in temperature and the nearest 10° in wind direction. Differences by those magnitudes between stations or between levels should not be considered significant.

Transmission of data by radio or telegraphy is another source of error. In the more obvious cases the error might be found, but most often it escapes detection unless the error introduces some absurdity into the report.

Plotting also introduces error in the data, for example, in wind directions where an average error of about 10° can be expected. There are other accidental errors made in converting the transmitted report into the final plotted form.

In plotting upper-air charts, the data obtained from balloon soundings are referred to the station of release irrespective of drift during ascent. Data at 100 mb might refer to some point as much as a hundred miles distant from the point at which it is plotted. With strong winds through the troposphere, significant error may result in analysis by this present method of plotting. In some cases at least a qualitative correction to geographic location might be made for temperature and humidity.

In view of the numerous ways by which errors can enter the data, one might conclude that accurate analysis is too much to expect. There are several means to avoid defeat, however. Many of the more obvious errors can be corrected. Errors in upper-air data can be found and corrected by reference to hydrostatic and stability limitations. Other inconsistencies in the data can be found and corrected by continuity checks in space and time and by exercise of a little judgment. The knack of trouble-shooting is one to be cultivated by the analyst.

Despite the fact that data might appear in order on first sight, an isolated report should be examined more carefully than others. This single report controls the analysis over a large area, but its error is less obvious in comparison. Many times it has been proved that analysis over large ocean areas can be completely reversed by a single error in a ship report!

While any report stands fault, the logical attitude to take is that a minimum number of reports are wrong. If a report does not agree with neighboring ones, it should be examined before applying a correction to all surrounding reports. A recommended approach in deleting inconsistent and otherwise erroneous data from the analysis is first to correct them, if that is profitable, by tracing the error to its source. The time and effort thus spent may be well rewarded.

b) Density of the data.—The relation of data density to possible accuracy in analysis was emphasized in connection with Figure 4.15. Only a few points are worth adding here. First is the ease with which errors in data can be detected with abundant data; inconsistencies stand out prominently. On the other hand, in regions of sparse data, errors must be more purposely sought.

An important way in which paucity of data restricts accuracy in analysis lies in the fact that possible scale and detail attainable is a function of the amount of data. The smaller details can be drawn accurately where data are sufficiently dense, but the same ones are open to question with data more widely scattered.

c) The nature of the particular variable.—Convenience and accuracy in drawing an analysis are functions of the uniformity and scale of the pattern of the variable. The large horizontal extent of topographic pressure features gives relatively smooth appearance to its patterns, and what makes the pressure field still more easily and accurately drawn is the close relation between pressure and wind fields. As a sharp contrast, the dimensions of relative humidity patterns are small and the shapes are extremely complicated.

d) Local conditions.—As there are as many local conditions as localities, this subject of small-scale meteorology can be mentioned only briefly here. When and how to account for local influences are everyday problems in analysis and forecasting. The decision is determined by the purpose.

The following short list should be extended and enlarged upon by the reader:

1. Orographic disturbances in the fields of most variables
2. Effects of localized convection cells and local winds
3. Modifications of stability, vertical wind shear, surface wind and pressure patterns, etc., at large temperature gradients of the underlying surface (coasts, snow lines)
4. Transport of a small-scale air mass produced by a rain shower or from a lake, river, or marsh
5. Containment of cold shallow air masses by mountain barriers
6. Differences in air trajectories
7. Differences produced at and beneath edges of cloud covers
8. Gradients in surface roughness and turbulence
9. Gradients of surface evaporation and condensation
10. Effect of differences in soil characteristics, soil cover, and convection currents in the sea surface

e) Scale of the phenomenon.—Any atmospheric pattern is the integral of perturbations ranging from infinitesimal to hemispheric in dimension. Thus the pressure field consists of a spectrum of waves having various lengths and amplitudes: the semipermanent wave whose hemispheric number ranges between one and ten, the shorter migratory waves corresponding to the traveling pressure centers observed at the surface, the semidiurnal pressure oscillations of more uniform amplitude and frequency, the more localized and short-lived pressure waves associated with short-period irregularities in the station barogram, and so on. Certainly an isobar is not a smooth curve at all but a very irregular line. The *average shape* of this isobar in large scale can be *approximated* by a smooth curve. It is impossible to account for all perturbations in the field of any variable, and analysis therefore must involve some selection.

Filtering the important from the unwanted features is one of those intangible "intricacies of analysis" which one appreciates by experience. One aspect bordering on the objective is our concept of scale significance. Since we are interested primarily in those features with periods equal to or greater than the synoptic time interval, and which are also credible results of data, we may suppress the minor irregularities. Generally, one should not attempt to bring out patterns of dimension smaller than the average distance between observing stations.

The small-scale "noise disturbances" in atmospheric patterns influence the station observations as well. Detailed analyses in micrometeorological networks show complicated structures in small scale; gradients are large, centers numerous, and lines eccentric in course. In effect, there is no assurance that an individual observation is entirely representative of the large-scale pattern to be drawn, no matter how accurately it was made and how idealized the station exposure. This error of sampling suggests that present-day analysis must have an equal or larger margin of error.

f) The type of base chart.—The scale of the map affects analysis technique both mechanically and psychologically, and it is a factor in any visual study of the analysis. The smaller the map scale, the more compact is the network of data, the more quickly the analysis can be done, the more composite features can be seen in a single glance, and the less is the opportunity for trivial details to enter. This is advantageous, since trivialities are more likely to be smoothed out, but it is harmful in that some important elements are more apt to be carelessly overlooked.

g) *The skill of the analyst.*—No matter how dense the data coverage might be and how learned one may be in the "academics," quality in analysis requires some skill or knack in the mechanics, in efficiently digesting the enormous mass of data, and in adapting the pertinent laws of nature.

4.24 *Special problems in the mechanics of analysis.*—Of the countless difficulties for the inexperienced, most can be remedied by a little thought. Owing to space limitations, only a few examples will be cited here.

One is analysis in the weaker areas of a pattern, where gradient directions are likely to be distributed haphazardly and where small disturbances and observational errors have their greatest relative effect. The procedure usually followed is a compromise between accuracy and effort. A more specific problem is analysis in the vicinity of cols, where several solutions are possible. In certain cases the advantage of clarity could even supersede drawing for detailed accuracy. A more trivial problem is excessive unrelated wiggles (sinuosities) in adjacent lines; where alternate lines are parallel, the intermediate line is seldom contradictory.

Figure 4.112 illustrates logical and synthetic approaches to analysis. Another common mistake is shown by Figure 4.24. Diagram (*a*) shows what might result by extrapolating patterns inward to a region of no data. In the process the gradient and curvature might be maintained all the way to the center. Although extrapolation is valid for *tentatively* locating the center, the process gives unnatural results when the entire pattern is extrapolated inward. Since the

unfavorable part of this pattern occurs with no data, we have liberty to change it to the more likely pattern shown in (*a'*). These examples show the need for thought while occupied by the mechanical procedure of drawing lines.

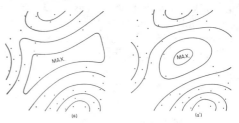

FIG. 4.24.—Patterns obtained from the same field of data.

4.25. *An outline for procedure in scalar analysis.*—There is no outline for analysis which is complete enough for different conditions and all types of analysis and which is not too unwieldy to be a handy guide. We give one here only to satisfy the usual demand for an "order of operations." The outline below serves as a start. It is designed as far as possible in chronological order of procedure for use only where reports are numerous enough without needing data determined indirectly: (i) fix in mind the history of the situation; (ii) bring continuity chart up to date; (iii) examine the plotted data in terms of (i) and (ii); (iv) sketch the patterns in a preliminary manner; (v) shape the patterns into final form; (vi) enter the final lines; and (vii) enter appropriate labels.

(i) Get all possible information from studying the history of the situation. Particular emphasis should be given to the broad-scale features, for they are more likely to be conserved (examine from a distance). During this step most of the present analysis is being evolved.

(ii) Prepare a continuity chart which includes all important components of the pattern. These might be locations, intensities, and trajectories of centers, cols, discontinuity lines (fronts, shear lines), zones of maximum gradient, etc. This step provides the preliminary objective procedure for locating those features existing in the past; new developments will have to be found in the current analysis. These first two steps are applicable only when previous charts in a sequence are available. The first chart in a sequence requires more study and trial and error.

The remaining procedure can be executed for the map as a whole or for successive parts of the map, depending on the area covered by the map and on particular preferences.

(iii) Gather additional information on motions and developments of the patterns from a survey of the plotted data. Before touching the pencil on the map, one can determine the general form of the subsequent analysis, especially in the most data. Where data are scattered, a little extra study should be given to individual observations.

(iv) Begin sketching the patterns, preferably first about the centers and in regions of largest gradient. Work outward from the area of most data, as a rule. The line interval used in sketching is usually a multiple of the final interval. In general, the amount of sketching should be minimized, and sketching should be done lightly. The required amount of sketching decreases with practice. Other charts for the same time (plotted or analyzed) should be consulted for guidance and continuity.

(v) With proper interpolation and reference to natural and preferred distributions in space, shape the patterns into more or less final form to clarify those of large scale as well as the important superimposed details. Check for internal consistency with other indications on the chart and with other charts in space and time and for possibility of unforeseen developments.

(vi) The final lines now can be entered—pencil preferred (for convenience in applying the second instrument, the eraser). (Ink is advocated for published charts, for charts used in weather briefings and conferences, or for those operations when additional drawing is performed on the same chart.)

Where sketching was done for multiples of the desired interval, intermediate lines should be drawn first. In the process care should be taken to insure that lines being entered conform with the data and that they are so located between sketch lines as to give the desired distribution of gradients over the area. Then the remaining lines are drawn over or in place of the sketches. For obvious reasons it is usually best to enter a line by pulling or drawing the pencil toward the body. One should also develop the skill for drawing smooth continuous lines over long distances without lifting the pencil from the paper. When distance is so great that a break is unavoidable, splicing is made at the more favorable places. Distracting details and errors can be removed by erasing long segments and redrawing the lines properly.

(vii) Label the lines and centers.[13] Each line should be labeled at the boundaries of the analysis and at a sufficient though limited number of places within the chart. Centers are usually designated by single capital letters: H or L for pressure, W or C for temperature and potential temperature, and M or D for moisture. Different color schemes are usually suggested for the designators of each type of center. For continuity and reference it may be desirable to designate ridge and trough lines in the pressure field, especially the more well-defined ones.

13. Conventions for labeling an analysis are outlined in weather service manuals (see "Reading References").

PROBLEMS AND EXERCISES

Following is a limited number of general exercises designed to supplement material in the text. Many other exercises and much practice are necessary using observed sets of data. An interesting exercise should be prepared based on the principle employed in Figure 4.15.

1. Give graphical examples of the various ways of representing the following:
   (i) Temperature variation at a point.
   (ii) Temperature distribution between two fixed points at a given time.
   (iii) Temperature distribution with time and distance between two fixed points.
2. List the various features of an isopleth for any (continuous) single-valued quantity.
3. In the vicinity of a point in the xy-plane, isotherms at intervals 2° C are oriented NNW–SSE (higher temperatures to the west) and are spaced 111 km apart. Find:
   (i) The direction and magnitude of the horizontal temperature ascendant.
   (ii) The eastward and northward gradients of temperature.
4. At a given point in space $\partial p/\partial x = 1\,\text{mb}/80\,\text{km}$, $\partial p/\partial y = -1\,\text{mb}/120\,\text{km}$, and $\partial p/\partial z = 1\,\text{mb}/20\,\text{m}$. Find:
   (i) The direction and magnitude of the gradient of p in each of the three coordinate planes and also the orientation and spacing of p lines in each of these three planes.
   (ii) The direction and magnitude of the resultant (space) gradient of p and also the spacing of consecutive p surfaces.
   (iii) The gradient of p along an axis with a slope of 1 in the xy-plane.
5. Using the same values given in (4) above, find:
   (i) The slope of p surfaces with respect to each coordinate vertical plane.
   (ii) The orientation of the vertical planes in which the p surfaces have maximum and minimum absolute slope, and give the maximum value of slope.
6. With reference to the curve in Figure 4.17b and its equation in terms of Z and x, find its actual slope in the atmosphere at $x = 3500$, $x = 4500$, and $x = 5500$ km.
7. At a given point on a horizontal chart the temperature ascendant is 1° C/100 km and directed to the south; a short distance east it is 4° C/100 km and directed toward 150°. If the gradient is discontinuous, determine graphically the orientation of the discontinuity line.
8. A discontinuity line in an isobaric surface slopes upward to the southwest with magnitude 1/10,000 and coincides with a trough in the surface. Draw the contours of the isobaric surface in the vicinity of this line if, in the region to the northwest, 100-foot contours are 200 km apart and, in the region to the southeast, they are 100 km apart.
9. Find the horizontal temperature gradient corresponding to isothermal slope 1/500 and vertical temperature lapse rate 7.5° C/km.
10. Determine the magnitude of the horizontal temperature gradient if the temperature gradient along the 700-mb pressure surface is 3° C/100 km and lies along a direction in which the pressure surface has a slope 1/8,000.
11. (*a*) Using Table 4.17b, determine Z at $x = 18$ and $x = 28$ by linear interpolation. Compute the departures from the actual values of Z found by the equation for Z.
    (*b*) Apply Bessel's formula with as many terms as possible to determine the interpolated values of Z at these two points. How do these results compare with the actual ones?
12. Find the value of Z for $x = 55$ in Table 4.17b by extrapolation from $x = 50$.

READING REFERENCES

BERRY, F. A., BOLLAY, E., and BEERS, N. R. (eds.). *Handbook of Meteorology*, pp. 160–76, 232–35. New York: McGraw-Hill Book Co., 1945.

BJERKNES, V., HESSELBERG, TH., and DEVIK, O. *Dynamic Meteorology and Hydrography*, Part II: *Kinematics*, pp. 1–8. Washington, D.C.: Carnegie Institution, 1911.

BJERKNES, V., and SANDSTROM, J. W. *Dynamic Meteorology and Hydrography*, Part I: *Hydro-statics*, pp. 19–23, 89–122, and related charts under separate cover. Washington, D.C.: Carnegie Institution, 1910.

PETTERSSEN, SVERRE. *Weather Analysis and Forecasting*, pp. 378–82, 441–59. New York: McGraw-Hill Book Co., 1940.

———. *Upper Air Charts and Analyses*. NAVAER 50-IR-148. Washington, D.C., 1944.

UNITED STATES DEPARTMENT OF THE AIR FORCE. *Preparation of Weather Maps and Charts*. Washington, D.C.: Hdqrs. A.W.S., 1950.

WHITTAKER, E. T., and ROBINSON, G. *A Short Course in Interpolation*. London: Blackie & Son, Ltd., 1923.

WILLERS, F. A. *Practical Analysis (Graphical and Numerical Methods)*, pp. 83–113. Translated from the German edition of 1928 by ROBERT T. BEYER. New York: Dover Publications, 1947.

CHAPTER 5

Graphical Analysis and Computations

Once a scalar pattern is drawn, there may be several more operations to perform from the analysis. Many are done for the most part visually in practice. In order not to skip over those aspects too lightly, we outline here the basic principles and mechanics involved.

This chapter, which is based largely on the discussion by V. Bjerknes, surveys the graphical techniques for obtaining secondary analyses and fields of space and time derivatives from given scalar patterns.[1] Familiarity with many of these will be assumed in succeeding discussions. Practice with each is necessary to understand them.

5.01. *Introduction to graphical analysis.*— From observed variables (p, T, r) it is possible to evaluate any number of the secondary variables $(\theta, \rho, T^*, \theta_E, U,$ etc.) for any observation point by use of formulas, special tables, or a thermodynamic chart. Scalar analysis of such a quantity then can be done by the same operations used for the observed variables, but that procedure is not always the most logical or the most accurate for deriving the analysis of the desired secondary variable.

Implicitly, if point values of a secondary variable can be obtained from measured p, T, and r, it should be possible to derive its pattern from previously determined patterns of p, T, and r without need for point computations or for duplicating the process

of scalar analysis. Merely as an example, assume that the analysis of saturation vapor pressure e_s is desired. One order of operations consists of computing e_s at each point from reported temperature,[2] plotting these values at the points, and drawing the pattern of e_s. But this is time-wasting if the temperature pattern has been drawn, and it may lead to inconsistencies in the patterns of T and e_s. Furthermore, any correction applied to point values during analysis of T must be applied in analysis of e_s. The more logical approach when the T analysis is given is to label each isotherm in the corresponding value of e_s. If additional lines are needed, these are located by interpolation. In this procedure of graphical analysis time and labor are saved, the result is consistent with and as accurate as the analysis on which it is based, and the method is objective and mechanical. In more complicated graphical analyses where two or more patterns are involved, it is useful to compute a few point values of the desired variable by which to guide the drawing of lines.

This objective method can be adapted to a variety of uses for obtaining fields of secondary variables and time-and-space derivatives. In each case the technique involves either converting values of lines or connecting certain intersections of scalar lines for the quantity in question. The process is addition, subtraction, and multiplication (or division) and is applicable to any space chart. When more than one pattern is employed in the process, use of a

1. See V. Bjerknes, Th. Hesselberg, and O. Devik, *Dynamic Meteorology and Hydrography*, Part II: *Kinematics* (Washington, D.C., Carnegie Institution, 1911), Chap. 8.

2. Assuming e_s is a function only of temperature.

light-table or transparent paper is required if all basic patterns are not on the same chart.

5.02. *Graphical addition.*—A simple type of graphical addition is indicated by

$$\Sigma = Q + \text{const.} \qquad (1)$$

An example is conversion of temperature in degrees centigrade to degrees Kelvin; 273 is added to the value of each isotherm. If isotherms are desired for integers other

intervals desired in the sum, integer values of the sum occur at each intersection of Q_1 and Q_2 lines. At point A in Figure 5.02a the sum of Q_1 and Q_2 is 16. Point B, diagonally opposite from A, also has $Q_1 + Q_2 = 16$. Line 16 for the sum therefore is drawn through these two points and extends in both directions through intersections beyond A and B. All other lines are obtained similarly. Where lines for Q_1 and Q_2 are mutually parallel, lines of the sum are parallel to them.

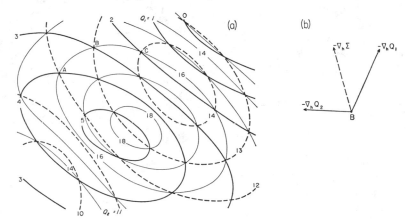

FIGS. 5.02*ab.*—Graphical addition of the patterns of Q_1 and Q_2

than so obtained, they are located by interpolation and drawn parallel to the initial pattern. In the process the gradient field is unchanged (seen by differentiating Eq [1]), and the two patterns are similar.

Graphical addition of two sets of lines is shown by referring to Figure 5.02a; given are the patterns of Q_1 (heavy continuous lines) and Q_2 (dashed lines). Q_1 and Q_2 may be the same variable on two different surfaces, the same variable on the same surface for different times, or two variables for the same surface and time. In any case the process is represented by

$$\Sigma = Q_1 + Q_2, \qquad (2)$$

and the objective is the pattern of Σ. If the lines for Q_1 and Q_2 are drawn for the same

Horizontal differentiation of Eq (2) gives

$$\nabla_h \Sigma = \nabla_h Q_1 + \nabla_h Q_2 .$$

Thus, the vector gradient for the sum of two patterns is the vector sum of the gradients for each of the two given patterns. Figure 5.02b shows the relation between the three descendant vectors for point B in Figure 5.02a.

It is possible to add three or more patterns in this way, only addition is done in two or more steps. If a third pattern, Q_3, is involved, the associative law of addition is used: $\Sigma = (Q_1 + Q_2) + Q_3$. In a first step Q_1 and Q_2 are added to give a third pattern, and adding the result to the pattern of Q_3 gives the desired sum.

In most uses of this technique the *aver-*

age pattern of a quantity is desired. The values of the sum lines then are divided by the number of patterns added. Thus in Figure 5.02*a* the average pattern for Q_1 and Q_2 is the third set of lines, but the average values are 7, 7.5, 8, 8.5, and 9 instead of 14, 15, 16, 17, and 18. The superfluous lines for fractional values of the average can be omitted initially.

Besides its use in deriving average patterns, graphical addition is used to obtain the fields of some secondary variables. Analysis of virtual temperature is one example, although it involves also division. If isotherms are drawn each 1° C and isopleths of moisture each 6‰, virtual isotherms for each 1° C are obtained by drawing through intersections in the two patterns. A similar procedure can be followed for the approximate pattern and values of θ_E from patterns of θ and r.

5.03. *Graphical subtraction.*—Subtracting

two sets of scalar lines is much like graphical addition, but it has many more uses in analysis because of its connection with space-and-time differentials. Through graphical subtraction are obtained the so-called change or *tendency charts* and *vertical differential charts* of varied sorts and uses.

In subtracting two patterns, a set of diagonal curves is drawn through points of intersection, like in addition; but now the opposite diagonals are used. In Figure 5.02*a* connecting A and B gives a sum line; connecting the other two vertices of that quadrilateral gives a line of constant difference between Q_1 and Q_2:

$$\Delta = Q_2 - Q_1 \qquad (\text{or } \Delta = Q_1 - Q_2) . \quad (1)$$

Figure 5.03 shows the pattern of $(Q_2 - Q_1)$ for the same patterns of Q_1 and Q_2 in Figure 5.02*a*. The pattern of $(Q_1 - Q_2)$ is the same as $(Q_2 - Q_1)$ except in sign. Finally, by subtracting two scalar patterns graphically, their gradient patterns are subtracted vectorially.

If analyses of a certain variable Q are given for the same surface for two synoptic times, their graphical difference yields directly the field of local change ΔQ integrated over the time interval Δt. As Δt is constant for the entire chart, the differential lines ΔQ are also lines for $\Delta Q/\Delta t$, which is assumed for many purposes the local tendency of Q. Particularly in the case of pressure, the

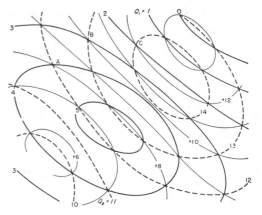

Fig. 5.03.—Graphical subtraction of the patterns of Q_1 and Q_2.

tendency chart is a basic tool in forecasting the fields of motion and weather. The ease with which the tendency field is derived by subtracting two successive analyses suggests that this graphical technique can be more efficient than drawing for point values of pressure difference. Graphical analysis of pressure change over periods 6, 12, or 24 hours is the desirable method, especially where analyzed pressure patterns give more reasonable pressures than do reports themselves. Also, the graphical method yields a tendency pattern over any region for which pressure patterns have been drawn, even where observed changes are so few as to preclude any reliable scalar analysis of them.

Tendency charts may be drawn graphically for any variable analyzed systematically. They are definite aids in weather analysis and forecasting and in extrapolating the latest scalar analysis where data are few and far between. Here the process applies in reverse: $Q_2 = Q_1 + \Delta$. From series of tendency charts in the past it is possible to extrapolate the next tendency pattern, at least approximately, based on the few scattered point differences available. Adding the extrapolated tendency pattern to the last analysis gives the desired analysis. This procedure provides an objective check on the analysis of remote areas of the chart, and meantime it is a useful introduction to prognosis.

Space differential analysis from one surface to the next in the vertical is useful because of its relation to hydrostatics. For example, differential analysis of pressure at two levels gives the mean isopycnics (lines of constant mass) for the layer, since $\bar{\rho} = \Delta p / 980 \, \Delta Z$ from the hydrostatic equation.

Differential analysis of temperature on two-level charts gives isopleths of $\Delta T/\Delta Z$; that is, lines of constant average temperature lapse rate for the layer (also approximate mean stability). Similarly, differential analysis of θ gives isopleths of $\Delta \theta/\Delta Z$; and, since horizontal variation of $1/\theta$ is small, these are almost exactly lines of average stability in the layer. Differential analysis of θ_E from two-level charts yields lines of $\Delta \theta_E/\Delta Z$; that is, average potential stability.

With respect to isobaric surfaces there are two principal types of differential analysis to be mentioned briefly now and discussed in detail later. First, the pattern of local height change of an isobaric surface is obtained by differential analysis of contour patterns on successive charts in time. Second, differential analysis of the topographies of two different isobaric surfaces for the same time gives the pattern of thickness for the layer; the differential lines thus obtained are *thickness lines* or *relative contours* for the layer. From Eq 3.05(2), $\bar{T}^* = \Delta Z/(K \ln p_1/p_2)$. Since they have the property $\Delta Z = $ constant, *relative contours for a given pressure layer are mean virtual iso-*

therms for the layer. If relative contours for the (700/1000)-mb layer are drawn for each 100 gpft, then by Table 3.05 these isotherms are spaced 2.9° C (Figs. 1.021). The 10,000-gpft relative contour for the layer is the mean isotherm for 292° K or 19° C.

Differential analysis of T, θ, or θ_E from synoptic isobaric surfaces gives approximately the same indications of stability as derived from differential analysis of level charts, but with slightly less accuracy.

From the definition of static stability one deduces that the stability of a layer between two isentropic surfaces varies inversely with its depth. Therefore, relative contours found by differential analysis of height on adjacent isentropic surfaces are lines of stability. It is then evident, by considering Eq 3.21(2), that the thickness tendency of such a layer will be related to lateral convergence of mass. Differential analysis of pressure from two adjacent isentropic surfaces gives the field of mass, and another step consisting of differential analysis in time gives the local accumulation of mass within that layer.

Differential analysis of the patterns of two different variables on the same surface to obtain the pattern of a third variable has limited application. The pattern of temperature-dewpoint difference can be determined quickly on any chart to give a crude but useful approximation to the pattern of relative humidity. For many purposes the pattern of $(T - T_s)$ suffices, but, if greater accuracy is desired, these differentials can be converted into the more exact pattern of relative humidity by considering the variation of temperature along the differential lines (Appendix Table J).

5.04. *Graphical multiplication and division.*—Multiplying (or dividing) in graphical analysis occurs in converting the analysis of any variable from one system of units to another, such as inches of mercury to millibars, Fahrenheit to centigrade degrees, meters to feet, and so on. To derive the second pattern the lines are labeled in the desired unit; other lines are drawn by interpolation. In each case the work is expedited if reference is made to a simple conversion table or nomogram for the particular variable. Tabular or graphical aids are necessary for most operations in multiplication or division.

Given the patterns of p and T in any plane, the pattern of θ can be obtained graphically with use of tables (Appendix Tables L and M), although a thermody-

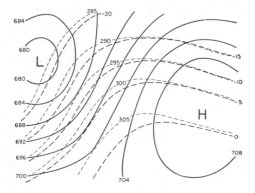

FIG. 5.04.—Potential temperature analysis from 10,000-foot patterns of pressure and temperature.

namic chart serves this purpose. As $\theta = 1000^{\kappa} T p^{-\kappa}$, Table M is constructed with the product (θ) and one independent variable (T) as arguments, and the remaining variable (p) is the tabulated quantity. That table is useful particularly in drawing cross sections.

Since on an isobaric surface θ is a function only of T, each isotherm is also an isentrope. On an isentropic surface isobars are also isotherms. Again, the analyses can be interchanged by using conversion tables and interpolation.

The pattern of saturation mixing ratio r_s can be determined from analyses of pressure and temperature (Appendix Table K). On an isobaric surface r_s is obtained direct-

ly from T. On an isentropic surface, r_s is given by p (or T). In each of the above, substitution of dewpoint temperature for actual temperature gives r instead of r_s.

To obtain the analysis of ρ or α, the patterns of p and T^* are used. If we neglect $T^* - T$, then at constant pressure the isopycnics and isosteres are derived from the analyzed temperature pattern only, and at constant potential temperature they are derived from isobars or isotherms.

Use of graphical multiplication with two different charts is quite restricted. Only one example is mentioned—analysis of $\bar{T}^*$ graphically from pressure patterns at two levels. As $\bar{T}^* = \Delta Z/(K \ln p_1/p_2)$, a line of constant p_1/p_2 is a mean virtual isotherm for the layer. If in particular the two levels are sea level and 10,000 feet, p_1/p_2 is roughly 3/2. With isobar intervals 3 mb at sea level and 2 mb at 10,000 feet, mean virtual isotherms are obtained approximately by graphical subtraction of the two patterns.

5.05. *Graphical differentiation from scalar patterns.*—For many purposes in analysis, particularly in connection with the wind field, it may be necessary to draw fields of space differentials (space derivatives). In some instances the magnitude of the total plane gradient, $G = \partial Q/\partial n$, is desired; at other times one may be concerned only with the component of gradient along a coordinate axis. In all cases use is made of finite differences in the quantity and in distance.

a) Analysis of the plane gradient field.— On horizontal or topographic charts the gradient can be analyzed by either of two general methods. First, hold δQ (which may be the scalar line interval) constant for the entire operation, so that the spacing of lines δn determines the value of gradient; or select a certain constant δn and thus determine gradients from measured values of δQ within the pattern.

(i) *Constant* δQ.—In this case $\delta n =$ constant/G. A scale is prepared on the edge of a sheet of paper, or on transparent material, which gives distance δn on the chart corresponding to certain desired values of the gradient. The scale shown in Figure 5.05a was prepared for use with Figure 5.05b and is valid only for a single-line interval ($\delta Q = 10$) and for a certain base chart.

are found similarly. The isopleths of gradient are then smoothed and perhaps adjusted slightly to a degree consistent with the accuracy of the initial pattern.

(ii) *Constant* δn.—This alternate method involves reference to a table, set up for a selected value of δn, which gives δQ as a function of desired integer values of G. To

Fig. 5.05a.—Example of a scale with constant δQ for locating the positions of desired values of the gradient.

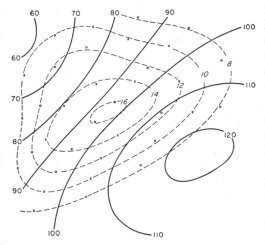

Fig. 5.05b.—Gradient analysis from a scalar pattern.

With this scale it is preferable to locate lines of constant G individually instead of first computing values at random points over the chart. Assume the line $G = 2$ is desired. The scale, held normal to the Q lines and with the origin (left end) on one Q line, is moved over the chart until scale value $G = 2$ coincides with the adjacent line δQ away. A mark is made between the two lines. The process is repeated between the adjacent pair of Q lines, and so on, until all points for $G = 2$ are located; line $G = 2$

is drawn through all the points. Other lines locate points for a certain value of G, first obtain the corresponding value of δQ from the table and then move the segment δn over the pattern until the segment coincides with that δQ.

Analysis of plane gradients by either method is restricted to synoptic maps; on cross sections, coordinate scales differ. However, caution may be necessary even with maps, since δn on the chart varies with map scale. If analysis is performed over a wide range of latitude, then it is preferable to use the second method (constant δn), perhaps defining δn in degrees-of-latitude units and using dividers for setting the value of δn in certain ranges of latitude.

Regardless of which method is used, some judgment must be exercised in measuring gradients and in drawing their patterns, and the technique should be modified to the particular condition. Where the scalar field is weak, smaller values of δQ or δn should be used. Where gradients vary rapidly, care is taken in assuring that a representative value of gradient is being assigned to a geographic point. In many cases it is advisable to shift the located point slightly as adjustment for variation of gradient between the Q lines. Measurements of gradient across discontinuity lines should be avoided, where isopleths of gradient are discontinuous, but measurements can be made to each side.

After some practice with gradient analysis, it becomes obvious that the prior scalar analysis must be done with great care. In fact, gradient analysis promotes accurate scalar analysis.

b) Analysis of the gradient component along a line.—For measuring a vertical gradient along a sounding, determine the vertical change of the quantity between successive levels and divide by the distance between. The value thus obtained is assigned to the midpoint of the layer. Integer values of gradient then are located between the levels by interpolation. The minor discontinuities at significant points are thus smoothed out, but it may be desirable to retain the sharper ones. If the operation is applied to several soundings in a vertical cross section, the pattern of vertical lapse rate for that variable can be drawn.

In vertical cross section $(s, -\log p)$ computations of $(\partial Q/\partial s)_p$ may be required. Either of the methods above is applicable. Work is facilitated if the cross section has constant horizontal scale. Another method for evaluating isobaric gradients in cross section involves use of vertical lines or columns of dots, equidistant geographically, and drawn either on the chart itself or on a transparent overlay. The difference $(\delta Q)_p$ evaluated between each pair of verticals for a desired number of isobars bears everywhere the same ratio to $(\delta Q/\delta s)_p$, and isopleths of isobaric gradient can be drawn from the tabulated values of δQ.

For evaluating south-north and west-east gradients on maps, we may use an analogous method with the latitude-longitude grid. Take 1° of latitude (or convenient fraction or multiple thereof) as the basic finite difference δs in north-south computations. For each meridian in turn compute δQ in each interval of latitude, and

from these values draw the patterns of south-north gradient, $\delta Q/\delta\phi$ or $\delta Q/\delta y$.

For west-east direction one of two methods may be used. Compute δQ between successive meridians along each latitude circle to obtain $\delta Q/\delta\lambda$, and then divide by the cosine of the latitude. A more direct and accurate method requires a prior grid layout on a transparent overlay consisting of a series of points equidistant on the earth along each latitude circle. The longitude difference $\delta\lambda$ between each point is $\delta\lambda = \delta x/k \cos \phi$, where k is 60 nautical miles. Differences δQ thus computed along each latitude circle are all consistent, and the pattern of $\delta Q/\delta x$ then can be drawn. Since in many cases comparison between south-north and west-east gradients is desired, it is advisable in setting up the grid to have $\delta x = \delta y$.

5.06. *Evaluating space derivatives by finite differences between chosen points.*—In some of the work in weather analysis and forecasting it may be necessary to employ numerical techniques for solving space differential equations of first and second order. These are the first derivatives $\partial Q/\partial x$ and $\partial Q/\partial y$, which were mentioned in the preceding section; second derivatives $\partial^2 Q/\partial x^2$ and $\partial^2 Q/\partial y^2$ and their sum, the horizontal Laplacian of Q, $\nabla_h^2 Q$; and also the cross derivative $\partial^2 Q/\partial x\partial y$.

In this type of evaluation from a given scalar analysis, we again substitute a finite differential for the partial (or total) derivative. For example, $(\delta Q/\delta x) \simeq (\partial Q/\partial x)$, $(\delta/\delta x)(\delta Q/\delta x) \simeq (\partial^2 Q/\partial x^2)$, $(\delta/\delta x)(\delta Q/\delta y) \simeq (\partial^2 Q/\partial x\partial y)$, etc. From elementary calculus one recalls that these approximations are true only as the distances δx and δy approach zero and only if the derivatives are continuous over the distance considered. Compliance with the second limitation usually is not difficult, since continuity of the function can be determined by visual inspection of its pattern. However, for convenience and accuracy, it is necessary to employ a distance (on the earth) δx or δy of *practical length*. To discuss all details of this problem would lead too far into theory of approximation.

We are aware of many sources of error in a completed analysis, owing both to ob-

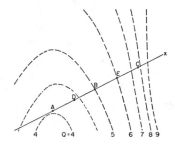

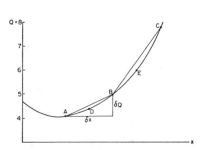

Fɪɢ. 5.06*a*

quantity along x at point D. We locate two points, A and B, equidistant from D which best satisfy requirements for distance.[3] Then, after subtracting Q_A from Q_B, divide by the distance δx from A to B to obtain $\delta Q/\delta x$, the approximation to $\partial Q/\partial x$ at D. If δx is denoted by L, Q_A by $Q_{-L/2}$, and Q_B by $Q_{+L/2}$, then

$$(\partial Q/\partial x)_0 \simeq (\delta Q/\delta x)_0$$
$$= (Q_{+L/2} - Q_{-L/2})/L . \quad (1)$$

In the lower part of Figure 5.06*a* is the corresponding Q, x diagram, in which the distribution of Q along x is shown by the curved profile. The slope of the straight

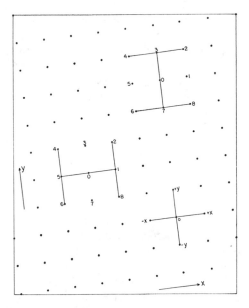

Fɪɢ. 5.06*b*

line connecting points A and B on the profile is $\delta Q/\delta x$; that is, the average variation of Q between the points. The quantity $\delta Q/\delta x$ we evaluated is $\partial Q/\partial x$ at D if this

servation and to mechanics and assumptions in drawing. We also know that this analysis contains many distracting small-scale disturbances which are real but nevertheless of little significance to the objective in mind. Choice of the suitable distance is an attempt to eliminate those two features. As a first criterion, *the distance used is large enough to minimize the errors inherent in analysis and to separate the erratic features of the pattern from the important ones.*

Now suppose we are given the pattern of Q in the upper portion of Figure 5.06*a* and wish to determine the variation of the

3. The points A and B do not have to be equidistant from the point through which the derivative is measured. In fact, where the *gradient* varies monotonically from A to B, more exact results are had if segment AB is shifted through D by an amount to compensate for variation of gradient.

straight-line segment parallels the tangent to the curve at D. In some cases $\delta Q/\delta x$ evaluated over a great distance is representative of $\partial Q/\partial x$ at the midpoint; in others, they are approximately the same only if δx is very small. The greatest length δx that may be used depends inversely on the curvature of the Q profile and therefore also on the nature of Q. To the first criterion for length units we may add: *The distance is small enough to insure that the average variation over the interval is representative of the variation in the vicinity of the point in question.* The best choice for δx is a compromise between the two.

Now consider measurement of a second derivative. If it is to be evaluated at point B in Figure 5.06a, the process becomes $(\partial^2 Q/\partial x^2)_B \simeq [(\delta Q/\delta x)_E - (\delta Q/\delta x)_D]/\delta x$. But, since $(\delta Q/\delta x)_E = (Q_C - Q_B)/\delta x$ and $(\delta Q/\delta x)_D = (Q_B - Q_A)/\delta x$, by substitution $(\partial^2 Q/\partial x^2)_B \simeq (Q_C - 2Q_B + Q_A)/(\delta x)^2$, or, in more general symbolic form,

$$(\partial^2 Q/\partial x^2)_0$$

$$\simeq (Q_L - 2Q_0 + Q_{-L})/L^2 . \quad (2)$$

This formula is valid for the second derivative along any axis.

The term $\partial^2 Q/\partial x^2$ is essentially the curvature of the Q profile along x. By Eq (2) this is determined from values of Q at three points along x. The truth of the approximation thus depends on how well these three values determine the curvature at the central point. It is apparent that, the smaller δx is, the greater the likelihood of this approximation. Therefore, for evaluating a second derivative another criterion for length is: *The distance must be small enough to insure that three values of Q give a representative curvature for its profile at the point in question.* The length L $(= \delta x)$ can be different from that used in computing a first derivative, and it is frequently taken smaller.

For evaluating all types and combinations of derivatives over a region of the chart it is convenient to use a rectangular grid illustrated by Figure 5.06b. This grid can be fixed geographically on the working chart, or, if printed on a transparent overlay, it can be adjusted to desired orientation about the point. To suit most types of computations, the unit L (distance between adjacent points along both coordinate axes) will have to be predetermined in accord with the principles above. Since over the entire grid L corresponds to constant distances *on the map*, for determining the necessary earth distance one might have to consider the map scale at the point for which the computation is made.

The method of evaluating first derivatives is evident. For obtaining $\partial^2 Q/\partial x^2$ and $\partial^2 Q/\partial y^2$, as well as their sum, refer to the rectangular cross in the lower right of Figure 5.06b. The term $\partial^2 Q/\partial x^2$ is approximated by $(Q_{+x} - 2Q_0 + Q_{-x})/L^2$, and $\partial^2 Q/\partial y^2$ by $(Q_{+y} - 2Q_0 + Q_{-y})/L^2$. The horizontal Laplacian is then

$$\nabla_h^2 Q = \partial^2 Q/\partial x^2 + \partial^2 Q/\partial y^2 \simeq (Q_{+x}$$

$$+ Q_{-x} + Q_{+y} + Q_{-y} - 4Q_0)/L^2 . \quad (3)$$

Less frequently the derivative $\partial^2 Q/\partial x \partial y$ may be required. With reference to the scheme in the left side of Figure 5.06b,

$$(\delta/\delta x)(\delta Q/\delta y) = [(\delta Q/\delta y)_1 - (\delta Q/\delta y)_5]/\delta x$$

$$= [(Q_2 - Q_8) - (Q_4 - Q_6)]/\delta x \delta y ,$$

or

$$\partial^2 Q/\partial x \partial y \simeq (Q_2 - Q_4 + Q_6 - Q_8)/4L^2 , \quad (4)$$

if L is still the distance between adjacent grid points. The same formula results by using the form $(\delta/\delta y)(\delta Q/\delta x)$ and referring to the scheme in the upper right of Figure 5.06b. The inconsistency of using $2L$ here compared to L in other operations is removed by taking a grid twice as dense for

$\partial^2 Q/\partial x \partial y$; the reference point again will be a grid point. However, if reference points are permitted in the center of each square, then $\partial^2 Q/\partial x \partial y \simeq (Q_1 - Q_0 + Q_7 - Q_8)/L^2$ at the center of square 1, 0, 7, 8 in Figure 5.06b.

5.07. Comments on local tendencies.—For all current meteorological applications of the local tendency $\partial Q/\partial t$ of any quantity Q, it is evaluated by finite differences $\delta Q/\delta t$. The interval δt is taken relatively large in most cases—3, 6, 12, or more hours. For many

by so doing, we depart from the true meaning of a tendency, which then is more appropriately a trend.

Use of long periods for δt may be beneficial particularly in the case of temperature near the ground. A net rise of 8° F between 1000 and 1300 local time has little significance because that is likely to happen daily. A net rise of 8° F between 1000 on one day and the same time on the next carries more meaning, even though the period is eight times as long. The distracting effect here is the diurnal temperature variation,

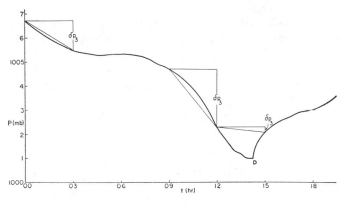

FIG. 5.07

atmospheric variables it can be evaluated at the surface over almost any time interval desired from recordings and hourly measurements. Above the surface the shortest possible time intervals correspond to the frequency of aerological observations.

The length of the time interval to be used is governed not only by the frequency of observations but also by the character of the element and the purpose. For obtaining a measure of *instantaneous* pressure tendency, for example, as short a period as possible should be used. Naturally, the undesirable short-period fluctuations influence and sometimes control the instantaneous tendency. By taking successively longer intervals for δt, more and more of the extraneous influences are removed. However,

which is large in comparison with inter-diurnal temperature variation (in continental areas outside high latitudes near the solstices). But the significance of a short-period temperature change is increased if the normal variation at that time of day is subtracted. The same is true for other elements with regular periodic variation. For most areas of the globe it is desirable to use periods shorter than 24 hours for δt and subtract the normal change for that period, for the reason that the "synoptically important" variations have periods of from one to just a few days.

To avoid possible misguidance by reported or computed tendencies, let us examine in more detail the nature of the tendencies obtained over finite time intervals.

Pressure is used as an example, but the reasoning is applicable to other variables. Consider a time unit 3 hours. This tendency is an average for the period, and in the usual condition it should be most representative of the tendency $1\frac{1}{2}$ hours before the observation. Suppose the t, p profile in Figure 5.07 represents an idealized station barogram from which most small fluctuations have been smoothed (partly by the instrument). The 3-hour change computed at time 0300 gives a good indication of average tendency and also of present tendency. The one computed at 1200 also gives a good indication of tendencies at the middle and end of the period but is less representative than earlier because the pressure tendency (slope of barogram) is changing with time (barogram curved).

The sharp change in the barogram between 1200 and 1500 is of special interest in view of the misleading information it can give for 3-hour tendency, and also most emphasis is given to pressure tendencies in just such a region. The slope of the barogram is discontinuous at D, and tendency cannot be defined there. Owing to sharp reversal of tendency and almost complete compensation of the pressure fall through most of the period by large rises near the end, the 3-hour change is small, and at best $\delta p/\delta t$ is a poor indicator of prevailing tendency. The important thing, however, is that the net pressure change is of sign opposite to the tendency at the end of the period. Over such a portion of the barogram the 3-hour change should not be interpreted as instantaneous tendency; it is useful only as a general trend. Areas of such discontinuous tendency can be located on the surface chart by the barometric symbols, in particular $\vee$, $\smile$, and $\diagdown$.

READING REFERENCE

Bjerknes, V., Hesselberg, Th., and Devik, O. *Dynamic Meteorology and Hydrography*, Part II: *Kinematics*, Chap. 8. Washington, D.C.: Carnegie Institution, 1911.

CHAPTER 6

Cross-Section Analysis

6.01. *Introduction.*—Although vertical cross sections are used little in routine work, there are several reasons for introducing analysis by that means, which is just one of the possible choices.[1] A primary reason is that, in preparing and using the common forms of analysis, we attempt to visualize patterns in three dimensions. Second, the simplicity of vertical atmospheric structure, as compared to horizontal, makes procedure in analysis more objective. Next, many important features found in the horizontal are more clearly identified and understood in vertical perspective. And, finally, cross sections are the logical step after studying vertical soundings in detail.

This description of cross sections is largely for introducing concepts basic to routine forms of analysis. For that reason it might appear unduly long. Further, it is realized that the full significance of some of the sections, particularly those on pressure and wind analysis, are not grasped entirely by restriction to a vertical plane. Therefore, those topics might be studied with related ones in the next chapter.

The cross-section chart used for most of the illustrations has for abscissa horizontal distance s and ordinate $-\log p$, and the average vertical scale is either 125 or 250 times the horizontal scale. Each was set up with constant horizontal scale. The layout of aerological networks (Fig. 2.20) restricts the choice of cross sections and gives difficulty in laying them out appropriately, unless the required data can be taken from

1. The value of a tool in learning cannot be judged solely by its use in practice.

analyses of a large number of synoptic horizontal charts.

6.02. *Identification of stable layers.*—An early step in thermal analysis is to distinguish the stable or inversion layers in each sounding. A stable layer is identified by small lapse rate of temperature or by rapid vertical increase of potential temperature. Since regulations for radiosonde observation require that data be given at significant changes in temperature lapse rate, *upper and lower boundaries of stable layers coincide with reported significant points* with few exceptions.

Stable layers of the free troposphere extend nearly along isentropic surfaces, as seen in Figure 6.02b. (Soundings for all stations appear in Fig. 6.02a.) The pT curve for Charleston shows a distinct concentration of stability between 315 θ and 320 θ. At Tampa this occurs between 316 θ and 321 θ; at Miami, between 318 θ and 321 θ. It is reasonable to assume that the same stable layer intersects all three soundings, and this layer was entered continuously. Another example is the layer between 294 θ and 299 θ at Miami, 293 θ and 297 θ at Tampa, and 297 θ and 300 θ at Charleston. There is good reason for entering this layer continuously in the cross section in spite of the higher values at Charleston, which may have resulted from warming by lifting and condensation (the layer is saturated). It is not always possible to extend stable layers over such large distances, as there are many localized processes to form or destroy them. Moreover, since reports limit the number of

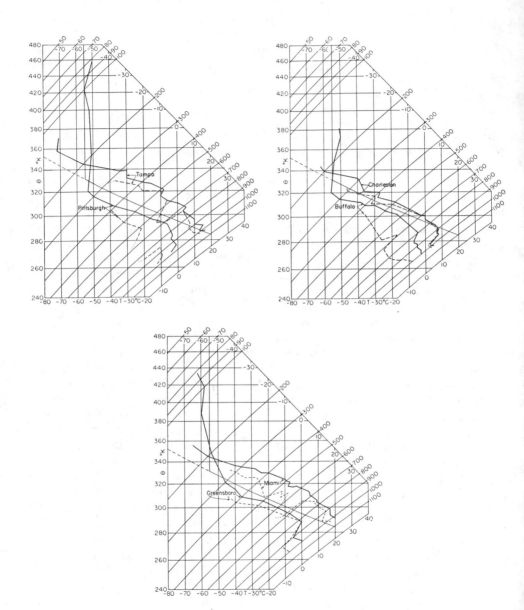

FIG. 6.02a.—Temperature and dewpoint soundings for 1500 GCT, 25 December 1947. (Fine continuous line is the U.S. Standard Atmosphere curve; fine dashed line is the extension of this curve upward from 234 mb with temperature lapse rate 6.5° C km⁻¹.)

significant points, there are many cases in which reports either minimize or accentuate contrasts in stability.

Near the ground, stability is principally a result of radiative and advective processes in the surface layer, with large diurnal and

in a way proportional to its height (Figs. 6.02*b* and 6.13*de*). The arctic tropopause has potential temperatures roughly 300° K, and the tropical tropopause roughly 400° K.

In the regions where the tropopause is steeply inclined, it is often difficult to as-

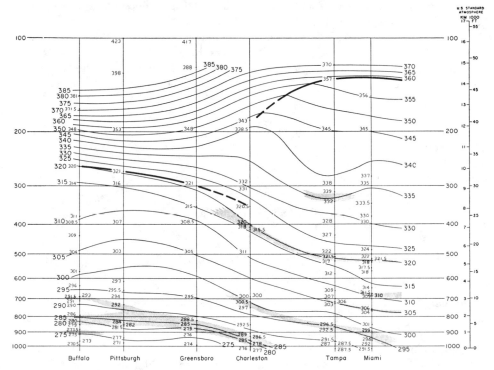

Fig. 6.02*b*.—Cross section of potential temperature (° K) for 1500 GCT, 25 December 1947. Heavy lines are the tropopause, and stippled areas represent stable layers. (Figures 6.06, 6.08, 6.09, 6.10, 6.11, and 6.13*abc* are for the same time. These cross sections have vertical scale 125 times the horizontal scale.)

local differences. Here there is no such stratification of stability along isentropic sheets.

6.03. *The tropopause.*—In many cases there is no difficulty in detecting the tropopause from the *change in lapse rate* along each sounding. The tropopause also tends to follow isentropic surfaces, at least in a general way over large areas (Fig. 6.02*b*). Where the tropopause is nearly flat, isentropic surfaces most nearly parallel it. Potential temperature at the tropopause is

sociate the tropopause with any one point in a sounding. Instead of the usual sharp transition to stratospheric stability, the tropopause may appear multiple in character, or it may not be distinguishable at all; and the troposphere merges more or less gradually with the stratosphere. Greensboro in Figure 6.02*b* is where the tropopause would have greatest slope if drawn as a continuous boundary. The Greensboro *pT* curve in Figure 6.02*a* shows several minor changes in stability upward from 400 mb, but some of these may be due only to reporting temperatures at mandatory levels and the last point of moisture measurement (near $-40°$ C). The true sounding curve might have been rounded smoothly throughout this region. In contrast, the tropopause is distinct at Buffalo, Pittsburgh, and Tampa. It is generally best defined in its highest and lowest regions and can be very diffuse where it would have the greatest slope. However, there are cases when the tropopause is difficult to define over any large extent of a cross section.

6.04. Identification of superadiabatic layers.

—It may be helpful to locate the rarely existing superadiabatic layers. In those the isentropic surfaces are inverted. Superadiabatic layers can be identified between two points in a sounding by drop of potential temperature with height or by excessive lapse rate of temperature.

6.05. Analysis of potential temperature.

—The potential temperature for any level in a sounding can be found from a thermodynamic chart or special tables, given pressure and temperature. Integer values of θ can be located along each sounding on the chart by interpolation between reported levels, but a useful procedure is to have those integer values located from the pT curve on the thermodynamic chart and entered at the proper pressures while plotting the cross section. This is equally applicable to other desired quantities, and scalar analysis then is a process of connecting the points by curves between adjacent verticals. To locate the points of intersection of isentropes with the ground, reference can be made to the surface chart.

Outside the region of indistinct or multiple tropopause, isentropes in the lower stratosphere are closely spaced, rather flat, and also quite uniform, and similarly in tropospheric stable layers (Eq 4.13[12]). Through most of the troposphere isen-

tropes are more irregular, but in broad sense they take the same general shape. Notice in Figure 6.02b that most isentropes in the troposphere take part in general north-south sweep from ridge to inflection to trough and that, except near the ground and tropopause, the isentropes are quite similar in shape.

Although boundaries of individual stable layers tend to be isentropic, some isentropic surfaces do intersect those boundaries especially when inclined. If the boundaries

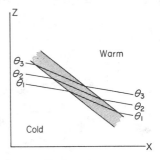

Fig. 6.05.—Cross-section pattern of potential temperature through an inclined stable layer.

are assumed discontinuities in stability, then isentropic surfaces normally are kinked along them as in Figure 6.05—pointed downward at the warm boundary and upward at the cold boundary. This pattern follows from Eq 4.13(12).

Notice in Figure 6.05 that $\partial\theta/\partial x$ is of the same sign everywhere but a maximum within the inclined stable layer (transition zone) and discontinuous at its boundaries. As a rule, *inclined stable layers or zones of transition in the free troposphere are concentrations of horizontal potential temperature gradient but not a reversal in direction of that gradient.* It follows that neither within nor at boundaries of such zones would we expect to find ridges or troughs in isentropic surfaces.

Now consider the tropopause, which is also the boundary of a stable layer. A significant difference is that in the region of appreciable slope in the tropopause there is also *reversal in direction* of horizontal θ gradient, with lowest θ at the tropopause. This gives a ridge in isentropic surfaces, and, where the tropopause is both distinct and inclined, the isentropes kink or bend rather sharply.

6.06. *Analysis of temperature.*—Procedure in temperature analysis is similar in many respects to potential temperature. Where temperature distribution is quite uniform, a sufficiently accurate analysis can be obtained merely by connecting interpolated points on adjacent soundings. That

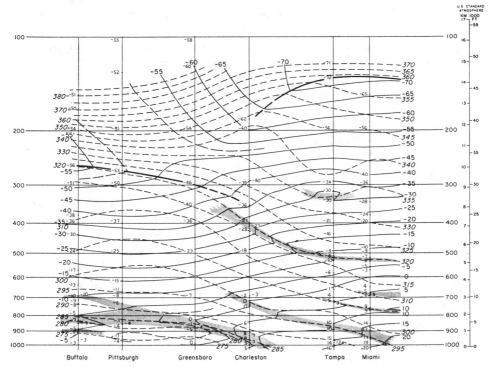

FIG. 6.06.—Cross section of temperature (°C) (*continuous lines*) and potential temperature (°K) (*dashed lines*) for same time as figure 6.02b.

In general, isentropic surfaces of the lower stratosphere are in phase with the shape of the tropopause—depressions in isentropic surfaces over depressions in the tropopause, and domes in isentropic surfaces over domes in the tropopause—but they have less amplitude than the tropopause. The phase relation between the tropopause and isentropic surfaces in the troposphere is different; depressions in isentropic surfaces are found beneath high tropopause and domes in isentropic surfaces beneath low tropopause.

procedure is most valid in the warm areas of the middle and upper troposphere.

At discontinuities in temperature gradient the isotherms kink (Fig. 6.06 and Eq 4.13[13]). If the zone of transition is an inversion, then isotherms within the layer are inverted relative to those outside. Regardless of magnitude or sign of the lapse rate through a stable layer, the kinks in isotherms point toward lower temperature at the warm boundary and toward higher temperature at the cold boundary. Since the vertical variations of θ and T are opposite in sign in the surrounding air masses, while their isobaric gradients are similar, the directions of kinks in isotherms and isentropes are opposite along the same discontinuity.

Especially when the discontinuity in the isotherms occurs at some distance from a sounding, isotherms can and should be determined from the potential temperature analysis. Within stable layers isentropes are likely to be most uniform, and here they can be relied upon to give the desired analysis of temperature which is most complex in this region. For proof of the validity of this procedure, one might draw isotherms first, then drawn isentropes consistent with the isotherms, and finally readjust both until logical patterns are obtained. In the lower stratosphere also, temperature analysis can be based primarily on a previous analysis of potential temperature, as suggested in Section 4.13(*e*).

At inclined tropopauses there is reversal of isobaric θ gradient with minimum of θ. The same is true for T, since isobarically the gradient of T is directly proportional to and in the same direction as that of θ:

$$(\partial T/\partial s)_p = (T/\theta)(\partial \theta/\partial s)_p . \qquad (1)$$

Thus, isotherms dip downward (kink) in crossing the tropopause. On the average, isotherms change almost from isobaric orientation in the troposphere to vertical orientation upon intersecting the tropopause.

To locate intersections of isotherms with the ground, reference is made to the synoptic surface chart. Particular attention can be given to adjusting isotherms in the surface layer of the cross section with zones of transition in the surface chart. These might be moving air-mass contrasts, or they might be maintained by coastlines, mountain barriers, or even edges of cloud shields.

6.07. *Interpretation from analyses of temperature and potential temperature.*—The vertical structure of the atmosphere can be read from a cross section in a manner similar to using thermodynamic charts. The amount of information derived from studying a cross-section analysis is a function of detail in the analysis.

Static stability is inversely proportional to the vertical spacing of isentropes (when $\partial \theta / \partial z > 0$). Adiabatic lapse rates are indicated where isentropes are vertical and superadiabatic lapse rates where θ decreases with height. From the temperature analysis, inversions are shown by inverted isotherms, and, when $\partial T/\partial z < 0$, the degree of stability is inversely proportional to the vertical spacing of isotherms. Superadiabatic lapse rates exist where the vertical spacing for $10°$ C is less than the vertical distance corresponding to 1 km on the attached height scale. This latter relation can be a useful guide in drawing the temperature pattern between soundings. The isobaric T gradient is inferred from the number of isotherms crossing an isobar per unit distance, or from the spacing of isotherms along an isobar, and similarly for other variables.

One should refer to Eq 4.13(14) and apply it to Figure 6.06. If slopes of isotherms and isentropes are taken relative to the isobars, the formula is exact.

Upon careful examination of the patterns for T and θ, one finds significant correlation between the T (or θ) distributions from level to level in the troposphere. Outside local modifications in the pattern by the ground, the general shapes of temperature or potential temperature surfaces remain much the same upward to the tropopause. In fact, with only slight reservation, the axes of thermal systems—troughs, ridges, and large isobaric gradient—in the troposphere tend to be almost vertical. Thermal systems above the tropopause are usually the reverse of those below.

6.08. *Moisture analysis.*—The process in analysis of humidity is considerably different from either temperature or potential temperature. Large gradients and small centers are common between levels and between stations. In Figure 6.08, for example, a complicated structure in moisture is found over Tampa and Miami, but this analysis is only a smoothed account. To locate centers as lar to that of temperature, and it may be opposite. Through sections of stable layers in which relative humidity is nearly constant, shapes of moisture lines are much the same as isotherms. Where relative humidity increases rapidly with height, inversions are more pronounced in moisture than in

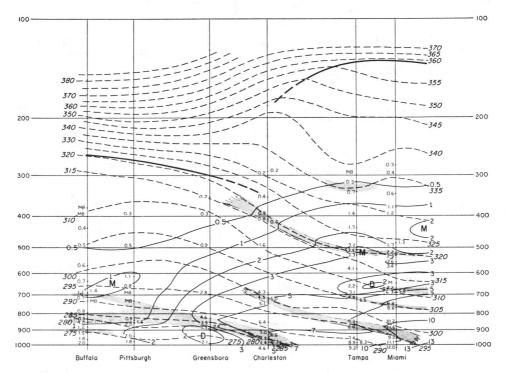

FIG. 6.08.—Cross section of moisture (‰) (*continuous lines*) and potential temperature (° K) (*dashed lines*) for same time as Figure 6.06.

well as approximate patterns between stations, reference was made to the various horizontal analyses of moisture. In sections of the chart cloud observations were used to determine approximate patterns of relative humidity, which then were translated into patterns of moisture.

Observe, on comparing Figures 6.06 and 6.08, the general positive correlation between temperature and moisture patterns and the large gradients of moisture at high temperature. At zones of transition in temperature the moisture pattern may be simi-

temperature. Other stable layers can have rapid decrease of moisture with height (e.g., over Buffalo and Pittsburgh). A significant feature in the moisture distribution through stable layers, as evidenced by Figure 6.08, is the geographic variation of vertical moisture gradient in the layer.

In drawing the analysis of moisture, it is usually advisable to use a varying moisture interval, that is, one in direct relation to the moisture itself.

Analysis of dewpoint temperature is similar to analysis of moisture, and their patterns are much alike. But, since dewpoint temperature is a function of pressure as well as of moisture content, a few differences occur in the vertical. The relation of saturation mixing-ratio lines and isotherms on a

midity may be obtained from the analysis of dewpoint temperature. Where isopleths of temperature and dewpoint are parallel, relative humidity is constant along either line; where the respective isopleths inter-

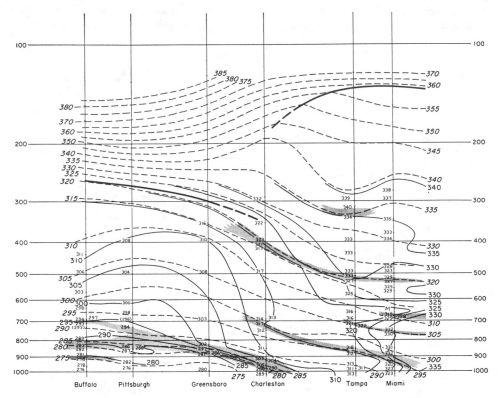

FIG. 6.09.—Cross section of equivalent potential temperature (° K) (*continuous lines*) and potential temperature (° K) (*dashed lines*).

thermodynamic chart is a basis for deducing the differences between patterns of moisture and dewpoint temperature. Where r is constant along an isobar, dewpoint is also constant; where r is constant with height, dewpoint decreases with height; and so on. Moisture inversions appear as weaker inversions in dewpoint temperature. Moreover, dewpoint maxima are displaced to higher pressure from moisture maxima, and dewpoint minima to lower pressure from moisture minima.

Point values or patterns of relative hu-

sect, relative humidity increases in the direction of decreasing dewpoint depression (see Appendix Table J).

6.09. *Analysis of equivalent potential temperature.*—The value of θ_E for each point in a sounding or the location of each desired integer value of θ_E can be approximated graphically from previously drawn patterns of potential temperature and moisture.

In Figure 6.09 isopleths of θ_E were not drawn in the upper region, where they are nearly identical with isentropes (Eq

4.13[15]). In the lower right portion of the chart the moisture and its gradient are both large, and values and pattern of θ_E differ widely from those of θ. Isopleths of θ_E kink at boundaries of stable layers oriented along θ lines if there is a moisture gradient along them. But in many examples θ_E lines are the pressure field, they are discussed briefly here. The virtual temperature pattern (Fig. 6.10) can be sufficiently approximated by graphical analysis, using the patterns of temperature and moisture. Since $(T^* - T)$ is always small, an accurate analysis also can be obtained from a few calculated point val-

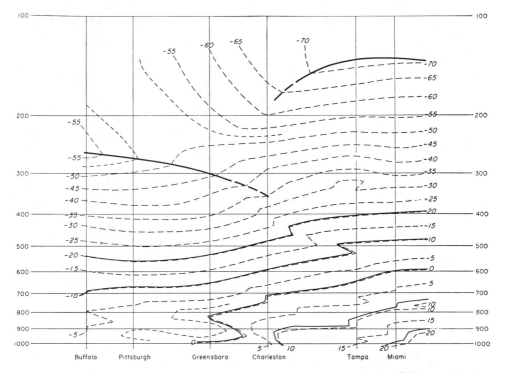

Fig. 6.10.—Cross section of virtual temperature (*continuous lines*) and temperature (° C) (*dashed lines*)

more nearly parallel to stable layers than are θ lines.

The pattern of θ_w is similar to that of θ_E. It is recalled from Figure 2.06 that each value of θ_E corresponds to a unique but smaller value of θ_w. Their surfaces are parallel, but, since 1° K change in θ_E gives less than 1° K change in θ_w, there are fewer θ_w surfaces, evident especially at high moisture.

6.10. *Virtual temperature.*—The patterns of T^*, ρ, and α are seldom drawn, but, because of their close relation to structure of

ues of T^* and using the temperature pattern as a guide. Figure 6.10 shows the difference in gradients of T^* and T is small, so that for most purposes the gradient of one can be substituted for the other.

Where virtual isotherms parallel isobars in cross section, the condition of *barotropy* exists in the plane. For barotropy in all three dimensions the same condition must be fulfilled in the vertical plane normal to the cross section considered, in which case surfaces of virtual temperature parallel isobaric surfaces in space. *Baroclinity* is the absence of barotropy.

6.11. *Density and specific volume.*—By differentiating the equation of state at constant pressure, $(\partial\alpha/\partial s)_p = (R_d/p)(\partial T^*/\partial s)_p$. Thus, isosteres (isopycnics) are steeply inclined to isobaric surfaces where the isobaric temperature gradient is large (Fig. 6.11). Barotropy exists where isosteres at the surface of the earth. These parallelepipeds are *isosteric-isobaric solenoids* or, more briefly, *solenoids*. The parallelograms seen in vertical cross sections are intersections of solenoid tubes at varying angles with the plane of the cross section.

Consider a plane defined along the re-

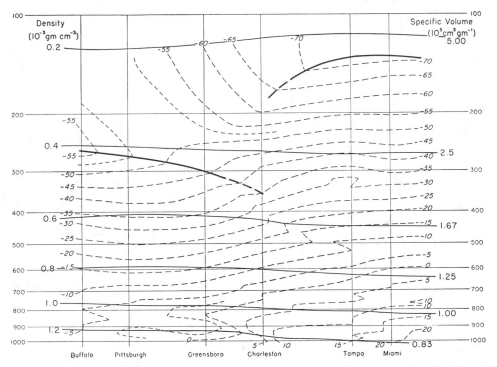

FIG. 6.11.—Cross section of density and specific volume (*continuous lines*) and temperature (° C) (*dashed lines*).

parallel isobars, and the atmosphere is most baroclinic where intersections of isosteres with isobars are numerous.

6.12. *Solenoids and baroclinity.*—If all unit isosteres are drawn on a cross section having also unit isobars, there results a pattern of parallelograms bounded on alternate sides by the respective isosteres and isobars (Fig. 6.12a). These parallelograms extend into space as parallelepipeds bounded by isosteric and isobaric surfaces and which either are closed rings or terminate

sultant pressure and specific volume ascendant vectors at a point in space (Fig. 6.12a). For all practical purposes this plane is a vertical cross section with azimuth along the isobaric ascendant of specific volume. It follows that, in the vicinity of the point, solenoid tubes are perpendicular to the plane. This we shall call the *solenoid normal plane*. In the isobaric pattern of α in Figure 6.12b the normal plane through B intersects the isobaric surface along ABC. If in this drawing the isosteres are drawn for unit intervals of α, then each strip of the isobaric surface bounded by two isosteres is the lower face of the solenoid tube above and the upper face of the one below.

The diagram in the right of Figure 6.12a

surface $\delta n = \delta a/(\partial a/\partial n)_p$. By substitution, $A = a(-\delta p \delta a)/g(\partial a/\partial n)_p$. The area A' per unit of p and a, $[A' = A/(-\delta p \delta a)]$, is then $A' = a/g(\partial a/\partial n)_p)$. The dimensions of A'

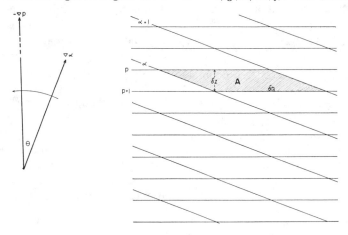

FIG. 6.12a

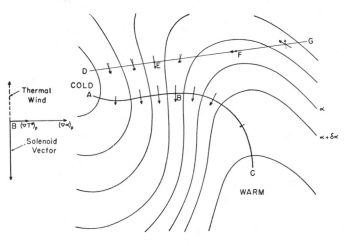

FIG. 6.12b

illustrates the pattern of isobars and isosteres in the normal plane for the vector scheme at the left. Shaded region A is the cross section of a *unit solenoid*. The area A is the product of the spacing δz of unit isobars and the isobaric spacing δn of unit isosteres; $A = \delta z \delta n$. From the hydrostatic equation $\delta z = (a/g)(-\delta p)$, and along an isobaric

are sec^2. A' has a geometric connotation and also a physical one of energy.

The importance of solenoids lies in the reciprocal of A', $N = 1/A'$, which is the number of unit solenoids per unit area in the normal plane, or the *solenoid density*. From the above formula for A' we have

$$N = (g/a)(\partial a/\partial n)_p. \tag{1}$$

The quantity N, of dimensions sec^{-2}, is specific energy per unit area of the normal plane and is always positive.

The solenoid vector $\mathbf{N}$ is the vector product of specific volume ascendant and pressure descendant:

$$\mathbf{N} = \nabla a \times -\nabla p . \qquad (2)$$

This vector is perpendicular to the normal plane with sense defined by right-handed rotation from ∇a to $-\nabla p$. The solenoid vector lies along the solenoid tubes; its magnitude is N. For the conditions indicated by Figure 6.12a, this vector is directed *from* the paper. Its magnitude

$$N = |\nabla a| \cdot |-\nabla p| \sin \theta , \qquad (3)$$

where θ is the acute angle between ∇a and $-\nabla p$, which is also the acute angle between isosteres and isobars in the normal plane. If ∇a had been to the left of $-\nabla p$ in Figure 6.12a, the solenoid vector would be *into* the paper. Derivation of Eq (1) from Eq (3) is left for the reader.

The most common reference to solenoids is in terms of baroclinity. In fact, baroclinity is best defined quantitatively as

$$\text{Baroclinity} \equiv N = (g/a)(\partial a/\partial n)_p .$$

Where baroclinity is zero, the atmosphere is barotropic.

It is only in the normal plane that we can define total baroclinity about a point. A vertical cross section tangent to curve ABC in Figure 6.12b gives the complete picture of baroclinity at the point of tangency. But the synoptic cross section is not parallel to the local normal planes everywhere in its horizontal and vertical extent. If a cross section is taken along $DEFG$ in Figure 6.12b, at this pressure the solenoid vector is normal to the cross section only at E, and isobars and isosteres in the cross-section plane indicate the true baroclinity at that one point. We can say nothing explicitly

about total baroclinity at other values of pressure, but at this one pressure baroclinity in the cross section underestimates true baroclinity at all points on either side of E. The partial baroclinity in the plane of the cross section is that component of the solenoid vector normal to the cross section, as indicated by the dashed arrows in Figure 6.12b. At F isobars and isosteres are parallel in the cross section, although the true baroclinity is greater here than at some places to the left of F.

The difference between partial baroclinity N_{sz} seen in sz cross section and true baroclinity N about a point is a function of the angle between the cross section and the local solenoid vector or normal plane.

$$N_{sz} = N \cos \theta_{sa} ,$$

if θ_{sa} is the angle between the azimuth s of the cross section and the isobaric ascendant of a (or T^*) through the point in question. Cross-section baroclinity is negative if $(\nabla a)_p$ has a component opposite to s.

From Eq (1) we may write

$$N_{sz} = (g/a)(\partial a/\partial s)_p . \qquad (4)$$

This can be transformed to expressions for baroclinity in terms of the commonly analyzed variables. From the equation of state, $(1/a)(\partial a/\partial s)_p = (1/T^*)(\partial T^*/\partial s)_p$. Also, $(1/T)(\partial T/\partial s)_p = (1/\theta)(\partial\theta/\partial s)_p$. Therefore,

$$N_{sz} = (g/a)(\partial a/\partial s)_p = (g/T^*)(\partial T^*/\partial s)_p$$
$$\simeq (g/T)(\partial T/\partial s)_p = (g/\theta)(\partial\theta/\partial s)_p . \qquad (5)$$

The analogous expression for total baroclinity about a point is

$$N = (g/a)(\partial a/\partial n)_p = (g/T^*)(\partial T^*/\partial n)_p$$
$$\simeq (g/T)(\partial T/\partial n)_p = (g/\theta)(\partial\theta/\partial n)_p . \qquad (6)$$

Baroclinity is the percentage or logarithmic variation of a, T, or θ at constant pressure, in each case multiplied by gravity. Notice the similarity between definitions of baroclinity $(g/\theta)(\partial\theta/\partial n)_p$ and static stability $(g/\theta)(\partial\theta/\partial z)$.

Cross-section baroclinity is inferred at a glance by noting the number of intersections of isosteres, isotherms, or isentropes in a given distance along isobars. With reference to Figure 6.11, the area just south of Pittsburgh between 700 and 400 mb is almost barotropic. Cross-section baroclinity is greatest between Greensboro and Tampa. The same conclusions are found from examining $p\theta$ intersections in Figure 6.02b. For estimating baroclinity visually by Eq (5), the temperature analysis is apparently the most convenient, owing to greater irregularity of the temperature pattern compared with θ or α.

A finite difference method is applicable for measuring baroclinity by Eq (5). As an example, consider a point midway between Greensboro and Charleston at 700 mb in Figure 6.06. The difference δT between stations is $5°$ C, and the mean value of T is $268.5°$ K. For $g = 980$ cm sec^{-2} and $\delta s = 430$ km (distance between stations), we obtain $N_{sz} = 2.4 \times 10^{-7}$ sec^{-2}.

A convenient way of visualizing baroclinity in a cross section is from the number of parallelograms per unit area formed by two intersecting sets of lines. In the sense that N refers to density of α, p solenoids, we take $N(T, p)$, $N(\theta, p)$, and $N(T, \theta)$ as the density of parallelograms formed by intersections of unit ($1°$ K) isotherms and isentropes and unit isobars. To each of these N bears the relation

$$N = (R_d/p)[N(T, p)] \; ; N = (R_dT/\theta p)$$
$$\times [N(\theta, p)] = (\alpha/\theta)[N(\theta, p)] \; ;$$
$$N = (c_p/\theta)[N(T, \theta)] \; .$$

The last form, which converts T, θ "solenoids" to baroclinity, is the most useful. As c_p is constant and $1/\theta$ is almost constant through a large part of the troposphere (for range of θ from $250°$ to $335°$ K, $1/\theta$ varies from 0.004 to 0.003), the density of T, θ solenoids has practically exact ratio to baroclinity. From the density of T, θ solenoids in Figures 6.06 and 6.13de the distribution of baroclinity (N_{sz}) is seen readily.

The equation $N_{sz} \simeq (g/\theta)(\partial\theta/\partial s)_p$ can be transformed into

$$N_{sz} \simeq - (g/\theta)(\partial\theta/\partial z) \tan_{sz} \beta_\theta$$
$$= -E_v \tan_{sz} \beta_\theta$$

by the slope formula[2] for θ surfaces and by definition of static stability. The above formula applies except where $E_v = 0$. Since stable layers tend to follow isentropic surfaces in the free atmosphere, the above equation suggests that *inclined stable layers are also concentrations of baroclinity.* Note how impressively this stands out in Figure 6.06 and expecially in Figures 6.13de. It is also evident that a horizontal stable layer can be barotropic even if its stability is large.

From hydrostatics it might be suspected that baroclinity can be interpreted in terms of vertical structure of the pressure field. Eq 3.14(1) can be written

$$\left.\begin{array}{c} \dfrac{g}{\delta Z}\left(\dfrac{\partial(\delta Z)}{\partial n}\right)_p = \dfrac{g}{T^*}\left(\dfrac{\partial T^*}{\partial n}\right)_p = N \, , \\[3mm] \dfrac{g}{\delta Z}\left(\dfrac{\partial(\delta Z)}{\delta s}\right)_p = \dfrac{g}{T^*}\left(\dfrac{\partial T^*}{\partial s}\right)_p = N_{sz} \, , \end{array}\right\} \quad (7)$$

where δZ is thickness of a unit pressure layer. As the *thermal wind vector* is (in northern hemisphere) opposite to the solenoid vector and directly proportional in magnitude, Eqs (7) and the scheme in the left of Figure 6.12b lead to the rule: *By facing toward warm air on an isobaric surface, the solenoid vector is directed to the right, the thermal wind vector to the left, and adjacent isobaric surfaces diverge most rapidly along*

---

2. The slope given in this equation is relative to the isobars in cross section. The isobaric reference is used in lieu of the horizontal in remaining discussions of cross sections.

the line of sight. (In the southern hemisphere the thermal wind has the same direction as the solenoid vector defined above.)

6.13. *Pressure analysis.*[3]—Since pressure is a coordinate of the base cross section, analysis of the pressure field must be done

the chart are the inverted images of the actual isobars in the atmosphere.

That type of pressure analysis is impractical, since it is difficult to draw a sufficiently accurate pattern of level surfaces giving a true picture of horizontal pressure distribution, although it is satisfactory for

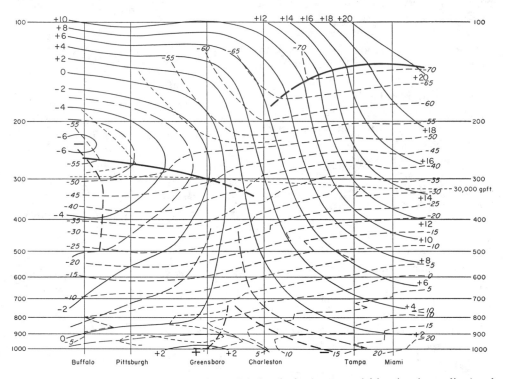

FIG. 6.13*a*.—Cross section of height departure *D* in hundreds of geopotential feet (*continuous lines*) and temperature (° C) (*dashed lines*), for same time as Figure 6.02*b*. Heavy dashed lines indicate pressure troughs and ridges. Contour patterns on the various isobaric charts were used in deriving this and also Figures 6.13*fg*.

indirectly by analysis of height (geopotential). From plotted heights for a number of pressure surfaces, it is possible to draw lines of equal height (isopotentials, equipotentials, isophyses). One such line, 30,000 gpft, is shown in Figure 6.13*a*. Notice its extremely small slope and curvature. A series of isophyses could be drawn, and the pattern is then interpreted in terms of pressure distribution. The curved isophyses in

3. It might be advisable to take this section concurrent with work in isobaric analysis.

pressure analysis along the vertical only. For any quantity whose vertical gradient is so dominant and relatively invariant horizontally, analysis of its anomalies (departures from a reference distribution) is a more suitable method of three-dimensional analysis. The reference pressure distribution can be an arbitrary sounding, an average sounding, an isothermal or adiabatic atmosphere, or even an average or reference cross section.

The reference used in the text is a modification of the U.S. Standard Atmosphere; its coordinates are given in the Appendix. All properties are the same as that standard except (i) height is in geopotential units (only for consistency with observations) and (ii) the temperature lapse rate is 6.5° C per gpkm up to 100 mb, which is the upper-most level analyzed (Fig. 6.02a). This was selected because (i) the sounding represents a suitable mean from which departures in temperature, lapse rate, and pressure are nearly a minimum for hemispheric conditions and (ii) a constant lapse rate of temperature removes the artificial discontinuity found at 234 mb in the U.S. Standard Atmosphere and thus provides for consistent thermodynamic analysis throughout the cross section. That reference is not being recommended for other purposes. Disadvantages in the selected reference compared to the standard atmosphere are the large departures in temperature and height in the stratosphere and the discrepancy there between departure in height (D) and the common altimeter correction.

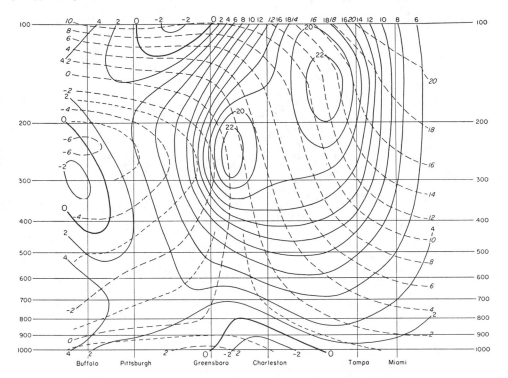

FIG. 6.13b.—Cross-section patterns of D (*dashed lines*) and of isobaric gradient of D (*continuous lines*) expressed in units 10 gpft per degree of latitude. Positive values for isobaric D gradient imply that the pressure surfaces slope upward to the right.

This analysis is similar in all respects to the D method suggested by Bellamy[4] except in the region above 234 mb. We may use the same symbol D to indicate the departure at a given pressure between actual height Z and height Z_r in the *reference atmosphere:*

$$D = Z - Z_r. \qquad (1)$$

4. J. C. Bellamy, "The Use of Pressure Altitude and Altimeter Corrections in Meteorology," *Journal of Meteorology*, II (1945), 1–79.

The quantity D, like any of the common variables, is a continuous function of space and time. *Contours on a pressure surface are intersections of* D *surfaces with that isobaric surface.* Surfaces of D therefore *connect* the contours in successive isobaric surfaces in the way that θ surfaces, for example, con-

surface-pressure and cross-section temperature analysis.

Figures 6.13*a* and 6.13*fg* illustrate D patterns in cross section. Along any isobaric surface, minima in D are troughs in those surfaces in the plane of the cross section, and maxima in D are ridges. From

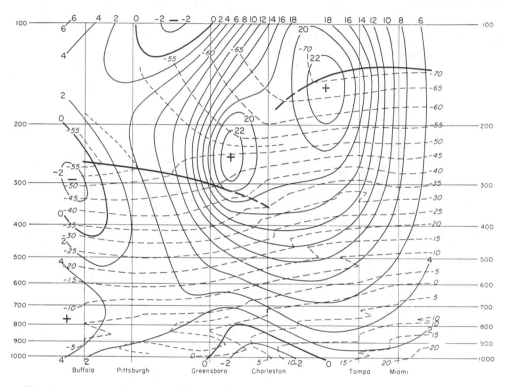

Fig. 6.13*c*.—Cross-section patterns of isobaric D gradient (*continuous lines*) and of temperature (° C) (*dashed lines*).

nect the θ lines on successive isobaric surfaces.

The values of D in any vertical sounding are obtained from Eq (1). They can be computed also for intermediate points along soundings by the D method of pressure-height evaluation (Sec. 3.09). Data between soundings on the cross section can be taken from analyzed isobaric charts, as contours of isobaric surfaces are isopleths of D. Values of D are also obtained along arbitrary verticals in the cross section from

Eq (1)

$$(\partial D/\partial s)_p = (\partial Z/\partial s)_p. \qquad (2)$$

Isobaric variation of D gives the slope of the isobaric surface uniquely. Thus the slopes of all isobaric surfaces can be determined visually by inspection of D patterns, and all details in isobaric topography lie in the isobaric variation of D. Curvature of an isobaric surface can be inferred from the variation of D gradient along that surface.

The height of any pressure (above sea

level) in any vertical is obtained by adding algebraically to the reference height of that pressure the value of D read from the pattern. Thus the height of 200 mb at the ridge line between Havana and Swan Island in Figure 6.13g is 38,630(Z_r for 200 mb) plus 1800 (D at that point), or 40,430 gpft.

From Eq 3.09(4) $\partial D/\partial Z_r = S^*$, or

$$\partial D/\partial Z = (T^* - T_r)/T^* . \quad (3)$$

The vertical gradient of D gives the anomaly of temperature from the reference. Where D lines are vertical or absent, the temperature is the same as reference for that level. When temperature is below reference, D decreases with height. Notice in the vicinity of Greensboro in Figure 6.13g the D lines are almost vertical in the troposphere; the Greensboro sounding is very similar to the reference sounding[5] (Fig. 6.20d). Since S^* is zero only where actual and reference sounding curves meet, maximum and minimum values of D in any vertical can occur only where $T^* = T_r$. In the cross section this is shown in the form of closed centers of D if $T^* = T_r$ in troughs and ridges of pressure and elsewhere as vertical sections of D lines.

Now examine the patterns of both T and D. Where isotherms parallel isobars (barotropy), there is little if any vertical change in *pattern* of D. In the middle and upper troposphere there is rather high positive correlation between values of D and T as well as in their patterns.[6] In comparison, notice that, in general, there is no correspondence of D and T patterns near the ground. Since D can be translated to height of pressure surfaces, and in turn to pressure on level surfaces, the same relations hold

5. It is generally the case that the zero D line is found in the region of strong baroclinity between the polar and tropical air.

6. Notice how the patterns of D and $T - T_r$ should correlate.

for pressure and temperature through the troposphere.[7]

By differentiating Eq (3) with respect to Z and making appropriate substitutions,

$$\partial^2 D/\partial Z^2 = (T_r/T^{*2})(\partial T^*/\partial Z - \partial T_r/\partial Z_r) . \quad (4)$$

$\partial T_r/\partial Z_r$ is constant at $-6.5°$ C gpkm^{-1}, and thus the quantity in the second parentheses is departure of vertical temperature lapse rate from that reference. As (T_r/T^{*2}) is practically constant, the vertical variation in spacing of D lines, given by $\partial^2 D/\partial Z^2$, is almost a unique function of lapse-rate departure from reference. This relationship is illustrated schematically in Figure 6.13h. In the first of these T, Z diagrams, in which the reference sounding is straight, the atmosphere is warmer than reference. Hence, D increases with height, as shown by the vertical order $D_1, D_2, D_3, \ldots$. In the lower part of the sounding T is only slightly greater than T_r (slow increase of D with height); in the upper part $T - T_r$ is greater and D increases with height more rapidly. In this relatively *warm* and relatively *stable* sounding, D increases upward at increasing rate.

In (b) the atmosphere is colder and more stable than reference. Thus D decreases upward at decreasing rate. Diagram (c) illustrates intersection of actual and reference soundings in a condition of relative stability. The portions of the sounding above and below the intersection are similar to (a) and

7. For statistical studies on correlations of pressure and temperature at various levels in the atmosphere see W. H. Dines, "Correlation and Regression Tables for the Upper Air," *Beiträge zur Physik der freien Armosphäre*, V (1913), 213; B. and E. Haurwitz, *Pressure and Temperature in the Free Atmosphere over Boston* ("Harvard Meteorological Studies," No. 3 [Cambridge, 1939]); C. M. Penner, "The Effects of Tropospheric and Stratospheric Advection on Pressure and Temperature Variations," *Canadian Journal of Research*, Vol. XIX, No. 1, Sec. A (1941).

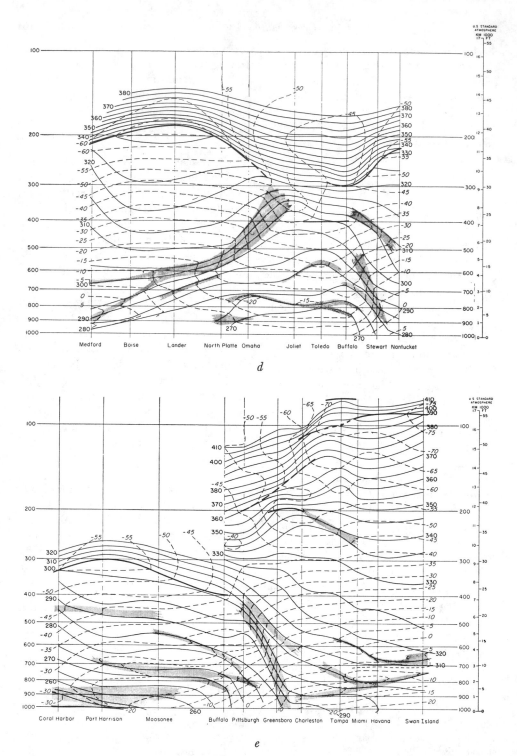

FIGS. 6.13*de.*—Cross sections of potential temperature (° K) (*continuous lines*) and temperature (° C) (*dashed lines*) for 1500 GCT, 1 March 1950. (Synoptic charts given in Chaps. 7, 8, and 9.)

162

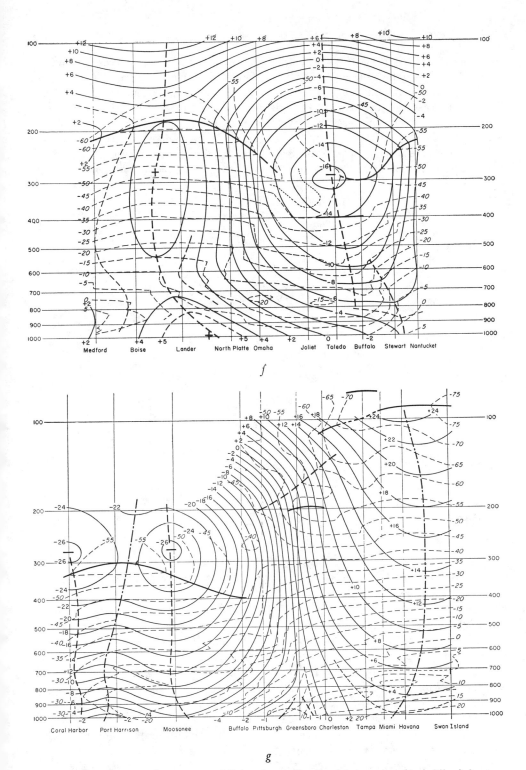

f

g

Figs. 6.13*fg.*—Cross-section patterns of D (*continuous lines*) and temperature (*dashed lines*) for same situation given in Figures 6.13*de.* (Same units as in Fig. 6.13*a* and Fig. 6.13*c*.)

(*b*), respectively. At the intersection, $T = T_r$ and $\partial D/\partial Z = 0$; but, since $\partial^2 D/\partial Z^2 > 0$, this is a point of minimum D in the vertical. Thus D decreases upward (relatively cold) up to that point and increases (relatively warm) above it.

The above relations between pressure and temperature can be verified readily by

in the lower part of the cross section near Greensboro with diagram (*b*) in Figure 6.13*h*, we conclude that the atmosphere is more stable than reference. In the stratosphere above Greensboro the pattern of D varies in the manner of scheme (*a*). One concludes from this analysis that, while the D pattern gives a picture of pressure distribu-

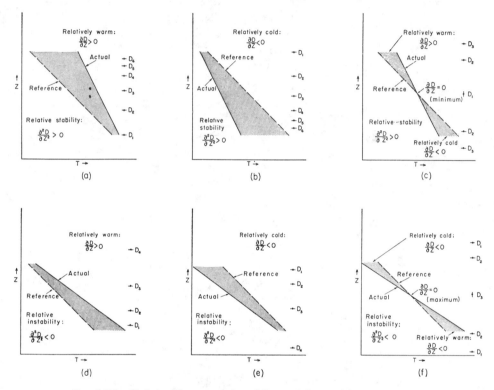

Fig. 6.13*h*.—Relation between static stability and the vertical variation of D

comparing the soundings in Figures 6.02*a* and 6.20 with the respective cross sections in Figures 6.13*afg*. The Greensboro sounding in Figure 6.02*a*, for example, is colder than reference up to about 280 mb and warmer than reference above. In the cross section, D decreases upward to 280 mb, where the D line for approximately -210 gpft is tangent to the Greensboro vertical, and above that D increases. On comparing the vertical variation in the pattern of D

tion and geometry of isobaric surfaces in space, it also gives a rather complete picture of relative temperature and temperature lapse rate.

The slope of a D line with respect to an isobar in cross section is

$$\tan_{sz} \beta_D = -(\partial D/\partial s)_p/(\partial D/\partial Z) . \quad (5)$$

The denominator on the right is considered continuous, since temperature is continuous (Eq [3]), and discontinuities in slopes of D

lines occur only at discontinuities in slopes of isobaric surfaces. However, from Eq (4), a discontinuity in temperature lapse rate gives second-order discontinuity in D. Thus D surfaces can have accentuated bends at well-defined tropopauses and boundaries of tropospheric stable layers (Fig. 6.13f).

6.14. *Isobaric slope.*—From the cross section D pattern it is possible to determine direction of slope, magnitude of slope, and positions of troughs and ridges in isobaric surfaces. The slope of an isobaric surface can be measured by finite differences for $(\partial D/\partial s)_p$. Such analyses are shown in Figures 6.13b and 6.14ab. Lines of zero slope coincide with pressure ridges and troughs, and the regions of maximum absolute isobaric slope along any isobar are the inflections in curvature of isobaric surfaces. These patterns could have been estimated visually by examining the patterns of D with some care.

These cross sections bring out several interesting facts about the three-dimensional pressure field. First, observe that maximum absolute slopes are to be found in the tropopause region. Slopes of isobaric surfaces are far from uniform; instead, the slope is likely to be concentrated in a few centers near the tropopause. Correlation of isobaric slopes between 500 mb and 100 mb is usually quite good; except in few conditions, pressure distribution at any level is good indication of that at another level in that region. One also detects rather large departures in pressure distribution near the ground from that through most of the region above. This poor correlation is not peculiar to these situations but is a fact observed generally and is a primary source of difficulty in relying on surface or sea-level pressure patterns for a three-dimensional picture of the atmosphere.

Discontinuities or sharp changes can occur in slopes of isobaric surfaces, and they would be found in the isopleths of isobaric slope. Although there should be some in the examples shown, they were omitted purposely to simplify the analysis and its legibility.

From the slope formula for isobaric surfaces (Eq 4.13[1]), the hydrostatic equation, and the relation between $\partial a/\partial s$ and $(\partial a/\partial s)_p$, one can show (as in Eq 6.12[7]) that

$$\frac{\partial}{\partial z}\tan_{sz}\beta_p = \frac{1}{T^*}\left(\frac{\partial T^*}{\partial s}\right)_p = N_{sz}/g. \quad (1)$$

Vertical variation of isobaric slope is directly proportional to isobaric virtual temperature gradient and equal to baroclinity divided by gravity. If the cross section is along the virtual temperature ascendant at constant pressure (e.g., tangent to curve ABC in Fig. 6.12b), then isobaric slope varies most rapidly in the vertical along this direction, slopes increase in positive sense in the vertical, and in the northern hemisphere the thermal wind vector is directed into the cross section. This relation is clarified by examining Figure 6.13c.

Figure 6.14b reveals an interesting feature of the distribution of isobaric slope in the atmosphere. Notice that the larger slopes of isobaric surfaces are concentrated in almost vertical zone within a narrow range of latitude. In this chart essentially barotropic conditions prevail in the tropical and polar portions, and the baroclinity is mostly between Buffalo and Charleston. The axis of maximum baroclinity is found near Greensboro, where it tilts slightly toward the cold air with height, and the reversal in sign of isobaric temperature gradient occurs just above 300 mb. Isobaric slopes have corresponding pattern.

6.15. *Tilt of pressure systems in the vertical.*—Consider two points a and b in s, $-\log p$ cross section (Fig. 6.15a). The height Z of each of the corresponding points in the atmosphere are functions of s and p,

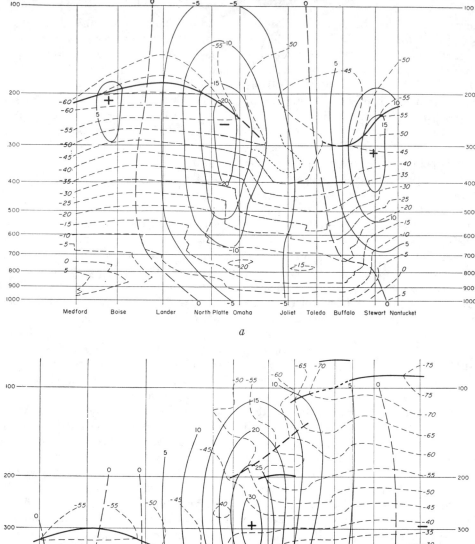

FIGS. 6.14*ab*.—Cross-section patterns of isobaric *D* gradient (*continuous lines*) and temperature (*dashed lines*) for the cross sections given in Figures 6.13*de, fg*. (Same units as in Figs. 6.13*bc*.)

where s is distance isobarically; thus, $Z = Z(s, p)$. By differentiation, and use of the hydrostatic equation, we find for the relation between actual isobaric slopes at b and a:

$$(\partial Z/\partial s)_b - (\partial Z/\partial s)_a = (\partial^2 Z/\partial s^2)\delta s$$
$$- (\partial a/\partial s)(\delta p/980) .$$

surfaces in s, Z plane. The denominator is the baroclinity in the plane (except for the factor g).

With reference to three-dimensional space, the *lines* of zero isobaric slope are *axes* of domes and depressions in isobaric surfaces. Ridges and troughs in successive isobaric surfaces are connected by *sheets*

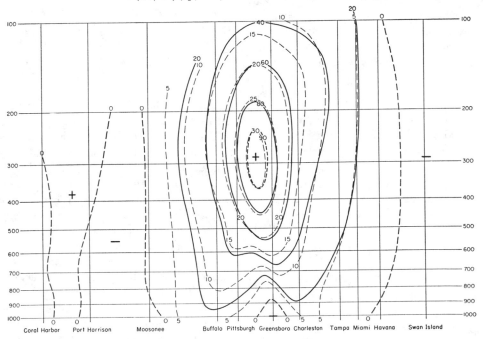

Fig. 6.14c.—Conversion of isobaric D gradient of Figure 6.14b into geostrophic wind normal to the plane (m sec⁻¹, continuous lines).

Along any line for which $(\partial Z/\partial s)_p = (\partial D/\partial s)_p = 0$, the left side is zero, and $\delta p/\delta s = 980(\partial^2 Z/\partial s^2)_p/(\partial a/\partial s)_p$. Upon substituting $-(980/a)\delta Z$ for δp,

$$\frac{\delta Z}{\delta s} = -\left(\frac{\partial^2 Z}{\partial s^2}\right)_p \Big/ \left(\frac{1}{a}\frac{\partial a}{\partial s}\right)_p$$
$$= -\left(\frac{\partial^2 Z}{\partial s^2}\right)_p \Big/ \left(\frac{1}{T^*}\frac{\partial T^*}{\partial s}\right)_p \quad (1)$$

for the slope $\delta Z/\delta s$ of the axis with property $(\partial Z/\partial s)_p = 0$. The numerator on the right is approximately the curvature of isobaric

normal to which there is no slope in isobaric surfaces; intersection of this sheet with the plane of the cross section gives a line for which $(\partial Z/\partial s)_p = 0$. In Figures 6.13fg these ridge and trough lines are the axes in the patterns of D, and in Figures 6.14ab they are the slope lines zero. Given only a vertical cross section, it is not possible to distinguish between dome and ridge nor between depression and trough.

A pressure ridge has the property $(\partial^2 Z/\partial s^2)_p < 0$, and a trough $(\partial^2 Z/\partial s^2)_p > 0$. So, for the slopes of ridges and troughs:

Ridge: $\dfrac{\delta Z}{\delta s} = \left| \dfrac{\partial^2 Z}{\partial s^2} \right|_p \Big/ \left(\dfrac{1}{T^*} \dfrac{\partial T^*}{\partial s} \right)_p$, (2a)

Trough: $\dfrac{\delta Z}{\delta s} = -\left| \dfrac{\partial^2 Z}{\partial s^2} \right|_p \Big/ \left(\dfrac{1}{T^*} \dfrac{\partial T^*}{\partial s} \right)_p$. (2b)

By the denominator of each formula, *the slope of a pressure system is inversely proportional to baroclinity.* As $(\partial T^*/\partial s)_p$ approaches zero, the axis of the system becomes vertical, and *in barotropic regions pressure systems are vertical.* Also, *pressure systems have the least slope (the most tilt)*

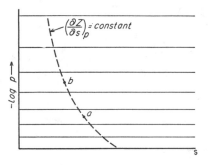

FIG. 6.15*a*

where the atmosphere is most baroclinic across their axes. Observe the ridge and trough in Figure 6.13*f* (or 6.14*a*) in the lower troposphere. Vertical axes are usual in the tropical and arctic portions of the globe. In middle latitudes typical pressure systems at the ground have temperature gradients, and these systems tilt in the lower troposphere.

Consideration of *direction* of isobaric temperature gradient leads to other important rules from Eqs (2*a*) and (2*b*). A pressure ridge is displaced upward in the *s* direction if temperature increases in that direction. Thus, other than in barotropic conditions, *a pressure ridge tilts toward the warm air with height.* The opposite holds for troughs; *a pressure trough tilts toward the cold air with height.*

Consider now the effect of curvature of isobaric surfaces. We may define *intensity* of

a pressure system in terms of vertical curvature of isobaric surfaces within it. Intense pressure systems are pronounced bulges in isobaric surfaces, upward for ridges and downward for troughs. "Flat" ridges and troughs have only small curvature of isobaric surfaces and thus also minor isobaric slopes on both sides. Eqs (2*a*) and (2*b*) show that, *the more intense the pressure system, the greater is its slope (the less its tilt)* for either trough or ridge. In weak systems the relative effect of temperature gradient is larger, and these have greater tilt for given baroclinity.

To show the effects of pressure system intensity and temperature gradient on axes of domes and depressions in isobaric surfaces, we may take the contour and temperature patterns on one surface and deduce the location of the center on an adjacent isobaric surface above (Fig. 6.15*b*), if the center is conserved. In all five diagrams only a depression is considered; a dome tilts the opposite way. The first two diagrams show schematically the effect of temperature pattern. The center shifts along the isobaric temperature descendant, and its displacement is directly proportional to the temperature gradient. Diagrams 3 and 4 reveal the same, but, when compared with 1 and 2, respectively, they show the effect of system intensity. Diagram 5 gives an asymmetric depression. By temperature distribution the center is displaced in direction *A* with height. From the standpoint of intensity, the center has additional displacement in direction *C*. The direction of displacement is neither but is an intermediate one *B*. *The direction of tilt is that along which the right side of Eq (1) is smallest,* and the rate of geographic displacement is inversely proportional to the magnitude of the ratio in that direction.

Equally as important are the zones of greatest isobaric slope. The *axis of inflection* in cross section connects points of maximum slope in consecutive isobaric surfaces. Points *A* in Figure 6.15*c* are at maximum slope in isobaric surfaces *p*. The vertical marks give positions of isotherms on this isobar. In all three cases temperature increases to the *height*. The physical relation is expressed in following form for the slope of the inflection axis:

$$\frac{\delta Z}{\delta s} = -\frac{\partial}{\partial s}\left(\frac{\partial^2 Z}{\partial s^2}\right)_p \bigg/ \frac{1}{T^*}\left(\frac{\partial^2 T^*}{\partial s^2}\right)_p. \quad (3)$$

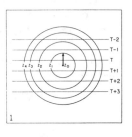

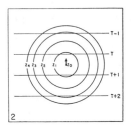

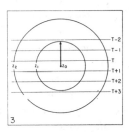

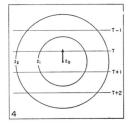

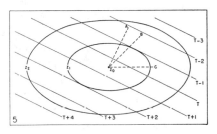

FIG. 6.15*b*.—Schematic illustrations of the geographic displacements of depressions in pressure surfaces with elevation, considering both the temperature distribution and the intensity of the pressure system.

right; thus isobars diverge from left to right. By Eq 6.14(1), in case (*a*) isobaric slopes increase most rapidly to the right of *A*, and the axis of inflection is thus deflected to the right of *A* with height. In (*b*) the axis of inflection is vertical (similar with barotropy). In (*c*) the axis of inflection tilts to the left with height. The general conclusion is that *the zone of maximum isobaric slope is tilted toward the greatest baroclinity with*

Figure 6.14*b* provides an excellent illustration for Eq (3). One axis of inflection begins at the ground just south of Charleston, is over Greensboro at 600 mb, and curves upward into the center marked +. Another begins in low levels near Buffalo and merges with the other near 400 mb. Up to this level both axes tilt toward greater baroclinity. Above, the rapid variation in isobaric curvature across the axis of inflection has become so large (numerator of Eq [3] so large) that this factor dominates, and the axis is almost vertical. Notice also that the denominator of Eq (3) decreases upward from about 600 mb.

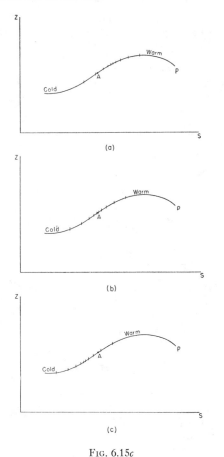

(a)

(b)

(c)

Fɪɢ. 6.15c

6.16. *Geostrophic wind.*—The geostrophic wind, which is an approximation to actual wind, is that fictitious motion representing equilibrium between the horizontal pressure force (per unit mass) and the deflecting force due to motion on a rotating earth. The concept of geostrophic wind, including departures from it, is discussed in more detail in Chapter 7. If we assume for the present that atmospheric motions are sufficiently

geostrophic above the surface layer, it is possible to deduce wind distributions in the cross sections illustrated.

Consider a unit cube of air in the horizontal pressure field of Figure 6.16. In hydrostatic equilibrium the net pressure force acting on the cube is along the horizontal pressure descendant; this force is of magnitude $-(\partial p/\partial n)\delta n\delta A$, if δn is the distance between sides and δA is unit area. Since the mass is $\rho\delta n\delta A$, the magnitude of the horizontal pressure force per unit mass is $-a(\partial p/\partial n)$, which is dimensionally an acceleration. This vector is represented by $-(1/\rho)\nabla_h p$ in the figure.

If the parcel of air is initially at rest, this acceleration gives the parcel motion toward lower pressure with increasing speed. However, the rotation of the earth gives it an apparent acceleration normal to its motion. In the northern hemisphere this deflecting force acts to the right of the motion; in the southern hemisphere, to the left. The deflecting force per unit mass, the *Coriolis acceleration*, is directly proportional to the speed c of the parcel, and its magnitude is $(2\omega \sin \phi)c$, where ϕ is latitude, and ω the earth's angular velocity ($2\omega = 1.458 \times 10^{-4}$ $\sec^{-1}$). The factor $2\omega \sin \phi$ is the *Coriolis parameter* and is denoted f; the Coriolis acceleration has magnitude fc.

The vector Coriolis force in the northern hemisphere is always 90° to the right of the vector wind. Since there exists no adequate simple notation for the Coriolis force in terms of resultant vector wind, we are at liberty to provide one here for this discussion. We are using symbol c for the speed of vector wind $\mathbb{C}$, and fc for the magnitude of the Coriolis force. For consistency, the notation $f * \mathbb{C}$ will be used for vector Coriolis force. The factor $f *$ is an operator which rotates direction 90° to the right of the vector it is multiplying; it has magnitude f. Thus, $f * \mathbb{C}$ is a vector of magnitude fc directed 90° to the right of $\mathbb{C}$. In the southern hemisphere $f < 0$, and there $f * \mathbb{C}$ is 90° to the *left* of $\mathbb{C}$.

In geostrophic motion the balance between pressure and Coriolis force implies that geostrophic wind blows along the isobars with low pressure to the left (in the northern hemisphere). In that state $f * \mathbb{C} + (-a\nabla_h p) = 0$, and

$$f * \mathbb{C} = a\nabla_h p , \qquad fc = a(\partial p/\partial n) . \quad (1)$$

In geostrophic motion the Coriolis force is

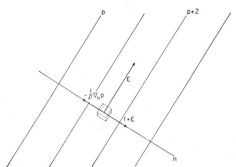

FIG. 6.16.—Relation betwen horizontal pressure gradient, the geostrophic wind ($\mathbb{C}$), and the Coriolis acceleration $f*\mathbb{C}$ in balanced motion (northern hemisphere).

merely equal to the horizontal ascendant of pressure multiplied by a.

For the present consider only the scalar form of Eq (1):

$$c = (1/\rho f)(\partial p/\partial n) .$$

Geostrophic wind speed is directly proportional to horizontal pressure gradient and inversely proportional to the sine of latitude. Since geostrophic wind blows parallel to isobars, and we took n along the horizontal pressure ascendant, Eq (1) states that *pressure increases to the right of the wind*.[8]

The geostrophic wind relation, Eq (1), is a simple method for relating pressure field and motion. Because of the convenient way

8. Here, and henceforth in our discussions of geostrophic wind, reference is made to the northern hemisphere.

in which the equation describes the field of motion from the pressure pattern, it is the basic tool for relating the two. However, it is very important to recognize that geostrophic wind is uniquely related to the pressure field rather than to the true wind field and that the geostrophic wind should be regarded as merely a representation of the pressure field. This is extremely useful because of the *empirical* fact that to a large extent the true winds are approximately geostrophic.

Figure P (Appendix) is a geostrophic scale for converting pressure gradient into wind speed, and conversely; this scale was devised for sea level (lower scale) and 10,000 feet (upper scale). Horizontal variation of ρ is neglected; the value of ρ for constructing the chart is that in a standard atmosphere. The distance scales at top and bottom give horizontal spacing Δn for an appropriate constant value of Δp. Wind speed is the ordinate of the chart, and latitude is accounted by the sloping lines. To find the spacing of 3-mb isobars at sea level corresponding to geostrophic wind 30 knots at 30° latitude, follow the 30-knot horizontal line to the intersection with the 30° latitude line and read Δn directly below: $\Delta n = 215$ km. The reverse process gives geostrophic wind from given Δn and ϕ.

From Eq 4.14(1) it can be shown that $\partial p/\partial n = 980\rho \, (\partial Z/\partial n)_p$. Therefore,

$$f * \mathbb{C} = 980 \, (\nabla Z)_p ,$$

$$fc = 980 \, (\partial Z/\partial n)_p . \quad (2)$$

The slope of an isobaric surface is related to geostrophic wind in the same manner as pressure gradient on a level surface, except that density is not a factor, and now a single geostrophic scale (Appendix, Fig. N) applies for all isobaric surfaces.

6.17. *Horizontal and vertical wind shear.*— Consider that vector $\mathbb{C}$ in Figure 6.17a(1) is the wind at position A. Direction n at A is

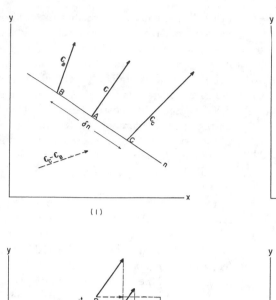

(1)

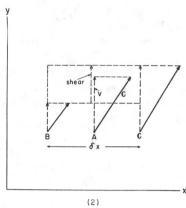

(2)

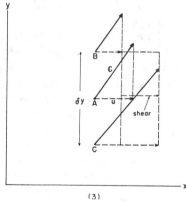

(3)

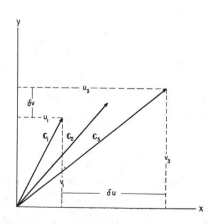

(4)

FIG. 6.17*a*

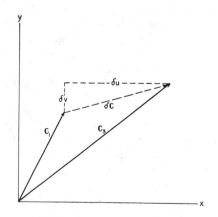

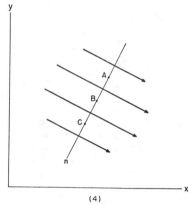

FIG. 6.17*b*

normal to the wind and in the horizontal plane through A and positive to the right of the wind. Two points, B and C equidistant from A, are located along n to the left and right of A, respectively, and a short distance δn apart. The vector difference $\mathbb{C}_C - \mathbb{C}_B = \delta\mathbb{C}$, when divided by δn, is taken as the horizontal wind shear at A. By this vector definition, horizontal shear is the change of vector wind per unit distance horizontally normal to the wind.

The scalar definition of horizontal wind shear, which considers change only in *wind speed* in the normal direction, is

$$\text{Horizontal wind shear} \equiv \partial c / \partial n .$$

If $\partial c / \partial n > 0$, there is *cyclonic shear;* if $\partial c / \partial n < 0$, there is *anticyclonic shear;*[9] if $\partial c / \partial n = 0$, the wind or motion is *nonshearing* horizontally.

In diagram (4) the shear at A is positive and the shear at C negative. If we suppose the wind is a maximum at B, then there is no shear at that point. *Wind shear is zero in a wind maximum, and it is positive to the left and negative to the right of the wind maximum.*

Later, use will be made of horizontal shear in the wind components (i.e., west-wind shear, south-wind shear, or shear of the wind component normal to a vertical plane). Diagram (2) in Figure 6.17a gives the winds at A and at two points δx apart and equidistant from A along x. The v component at each point is that component along y. Horizontal shear of v, $\partial v / \partial x$, is approximated by $(v_C - v_B)/\delta x = \delta v / \delta x$. The same applied to the u components in (3) yields the component shear $-\partial u / \partial y$. (The negative sign is necessary here, since y decreases to the right of the positive u direction.)

The vertical wind shear $\partial \mathbb{C}/\partial z$ may be described by reference to winds for three successive levels in a hodograph (Fig. 6.17b).

9. Nomenclature valid for the northern hemisphere.

Let vectors $\mathbb{C}_1$, $\mathbb{C}_2$, and $\mathbb{C}_3$ represent the winds at levels 1, 2, and 3 above a geographic point. By the method shown in Figure 6.17a, the shear at level 2 is the difference $\mathbb{C}_3 - \mathbb{C}_1$ divided by the vertical distance δz.

Vertical wind shear can be resolved into its coordinate components, as indicated in Figure 6.17b (left). Vertical shear of v, $\partial v/\partial z$, at level 2 is approximated by $\delta v/\delta z$, where $\delta v = v_3 - v_1$ and $\delta z = z_3 - z_1$. For the vertical shear of u a similar approximation $\delta u/\delta z$ applies, and this shear appears in the vertical plane along x. If $\mathbb{C}_1$ and $\mathbb{C}_3$ are *geostrophic* winds, $\delta\mathbb{C}$ is the *thermal wind* vector for the layer. Under the same provision δu and δv are the x and y components of thermal wind in the layer, and $\partial u/\partial z$ and $\partial v/\partial z$ each express geostrophic vertical shear and thermal wind per unit vertical distance.

6.18. *Geostrophic wind normal to a cross section.*—

a) *Wind speed.*—The geostrophic wind component in an arbitrary direction is expressed in terms of pressure gradient normal to that direction (Fig. 6.18). The x component of horizontal pressure force is $-(1/\rho)(\partial p/\partial x)$, and the related component of geostrophic wind $\mathbb{C}$ is indicated by v. Thus

$$v = (1/\rho f)(\partial p/\partial x)$$
$$= (980/f)(\partial Z/\partial x)_p . \quad (1a)$$

Similarly for the other component of geostrophic wind:

$$u = -(1/\rho f)(\partial p/\partial y)$$
$$= -(980/f)(\partial Z/\partial y)_p . \quad (1b)$$

In cross sections it is possible to determine only the component of pressure gradient (isobaric slope) along the azimuth of the section and thus only the component of geostrophic wind *normal to the cross section*. From the preceding it is apparent that, where pressure increases to the right, this wind component is *into* the cross section. Let c_n denote this wind component (positive into the cross section). Then the geostrophic value of c_n is

$$c_n = (980/f)(\partial Z/\partial s)_p . \qquad (2)$$

For west-east cross section f is constant, and the normal geostrophic component varies only with the slope of isobars in that plane. Lines of constant isobaric slope in Figure 6.14a are isopleths of c_n. Latitude is

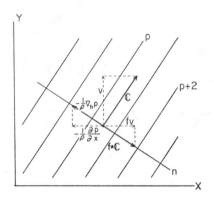

FIG. 6.18

about 41°, where $f = 0.96 \times 10^{-4}$ sec^{-1}; multiplying isobaric slope in the given units by 2.8 gives c_n in m sec^{-1}. Thus, line *20* is the isopleth of wind speed (*isotach*) for 56 m sec^{-1}, or 109 knots. Where slope is positive (upward to east), the normal wind is from south, and, where slope is zero, there is no meridional geostrophic wind.

In north-south cross section the pattern of wind speed is shifted southward relative to the pattern of isobaric slope, except that lines for zero slope remain as lines for zero normal wind. Figure 6.14c gives the result of converting Figure 6.14b into geostrophic wind. In the region of large isobaric slope the pattern undergoes visible southward shift.

  b) Isobaric shear.—Isobaric shear of nor-

mal geostrophic wind is obtained by differentiating Eq (2).

$$(\partial c_n/\partial s)_p = (980/f)(\partial^2 Z/\partial s^2)_p$$
$$- (c_n/f)(\partial f/\partial s) . \qquad (3)$$

The first term on the right is the vertical curvature of the isobaric surface. A surface curved convex upward in the plane indicates negative or anticyclonic shear in the normal geostrophic wind, and one curved concave upward has positive shear in that component. Thus, ridges in the isobars are attended by anticyclonic shear of c_n and troughs by cyclonic shear.

The last term in Eq (3) is zero for a west-east cross section and of maximum importance along a meridian. For meridional cross section with s southward, $\partial f/\partial s = (2\omega \cos \phi)(\partial \phi/\partial s) = (2\omega \cos \phi)(-1/a)$, where a is the earth's radius, and the west-wind shear is

$$(\partial c_n/\partial s)_p = (980/f)(\partial^2 Z/\partial s^2)_p$$
$$+ (c_n/a) \cot \phi . \qquad (4)$$

Latitude variation effectively increases cyclonic shear (decreases anticyclonic shear) of west winds and decreases cyclonic shear (increases anticyclonic shear) of east winds. If the cross section is other than north-south, the second term on the right of Eq (4) is multiplied by the cosine of the acute angle between the cross section and south. Eq (4) should be analyzed thoroughly with reference to Figure 6.14c.

As a rule, the second term on the right of Eq (4) is only about one-tenth the preceding term. With relatively light winds (c_n small) or high latitude, the second term can be neglected. Only where c_n is very large or where $(\partial^2 Z/\partial s^2)_p$ is near zero is this term of significance.

Figure 6.14c indicates large horizontal shear on both sides of the maximum current. Maximum shear of west wind, averaged between the 60 and 80 m sec^{-1} lines, is about

-1.0×10^{-4} sec^{-1} on the south side and 1.3×10^{-4} sec^{-1} on the north side of the current, both near the magnitude of the Coriolis parameter. Shear of 1.0×10^{-4} sec^{-1} represents change by 1 m sec^{-1} in 10 km. The shear is obviously a minimum in the center of the current, but it is also apparent that horizontal wind shear is relatively small outside the general region of strong winds.

c) *Vertical shear.*—The close relation between change of wind with height and temperature distribution is clearly evident in Figure 6.14b. If we differentiate Eq (2) above with respect to height and substitute from Eq 6.14(1),

$$\partial c_n / \partial z = (g/fT^*)(\partial T^* / \partial s)_p = N_{sz}/f . \quad (5)$$

Vertical geostrophic wind shear is directly proportional to the isobaric temperature gradient and bears the same relation to baroclinity as does geostrophic wind to slopes of isobaric surfaces. The normal wind component *increases* with height where *warmer* air is to the right of the wind and *decreases* with height where *colder* air is to the right of the wind (in the northern hemisphere).

Eq (5) is the *thermal wind equation.* It shows how important temperature distribution is to the structure of atmospheric motion. The thermal pattern governs hydrostatically the pattern of pressure, and therefore the pattern of motion is controlled by the pattern of temperature. Since near the ground the geostrophic wind is comparatively light, strong winds aloft depend almost entirely on thermal wind. In Figure 6.14b geostrophic west wind increases rapidly upward in the strongly baroclinic region between Pittsburgh and Greensboro and then decreases almost as rapidly above where baroclinity is large but of opposite sense. At the extremities of the cross section, wind is relatively constant and light throughout the troposphere with the nearly barotropic conditions.

The most impressive feature of Figure 6.14c is the concentration of wind in a narrow zone of small meridional and vertical dimensions. The 80 m sec^{-1} isotach incloses an area less than 20,000 feet in depth and less than 5° latitude in width. Shear about the current is large horizontally and vertically. For the narrow extent of this wind system, its wind speed, and the strong shear around it, a suggestive name is *jet stream.*

6.19. *Tropospheric baroclinity and the polar-front region.*—Figure 6.13e extends about along the 80th meridian from 65° N to 17° N. In the main, this picture is typical of winter conditions over eastern North America, from both the properties of the atmosphere and the geographic locations of important phenomena. If attention is given only to the troposphere below about 400 mb, we can segregate the pattern into three rather distinct regions on the basis of temperature and baroclinity; these are (i) the *polar air mass;* (ii) the strongly baroclinic *polar-front region;* and (iii) the *tropical air mass.*

In the north are temperatures characteristic of polar regions in this season. The lowest 500-mb temperature is $-43°$ C (at Coral Harbor), which is nearly the average daily minimum temperature for that level and season.[10] Within the surface layer are the very low temperatures and strong low-level stability characteristic of continental polar regions in winter. In fact, because of its low temperatures and its separation from the milder polar air above, this shallow region has been called the "arctic air mass."

With exception of the extreme northern end, and also excluding the shallow arctic air near the ground, the large pool of air extending southward to the vicinity of

10. On this day the lowest reported 500-mb temperature in arctic North America was $-48°$ C at Eureka (80° N 86° W).

Buffalo is relatively barotropic. Between Port Harrison and Buffalo the average temperature gradient at 500 mb and at 850 mb is 0.3° C/° ϕ, many times smaller than south of Buffalo. Because this broad region is so nearly homogeneous at most levels, it may be treated as an entity; and, further, since temperatures are polar in character, it is the *polar air mass*.

South of Charleston essentially the same barotropic conditions exist. Average temperature gradient at 500 mb between Charleston and Swan Island is 0.2° C/°ϕ. At the surface the average gradient through this region is 0.5° C/°ϕ, yet for the surface this is unusually barotropic relative to the rest of the pattern to the north. The deep and relatively barotropic pool of air shown in this picture extends southward across the tropics; it is the *tropical air mass*.

Within both polar and tropical air masses there are several more or less horizontal stable layers or transition zones imparting local irregularity to the otherwise homogeneous aspects of the two broad masses. If only a single level is considered, for example, 750 mb, the over-all homogeneity may not be so evident. Thermal structure near the ground generally is not a representative picture of the air masses, as there are local patterns at the surface which can overshadow the large-scale temperature pattern. Therefore, *our descriptions of the different air masses should not be confined to indications at one level but should be based on characteristics and structure of a large volume of air*.

Figure 6.13*d* intersects 6.13*e* near Buffalo.[11] The coldest air is centered over Joliet and Toledo. It is slightly warmer than the polar air to the north, but the tendency for polar air to warm in moving southward allows this air to be classified polar also. Over

Nantucket temperatures approach tropical values except near the surface, as also through the middle and upper troposphere in the west. But again, with due allowance for difference in latitude, we are justified in calling this air "tropical." The three-dimensional picture constructed from the two cross sections shows a cold mass of air extending southward over the central United States and surrounded by the tropical air.[12]

Between these two air masses most of the temperature contrast is concentrated in quite a narrow region, the *polar-front region*. The baroclinic region between the two masses in the western part of Figure 6.13*d* is greatly tilted. At 600 mb the contrast exists almost entirely between Lander and North Platte, while at 400 mb it is all between Omaha and Joliet. In comparison, the baroclinic region in the east extends almost vertically from the surface to 400 mb in both cross sections. The difference is due to different motions and also to difference in angles between temperature gradient and cross section.

We found previously from Figure 6.13*e* that isobaric temperature gradients were on the whole less than 0.5° C/°ϕ within the two air masses. Through the polar-front region the average gradient at 850 mb between Buffalo and Greensboro exceeds 2° C/°ϕ; at 500 mb the average between Pittsburgh and Greensboro is 3.3° C/°ϕ. Values are larger over shorter distances. These figures bring out the additional fact that temperature gradients are frequently larger aloft than near the surface.

The importance of the polar-front region is evident in terms of temperature—at the ground and throughout the troposphere—as well as in those variables controlled by temperature. This region is critical in the vertical structure of pressure and wind systems. Because of its relation to the field of motion, the polar-front region must have an im-

---

11. Actually, the planes intersect between Buffalo and Pittsburgh, but only the Buffalo sounding was used in analysis of the west-east section.

12. Now check with Figure 7.08.

portant bearing on the weather pattern. But, as already seen, this region is subject to varying degrees of baroclinity and of total temperature contrast.

6.20. *The polar-front zone.*—In Figures 6.13*de* the shaded areas indicate stable layers (zones of transition). These were located and drawn from indications in the cross sections and from temperature patterns on isobaric charts. Some are virtually barotropic and have no meaning except in static stability. Others represent concentrations in baroclinity as well as stability, and they are of greater significance in atmospheric motions and weather.

From these cross sections it appears that in *most* areas a large *part* of the baroclinity in the polar-front region is concentrated within one of the inclined transition zones. It is such a zone that is presently called a "frontal zone," a "front," or *the polar front.* There are many obvious difficulties presented by these pictures for defining and describing a frontal zone in a manner as simple as one would like. We resort to the usual procedure of describing a frontal zone theoretically and then point out the more glaring discrepancies between that and what is observed.

In the classical model we think of the polar front as an inclined transition zone (Fig. 6.05) separating relatively barotropic polar and tropical air. This zone contains all (or almost all) the baroclinity at each level. The warm boundary of the transition zone is the more prominent dynamically, and it is that boundary we draw as a line to indicate the front in a horizontal chart. If the front is not moving in a certain level, it is a *stationary front* in that level. If the front is moving such that warm air replaces cold air locally, it is a *warm front.* A *cold front* indicates the front is moving such that cold air replaces warm air locally. In all cases the cold air underlies the warm air in the shape of a wedge. The frontal zone is shown in the vertical by a layer of strong stability—the "frontal inversion"—which marks transition from cold air of low humidity to warm air of high humidity. Hence, a frontal inversion should be a moist inversion. All these properties of a frontal zone are discussed at greater length in the reading references.

First, we indicate some caution in attempting to associate frontal inversions with certain vertical lapse rates of moisture. It is true that through many frontal zones (warm fronts in particular) moisture increases with height, but it is also true that through many real frontal inversions (cold fronts in particular) the moisture does not increase significantly with height and may decrease more rapidly there than elsewhere in the sounding. The notion that a frontal inversion should be marked by vertical increase of humidity has led to many erroneous deductions from individual soundings. Except in certain situations, it is not safe to deduce from a single ascent the three-dimensional thermal structure of the atmosphere without considering several other factors.

A frontal zone is a layer of locally greater stability in a vertical sounding, but there are frequent occasions in which the variation of stability is so slight that the zone of transition is not clearly evident from one sounding even though baroclinity might be extremely great.

A third difficulty encountered in applying the single model of a front to actual conditions is the empirical fact that most often the frontal zone identified as a stable layer embraces only a small fraction of the total baroclinity found in the polar-front region. On occasion the zone of transition is an insignificant part of the temperature contrast between air masses, and the frontal zone itself is minor in the vertical structure of the atmosphere and in the weather (in spite of the emphasis placed on it). In contrast, situations are occasionally found in which almost all the baroclinity between air masses is concentrated in a sharply bounded zone of transition.[13]

In the cross sections there are several examples of frontal zones worthy of some study. These are a fair sample in principle of what can be expected in daily conditions. The extended zone of transition in Figure 6.13*d* is, from Lander to Joliet, a deep and almost isothermal layer containing most of the baroclinity. Above it, isotherms slope downward gently in approaching the upper discontinuity, indicating slight baroclinity, which is seen also below. This zone is a good example of the frontal model as far as temperature pattern is concerned, which is the primary consideration.

This frontal zone appears to be continuous with the stability of the stratospheric depression over Joliet and Toledo. The position of intersection of the frontal zone with the tropopause region is questionable, as it is in most cases. On horizontal or isobaric charts in the region from 600 mb to 400 mb the frontal zone should appear as a channel or "ribbon" of closely spaced isotherms, with temperature gradient discontinuities at its boundaries and surrounded by relatively homogeneous pools of air in which spacing of isotherms increases gradually outward.[14]

Additional evidence shows that this front is moving slowly to the right with time; it is a warm front. Here its displacement is due partly to advection and partly

13. This type of frontal zone is described in recent publications by Palmén and associates; cf. E. Palmén, "On the Distribution of Temperature and Wind in the Upper Westerlies," *Journal of Meteorology*, Vol. V, No. 1 (1948); E. Palmén and C. W. Newton, "A Study of the Mean Wind and Temperature Distribution in the Vicinity of the Polar Front in Winter," *Journal of Meteorology*, Vol. V, No. 5 (1948), and "On the Three-dimensional Motions in an Outbreak of Polar Air," *ibid.*, Vol. VIII, No. 1 (1951).

14. See Figure 7.08.

to subsidence, which accounts for a dry frontal inversion and absence of cloudiness. This type of frontal zone is characteristic of the rear (western) side of a subsiding polar air mass.

Figure 6.20*a* gives soundings for Omaha, Lander, Rapid City, and Dodge City. Notice in the Omaha sounding that there is a sharp inversion of 5.5° C in a 10-mb layer near 730 mb of stability many times greater than in the frontal zone above, but (from considering north-south temperature distribution also) its total baroclinity is less than in the frontal zone which is less clearly defined in the sounding.

At Lander the upper boundary of the transition zone appears distinctly at 575 mb. The lower boundary is masked by stability near the ground, and local modifications there are also expected by the rugged orography. But the slight change in stability at 635 mb suggested this point as the lower boundary of the stable zone, since it also gave consistency with stations to the east. In comparing the soundings for Omaha and Lander, notice that this frontal zone extends almost isentropically between stations. Usually, however, there is some increase of θ upward along the front. In addition, there is meridional variation of θ along the frontal zone, as seen by comparing the soundings for Rapid City and Dodge City.

Comparison of the soundings below 700 mb with the region above the frontal zone reveals a commonly observed feature of temperature distribution in the polar-front region—there are usually large temperature contrasts near the ground in the polar air mass itself. From Omaha to Dodge City the temperature contrast is about 10° C (in slightly more than 500 km) near the ground. This temperature difference decreases upward, until the vicinity of the frontal zone, and above the temperature contrast between stations is comparatively small. In

Figure 6.13*d* the frontal zone embraces almost all the baroclinity near Omaha, but westward the frontal zone—as identified by stability—begins to detach itself from the greater part of the baroclinic field. Farther west we reach a common dilemma in identifying a front; time, effort, and argument can be wasted. At Boise and Medford the upper of the two stable layers is about 20 mb in depth and is essentially barotropic (compare soundings for Oakland, Boise, and Tatoosh). To call this layer a "frontal

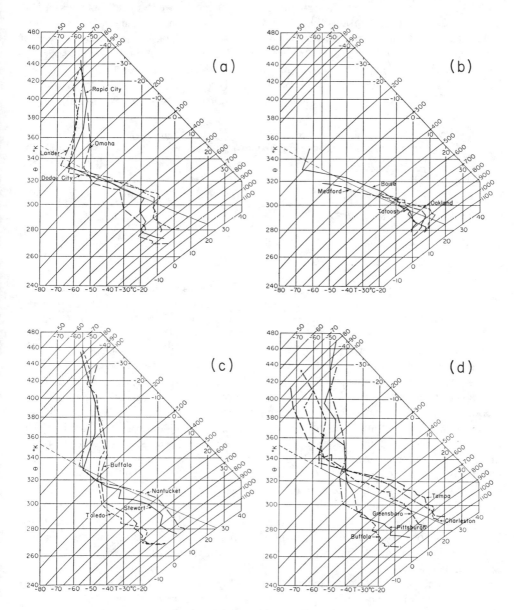

Fig. 6.20.—Pressure-temperature soundings for 1500 GCT, 1 March 1950

zone" is misleading in terms of temperature contrast. The next stable layer below was drawn also as a continuation of the deeper one at Lander, although there is room for doubt. The stable layer extending from near 900 mb at Medford to near 750 mb at Boise does have a reasonable *temperature gradient*, as evidenced by all illustrations, but it accounts for only a small fraction of *temperature contrast* between adjacent stations. Therefore, this stable-layer "frontal zone" is really insignificant. We should be more concerned in such cases with low-level

frontal zone) was about midway between Greensboro and Charleston. At Pittsburgh (Fig. 6.20*d*) data were missing between 490 and 400 mb. In view of the strong average stability between the two levels, and also in view of the large temperature gradient at 500 mb between Greensboro and Pittsburgh, we can assume the frontal zone intersects Pittsburgh within the layer of missing data. At Buffalo the atmosphere is very stable everywhere above 400 mb, but in this region it is difficult to distinguish between low stratosphere and frontal zone.

TABLE 6.20

VERTICAL TEMPERATURE LAPSE RATES (° C/KM) FOR GIVEN ISOBARIC TEMPERATURE GRADIENT (° C/100 KM) AND ISENTROPIC SLOPE

$$(\gamma = -\partial T/\partial z)$$

$(\partial T/\partial s)_p$	1/25	1/50	1/100	1/150	1/200	1/250	1/300
4	8.8	7.8	5.8	3.8	1.8	− 0.2	− 2.2
8	7.8	5.8	1.8	− 2.2	− 6.2	−10.2	−14.2
12	6.8	3.8	− 2.2	− 8.2	−14.2	−20.2	−26.2
16	5.8	1.8	− 6.2	−14.2	−22.2	−30.2	−38.2

baroclinity in the polar air. The important thing here is baroclinity, which cannot be placed into a simple transition zone.

Now examine the eastern side of the cold-air dome. The baroclinic region found here in Figure 6.13*d* appears from a better angle in the next figure. It is oriented SW-NE in low levels and veers upward into WSW-ENE (Fig. 7.08). In Figure 6.13*e* two large transition zones within the polar-front region were located during analysis. The upper one, centered near 550 mb over Greensboro, was entered from the sounding, and its tilt was decided by reference to the isentropic surfaces. The significance of this zone is less than can be attached to the broader and more baroclinic frontal zone below. The frontal zone is moving southward, and thus it is a cold front.

The frontal zone was located first on the surface chart from the temperature and wind fields. The front (warm boundary of

As checks on locating this frontal zone two additional operations were used. First, it was checked against the temperature analysis on isobaric charts. The second involved a likely pattern of potential temperature in the cross section. The "bundle" of θ lines between 500 and 400 mb at Pittsburgh intersects Greensboro below 650 mb, with exception of a few lines lost to the tropical air and a few gained from the polar air. As to the exact shapes of the isentropes between Pittsburgh and Greensboro we are not certain, but it is logical to assume that they trend mutually parallel.

Reference to the Greensboro sounding in Figure 6.20*d* shows a glaring apparent discrepancy between stability in the sounding and location of the frontal zone in the cross section. The zone is drawn to include most of the layer below 660 mb. In the sounding there is a 10-mb isothermal layer bounded above by 660 mb and the deeper one from 565 to

535 mb. If it is suggested that the upper one is continuous with the front at the surface, there would be difficulty explaining the small temperature differences aloft between Charleston and Greensboro compared to those between Greensboro and Pittsburgh. The 10-mb stable layer is 2° deep in potential temperature, which means that intersection of this zone with an isobaric surface gives about 2° in temperature contrast. Since temperature differences between Charleston and Pittsburgh exceed 15° C, the possibility of associating the frontal zone entirely with this stable layer is discarded too. Then, where is the stable layer at Greensboro by which to identify the front?

To show that a frontal zone need not be a layer of great stability, we might examine the slope formula for θ surfaces, Eq 4.13(12). That can be written

$$\tan_{sz} \beta_\theta = -(\partial T/\partial s)_p/(\partial T/\partial z + \Gamma_d) \, .$$

As $-\partial T/\partial z = \gamma$,

$$\gamma = (\partial T/\partial s)_p \cot_{sz} \beta_\theta + \Gamma_d \, . \qquad (1)$$

Table 6.20 gives γ from Eq (1). It is seen that, for small isobaric temperature gradients (weak fronts) and/or large isentropic slopes (steep fronts), *vertical temperature lapse rate in the frontal zone can be almost dry adiabatic.* Thus cold fronts, which are quite vertical near the ground, are not always the most stable layers. With more level isentropic surfaces and large isobaric temperature gradients, strong inversions are required. For the "average" frontal zone, consider a temperature gradient 5° C/100 km and isentropic slope 1/150; this gives a nearly isothermal layer. Since fronts are usually somewhat steeper than isentropic surfaces by a factor between 1 and 2, the slopes to be used in the table are slightly smaller than the actual slope of the front.

The results above do not contradict our statement made with reference to Figure 6.05 that inclined stable layers are baroclinic

zones. However, the converse does not hold, since strong baroclinity is possible with θ surfaces vertical.

Slopes of θ surfaces as drawn in the frontal zone over Greensboro in Figure 6.13e are about 1/100, and isobaric temperature gradient measured in this zone is about 5° C/100 km. From the table this implies that γ is half the dry adiabatic. The sounding has this lapse rate from 910 mb to the shallow isothermal layer near 670 mb.

Now we might examine other important aspects of the broader polar-front region. Observe that, while the baroclinic region extends almost vertically and narrows appreciably between the surface and 400 mb, the front tilts from the warm side of the baroclinic region at the surface to its cold side aloft. In low levels the cold air near the front is quite baroclinic and the tropical air barotropic, and the reverse is usually true above about 700 mb.

In the right-hand portion of Figure 6.13d is the same frontal zone with somewhat less intensity in that plane. The small temperature contrast at the ground is due to stagnant modified polar air in low levels to the east, and the front is not drawn down to the surface. At higher levels the cross section runs fairly parallel to the frontal zone, and it is not sufficiently defined in this plane. Furthermore, Buffalo's position north of the other stations in the chart adds difficulties.

Soundings for this portion of the chart are given in Figure 6.20c. In the Stewart[15] sounding the upper of the two distinct inversion layers can be considered the continuation of the upper one at Greensboro; both stations give similar potential temperatures. The lower inversion at Stewart can be associated with the frontal zone, of which it might be all or part. The inversion at Buffalo between 280 and 284 θ can be drawn in any of a number of ways in the

15. Stewart Air Force Base, Newburgh, N.Y.

cross section. Such inversions are common in the subsiding cold air behind most cold fronts and might be entirely separate from the frontal zone. As the pattern is drawn, this inversion appears dubiously as a "branch" of the frontal zone over Stewart. Neither alternative can be certain because of insufficient data in the critical area near the front. Past practices might have favored drawing the principal frontal zone from the surface position through the Stewart inversion at 282–90 θ and through the colder inversion at Buffalo, as a means of accommodating inversions, with disregard for the greater baroclinity above. But the logic of such an objective procedure defeats the purpose for which fronts are intended in analysis; the strong temperature contrast between Buffalo and Stewart above 700 mb would be masked in that type of analysis, while the front would be dismissed from its connotation of baroclinity to a less meaningful one of stability.

In all three frontal zones discussed, potential temperature increases upward along the boundaries of the front. Since moisture content normally decreases upward along the front, it can be expected that θ_E surfaces conform more with the front than do θ surfaces. In Figure 6.13e the 310 θ_E line extends (not shown) from near 400 mb at Pittsburgh to 560 mb at Greensboro to the surface position of the front. A careful θ_E analysis is often an objective guide for locating fronts in three dimensions.

Another example of frontal structure in cross section is afforded by Figure 6.06. Here again the baroclinic region extends vertically to the tropopause region. South of Charleston the troposphere is tropical and north of it the temperatures are freezing near the ground—polar air—but above 800 mb there is no real polar air anywhere in the picture. Temperatures −25° C at 500 mb over northern United States in winter are more nearly tropical than polar (Fig.

6.13e). The polar air mass deepens some distance north of Buffalo.

A weak frontal zone can be located just south of Charleston at the surface. It slopes upward to the north with average rate about 1/300 between Charleston and Greensboro, but northward it becomes horizontal and thereby loses its true frontal character, even though it still separates polar air near the ground from mild tropical air above. This case might be extreme in showing separation of the baroclinic field from the stable zone which might be identified as the polar front, but many situations approach it.

6.21. *Pressure and wind discontinuities at zones of transition.*—Inclined zones of transition (fronts) in temperature and potential temperature fields are basically concentrations of horizontal gradient, at the boundaries of which we usually can assume that, for practical purposes, the gradient is discontinuous, although in the average case it is only "nearly discontinuous" and in some cases not even that. If there are discontinuities in the temperature field, then from hydrostatics corresponding effects exist in the pressure field near fronts, which in turn are reflected in the wind field. So much practical and theoretical significance is attached to the relation between a front and horizontal pressure and wind fields that we must examine this question in some detail.

We approach the problem first from the classical viewpoint,[16] in which instead of a finite *zone* of transition we assume the change in temperature between air masses occurs abruptly at a *surface of discontinuity* (Fig. 6.21a). Across this surface T and θ are discontinuous, but pressure must be continuous, as otherwise pressure forces are infinite and the surface is not conserved. On

16. M. Margules, "Über Temperaturschichtung in stationar bewegter und in ruhender Luft," *Meteorologische Zeitschrift*, Hann-Band, 1906.

this basis it can be shown that, with prime symbols for the warm mass and double-prime symbols for the cold mass,

$$-\frac{\delta z}{\delta s} = -\frac{980}{g}$$

$$\times \frac{T''\,(\partial Z/\partial s)'_p - T'\,(\partial Z/\partial s)''_p}{T'-T''}$$

$$= \tan_{sz}\beta_F \quad (1)$$

for the slope of the line from 1 to 2, and therefore the slope of the frontal surface. The effect of differing temperatures in the numerator is small, and for simplicity we

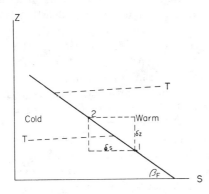

FIG. 6.21*a*

may use a value for T the average for the two air masses. Thus,

$$-\frac{\delta z}{\delta s} = -\frac{980}{g}$$

$$\times \frac{\bar{T}\,[\,(\partial Z/\partial s)'_p - (\partial Z/\partial s)''_p\,]}{T'-T}$$

$$= \tan_{sz}\beta_F. \quad (2)$$

The denominator of Eq (2) is positive. Also, the frontal surface slopes upward toward cold air; in Figure 6.21*a* the slope is negative. Therefore the slopes of isobaric surfaces are greater algebraically in the warm air than in the cold. This implies sharp *troughlike kinks* in isobaric surfaces at the front, which may or may not be actual troughs as defined by minimum elevation. Three possible isobaric configura-

tions in the vicinity of this discontinuity are shown in Figure 6.21*b*.

If geostrophic wind is introduced into Eq (2),

$$-\frac{\delta z}{\delta s} = \tan_{sz}\beta_F = -\frac{f\bar{T}}{g}\frac{c'_n - c''_n}{T'-T''}. \quad (3)$$

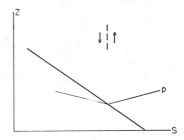

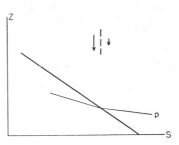

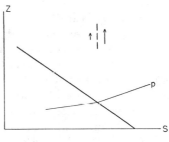

FIG. 6.21*b*

The wind c_n is the component of geostrophic wind into the cross section. It follows that *horizontal wind shear across the frontal surface is cyclonic*. In Figure 6.21*b* is given the distribution of this component in the vicinity of the front for each indicated pressure pattern; an upward vector implies wind into the cross section.

Excluding the small variation in $\bar{T}$, Eq (3) shows the magnitude of the frontal slope is (i) directly proportional to wind shear; (ii) inversely proportional to the temperature contrast; and (iii) directly proportional to the sine of latitude. The formula does not apply at the equator.

In isobaric surfaces (Fig. 7.03h) the front appears as a troughlike discontinuity line. Contours kink to "point" toward higher elevations. A similar pattern is observed in pressure in horizontal charts. It is found also that the geostrophic wind component normal to the discontinuity line must be the same on both sides.

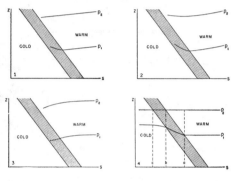

FIG. 6.21c

The above theory is an elementary basis for associating pressure and wind fields with fronts. Pressure patterns so deduced are in many ways similar to those observed with fronts near the ground and frequently through the troposphere. Important departures from the model are taken up shortly, but a word of caution should be injected here. Although we find that cyclonic shear and troughlike pressure patterns might be required for existence of fronts, the converse—cyclonic shear and pressure troughs *requiring* fronts—does not hold. Wind shifts, pressure troughs, and baroclinity are not all synonymous.

Even though the above description is a simple working basis and has applied reasonably in practice, there is some skepticism in accepting it, since it is based on a zero-order discontinuity in temperature—something not observed. To visualize this

concept factually, we now consider a *frontal zone* at the boundaries of which are first-order discontinuities in temperature. With the warm boundary first, we deduce the change in slope of a given isobaric surface p_1 at the front from hydrostatics by referring to (i) the slope of a free isobaric surface p_2 in the warm air above and (ii) the discontinuity in gradient of mean virtual temperature at the zone of transition.

In the first diagram in Figure 6.21c it is assumed that, in the plane of the cross section, p_2 is straight and the warm air barotropic. Where p_1 intersects the warm boundary of the frontal zone, the depth of layer p_2/p_1 changes from constant in the right to progressively decreasing values to the left. Thus the shape of p_1 within the frontal zone is different from p_2 directly above it. The slope in p_1 changes at the frontal boundary and gives a troughlike shape in p_1. There is thus cyclonic shear of geostrophic wind across the frontal boundary. (Incidentally, observe that p_1 must be *curved* concave upward within the frontal zone if p_2 is straight.) The discontinuity at the front is one of second order in pressure (i.e., discontinuity in curvature).[17]

From the second diagram one deduces the effect of concave curvature (cyclonic shear) in the upper isobaric surface is a bend in p_1 across the frontal boundary more accentuated than in the first case. The third shows the opposite type of curvature in p_2. Here the curvature in p_1 contributed by change in baroclinity at the front is diminished by the opposite effect aloft.

The general expression for pressure and wind distribution at the warm boundary of a zone of transition can be found by considering the variation of isobaric slope

17. We should now speak of "bends" rather than "kinks" in pressure surfaces, since defining a front as a first-order discontinuity in temperature gives that distinction in the type of discontinuity in pressure. Thus we are technically wrong in kinking isobaric surfaces at fronts, but there is some justification in continuing to do so in view of scale.

around the upper rectangular circuit *1, 2, 3, 4, 1* in Figure 6.21*d*. The object is to determine the difference in isobaric slopes between points *1* (in the warm air) and *4* (in the frontal zone), which are equidistant horizontally from the frontal bounary at *a*. The result is

$$\left[\frac{\delta\,(\tan\beta_p)}{\delta s}\right]_{FW}$$

$$=\frac{1}{2g}(N_F - N_W)\,|\tan\beta_F|$$

$$+\frac{980}{g}\left[\frac{\partial}{\partial s}\left(\frac{\delta Z}{\delta s}\right)_p\right]_W, \quad (4)$$

if the term on the left is the variation of isobaric slope from *4* to *1*, N_F and N_W are the baroclinity in the frontal zone and warm mass, and the last term refers to isobaric curvature at *b* directly above *a*. We may convert isobaric slope into geostrophic wind by reference to Eq 6.18(3) and by neglecting latitude variations:[18]

$$(\delta c_n/\delta s)_{FW} = (1/2f)(N_F - N_W)$$

$$\times\,|\tan\beta_F| + (\partial c_n/\partial s)_W. \quad (5)$$

According to the distribution of temperature in Figure 6.21*d*, $N_F > 0$. Warm-air baroclinity is either positive or zero, but $N_W < N_F$ is required to define a front. From Eq (4), the intensity of troughlike bend in the isobaric surface through *a* is proportional to (i) the contrast in baroclinity at the frontal boundary; (ii) the slope of the front; and (iii) the concave curvature in isobaric surfaces above the point in question. Eq (5) expresses the same in terms of wind: Cyclonic shear across the frontal boundary is proportional to the contrast in baroclinity, to the slope of the front, and to cyclonic shear in the warm air above. Thus, *when horizontal wind shear in the warm air is negligible, there is cyclonic shear across the front; the cyclonic shear is proportional to*

18. Eqs (4) and (5) refer to arbitrary *sz*-plane. The slopes and baroclinity should properly have *sz* subscripts, but these are being omitted intentionally to avoid confusion in symbols.

the slope of the front and to its contrast in baroclinity. When the last term in Eq (5) is positive, there is dual contribution to cyclonic wind shear at the front. *If there is cyclonic wind shear in the warm air above the front, the cyclonic shear across the frontal boundary is the sum of the warm-air shear and the contribution by the front.* If shear in the

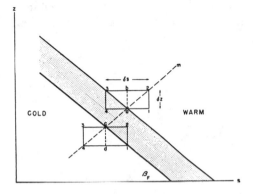

Fig. 6.21*d*

warm air is negative, the cyclonic shear across the front is less than contributed by the front but greater than in the warm air. There are three cases to consider. (i) *If horizontal shear in the warm air is anticyclonic but less in magnitude than contributed by the front, horizontal shear across the front is cyclonic.* This is revealed as a slight dip in isobaric surfaces at the front. (ii) *If horizontal shear in the warm air is anticyclonic and of magnitude equal to the contribution by the front, there is no horizontal shear across the front.* There is no bend in the isobaric surface. (iii) *If horizontal shear in the warm air is anticyclonic and of magnitude greater than contributed by the front, there is anticyclonic shear across the front.* However, for any condition with anticyclonic shear in the warm air, anticyclonic shear decreases from the warm air to the frontal zone, as the change in baroclinity is a maximum at the frontal boundary.

Horizontal distribution of wind and pressure across the *cold boundary* is quite different. To determine qualitatively how the patterns differ from those found for the upper boundary, refer to Diagram 4 in Figure 6.21*c* and again reason by hydrostatics. It is found that *at the cold side of a frontal zone there is anticyclonic shear relative to the wind shear at the warm boundary directly above.* This at least suggests that horizontal wind shear increases cyclonically upward through the frontal zone. Observe also by the drawing that horizontal shear

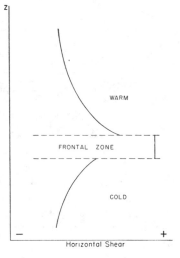

FIG. 6.21*e*

decreases isobarically from the warm boundary through the frontal zone.

For the cold boundary of the frontal zone, expressions for the bend in isobaric surfaces and for horizontal wind shear across a point c in relation to d (Fig. 6.21*d*) are the same as Eqs (4) and (5) if CF is substituted for FW and C for W. What was said for point a in relation to b is valid for point c in relation to d. But, from information deduced by Figure 6.21*c*, cyclonic shear and intensity of the troughlike bend in isobaric surfaces must be greater at the warm boundary than at the cold boundary below.[19] All this leads to the vertical distribution of horizontal wind shear given schematically in Figure 6.21*e*.

To see why horizontal wind shear increases upward in the cold air and then decreases upward in the warm air, as well as why there should be the indicated type of discontinuity in shear at the frontal boundaries, analyze

$$\frac{\partial}{\partial z}\frac{\partial c_n}{\partial s} = \frac{g}{f}\frac{\partial}{\partial s}\left(\frac{\partial}{\partial z}\tan\beta_p\right) = \frac{1}{f}\frac{\partial N}{\partial s}, \quad (6)$$

which excludes latitude variations. This single expression embraces most of what has been said about pressure and wind distribution at frontal discontinuities. The term on the left represents the slope of the profile in Figure 6.21*e*, and $\partial N/\partial s$ is the horizontal variation of baroclinity in the cross section. After careful analysis that drawing should be compared with the lower cross section in Figure 6.25.

The distribution of vertical wind shear $\partial c_n/\partial z$ upward through a frontal zone is similar in most respects to the profile for $\partial c_n/\partial s$ in Figure 6.21*e*, if the vertical line for zero shear is placed to the left of the profile. From Eq 6.18(5) it follows that vertical shear of c_n is a maximum in the frontal zone, is discontinuous at frontal boundaries, and decreases upward and downward in the warm and cold air masses from the frontal zone in the same manner as does the baroclinity.

While we have examined both horizontal and vertical wind shear at *boundaries* of a frontal zone, little attention was given to the horizontal wind shear *within* it. This was described earlier by Palmén.[20] With reference to the upper circuit *a, b, 3, 4* in Figure 6.21*d*, wind shear between *4* and *a* can be expressed in terms of wind shear between *3* and *b* and difference in baroclinity between the frontal zone and the warm air. That

19. This plus the effect of friction with the earth, and vertical mixing in the surface layer, would suggest that the cold boundary of a frontal zone is diffuse near the ground.

20. *Op. cit.*

derivation and analysis of results are left as a problem for the reader.

6.22. *Thermal distribution at zones of transition.*—The structure of a front should be analyzed further in its horizontal and vertical temperature distribution, since, in locating fronts daily, we must rely primarily on the thermal properties. This subject was discussed by Palmén and collaborators[21] and chiefly by Godson[22] among Canadian meteorologists.

As shown by Godson,

$$\delta z/\delta s = [(\partial\theta/\partial s)_F - (\partial\theta/\partial s)_W]/$$

$$[(\partial\theta/\partial z)_W - (\partial\theta/\partial z)_F] = -\tan_{sz}\beta_F , \quad (1)$$

where F and W refer to frontal zone and warm air. If potential temperature distribution is known in the vicinity, it is possible to determine the slope of the front. For example, take (i) $\gamma = 0$ in frontal zone; (ii) $\partial\theta/\partial s$ in frontal zone $10°$ C/100 km; (iii) $\partial\theta/\partial s$ in warm air near front $4°$ C/100 km; (iv) $\gamma = 5°$ C km^{-1} in warm air; and (v) pressure about 700 mb, where $\theta/T = 1.1$. The slope of the front is 1/92. Larger values are obtained by increasing the contrast in baroclinity or decreasing the contrast in stability.

By approximate substitution of $(\partial\theta/\partial s)_p$ for $\partial\theta/\partial s$ and reference to Eqs 6.12(5) and 3.17(3), Eq (1) becomes

$$\delta z/\delta s = (N_F - N_W)/(E_W - E_F)$$

$$= -(\tan_{sz}\beta_F) , \quad (2)$$

where N is baroclinity in the plane of the cross section, E is stability, and subscripts

21. See previous references in n. 13 and E. Palmén, "Discussion of Problems concerning Frontal Analysis in the Free Atmosphere," *Soc. Scient. Fenn., Com. Phys.-Math.*, Vol. XIII, No. 8 (1946).

22. W. L. Godson, "Synoptic Properties of Frontal Surfaces," *QJRMS*, Vol. LXXVII, No. 334 (1951), and "The Structure of North American Weather Systems," *Centennial Proceedings of the Royal Meteorological Society, 1950*.

F and W refer, respectively, to the frontal zone and warm air. As the sign of the frontal slope is governed by the distribution of baroclinity in the section, we may write

$$|\tan_{sz}\beta_F| = (N_F - N_W)/(E_F - E_W) , \quad (3)$$

and

$$E_F - E_W = (N_F - N_W)/|\tan_{sz}\beta_F| . \quad (3a)$$

For the cold boundary of the frontal zone, subscript C (cold mass) is substituted for W.

From Eq (3) frontal slope varies directly with the change in baroclinity and inversely with the change in stability. Since $N_F > N_W > 0$ and $N_F > N_C > 0$, stability in the frontal zone must be greater than in adjacent air masses, but no magnitudes are specified. The greater the contrast in stability, the smaller is the frontal slope; that is, fronts of small slope, including those approaching horizontal stable layers, are relatively well marked in soundings.

We may now eliminate frontal slope from Eqs 6.21(4) and (5) and replace it by contrast in stability by substitution from Eq (3) above. Then for the warm boundary

$$\left[\frac{\delta(\tan\beta_p)}{\delta s}\right]_{FW} = \frac{(N_F - N_W)^2}{2g(E_F - E_W)}$$

$$+ \frac{980}{g}\left[\frac{\partial}{\partial s}\left(\frac{\delta Z}{\delta s}\right)_p\right]_W . \quad (4)$$

$$\left[\frac{\delta c_n}{\delta s}\right]_{FW} = \frac{(N_F - N_W)^2}{2f(E_F - E_W)} + \left(\frac{\partial c_n}{\partial s}\right)_W . \quad (5)$$

For the cold boundary substitute C for W.

6.23. *Special geometric properties of frontal zones.*—In the lower atmosphere the shape of a front in cross section is governed largely by the retarding effect on air motion by friction with the ground. A cold front assumes the type of curvature shown in Figure 6.23a: *convex in the direction of motion.* A warm front drags behind at the ground and takes the opposite shape but is also convex in the direction of motion. Near

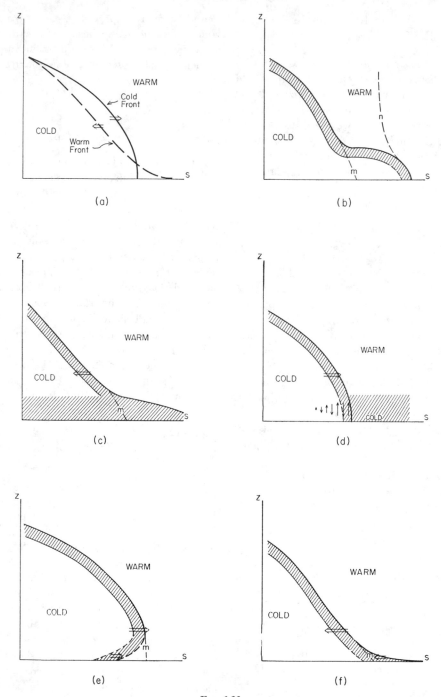

Fig. 6.23

the ground cold fronts tend to vertical orientation, and warm fronts tend to the horizontal. At levels sufficiently above there is no apparent difference in shape between warm and cold fronts.

Since warm and cold fronts differ so radically in slope near the ground, there must be corresponding differences in pressure and wind distributions at the surface. Section 6.21 suggests that on the surface chart cold fronts should be, and actually are, more clearly shown in pressure and wind fields than are warm fronts.

A front which at one time is moving as a cold front may retard, become stationary, and eventually return as a warm front, and conversely. A front may be a cold front in one level and a warm front in another, and the shape of the front changes accordingly. There is no real difference between cold and warm fronts except in the direction of motion of the temperature pattern and in the consequences of that motion.

Figure 6.23*b* depicts a vertical structure[23] frequently found in both warm and cold fronts. The frontal zone may be wave shaped in a vertical plane. Along with the alternating regions of steep and flat slope there are corresponding irregularities visible in the pressure and wind fields above or below. If the region of greatest frontal slope lies in a pressure trough, then as hydrostatic necessity this trough is reflected downward through the cold air along an axis *m*, whose tilt depends on the baroclinity of the cold air. Thus, a nonfrontal trough beneath the sharp bend in the front may be found at the ground some distance to the cold side of the surface front, which itself might lie in a pressure trough of greater or less intensity. If the part of the front near the ground is steep and lies in a sufficiently intense pressure trough at the surface, the chances are

23. Figure 6.02*b* can be considered a detailed view of the right-hand portion of this one.

this trough veers upward into the warm air, in accord with hydrostatics, and does not follow the profile of the flat portion of the front; it is a nonfrontal trough aloft. A common feature of this pattern is a double field of baroclinity, one extending upward from near the surface position of the front and the other extending upward (and frequently also downward) from the region of greatest slope in the front.

Quite frequently over continents in winter, fronts are distinct at upper levels but apparent at the ground only in the hydrostatically reflected pressure and wind fields (Fig. 6.23*c*). A stable cold film of air at the ground, maintained by net radiation cooling or by advection of milder maritime air above the stagnant surface layer, is difficult to dislodge because of its resistance to vertical mixing and its subnormal surface winds. Advance of a warm front further stabilizes the surface layer, and, except when attended by sufficiently strong wind to give the necessary turbulent mixing, the warm front glides over the surface layer without evidence as a large temperature contrast at the surface but only as a weak or moderate pressure trough and wind contrast extending downward through the stable layer along axis *m*. This is true also in the case of cold fronts followed by relatively mild air at upper levels compared to the temperatures in the cold film at the surface. If the air following the cold front is considerably colder than the surface, instability leads to vertical mixing, and the surface shielding layer is "swept away" by the front (Fig. 6.23*d*). Quite frequently over land such conditions as in diagram (*d*) lead to higher surface temperatures to the rear of the cold front than ahead of it. This occurs most often at night or in early morning with light winds and clear skies in the warm air. By comparing diagrams (*c*) and (*d*), one will conclude that, for equal frequencies of warm and cold fronts at upper levels, at the surface cold fronts are observed more often than warm fronts.

We now come to the question of the "overrunning" cold front illustrated by (e). Surface friction retards the air in contact with the ground while above the air moves less impeded, and a material surface with the ordinary slope of a cold front becomes deformed as indicated. Thus, if a front is a material surface or zone, it soon develops an inverted slope near the ground, and cold air at intermediate levels overruns its surface position. Certainly, such overrunning does happen, but we cannot assume a front is a substantial surface, especially in this region of mixing. The condition shown by the drawing is an unstable one. The warm air trapped below the front is mixed rapidly with the cold air above, producing surface cooling as far as line m—the approximate axis of the pressure trough—and the inverted portion of the front is wiped out. Overrunning and vertical mixing are characteristic of cold fronts, but the inverted front is destroyed in a continuous process. Air is transferred from the warm mass to the cold mass where it loses its properties through mixing; as a result the front and the adjacent cold air approach the temperature of the warm air mass unless other processes compensate.

By the same token we should suspect that "overrunning" warm fronts are a rule near the ground; this possibility has not had sufficient emphasis in the past. Through surface friction the stable cold air in contact with the ground (shown in [f] as part of the frontal zone) may lag far behind the front aloft. Often in winter over land and in summer over cold water, this thin film or *schleppe* of cold air lags behind the steeper part of the front by several hundred miles and may be just a few hundred feet deep. In such cases the surface boundary between warm air and transition zone usually is not found in a distinct pressure trough, as the

slope of that boundary is insufficient. Instead, a more pronounced trough is likely to be found below the sharp bend in the front (in the vicinity of axis m), where the product $(N_F - N_W) |\tan \beta_F|$ is much larger. The surface front can re-form at somewhat lower temperature near m. This too is expected to be a continual process of overrunning, redevelopment of the front at the surface, and transfer of surface air from one air mass to the other. The air thus transferred loses its properties by mixing with the warmer mass, but mixing is usually slower in this case because of stability. Consequently, behind a warm front there is generally a stable layer near the ground (often with fog or stratus if the moisture is sufficient) and temperatures of more tropical nature above. The usual explanation offered for this stable layer is contact cooling of tropical air with the ground, a process which no doubt can play a part. That theory probably arose from assuming a front is a substantial surface.

An interesting aspect of this thin film of cold air is its apparent horizontal accelerations over land with diurnal changes in surface radiation. With favorable skies, daytime heating mixes this air with the warm mass above, and the boundary of the warm air retracts toward m. During nighttime cooling the film extends itself horizontally, and the cold surface air mass retards.

There are a few points worth adding in connection with locating frontal zones in the lower portions of soundings. In the case of a cold front, the usual type of sounding through this region is a surface mixed layer with relatively unstable conditions and, except perhaps in the frontal zone itself, bounded above by a more stable "turbulence inversion" (Toledo, Buffalo, Stewart, Pittsburgh, and Greensboro in Fig. 6.20). The turbulence inversion is easily mistaken for the frontal zone, and the mixed layer may not be accepted at first hand as part of the frontal zone. Subsidence inversions are common some distance behind the cold front (Figs. 6.13*e*). With warm fronts, the surface stability behind the front, and frequently also ahead of it, is often an inducement for association with the warm-front zone.

Besides being a zone of large temperature gradient, the polar front is also characterized by certain preferred values of potential temperature which vary somewhat with elevation and latitude and more with the season. Roughly, the polar front should be embraced within the range 280–95 θ near the ground in latitudes of the United States in winter (Figs. 6.02*b* and 6.13*e*). In summer the values of θ are some 10° higher. Average annual values correspond nearly with those in the U.S. Standard Atmosphere. The ranges in θ given above correspond to certain ranges in θ_E (or θ_w) and also to a certain band of temperature at each level. Statistics of this type will be an objective aid for locating fronts in any type of chart, but still the greatest difficulty comes in analysis near the ground.

6.24. *Discontinuities at the tropopause.*— The tropopause is assumed a first-order discontinuity of temperature in the vertical and is similar in this respect to lower boundaries of tropospheric stable layers. As evident from the illustrations, the tropopause is also a discontinuity in horizontal (isobaric) temperature gradient. The slope of the tropopause is a function of these discontinuities, and also the tropopause must be featured by certain changes in pressure and wind field. This cannot apply to the indistinct or multiple sections of the tropopause.

We may determine the tropopause slope by the method used in obtaining Eq 6.22(2). The small rectangle b, a, 3, 4 in Figure 6.24 has dimensions δz and δs, and the slope of the tropopause ($\tan_{sz} \beta_t$) between 4 and a is $\delta z/\delta s$. First, the temperature on the

stratosphere (S) face of the tropopause at a is related to the temperature at 4, and then similarly for the troposphere (T) face. As temperature is continuous, $T_a - T_4$ is the same on either face, and

$$\delta z/\delta s = [(\partial T/\partial s)_T - (\partial T/\partial s)_S]/$$
$$[(\partial T/\partial z)_S - (\partial T/\partial z)_T] = \tan_{sz} \beta_t \quad (1)$$

is the slope of the tropopause. The formula is the same with potential temperature in

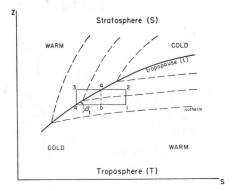

FIG. 6.24

place of temperature. Substitution of baroclinity (N) and stability (E) gives

$$\tan \beta_t = (N_T - N_S)/(E_S - E_T) . \quad (2)$$

The denominator is always positive, and the sign of tropopause slope is governed only by distribution of baroclinity. In Figure 6.24 slight positive baroclinity is indicated in the troposphere ($N_T > 0$), and in the stratosphere baroclinity is somewhat greater but negative. Hence, ($N_T - N_S$) > 0, and $\tan \beta_t > 0$. Near Omaha in Figure 6.13*d* tropospheric baroclinity near the tropopause is small, and, since $N > 0$ in the stratosphere, the tropopause slopes downward to the right. In areas where $N_T = N_S$ the tropopause is horizontal irrespective of the relative stability distribution. As E_T approaches E_S, $\tan \beta_t \to \infty$; recall that earlier we said the tropopause is ill defined in regions where it otherwise would have greatest slope.

The features in the pressure and wind fields at the tropopause can be found by the procedure used in developing Eqs 6.21(4) and (5). With reference to rectangle *1, 2, 3, 4* in Figure 6.24, the dimensions of which are δz and δs, the following expressions are obtained:

$$\left[\frac{\delta(\tan\beta_p)}{\delta s}\right]_t = (N_T - N_S)(\tan\beta_t)/2g$$
$$+\frac{980}{g}\left[\frac{\partial}{\partial s}\left(\frac{\delta Z}{\delta s}\right)_p\right]_T, \quad (3)$$

$$\left[\frac{\delta c_n}{\delta s}\right]_t = (N_T - N_S)(\tan\beta_t)/2f$$
$$+\left[\frac{\partial c_n}{\partial s}\right]_T. \quad (4)$$

By substitution from Eq (2), the slope of the tropopause can be eliminated from Eqs (3) and (4).

$$\left[\frac{\delta(\tan\beta_p)}{\delta s}\right]_t = \frac{(N_T - N_S)^2}{2g(E_S - E_T)}$$
$$+\frac{980}{g}\left[\frac{\partial}{\partial s}\left(\frac{\delta Z}{\delta s}\right)_p\right]_T. \quad (3a)$$

$$\left[\frac{\delta c_n}{\delta s}\right]_t = \frac{(N_T - N_S)^2}{2f(E_S - E_T)} + \left[\frac{\partial c_n}{\partial s}\right]_T. \quad (4a)$$

The left side of each equation refers to point *a* in Figure 6.24, and the last term in each refers to *b* directly below *a*. From these equations it is readily apparent that the first term on the right is always a positive contribution to a dip in isobaric surfaces (cyclonic shear) at the tropopause. This dip is expected where horizontal shear in the upper troposphere is either positive or zero.[24] Where this shear is strongly negative, as on the right side of fast wind currents, such a dip need not appear, and horizontal shear across the tropopause can be of either sign or zero. When there is no contrast in baroclinity across the tropopause (tropopause horizontal), there is no discontinuity

24. Such features in the pressure pattern at the tropopause might extend upward into the stratosphere or downward into the troposphere.

in the pressure or wind field. These expressions are not valid when E_T approaches E_S. In summary, the tropopause is characterized by a relative maximum of cyclonic shear.

The *vertical* wind shear across the tropopause is different from that through fronts. It is theoretically discontinuous at frontal boundaries and a maximum in the frontal zone. The tropopause and lower stratosphere are not zones of concentrated baroclinity in that sense, and they are not characterized by maximum vertical wind shear. In fact, *where the tropopause is a reversal in horizontal temperature gradient, it is a boundary of zero vertical wind shear.*

6.25. *Summarizing remarks.*—In the upper diagram of Figure 6.25[25] are wind and temperature distributions in eastern North America averaged meridionally with respect to the polar front at 500 mb in the twelve more distinct situations in December, 1946. There are several features indicated which either have been discussed or are apparent in Figures 6.13*de*. Above and below the sloping frontal zone the two air masses are strongly baroclinic. The upper wind maximum has almost vertical axis above about the 600-mb position of the front. Approximately at the intersection of the tropopause with a vertical axis through the center of tropospheric baroclinity is found the maximum geostrophic wind. To both sides of the maximum wind, or jet stream, there is large horizontal wind shear (lower cross section). A few degrees of latitude to the left of the current center is found a stratospheric temperature maximum. Within the frontal zone the vertical wind shear is a maximum. Across the surface position of the front there is cyclonic wind shear, and the same is true in the uppermost levels of the front. In the intermediate region (700 mb) cyclonic shear at the front is a mini-

25. See also Figures 1.07.

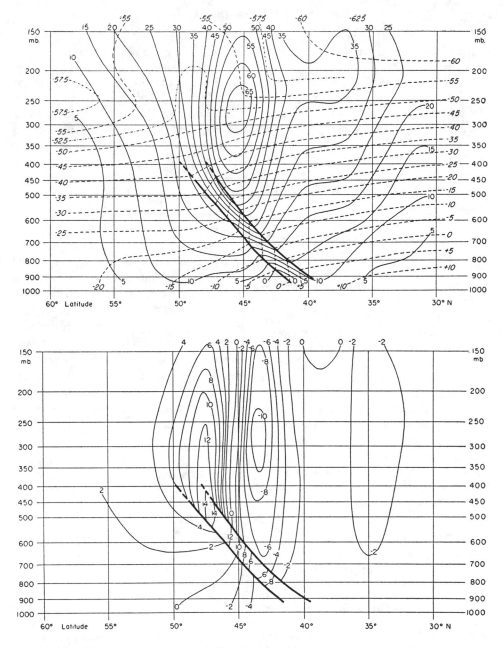

Fig. 6.25.—*Upper:* Mean temperature (° C) and geostrophic west-wind component (m sec^{-1}) across the polar front, computed along the meridian 80° W from twelve cases in December 1946. Heavy lines are frontal boundaries. Means were computed with respect to the polar front. *Lower:* Mean horizontal wind shear computed from the cross section above. Units are m sec^{-1} per 100 km, or 10^{-5} sec^{-1}. Positive numbers indicate cyclonic shear; negative, anticyclonic shear. (After E. Palmén and C. W. Newton, "A Study of the Mean Wind and Temperature Distribution in the Vicinity of the Polar Front in Winter," *Journal of Meteorology*, Vol. V [1948].)

mum, and this occurs below the strong anticyclonic shear on the right flank of the jet stream.

A striking feature of the wind distribution is the poor correlation of the pattern in the lower troposphere with the pattern above. In the more barotropic regions the wind varies little with height, both in magnitude and in pattern, but in the baroclinic region it increases rapidly with height; and the pattern at the surface is rapidly and radically modified with height. Notice also the tendency for the upper wind maximum to branch out downward to north and south.

Another interesting feature attending the stronger wind currents of middle latitudes is the sort of "pseudo-tropopause" located to the upper right of the wind maximum in Figure 6.14*b* and apparently imbedded in the tropical troposphere. This ridge-shaped structure follows the axis of the wind current and appears as the only tropopause over Nantucket in Figure 6.14*a*. It slopes downward and becomes diffuse toward the current center, but in the opposite direction it occasionally slopes gradually downward as a sheet or thin stable layer for hundreds of miles from the current center. The ridge in this canopy has a temperature minimum[26] similar to the common tropopause.

The temperature patterns in the cross sections we examined indicate that the principal regions of baroclinity in the troposphere extend through the troposphere. This suffices as a general picture, but, as in anything else, departures in details are frequent. There may be large temperature gradients only at the ground, maintained by varying controls at the surface, such as by coastlines, by differences in radiation over land, or by shallow circulations at the ground, all quite independent of the broader picture. Thus, the surface temperature pattern is not always reliable in giving the patterns of temperature and motion through the troposphere. Likewise, there is increasing empirical evidence of strongly baroclinic conditions (and attendant wind systems) in the upper troposphere with little if any traces in the lower troposphere. Such a structure is frequent in tropical air masses, which by nature are quite homogeneous near the ground, and is observed over the United States and similar latitudes in summer.

26. This feature was described earlier by H. Riehl, "Jet Stream in Upper Troposphere and Cyclone Formation," *Transactions of the American Geophysical Union*, Vol. XXIX, No. 2 (1948).

PROBLEMS AND EXERCISES

1. Find and compare the slopes of isotherms and isentropes in a vertical cross section for the sets of values given below:

$-\partial T/\partial z$ (° C/km):   4   0 -4 4   0 -4
$(\partial T/\partial s)_p$ (° C/100 km): 1   1   1   5   5   5

2. Two verticals are 100 km apart. Along the left one the (virtual) temperatures are $-13°$ C and $-20°$ C at 600 and 500 mb, respectively. Along the right one they are $-8°$ C and $-17°$ C at those pressures. Assuming constant vertical lapse rates of (virtual) temperature from bottom to top of the layer,

   (i) Determine the average baroclinity over this cross-section area.

   (ii) Determine the number of unit isosteric-isobaric solenoids in this element of cross section.

   (iii) What is the true baroclinity if the plane of the cross section makes an angle 60° with virtual isotherms at both 600 and 500 mb.

   (iv) With the results of (i) above, find the average change in geostrophic wind normal to the cross section from 600 to 500 mb ($f = 10^{-4}$ sec^{-1}).

3. In a pressure ridge line in a cross section the 500-mb height is 5700 gpm. On both sides the isobar slopes downward with average rate 2 gpm in 100 km. Using virtual temperature 250° K, determine the geographic displacement of the ridge per 1 gpkm elevation if $(\partial T^*/\partial s)_p$ across the ridge line is (i) 0.5° C/100 km; (ii) −4° C/100 km. How would the results differ if the pressure surface sloped downward to both sides only one-fourth as much?

4. Find the geostrophic wind (m sec⁻¹ and knots) normal to a plane for the given sets of latitude and isobaric slope $(\partial Z/\partial s)_p$ in that plane.

$(\partial Z/\partial s)_p$ (gpm/100 km):	5	5	20	20	80	80
ϕ (° North):	30	60	30	60	30	60

5. Give the expression for *horizontal* geostrophic wind shear in terms of *isobaric* geostrophic wind shear (Eq. 6.18[4]). Under what conditions do they differ most?

6. Find by integrating Eq 6.18(5) the increase of geostrophic wind speed between sea level and 10 km at 40° latitude for mean temperature 250° K and isobaric temperature gradient 1° C per 100 km.

READING REFERENCES

BYERS, H. R. *General Meteorology*, pp. 184–99, 211–13, 217–18, 245–46, 298–308, 350–62. New York: McGraw-Hill Book Co., 1944.

HOLMBOE, J., FORSYTHE, G. E., and GUSTIN, W. *Dynamic Meteorology*, pp. 220–32, 300–306. New York: John Wiley & Sons, 1945.

PETTERSSEN, SVERRE. *Weather Analysis and Forecasting*, pp. 274–86, 482–83. New York: McGraw-Hill Book Co., 1940.

CHAPTER 7

Isobaric Analysis

7.01. *Introduction.*—Analysis at constant pressure affords the nearest approach to three-dimensional analysis of the physical and kinematic variables all at once. Data entered on isobaric charts are observed temperature, humidity, height, and wind. These give a complete analysis in each isobaric surface, and, when charts for successive surfaces are examined together, a picture of three-dimensional structure in those variables can be obtained in addition to certain other indications (such as stability) from the vertical variations.

Even more value perhaps could be derived from isobaric analysis in understanding the daily weather by modernizing those phases of analysis dating to the times when data were obtained only from the surface. As one example, cloud data giving visual indications of conditions rather high in the atmosphere are entered exclusively on the surface chart. Of course, rain falling from clouds at 700 mb is more intimately associated with atmospheric processes and state at 700 mb than at the surface.

7.02. *Broad-scale pressure patterns in the troposphere, 1 March 1950.*—Figure 7.02a is the hemispheric sea-level pressure analysis for 1230 GCT, 1 March 1950, and Figure 7.02b the pattern at 500 mb. The sea-level pattern can be interpreted in terms of topography of pressure surfaces near sea level. In average conditions the 1000-mb surface slopes upward 100 feet for each $3\frac{3}{4}$-mb increase in sea-level pressure.

If the average pressure along each latitude circle is computed from Figure 7.02a,

one should find that the meridional pressure profile is in good agreement with the winter mean for the hemisphere (Fig. 1.031a) and typical of daily winter conditions. In Figure 7.02a the Siberian HIGH is split, and near the location of the intense HIGH center in the winter mean there is a deep LOW. That is perhaps the greatest departure, but it does show that climatic charts are not completely indicative of daily patterns.

Figure 7.02a shows that, with possible exception of the more semipermanent features, low-level circulation really consists of a large number of eddies of varying dimension and somewhat erratic distribution. Also, the pattern varies considerably with latitude; it is difficult to establish in this pressure field an organized wave form holding for a large range of latitude. For example, the subtropical HIGH of the eastern Pacific Ocean is in the longitudes of the low pressure near the Aleutians. Therefore, though we may speak of wave forms in sea-level pressure over a limited area or along a certain latitude, such waves generally vary in amplitude, location, shape, and number over the map.

On comparing Figure 7.02b with Figure 1.032a, several common features are found. Perhaps the most striking is the concentration of isobaric slope within a relatively narrow range of latitude. When interpreted by geostrophic wind, this shows that "zonal" (midlatitude) westerlies are far more concentrated and stronger aloft than near the surface. Except for the split in the European sector, this westerly flow appears more as a single current with greater speeds and waves of greater length than at the surface. As a rule, the pressure pattern aloft is organized in more simple wave pattern than below, which is mostly an indication of the similar wave pattern of temperature.

The subtropical high-pressure belt is displaced southward by roughly 10° and is generally more continuous than at sea level, and its pressure cells have greater west-east elongation aloft than at the ground. There is little isobaric slope between the subtropical high-pressure belt and the equator, which suggests that tropical easterlies aloft are relatively weak.

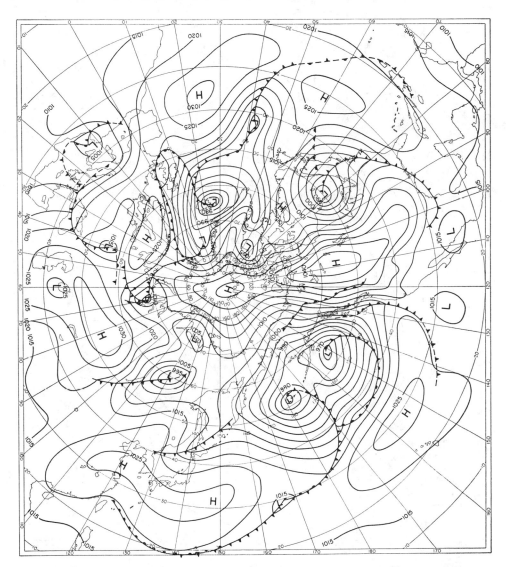

FIG. 7.02a.—Sea-level pressure and frontal analysis (based on *Daily Series, Synoptic Weather Maps, Northern Hemisphere Sea Level and 500-Millibar Charts*) for 1230 GCT, 1 March 1950. (Units of pressure are millibars; fronts are entered in standard symbolic form.)

The polar regions are another weak section of the pattern, but here the flow is dominated by several depressions situated for the most part just north of the principal westerly current. The topographic features at 500 mb correspond in general to those near sea level, and their shift in location as well as change of shape and intensity upward are dependent on the temperature distribution. Depressions have westward tilt with height when their western sectors are colder than their eastern sectors. The

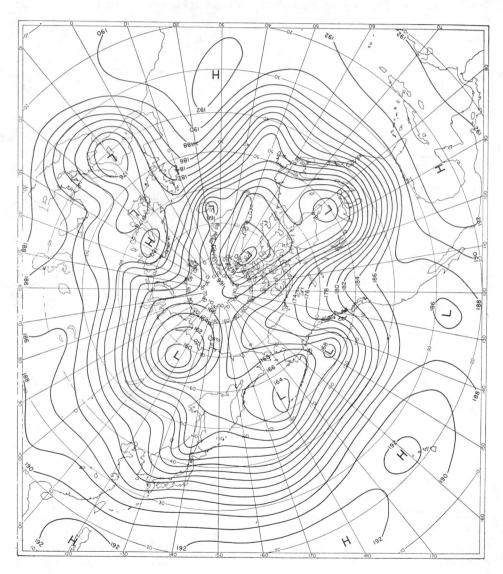

Fig. 7.02b.—500-mb analysis (based on *Daily Series, Synoptic Weather Maps, Northern Hemisphere Sea Level and 500-Millibar Charts*) for 1500 GCT, 1 March 1950. (Contours are labeled in hundreds of geopotential feet.)

large northwestward shift of the main Icelandic LOW across Greenland between sea level and 500 mb agrees with the northwestward temperature drop and weak sea-level pressure difference across Greenland. Just west of Greenland is the cold center of the North American sector for the troposphere as a whole, and air temperatures in the oceanic area southeast of Greenland are controlled by a comparatively warm water surface.

7.03. *Topographic analysis of constant pressure.*

—From reported heights of an isobaric surface the topography is analyzed to determine the field of motion and also to be used as a parameter in forecasting it, which are the primary reasons for performing any pressure analysis. Figure 7.03 gives topographic analyses of several surfaces at 1500 GCT, 1 March 1950. On the 1000-mb chart the winds given are the third level in the wind reports, which is by custom the "gradient level."

At all levels there is reasonably good agreement between actual winds and isobaric topography. Geostrophic motion is parallel to contours, with downward slope to the left of the wind (northern hemisphere), and with speed proportional to isobaric slope and inversely proportional to to the sine of latitude (Eq 6.16[2]). At most levels actual winds agree sufficiently well with geostrophic to be an objective guide in drawing the pressure pattern. Even near the ground, where the wind ordinarily is subgeostrophic with a cross-contour component toward lower pressure, observed winds are of help, particularly when there are more wind than pressure reports.

Besides the use of the pressure-wind relation, which definitely adds an objective element, there are other useful principles in pressure analysis. All rules of ordinary scalar analysis are applicable. Space and time continuity, and all indirect evidence given by other forms of observation, must be employed in the areas of fewer data. Vertical continuity will be covered in detail later. Time continuity was employed over ocean areas in Figure 7.03 and, with less emphasis, over regions with more data.

7.04. *The geostrophic equation in pressure and wind analysis.*

—Though the geostrophic wind equation is not always reliable in relating the actual wind to the pressure pattern, it serves a useful purpose. With reference to constant pressure, the geostrophic equation is

$$fc_g = 980(\partial Z/\partial n)_p , \qquad (1)$$

if n is directed upslope along the surface and c_g is geostrophic wind speed. In finite differences

$$c_g = (980/f)(\delta Z/\delta n)_p , \qquad (1a)$$

or

$$\delta n = 980\delta Z/fc_g , \qquad (1b)$$

with constant pressure implied in Eq (1b). Either equation is solved readily for the quantity on the left from given values for the three independent variables by a geostrophic scale.

We proceed first with pressure analysis, given observed winds, using the example in Figure 7.04a. At a point near 57° N, 700-mb height is 9830 feet, and the wind is southwest 30 knots. The problem is to approximate the topography of the 700-mb surface in the vicinity, assuming the wind is geostrophic. For that latitude and wind speed the geostrophic scale[1] (e.g., Appendix, Fig. N) gives 1.4° latitude for spacing of 100-foot contours. The 9800-foot contour is evidently 0.4° latitude to the left of the wind. The 9900-foot contour then is located also. Additional contours may be entered to

1. There are many nomograms and transparent overlays available for determining the spacing of contours about a given wind or, conversely, for determining the wind from the spacing of contours.

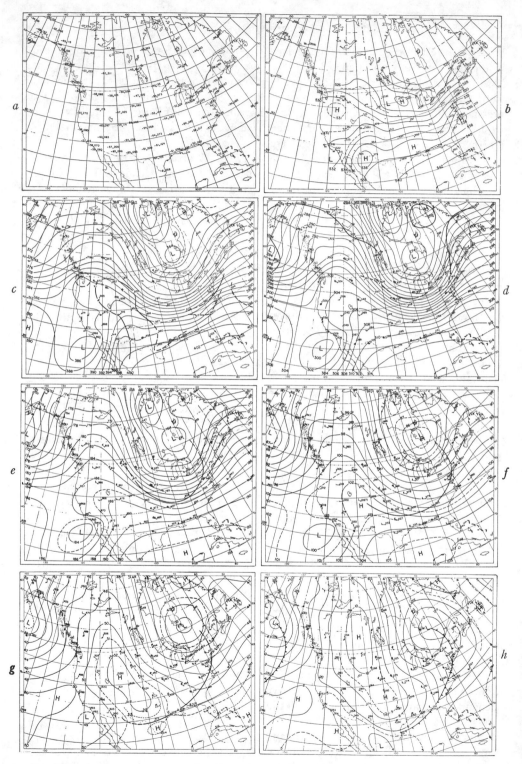

FIG. 7.03.—Analyses of isobaric topography for 1500 GCT, 1 March 1950. (Contours labeled in hundreds of geopotential feet.) (a) Temperature and pressure of the highest point reported in the radiosonde reports received; (b) 100 mb; (c) 200 mb; (d) 300 mb; (e) 500 mb; (f) 700 mb; (g) 850 mb; and (h) 1000 mb.

either side with the same spacing. But remember that the spacing obtained is geostrophic and also applicable only in the vicinity of the given wind. The "vicinity" may or may not be larger than a 100-foot interval, depending on wind shear. Caution here is necessary. Although the contours may have been entered straight, chances are that they are curved. The curvature of contours and their variation in spacing are determined from other indications. It is evident that a reasonable pressure analysis

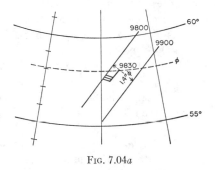

Fig. 7.04a

can be derived by this method from a good network of wind reports, provided not too large a percentage of the winds depart radically from geostrophic.

The reverse procedure is taken to determine geostrophic wind from a given pressure analysis. For practice in this operation and for comparing meantime geostrophic and actual winds, the following exercise is suggested: In a region of dense pressure data draw the isobaric topography without reference to observed winds. Then, using a geostrophic scale, enter winds at each observation point. Compare the results with reported winds by the vector differences between computed and reported winds.[2] For deriving the pattern of geostrophic speed in the isobaric topography,

2. A discussion of this type is given by M. Neiburger, L. Sherman, W. W. Kellogg, and A. F. Gustafson, "On the Computation of Wind from Pressure Data," *Journal of Meteorology*, V (1948), 87–92.

a method based on Section 5.05*a* is applicable.

The vector difference $\mathbb{C}'$ between actual wind $\mathbb{C}$ and geostrophic wind $\mathbb{C}_g$ at the same point (Fig. 7.04*b*) is the "geostrophic departure" or "ageostrophic wind."

$$\mathbb{C}' = \mathbb{C} - \mathbb{C}_g . \qquad (2)$$

Ageostrophic wind is the vector added to geostrophic wind to obtain actual wind. Outside equatorial regions and except near the ground, the ageostrophic vector is usually much smaller than geostrophic or actual wind, as shown by careful examination of all charts in Figure 7.03. In certain large areas it appears negligible. Comparison of actual and geostrophic winds might be conditioned by the accuracy of observations, by the accuracy of pressure analysis, by the fact that pressure analysis was based on actual winds, and also by the difference in scale between pressure patterns to which we conform in analysis and those affecting the winds locally.

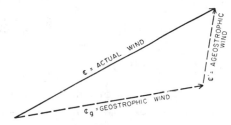

Fig. 7.04*b*.—Vector relation of actual, geostrophic, and ageostrophic wind.

We should expect less agreement between actual and geostrophic winds near the ground (1000 mb or sea level) because of surface friction, land obstacles modifying the flow, and the difficulties with pressures below ground. In view of all this, agreement rather than disagreement between wind and topography of 1000 mb over land is remarkable. Over the Pacific coast of Canada and southern Alaska in Figure 7.03*h*, winds seem to follow more nearly the contours of land than contours of the pressure surface, with a component toward lower pressure, as is common where strong geostrophic wind impinges on a steep land barrier. Some air is displaced upward, but some flows parallel to the slope. Strong downslope motion also departs from the pressure pattern.

Over parts of western United States and Canada, 1000 mb is several thousand feet below ground and a still greater distance below the levels of the plotted winds. There, and also in flat pressure regions (e.g., the California sector) where thermal and orographic circulations might contradict the large-scale pressure pattern, we should not relate actual to geostrophic wind. Over an area extending from south-central Texas, along the Appalachians, and into the eastern Great Lakes region, there is pronounced ageostrophic flow toward lower pressure which apparently is due to other factors.

The pressure-wind relation improves substantially upward to 850 mb. Departures in some areas are still apparent, however, along the Pacific coast, in regions where 850 mb is near or below ground, and in some areas where departures are controlled by other than surface friction and physiography and which therefore might appear at still higher levels also, either larger or smaller than near the ground.

In the northern hemisphere, air currents curved clockwise are said to be *anticyclonic;* those curved counterclockwise, *cyclonic.* A complete clockwise circulation is called an *anticyclone* and the opposite a *cyclone,* which correspond to HIGHs and LOWs in the pressure field, respectively.

7.05. *Analysis of the 1000-mb chart.*— The topography of 1000 mb shown in Figure 7.03*h* was determined from heights given by radiosonde reports, from gradient-level winds, and from other considerations. The pressure data can be supplemented by sea-level pressures reported hourly, three-hour-ly, or six-hourly. These are converted to 1000-mb heights by a barometric conversion table or formula. Table F (Appendix) gives sea-level pressures for the usual range of 1000-mb heights as a function of temperature between sea level and 1000 mb. That temperature can be estimated from surface reports to within the necessary accuracy. For many purposes conversion of sea-level pressure to 1000-mb height based on standard atmosphere temperatures will suffice, and it can be done mentally for each station in the process of drawing contours. With some practice this can be accomplished as quickly as drawing sea-level isobars. Or, more roughly, the completed sea-level pressure analysis can be converted into 1000-mb topography by labeling the 1000-mb isobar zero and locating other contours by interpolating between isobars.

7.06. *Relative topography between pressure surfaces.*—Contours of thickness (relative topography) of the layer between two pressure surfaces can be drawn either by the graphical subtraction method described in Section 5.03 or from scalar analysis of point values of thickness. The former method was used almost exclusively in deriving the analyses shown in Figure 7.06 for consistency with the isobaric patterns.

Eq 3.05(2) shows that relative contours (thickness lines) are mean virtual isotherms for the layer. For each 100-foot increment of thickness in the (500/700)-mb layer, for example, there is change by 3.1° C in $\bar{T}^*$.

Parts of the analyses in Figure 7.06 actually preceded the analyses in Figure 7.03. Reports over Alaska, western Canada, and the United States led to sufficiently accurate isobaric analyses over those areas. From topographic patterns the thickness patterns were determined there, from which thickness patterns were extrapolated laterally, using such information as obtained from time continuity, certain geographic

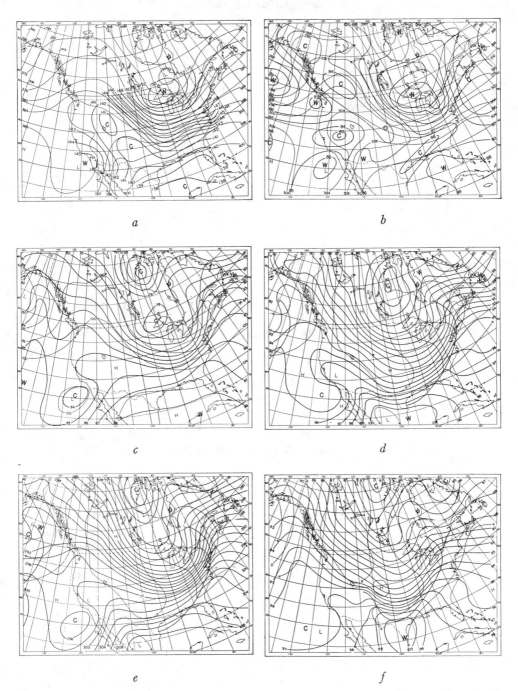

FIG. 7.06.—Relative topography of the various pressure layers for: (i) 1500 GCT, 1 March 1950, (a) 100/200 mb, (b) 200/300 mb, (c) 500/700 mb, (d) 700/1000 mb, and (e) 300/1000 mb; (ii) 0300 GCT, 1 March 1950, (f) 700/1000 mb.

controls on temperature, relations of temperature to surface flow patterns, etc. Except at 1000 mb, the pressure analyses given over the Pacific Ocean and northeastern Canada are due largely to this procedure. For evident reasons, the 100-mb chart had to be either derived or adjusted by this means.

7.07. *Thermal wind and relative topography.*—If geostrophic wind at Z_1 is $\mathbb{C}_{g1}$, and at Z_2 is $\mathbb{C}_{g2}$, where $Z_2 > Z_1$, then

$$|\Delta\mathbb{C}_g| = (980/f)(\partial\Delta Z/\partial n), \quad (1)$$

where n is normal to thickness lines ΔZ of the pressure layer. In vector form:

$$f * \mathbb{C}_{g2} - f * \mathbb{C}_{g1} = f * \Delta\mathbb{C}_g$$
$$= 980 \, \nabla(\Delta Z) . \quad (2)$$

Thus, the thermal wind vector $\Delta\mathbb{C}_g$ for a given pressure layer is related to the thickness pattern *exactly* as geostrophic wind is related to pressure topography. A geostrophic wind scale applies to the thermal wind and relative isobaric topography just as it applies to geostrophic wind and isobaric topography. Eq (2) is demonstrated graphically in Figure 7.07. Thickness lines are dashed, and contours for the two pressure surfaces are shown parallel to their geostrophic vectors; all three sets of lines are drawn for the same interval of height.

From Eq 7.04(2) it can be shown that vertically through a pressure layer $\Delta\mathbb{C} = \Delta\mathbb{C}_g + \Delta\mathbb{C}'$. Then, the thermal wind in the layer equals the actual vector change of wind if (i) $\mathbb{C}_1$ and $\mathbb{C}_2$ are both geostrophic or if (ii) ageostrophic wind is the same at both pressures, no matter how large it might be. This suggests that actual vector change of wind with height is likely to be a good indication of thermal wind and therefore also an indication of horizontal temperature distribution. Thus, where either of the two above conditions is reasonably ful-

filled, the local temperature distribution in various pressure layers can be determined solely from analysis of the hodogram. This information is valuable especially near isolated wind stations.

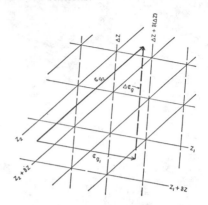

Fig. 7.07.—Relation of the thermal wind vector in a layer to the vector geostrophic winds at the same geographic location on the bounding isobaric surfaces and relation of the relative contours (*dashed*) to the contours of the isobaric surfaces.

7.08. *Temperature analysis at constant pressure.*—The series of temperature charts is given in Figure 7.08. Little reliance could be placed on the analyses unless we mentioned how each was derived; common procedures might have led to quite different results in certain areas.

Since thickness charts were prepared beforehand, they were the objective guide in temperature analysis, although they could not be relied on for that purpose near the tropopause and near the ground. From the thickness pattern for the (500/700)-mb layer alone it is possible to derive the temperature patterns (and also the temperatures) at 700 and 500 mb with some accuracy. This follows from observation that vertical temperature lapse rates usually change slowly both vertically and horizontally in the troposphere above the surface layer. Large changes in lapse rates often are associated with frontal zones, but such zones can be detected in the thickness

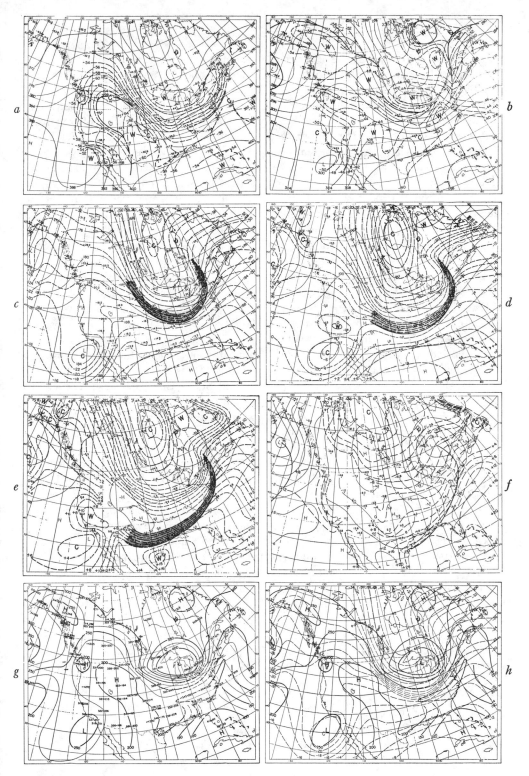

FIG. 7.08.—Isotherms and contours at 1500 GCT, 1 March 1950, for (a) 200 mb; (b) 300 mb; (c) 500 mb; (d) 700 mb; (e) 850 mb; and (f) 1000 mb. (g) Pressure of the tropopause, and (h) pressure of the tropopause and 500-mb isotherms.

pattern. Notice the resemblance of the temperature patterns at 700 and 500 mb with the thickness pattern for the layer between, both over the continent with many data and over the oceans with few.

Another aid in drawing the temperature analyses were the cross sections of temperature and potential temperature, including Figures 6.13*de* and others. Their greatest help was in frontal regions, where we have seen that proper cross-section analysis gives more accurate results than does interpolation of temperature between stations. In some instances, however, temperatures at constant pressure suggested adjustments in the cross-section patterns.

Examine the analysis of temperature at 700 mb in Figure 7.08*d* over the eastern United States. By ordinary interpolation from plotted temperatures there would be no such concentration of isotherms over Washington ($-9.2°$ C) and Stewart ($-8.8°$ C) and, therefore, no evidence of a front in that area. Concentrations of isotherms are made more evident by reference to thickness charts, cross sections, complete analysis at the ground, and time continuity than from indications by the few plotted temperatures in one chart alone.

Temperatures shown on the 1000-mb chart are for the surface. The analysis was based on these data and on six-hourly surface reports. The pattern shown deserves less claim for accuracy than those at some of the higher levels, because of more reasons for surface irregularities in temperature over land.

7.09. Topography of the tropopause.—The intersection of the tropopause is indicated by the broad lines on the 200-mb and 300-mb charts in Figure 7.03. A more complete picture of the shape of the tropopause is given by Figure 7.08*g*, where the curves are isobars of tropopause surface, and the figures at each station are potential temperature and pressure at the tropopause. The *H*'s and *L*'s refer to height instead of pressure.

The analysis given should be taken only as an approximate indication of tropopause topography. In several areas the tropopause is discontinuous and overlapping (Figs. 6.13*de*). There the isobars really should be discontinuous, and two or more pressure patterns might be indicated. However, that is difficult to draw sufficiently accurately, and the analysis shown is more easily interpreted. Incidentally, the pressure pattern given is not strictly similar to the contour pattern, even though both sets of lines would tend to parallel.

The area of the analysis between the Great Lakes and Florida is the most unreliable. In this region many soundings indicated either no distinct tropopause or a number of possible ones. Over the western United States the tropopause was distinct in most soundings. The Tatoosh sounding gave some difficulty; the analysis there is only an attempt to account for the data.

The same pattern is shown in Figure 7.08*h* superimposed on the 500-mb temperature pattern. A positive correlation is indicated between the height of the tropopause and temperatures in the troposphere, a relation known since the studies by Dines[3] and Schedler.[4] Since it is also found that temperature and pressure patterns are much alike in the upper troposphere, there is similar agreement between configuration of the tropopause and upper tropospheric pressure patterns. Locations of troughs and ridges in pressure, in the tropopause, and in tropospheric temperature all agree fairly well in longitude, while contours of the tropopause have generally larger north-south

3. W. H. Dines, "Correlation and Regression Tables for the Upper Air," *Beiträge zur Physik der freien Atmosphäre*, V (1913), 213.

4. A. Schedler, *Beiträge zur Physik der freien Atmosphäre*, VII (1915), 88–101.

amplitude than do isobaric contours or isotherms.

In Figure 7.08*g* notice the correlation between height and potential temperature of the tropopause. Over the tropical part of the pattern the tropopause is near 100 mb and 400 *θ*, while in the more arctic-like area the tropopause is nearer 300 *θ* and even below 400 mb. (In some extreme arctic depressions the tropopause at times appears still lower.) Another interesting observation from this chart is the tendency for the tropopause to maintain itself leaflike parallel to isentropic sheets.

7.10. *Temperature field near the tropopause.*—In Figure 7.08 there is general simplification of the temperature field upward from the surface to 500 mb (and perhaps even higher). Local irregularities vanish, and areas of large and small gradient become organized. Still higher toward the tropopause this evolution breaks down almost suddenly, and the temperature pattern becomes quite confused in places. At 300 mb the tropopause can be found as a cold discontinuity in the temperature field, but at some distance south of that the temperature gradient generally lessens, and weak cells appear in temperature. This region has small vertical and horizontal potential temperature gradients which favor the disorganized pattern shown. There is also justification for both vertical and horizontal mixing of air in this region.

Still higher there is semblance of simplicity in temperature pattern once more. At 200 mb the tropopause remains a cold discontinuity, but the isotherms in the increased stratospheric area of the map are assuming smooth arrangement of larger scale. Notice also the peculiar temperature pattern on the 200-mb chart just equatorward from the strongest west winds (see Sec. 6.25). There is probably continued simplification of the temperature pattern

some distance upward from 200 mb, as indicated by the thickness pattern for the (100/200)-mb layer (Fig. 7.06*a*).

Over all the United States except the extreme Southeast and an area over the Rocky Mountains the tropopause is below 200 mb (Fig. 7.08*g*). Thus the thickness pattern for the (100/200)-mb layer should indicate horizontal temperature distribution in the lower stratosphere. This pattern and the tropopause topography are illustrated together in Figure 7.10*a;* there is great similarity.

7.11. *Vertical change in pressure and wind patterns.*—Knowledge and application of a few simple rules governing the vertical variation of wind are fundamental in weather analysis and forecasting. For example, one objective of surface synoptic analysis is to represent patterns of motion and temperature at that level. But from this alone some indications of motions at higher levels can be obtained. Conversely, patterns of motion and temperature at the upper levels give indication of the respective patterns at the surface. Further, comparison of flow patterns at two levels is an objective method of proper thermal (frontal) analysis.

Material in this section is outlined by a number of rules. A way to learn to apply them is by practicing graphical analysis of each, using three winds or three sets of lines, as illustrated in Figure 7.07, for several geographic points in the area.

(1) *The vertical difference in geostrophic vector wind is the thermal wind, which bears a similar relation to relative topography for the layer between as does geostrophic wind to isobaric topography.*

(2) *The vector wind and the entire flow pattern are invariant with height in barotropic regions.*

(3) *Wind maintains constant direction but increases in speed with height where pressure and temperature patterns are similar.*

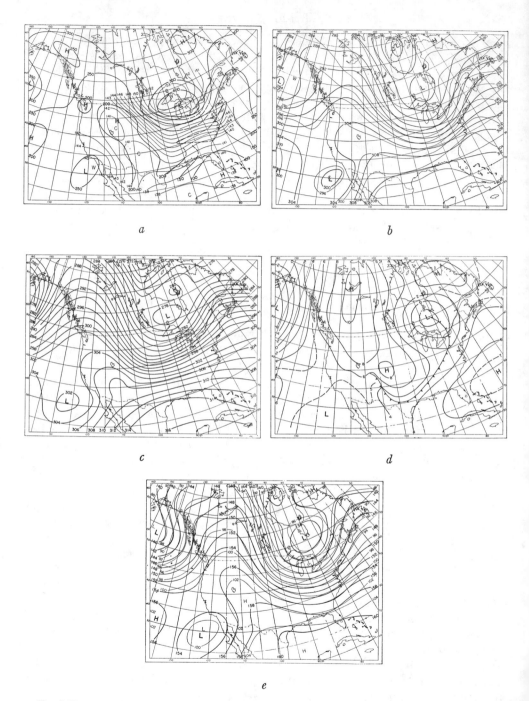

a

b

c

d

e

Fig. 7.10.—Miscellaneous charts for 1 March 1950. (a) Tropopause pressure and (100/200)-mb thickness, 1500 GCT; (b) 300-mb contours (*heavy lines*) and (300/1000)-mb thickness; (c) 300-mb contours, 0300 GCT; (d) 1000-mb contours, 0300 GCT; (e) average contours between 300 and 1000 mb (*heavy lines*) and 700-mb contours, 1500 GCT.

Wind and thermal wind have the same direction. Notice Lower California and the Dakotas between 700 and 500 mb in the series of charts.

(4) *Wind from a given direction decreases in speed with height, and may reverse its direction, where pressure and temperature patterns are opposite.* The lower-level wind and the thermal wind have opposite directions. Examine the wind variation from 1000 to 700 mb over central Texas and James Bay.

(5) *Wind backs with height (decreases direction) where it blows from cold to warm* (northern hemisphere). The thermal wind then has a component to the left of the wind. The rate of turning is directly proportional to the temperature gradient downwind and inversely proportional to wind speed and sine of latitude. Examples are found up to 500 mb over the eastern United States.

(6) *Wind veers with height (increases direction) where it blows from warm to cold* (northern hemisphere). Here the thermal wind has a component to the right of the wind. Examine the (700/1000)-mb layer over western and extreme southeastern Canada and in the lower stratosphere east of the Great Lakes. It is useful to observe that the wind usually changes most *in direction* near the ground and in the lower stratosphere. In the deep intermediate layer thermal wind has in most cases almost the same direction as the wind.

Now we consider the vertical variation of wind patterns as contrasted with variation of wind along individual verticals. Summarizing applicable material introduced in Chapter 6, we consider first the geographic shift of pressure centers, ridges, and troughs with height. From Eq 6.15(1),

$$\delta Z/\delta x = -(\partial^2 Z/\partial x^2)_p/(1/T^*)(\partial T^*/\partial x)_p\,,$$

$$\delta Z/\delta y = -(\partial^2 Z/\partial y^2)_p/(1/T^*)(\partial T^*/\partial y)_p\,,$$

$$(1)$$

for the coordinate components in slope of the axis of a pressure center (and of the sheet connecting a trough or ridge line in successive pressure surfaces). Consider the origin of the coordinate system is the center of a circular pressure system with x parallel to isotherms in the direction of the thermal wind in the northern hemisphere. Then $(\partial T^*/\partial x)_p = 0$, and the axis of the center lies in the yz-plane. Thus, from Eqs (1) and Section 6.15,

(7) *A circular cyclone (anticyclone) is displaced toward cold (warm) air with height at a rate directly proportional to the baroclinity and inversely proportional to the intensity of the system.*

In general, a pressure system does not have circular contours; asymmetry may have significant effect on its geographic displacement with height. To examine this, place y along the temperature descendant and another axis along the major axis of the pressure system, determine displacement of the center along each axis in given δZ, and obtain the resultant displacement by the method in Figure 7.143. Displacement is in the direction along which $(1/T^*)(\partial T^*/\partial s)_p/(\partial^2 Z/\partial s^2)_p$ has maximum absolute value, if s is the direction of displacement. From this follow the more general rules for geographic displacement of pressure centers with height:

(8) *A cyclone (anticyclone) is displaced geographically with height with a component toward the cold (warm) air, the magnitude of which is proportional to the temperature gradient, and with a component along the longest symmetry axis toward colder (warmer) air, the magnitude of which is proportional to the ellipticity of the pattern.* Existence of the latter component depends on the first.

Rule (7) is illustrated by Figure 6.15*b*, by the first two diagrams in Figure 7.11, and by the LOW in southeastern Canada between 1000 and 700 mb in the map series. The shift in location of the large HIGH over western North America from 1000 to 850 mb conforms with Rule (8). From temperature distribution only, it shifts southwestward with height, but, owing to north-south elongation, it has an added component of displacement southward.

Displacements of troughs and ridges in the pressure pattern are governed by the same principle, but we speak only of dis-

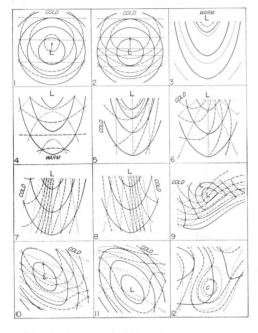

FIG. 7.11.—Variation of pressure patterns with height; *continuous lines*, lower pattern; *dashed lines*, upper pattern; *dotted lines*, thermal pattern.

placements normal to trough and ridge lines. With y along the trough or ridge line, we need consider only the first expression in Eqs (1).

(9) *Pressure troughs (ridges) are displaced toward cold (warm) air with height at a rate directly proportional to the temperature gradient and inversely proportional to the intensity of the trough (ridge).* The intensity of the trough or ridge may be defined either as vertical curvature of the pressure surface along x or as vector change of wind horizontally across the trough or ridge. The

above principles are clearly illustrated by (5) and (6) in Figure 7.11.

We now reach the lengthy and more complex subject of changes with height in intensity and shape of pressure patterns.

(10) *Cold cyclones (troughs) and warm anticyclones (ridges) intensify with height* if isotherms and pressure contours are for the most part mutually parallel. Here the thermal flow has in each case direction consistent with the flow at the lower level, and circulation increases upward. All these conditions are commonly observed through the upper troposphere.

(11) *Warm cyclones (troughs) and cold anticyclones (ridges) weaken with height.* The thermal flow is opposite to the flow at the lower level, and the circulation may even reverse with height. Such systems are found in the lower stratosphere and less frequently near the ground. Examine (3) and (4) in Figure 7.11.

With reference to Figure 6.15c and the above statements, we may add:

(12) *The axis of maximum geostrophic wind intensifies in wind speed with height if there is a component of thermal wind along the geostrophic wind, and it is displaced laterally with height to the side with the largest thermal wind component parallel to that wind.* The wind axis rotates horizontally with height in the sense from pressure ascendant to temperature ascendant at a given level.

From comparing charts for upper and lower troposphere, one sees that, besides vertical changes in location and intensity of pressure systems, there are also pronounced changes upward in shape of the pattern and in number of centers. Some systems completely vanish upward, while others vanish downward, and all the while there are related changes in shape and spacing of contours. These vertical changes evidently are due to distribution of temperature and to peculiarities in the pressure pattern.

The general tendency for vertical change in shape of pressure patterns is indicated in most of the diagrams in Figure 7.11. Note in particular the first one, where a uniform temperature gradient superimposed on a circular pressure pattern leads to an asymmetric pressure pattern above. In order for the pressure pattern to remain circular with height, there either must be barotropy or the temperature pattern must be concentric with the pressure pattern. This is evidence for: *The shape of the pressure pattern has preference to vary with height.*

To explain vertical variations in pressure and wind patterns, we use here the concept of vorticity. If we express the wind at the upper level by $\mathbb{C}_{g2} = \mathbb{C}_{g1} + \Delta\mathbb{C}_g$, then its vector cross product (vorticity) is

$$\nabla \times \mathbb{C}_{g2} = \nabla \times \mathbb{C}_{g1} + \nabla \times \Delta\mathbb{C}_g .$$

(13) *The (geostrophic) vorticity at the upper level is the sum of the vorticity at the lower level and the thermal vorticity between levels.* Vorticity at any level can be expressed by the sum of the shear $(\partial c/\partial n)$ and a curvature term (c/R), in which R is the radius of curvature of contours or streamlines.[5] Thus,

$$(\partial c_g/\partial n + c_g/R)_2 - (\partial c_g/\partial n + c_g/R)_1$$
$$= \partial \Delta c_g/\partial n + \Delta c_g/R_T , \quad (2)$$

where R_T is the radius of curvature of isotherms (thickness lines) and each other R refers to the contours of the respective pressure surfaces.

(14) *If the sum effect of shear and curvature of thermal flow is positive, there is vertical increase cyclonically in the shear or curvature terms, or in both.*

(15) *If the sum effect of shear and curvature of thermal flow is negative, there is vertical increase anticyclonically in the shear or curva-*

5. c/R is seen physically as angular velocity of an air parcel following a path of curvature $1/R$. The term $\partial c/\partial n$ has similar meaning. (See Sec. 10.24.)

ture terms, or in both. These are merely a rewording of (13), and, though they might appear too general, to make them more specific introduces a number of conditions.

The above rules and Eq (2) may be used in explaining many vertical changes in shape of pressure and wind patterns. In the first example in Figure 7.11 the uniform temperature gradient gives vertical change of wind and geographic displacement of the pressure pattern, but thermal wind vorticity is zero over the entire region. As a result, certain vertical changes may occur in shear and curvature, but they compensate one another at any geographic point. Owing to shift of the center, cyclonic curvature decreases with height in the lower part of the diagram and is compensated by either increased speed or increased cyclonic shear, or both. In the upper part the increase in cyclonic curvature of contours is compensated either by decreased speed (as appears the case) or by decreased cyclonic shear.

In diagrams (7) and (8) cyclonic thermal shear gives increased cyclonic shear and curvature upward, and anticyclonic thermal shear the opposite. The patterns shown schematically in diagrams (7) and (8) are combined in the next drawing, which represents an idealized frontal wave near the ground. The warm sector in the lower right is virtually barotropic, and upper flow agrees closely with surface flow. Disappearance of the warm-front trough with height occurs through strong anticyclonic thermal shear over the surface trough and to some extent through anticyclonic curvature of thermal flow. In contrast, cyclonic curvature is found above the surface cold-front trough, in spite of large anticyclonic thermal shear, and reflects some contribution by cyclonic curvature of thermal flow near the cold front. The trough is displaced over the surface cold air with height nearly to a position at which the sum of thermal and lower-level vorticity is a maximum.

Diagram (10) in Figure 7.11 illustrates how two troughs can appear above a single one. This pattern turned 180°, and with temperature and pressure gradients reversed, gives a double ridge aloft from a single ridge below. Diagram (11) reveals how a pressure center gives way with height to semisaddle structure due to superposition of certain thermal vorticity on opposite vorticity below. By turning the pattern 180° and reversing all gradients, a structure is found similar to the anticyclone over the western United States between 850 and 500 mb, 1 March 1950. In vertical transition from a large pressure cell below to the semisaddle above, a well-developed col appears in an intermediate level.

In diagram (12) is an arrangement frequently found some distance equatorward of the main westerly current aloft. This is the "cold LOW" which often exists above a flat or indifferent pressure pattern at the ground. Examine its analogue off the California coast in the map series.

7.12. *Additional remarks on vertical variation of pressure patterns.*—Figure 7.12 gives the profiles of height and temperature along isobaric surfaces at 45° latitude prepared from Figures 7.03 and 7.08. Much of the information shown here could also be derived from Figure 6.13*f*. Troughs and ridges in successive isobaric surfaces are connected by broad dashed and dot-dashed axes, respectively, and troughs and ridges in the temperature similarly by finer axes. The dotted lines connect positions of local maxima in isobaric slope and isobaric temperature gradient.

The axis of minimum temperature in the troposphere is almost vertical. In the lower stratosphere an axis of maximum temperature lies almost exactly above. In the troposphere the axes of high temperature in this case are not strictly vertical, possibly owing to errors of analysis in these broad, flat

regions of temperature. Nevertheless, regions of low temperature in the stratosphere are situated very nearly above high temperatures in the troposphere. The reversal occurs near the tropopause, which in this drawing would have shape in phase with, but of greater amplitude than, the isobars.

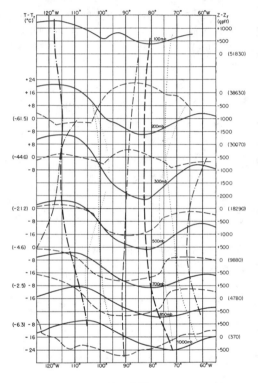

Fig. 7.12

Near the ground, pressure and temperature patterns are out of phase by roughly 90°; this figure varies in wide limits depending on land-sea temperature differences (notice the extreme west), mountain barriers and orography, and such other factors as cloudiness, precipitation, and stability, which modify diurnal variations of temperature at the ground. Upward through the troposphere, pressure and temperature patterns approach perfect phase coincidence, with cold troughs and warm ridges. In the lower stratosphere pressure and temperature patterns are just about 180° out of phase. All this leads to certain vertical variations in the pressure pattern which may be studied along with Figure 6.13*f*.

The frontal zones are more clearly shown by the temperature profiles at 850, 700, and 500 mb than at 1000 mb. Occasionally the reverse occurs with cold fronts. However, the differences shown here between temperature profiles at the ground and representative ones aloft are a rule over continents in winter, and they are sufficient caution not always to expect the clearest indications of fronts by the temperature pattern near the ground.

Figure 7.10*e* gives the 700-mb topography with the average of the 300-mb and 1000-mb topographies. That average isobaric surface is midway vertically between 1000 and 300 mb and about 5000 feet above 700 mb. The amazing parallelism between the two sets of lines suggests that wind *direction* at 700 mb is on the whole a good indication of the mean in the troposphere, at least for middle latitudes. Interpreted in other ways, this means that the change of wind direction and geographic displacement and vertical changes in orientation of the pressure pattern are as great in the lower 10,000 feet as in the 20,000 feet above. There is one important difference between the patterns in Figure 7.10*e*, however. In the main westerly current the mean flow is almost everywhere stronger than at 700 mb.

7.13. *Height tendencies at constant pressure.*—Local changes in height of various pressure surfaces are given in Figure 7.131. The lines are labeled in hundreds of feet.[6] These local changes are commonly called "height changes" of the pressure surfaces or "pressure-height changes."[7]

Casual examination of the charts shows

the larger tendencies are at 300 mb, which is brought out more clearly by Figure 7.132. The main centers of fall and rise are located in the tropopause region, but the ground is a secondary maximum. In some cases tendencies near the ground exceed in magnitude those in the centers near the tropopause (e.g., a tropical storm). The charts reveal that some surface between 800 and 600 mb (say, 700 mb) has minimum average tendency, and it is conceivable that another pressure surface of minimum tendency is found in the stratosphere. In certain conditions the largest tendencies are situated at intermediate levels in the troposphere, but these are weak ordinarily and exist in flow patterns less pronounced than this one.

Height tendencies $\partial Z/\partial t$ of pressure surfaces may be converted to pressure tendencies $\partial p/\partial t$ at fixed levels through the hydrostatic equation

$$\partial p/\partial t = -(\delta p/\delta Z)(\partial Z/\partial t)_p. \quad (1)$$

Thus, where isobaric surfaces are spaced vertically 25 feet/mb, for each 100-foot height change the local pressure change is 4 mb. Local pressure tendency is the number of unit isobaric surfaces passing the point in unit time.

Without attempting to explain the ultimate cause of height or pressure tendencies, it is possible to analyze the *distribution* of tendencies in the vertical in terms of temperature tendencies only. From Eq 3.13(1) the tendencies of two isobaric surfaces are related to the tendencies of temperature and thickness ΔZ by

$$(\partial Z/\partial t)_2 - (\partial Z/\partial t)_1$$
$$= K (\ln p_1/p_2)(\partial \bar{T}^*/\partial t) = \partial \Delta Z/\partial t. \quad (2)$$

Warming gives increasing rises (decreasing

---

6. These are also values for change of height departure *D*.

7. A suitable name for an isopleth of height change is not agreed upon. R. C. Bundgaard (in American Meteorological Society, *Compendium of Meteorology* [Boston, Mass., 1951]) suggests "isallohypse,"

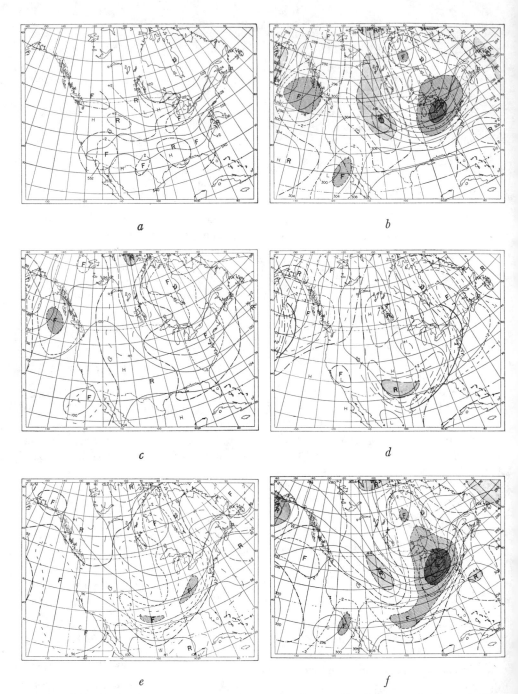

FIG. 7.131.—Height changes of pressure surfaces during the preceding 12 hours and contours for 1500 GCT, 1 March 1950: (*a*) 100 mb; (*b*) 300 mb; (*c*) 700 mb; and (*d*) 1000 mb. (*e*) Change at 700 mb minus change at 1000 mb, and 700-mb contours; (*f*) change at 300 mb minus change at 1000 mb, and 300-mb contours.

falls) of isobaric surfaces upward, and cooling gives decreasing rises (increasing falls) upward. It is thus possible to infer local time variations of temperature from the vertical

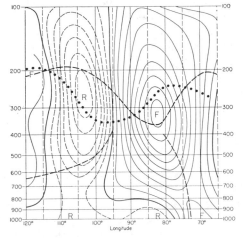

Fig. 7.132.—Cross-section pattern of pressure-height change at 41° N during the 12 hours preceding 1500 GCT, 1 March 1950. (The dashed line is the approximate position of the present tropopause; the line of dots, the position 12 hours previously.)

variation of pressure-height tendencies, and conversely.

In Figure 7.133 the vertical scale in the chart is $-\log p$, and therefore by Eq (2) the slope of each profile is inversely proportional to local temperature change. At Lander decreasing rises upward from the ground indicate cooling. Near 700 mb there was no temperature change, thence upward to 200 mb slight warming, and, finally, cooling above 200 mb. Similar variations of temperature are indicated at Rapid City and Omaha but with different magnitudes. The Joliet and Pittsburgh profiles indicate strong cooling in the troposphere and approximately equal warming in the stratosphere. The almost vertical profile at Nantucket indicates little net temperature change. A similar series of profiles can be derived from Figure 7.132.

When comparing Figure 7.133 with the series of temperature charts (Fig. 7.08), certain important deductions follow immediately. Cooling near the ground over the

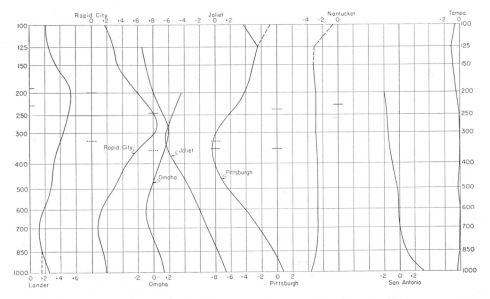

Fig. 7.133.—Profiles of pressure-height change for the 12-hour period ending at 1500 GCT, 1 March 1950. (Present and previous positions of the tropopause are indicated by the short horizontal continuous lines and dashed lines, respectively.)

central United States suggests southward (and eastward) advance of the cold air mass and that the front in the east is a cold front. The low-level cooling just east of the Rocky Mountains indicates the net diurnal cooling during the previous 12 hours and perhaps some westward spreading of the cold mass. In the troposphere above 700 mb, strong cooling over the Great Lakes region and strong warming near Rapid City indicate clearly the eastward movement of the cold air mass and of its bounding fronts at those levels. Similar progression is indicated in the lower stratosphere. The constant pressure-height changes over Nantucket indicate either that motion of the temperature pattern is nearly parallel to the isotherms or that isobaric temperature gradients are small; the former appears justified by the temperature charts.

The relation of the main centers of pressure-height change to the shape of the tropopause in Figure 7.132 is quite typical. With respect to the *present* tropopause, the center of negative change is located above and the center of positive change below. The maximum fall along any vertical must be located at the level separating net cooling below from net warming above, and conversely for the center of rise (Eq [2]).

The last two charts in Figure 7.131 give the patterns of 12-hour thickness change for the indicated layers. In (*e*) the near-coincidence of the zero line in the east with the surface (1000-mb) position of the front indicates, along with small gradients to the east, nearly barotropic conditions in the warm air mass for the layer. The large gradients to the rear of the zero line indicate the large temperature contrast across the moving front. The -4 line $(-400$ feet) implies 12-hour local temperature drop about $12°$ C for the layer. For the $(300/1000)$-mb layer the zero line in the eastern United States is farther advanced toward the warm air than in the $(700/1000)$-

mb layer, which indicates that near cold fronts the cooling at high levels can precede cooling near the ground and consequently leads to decreasing stability in air columns with approach of the cold front. The location of the area of tropospheric cooling in relation to the center of cold air off California indicates northeastward motion of that center and future developments in the weather of the southwestern states. Several other features on this chart are deserving of analysis, and correlation between patterns in this chart and in others in Figure 7.131 should be examined.

7.14. *Relation of pressure tendency field to changes in pressure pattern.*—In a complete analysis it is useful to determine as much as possible about changes in the pressure pattern. This is important in forecasting the field of motion and thus in forecasting weather. It is apparent that the field of pressure change indicates evolution in the topographic field of a pressure surface, viewed as combined effects of the motions of contours and features described by contours, of deformation, and of intensification and deepening in the individual systems. As these usually occur in combination, to explain the local pressure change in terms of motion of the pressure pattern, for example, it is necessary to account for all other concurrent processes.

The problem may be approached by considering motions of individual contours in the pattern and relative motions of adjacent contours. The speed of a contour can be derived by expanding the total derivative of height at constant pressure as a function of horizontal distance x (normal to the contour) and time t. Then $(dZ)_p = (\partial Z/\partial x)_p dx + (\partial Z/\partial t)_p dt$. Division by dt and substituting U_x for dx/dt gives

$$(dZ/dt)_p = U_x (\partial Z/\partial x)_p$$

$$+ (\partial Z/\partial t)_p . \quad (1)$$

If we maintain the point of reference on the contour as it moves isobarically, then $(dZ/dt)_p = 0$, and

$$U_x = -(\partial Z/\partial t)_p/(\partial Z/\partial x)_p . \quad (2)$$

If the pressure surface falls, its contours move toward higher topography. From this point on we drop the subscript p with the understanding that constant pressure is implied.

From Eq (2) it is possible to compute the motion of any contour if the local tendency

concept of deepening as applied to individual systems in the topographic field—centers, troughs, and ridges—describes the change in height at a point fixed relative to the system. A depression in a surface "deepens" or "fills" according as its central height decreases or increases, and corresponding definitions hold for anticyclones, ridges, and troughs. In all cases we are concerned *only with the change of height per unit time within the system;* whether the system moves or not and whether pressure gradients are increas-

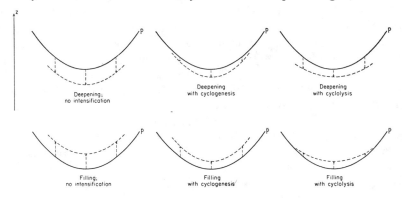

and the topographic pattern are given. As an example, for a contour oriented south-north with higher values east, if the tendency is -50 feet/12 hours and (resultant) pressure slope 100 feet per 100 km, then the contour moves eastward 50 km per 12 hours.

It is not necessary to restrict Eq (2) to displacements only along an axis perpendicular to the contour. The x-axis may be placed any finite angle with the contours. In the above example if contours are parallel and x is southeast instead of east, U_x in that direction is 70 km per 12 hours. For given pressure pattern and tendency, the smaller the angle between the contours and the direction in question, the greater is the speed of contours in that direction.

a) Deepening of pressure systems.—The

ing or decreasing are of no concern in this definition.

For a center in topography, deepening may be considered the rate at which new contours are forming in the center and expanding outward or at which contours are shrinking in diameter and disappearing into the center. It also may be visualized in three dimensions as the rate at which successive unit surfaces are intersecting fixed levels by displacements of the surfaces upward or downward along the axis of the center (Fig. 7.141). In troughs and ridges deepening has similar meaning; deepening involves motion of contours along the trough or ridge line in that surface.

Consider a coordinate system fixed to a moving pressure system with the origin of coordinates at the center and x along the motion of the system. A hypsometer fixed anywhere in the coordinate system will measure height change due to internal changes in the system itself. Internal change is deepening. On the other hand, a similar device fixed geographically on the pressure surface records changes due both to that and to horizontal displacement of the pattern. To interpret the rate of deepening from height tendencies at fixed observing points requires correcting reported tendencies for motion of the system. The rate of deepening, $(dZ/dt)_S$, is the difference between local (observed) tendency $\partial Z/\partial t$ and contribution $(\partial Z/\partial t)_M$ due to movement of the system. Thus, $(dZ/dt)_S = \partial Z/\partial t - (\partial Z/\partial t)_M$.

The local tendency of an isobaric surface resuting from motion of the system is the product of system speed U_x and pressure slope $(-\partial Z/\partial x)$ in the direction of motion. Thus, $(dZ/dt)_S = \partial Z/\partial t + U_x(\partial Z/\partial x)$.

Around a given pressure center the second term ("advective" term) is a maximum where contours are perpendicualr to x. For circular cyclones and anticyclones this is along the direction of motion. Where contours parallel the motion of the system, $\partial Z/\partial x = 0$, and, regardless of the speed of the center, local tendencies there give the rate of deepening uniquely. It follows that *at centers, troughs, and ridges in a pattern deepening is given by the local tendencies.*

Because of the confusion that usually results from applying the terms "deepening" and "filling" to both LOWs and HIGHs, different terminology is introduced for our discussion. The term *katabaric* denotes *pressure decreasing*, and *anabaric* denotes *pressure increasing*. Thus a katabaric LOW is one whose central pressure height is decreasing with time; it is a "deepening LOW" in the usual sense. The opposite is an anabaric or "filling" LOW. An anticyclone whose central height decreases is katabaric, and one whose central height increases is anabaric. The same analysis can be applied to deepening of troughs and ridges. For example,

if the tendency is negative in a trough line (katabaric trough), contours are moving along the trough line toward higher values.

In applying the above to height change and topographic patterns on the pressure chart, it is well to bear in mind the distinction between instantaneous tendency and net change over a finite time interval. In a moving cyclone whose central height is constant, the 6-hour or 12-hour local change is negative at the center due to displacement. Thus, in traveling cyclones or troughs, the line of zero change lies to the rear of the system if there is no change in central height or if the center is katabaric; only if the system is filling beyond a critical rate does that line lie in or ahead of the center or trough line. Analogous deductions are made for anticyclones and ridges. In any case, the distance between zero change and the axis of the moving pressure system is a direct function of the time interval over which the change is determined. In the special case of a symmetric trough or ridge with no deepening, the line of zero change lags behind by a distance half that over which the system has moved during the tendency interval.

Deformation in the pattern is a problem in analysis and forecasting; the various systems followed from map to map are constantly changing in shape. Centers may form and others disintegrate. Development of secondary offshoots from cyclones and anticyclones must be watched carefully. It is not possible to discuss here in detail all types of deformation in the pressure pattern, but it can be viewed as the gradient of deepening.

b) Intensification of pressure systems.— Although deepening and intensification of systems are often taken synonymously, they are different (Fig. 7.141). A cyclone may be deepening while weakening, or it may be intensifying and filling. While deepening refers only to change of pressure height in the system, *intensification refers to the increase of pressure gradient or increase of circulation with time.*

Intensification is seen in terms of relative speeds of adjacent contours. If contours are spreading apart, the pattern is *weakening.* With the opposite the pattern is strengthening or *intensifying.* For pressure slope to increase, the isallohyptic descendant must have a component downslope along the pressure surface, and with weakening it has a component directed upslope.

If pressure-height falls are largest in the center of a cyclone, or if rises are smallest in the center, the cyclone is intensifying; it is cyclogenetic or undergoing *cyclogenesis.* For the opposite the cyclone undergoes *cyclolysis.* An anticyclone is in the process of *anticyclogenesis* (intensifying) or *anticyclolysis* (weakening, or breaking down) according as the isallohyptic descendant is directed outward from or inward to the center of the anticyclone. Similar principles hold for intensifying and weakening troughs and ridges.

c) Motion of pressure systems.—The relation of the isallobaric field to horizontal displacements of various features in the pressure pattern was introduced into forecasting on a quantitative basis by Petterssen, whose contribution has become more widely known as the *Petterssen Formula and Rules.*[8] By a somewhat different approach Byers later gave a simplified displacement formula.[9] Although use of the formulas and related rules lies more properly in the realm of forecasting, we use them here for analyzing what occurs (diagnosis) rather than for forecasting (prognosis) what will occur. In

fact, the formulas and rules describe only in an approximate way what has occurred over the period from which changes were computed and less accurately what is occurring at present. Conclusions drawn from past indications are applicable to prognosis only if the same pressure and tendency patterns persist into the future or if departures can be anticipated.

(i) *The Byers method.*—Return for a moment to the relation of the tendency field to displacements of individual contours. From that approach it is possible to determine displacements or speeds of all features in the pressure pattern. From existing indications, all contours can be displaced through the desired time interval by use of Eq (2), and the final positions of contours determine the final positions of centers, troughs, and ridges in the pattern. The problem can be simplified by computing displacements of a minimum number of contours in the more critical areas of the map and then constructing the remainder of the topographic pattern by geometry. This is in essence the Byers method for displacement.

For a laterally symmetric trough (or ridge) moving without deepening, the speed of its axis can be approximated by the speed of any contour in the system measured along a direction normal to the trough (or ridge) line. Suppose the pattern in Figure 7.142*a* satisfies these conditions. The speed along x of the contour through A can be determined from the height tendency at A and the contour gradient along x through A. Or the same computation could be made for B. Either result approximates the speed if the trough is not deepening.[10] Thus,

$$U_x = -(\partial Z/\partial t)/(\partial Z/\partial x)_A$$
$$= -(\partial Z/\partial t)/(\partial Z/\partial x)_B .$$

If the trough is deepening, contours move relative to the trough line. If tendencies in

8. Sverre Petterssen, *Weather Analysis and Forecasting* (New York: McGraw-Hill Book Co., 1940), pp. 378–407.

9. H. R. Byers, *General Meteorology* (New York: McGraw-Hill Book Co., 1944), pp. 450–62. Cf. J. F. O'Connor, in F. A. Berry, E. Bollay, and N. R. Beers (eds.), *Handbook of Meteorology* (New York: McGraw-Hill Book Co., 1945), pp. 669–74.

10. This is true using instantaneous *tendencies* but may be in error using common pressure *changes.*

the trough are negative, contours move outward. The speed of the contour at A then overestimates, and the speed of the contour at B underestimates, the speed of the trough. The converse is true with rising tendencies in the trough. As long as the system remains symmetric, the effect of deepening on the tendencies, and thus on the computed speed of the trough, can be

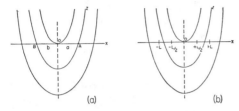

FIG. 7.142.—Coordinates used in (*a*) the Byers method and (*b*) the Petterssen method for computing displacements of troughs and ridges in the pressure pattern.

accounted by taking for the speed of the trough line the average of computed speeds at A and B:

$$U_x = (U_A + U_B)/2 . \qquad (3)$$

The same result should be obtained if displacements of the contours at A and B are computed individually and the final position of the trough line obtained by reconstructing the trough geometrically from the two contours.

Troughs and ridges in the pressure pattern usually have some asymmetry, but, if asymmetry is not too great, the above formula for speed is adequate. Byers shows that symmetry of a trough or ridge can be expressed as the ratio of a to b in Figure 7.142*a*. Introducing this modification into the last formula, he obtains

$$U_x = (bU_A + aU_B)/(a + b) .$$

For a trough $(\partial Z/\partial x)_A > 0$ and $(\partial Z/\partial x)_B < 0$. The signs of $(\partial Z/\partial t)_A$ and $(\partial Z/\partial t)_B$ are usually opposite also, and both terms in parentheses of Eq (3) contribute

in similar sense. The trough speed is intermediate to the speeds of the contours about the trough. The direction of motion is consistent with the sense of the tendency gradient along x; the trough moves from rises to falls. Similarly for the symmetrical ridge, but it moves from falls to rises. If the trough or ridge is deepening, its motion is different from the motion of contours in the system.

The formulas given above are also applicable to centers. Computations are made along two axes (Fig. 7.143), not necessarily at right angles. The actual displacement is given by the intersection of perpendiculars drawn through the final positions computed for each axis. (This differs from addition of two vectors, since speed along either axis is the *projection* of the velocity upon it.)

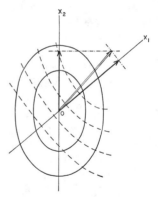

FIG. 7.143.—Computation of displacement for a pressure center; solid lines are contours, dashed lines are isallobars (or isotherms, for displacement in the vertical).

(ii) *The Petterssen method.*—While the method above may be more convenient to apply for actual computations of displacement, the original Petterssen formula gives a more complete interpretation of present motions of systems in terms of the geometry of pressure and tendency patterns.

By placing x normal to the trough or ridge line (Fig. 7.142*b*), the trough or ridge is defined through the point at which $\partial Z/\partial x = 0$. If the coordinates move with the trough, the speed of that point is also the speed of the trough along x. This can be expressed in a manner similar to that shown earlier for an individual contour. The differential operator,

$$\left(\frac{d}{dt}\right)_s = \frac{\partial}{\partial t} + U_x \frac{\partial}{\partial x},$$

can be applied to the characteristic point in question:

$$\left(\frac{d}{dt}\right)_s \frac{\partial Z}{\partial x} = \frac{\partial}{\partial t}\left(\frac{\partial Z}{\partial x}\right) + U_x \frac{\partial}{\partial x}\left(\frac{\partial Z}{\partial x}\right).$$

But, since $\partial Z/\partial x$ at the trough is constant (in fact, zero), its substantial derivative $(d/dt)_s$ is zero if the coordinates are fixed relative to this moving point. Upon reversing the order of differentiation in the first term on the right and solving for U_x,

$$U_x = -\frac{\partial}{\partial x}\left(\frac{\partial Z}{\partial t}\right) \Big/ \frac{\partial^2 Z}{\partial x^2}. \qquad (4)$$

The numerator in Eq (4) is the tendency gradient along x, and the denominator is merely the variation of pressure slope across the trough. Let distance OL in Figure 7.142b be unit length δx; Eq 5.06(2) applies. If we assume the isobaric slope measured between $-L$ and O is representative for the midpoint $-L/2$, and similarly the pressure slope between O and L is representative for $L/2$, Eq (4) is

$$U_x = -L[(\partial Z/\partial t)_{L/2} - (\partial Z/\partial t)_{-L/2}]/$$
$$(Z_L - 2Z_O + Z_{-L}). \qquad (5)$$

This is the Petterssen formula for speed of a trough or ridge line. This equation multiplied by the unit time interval δt gives the displacement formula. Eq (5) gives about the same results as Eq (3) if points A and B are located at $L/2$ and $-L/2$, respectively, and if isobaric slopes are determined from heights read at points L, O, and $-L$ in both cases. In applying this formula to computation of speed or displacement, the critical element is the choice of L.

The formula is equally valid for pressure centers, and computations for their displacement are again made along two axes (Fig. 7.143). One axis is placed along the tendency gradient, along which the numerator of Eq (4) is a maximum; the other is placed along the major axis of the system, along which the denominator is a minimum. If the closed system is symmetric, contribution by the denominator of Eqs (4) and (5) is the same in all directions about the center, and only one computation is necessary.

From Eqs (4) and (5) we deduce useful principles governing the motions of pressure systems. It is apparent that cyclones and troughs move toward falls and that anticyclones and ridges move toward rises. Therefore we need examine only the magnitudes of the numerator and denominator without confusion by signs. By the numerator, the system speed varies directly with the tendency gradient across it. This is illustrated by comparing Diagrams 1 and 2 or Diagrams 3 and 4 in Figure 6.15b.[11] The effect of the denominator is such that intense pressure systems move slowly and weak systems rapidly for given tendency gradient. Compare Diagrams 1 and 3 or 2 and 4 in Figure 6.15b.

It is useful to remember that cyclones move in the direction of largest falls (isallobaric minimum) and anticyclones in the direction of largest rises (isallobaric maximum) *only when the systems are nearly circular*. The path of any center is always of direction along which the combined effect of tendency gradient and isobaric curvature is a maximum. If the system is elongated, the center takes a course intermediate between the tendency gradient and the major axis of the system, except when isallohypses parallel the major axis, in which

11. Figure 5.16b is applicable for displacement of pressure systems with *time* if change in height is substituted for temperature T in each drawing.

case the system moves along the tendency gradient. The more elongated the system, the greater is the departure of actual path from tendency gradient.

In computing displacements by either method above, care and judgment are needed in applying them properly. Any tool so sensitive can yield unreliable or misleading results if not applied with proper discretion or if required conditions are not satisfied. For the Byers formula the points selected should have tendencies and pressure slopes representative of a broad region

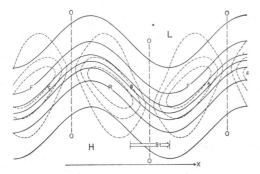

FIG. 7.144.—Isallobaric field resulting from longitudinal displacement of a sinusoidal pressure wave pattern.

to each side of the trough. Tendencies should be read at points between the axis of the system and the adjacent tendency center and never beyond it. With the Petterssen formula the tendency gradient used must agree with the tendency gradient across the axis of the system, and the three values of height should be representative for curvature of the pressure surface also across the axis of the system. In displacing trough and ridge lines, it is frequently advisable to make computations across several parts of the system as insurance against unreliable analysis and improper choice of length units.

In setting up computations across frontal or other discontinuities in the pressure pattern, the points at which changes are read should be at least as far removed from the

present position of the front as the barometric time interval in use. Although Eq (4) is not applicable at discontinuities in the pressure pattern, the finite-difference formula still can be applied there.

While Eq (4) is an exact expression for speed of a pressure system in terms of tendency and intensity, it is difficult to apply with the same accuracy from data now obtained. The denominator can be evaluated with some accuracy, but the numerator is difficult to evaluate, since there are no reports of *instantaneous* tendency. Inconsistency in the time for which the numerator and denominator in Eq (4) are evaluated can make results quite unreliable.[12]

To see what effect is due to the length of time interval over which changes are computed, consider the sinusoidal wave pattern of contours in Figure 7.144. It is moving with speed U_x to the right without deformation. Since U_x is uniform, Eq (2) shows the local (instantaneous) tendency is largest at the inflection points (A, B, and C). Zero tendency occurs at trough and ridge lines.

Now, instead of dealing with true tendencies, consider net change over a period of hours; say, δt is 12 hours. The result is the given pattern of change. The zero lines and centers all lag behind the positions for instantaneous tendency. The lag in distance d_x can be expressed in simple form by

$$d_x = (L/2T)\delta t , \qquad (6)$$

where L is the wave length of the pressure pattern, T the period of the wave, and δt the time interval over which the change is computed. As $\delta t \rightarrow 0$, the true tendency is approached, and $d_x \rightarrow 0$. The formula is not valid for $T = \delta t$, for then there are no net changes. When $T = 2\delta t$, centers of net change are a quarter wave length behind centers of true tendencies and therefore in the troughs and ridges of the pressure pat-

12. Average pressure pattern for the period might be used.

tern. For the example illustrated, $d_x = L/8$; centers of net change are midway between inflection points and the succeeding troughs or ridges. This corresponds to $T = 24$, $\delta t = 6$; $T = 48$, $\delta t = 12$, etc. A similar analysis is applicable to moving HIGHs and LOWs in the pressure pattern.

(iii) *Objective methods for forecasting displacement.*—In the preceding paragraphs we were concerned with methods for detecting the present state of motion of pressure systems. In forecasting for periods 24 hours or more, those methods based only on geometry of pressure and change patterns at one level are not sufficiently reliable. For that purpose, several other "objective" techniques have been suggested and practiced. The most simple is the *path method*, which is essentially extrapolating the path of a pressure or tendency system knowing its previous track (Fig. 9.02c); but the value of the method depreciates rapidly for long forecast periods (slightly less for tendency patterns). Other methods include steering (of pressure and tendency patterns), temperature advection, the Rossby long-wave and constant-vorticity principles, and more recently numerical prognosis. Some of these are covered in later chapters; the references below are offered for reading now.

J. M. AUSTIN, "An Empirical Study of Certain Rules for Forecasting the Movement and Intensity of Cyclones," *Journal of Meteorology*, IV (1947), 16–20.

S. P. CHROMOV, "Rules for Forecasting Synoptic Situations" (translated by I. I. SCHELL), *Bulletin of the American Meteorological Society*, XVI (1935), 21–22, 71–73, 108–10.

J. J. GEORGE, *On the Relationship between the 700-Mb Surface and the Behavior of Pressure Patterns at the Ground* (Atlanta, Ga.: Department of Meteorology, Eastern Airlines, 1949).

J. J. GEORGE et al., *On the Relationship between the Fields of Atmospheric Pressure and Temperature at Upper Levels and the Behavior of Surface Pressure Systems* (Reports Nos. 2 and 3 [Atlanta, Ga.: Department of Meteorology, Eastern Airlines, 1951]).

R. W. LONGLEY, "A Study of the Relationship between the 700-Mb Flow and the Movement of Surface Pressure Centers," *Journal of Meteorology*, IV (1947), 202–4.

V. J. and M. B. OLIVER, "Forecasting the Weather with the Aid of Upper-Air Data," in F. A. BERRY, E. BOLLAY, and N. R. BEERS (eds.), *Handbook of Meteorology* (New York: McGraw-Hill Book Co., 1945), pp. 813–57.

W. C. PALMER, "On Forecasting the Direction of Movement of Winter Cyclones," *Monthly Weather Review*, LXXVI (1948), 181–201.

P. E. WASCO, "The Control-Line Method of Constructing Prognostic Charts," *Bulletin of the American Meteorological Society*, XXXIII (1952), 233–36.

7.15. *Interpretation of developments in pressure patterns in terms of tendencies.*— When viewed in connection with the pressure field, there are many indications in the tendency field which give evidence of past, present, and future developments in pressure and wind patterns. While it is not possible to cover all details here, we can examine just a few points illustrated by the case of 1 March 1950.

By comparing Figure 7.131b with Figure 7.144, there is immediate evidence that the primary westerly wind pattern is moving eastward. For each trough and ridge there is an attendant change center a short distance downstream. The fall centers are to the left of the main current and the rise centers to the right, similar to Figure 7.144. Changes in amplitude of the pressure pattern (deepening of systems) as well as changes in wave length are indicated where motion of the pressure pattern would give changes different from those observed. Some evidence of deepening is found over Hudson Bay; motion of that trough could not explain the large falls or the near-coincidence of the fall center with the trough line. There is also evidence that the LOW has drifted (is drifting) southeastward. Deepening east and south of the Great Lakes is shown by the location of the maximum falls in the trough.

The LOW center off California is detached from the main isallobaric train, but the distribution of tendencies about it indicates eastward or northeastward motion. The center of small falls over northwestern Canada deserves particular attention, since it is situated peculiarly in the broad ridge. This suggests local deepening and consequent trough formation.

Most deductions made from this chart can be verified by reference to Figure 7.10c, which gives the pattern 12 hours earlier. Additional points to notice are the local intensification of the flow where pressure and change gradients have components in the same direction and local weakening where they are of opposite direction.

A significant item illustrated by Figure 7.131b is the relation of magnitude of pressure change to intensity of the current (Eq 7.14[1]). Centers of large change, aligned in an "isallobaric train," are associated with the pronounced currents. In low latitudes, where pressure surfaces are nearly flat, the changes must be small, and the same is true in the weak regions north of the westerly current. We excluded the effect of deepening, but the major part of the tendency field is almost always observed to result from motion of the pressure pattern.

The elongated region of falls at 1000 mb[13] in the east implies elongation of the cyclone during the period; the large isallobaric gradient near the east and south coasts, the passage of a sharp trough; the axis of rises extending eastward from Texas, the increase of anticyclonic curvature; the axis of minimum rise extending westward from the Great Lakes, the increase of cyclonic curva-

13. See also the sea-level isallobaric chart in Figure 9.08c.

ture (trough development); the fall center over California, the formation or deepening of a LOW; the fall center in northwestern Canada, the formation or intensification of a trough; the fall center off Vancouver, the advance of a trough and local increase of cyclonic curvature; and so on. More of these should be analyzed before referring to Figure 7.10d.

7.16. *Horizontal (isobaric) advection of an air property.*—For a function Q of space and time its total time derivative is

$$dQ/dt = \partial Q/\partial t + u(\partial Q/\partial x) + v(\partial Q/\partial y) + w(\partial Q/\partial z) ,$$

where u, v, and w are the x, y, and z components of velocity. The local tendency of Q is $\partial Q/\partial t$, and the remaining terms on the right are advective changes due to motion of air normal to Q surfaces. By combining the two central terms on the right and rearranging the equation, we get

$$\partial Q/\partial t = dQ/dt - \mathbb{C} \cdot \nabla_h Q - w(\partial Q/\partial z) . \quad (1)$$

The nonconservative effect dQ/dt and the vertical advection term were both discussed for temperature in Section 3.28. Here we are concerned with the local tendency of a quantity due only to horizontal advection:

$$(\partial Q/\partial t)_{\mathrm{adv}} = - \mathbb{C} \cdot \nabla_h Q .$$

The term on the right can be expressed and evaluated in either of two forms. As $-c(\partial Q/\partial s)$, it is merely the product of wind speed and horizontal variation of Q downstream. Thus, with west wind 20 m sec^{-1} and temperature increase eastward 1° C in 100 km, horizontal advection is $-2°$ C per 10^4 sec, or $-0.7°$ C per hour, and is the true local tendency of temperature if there is no net contribution from the other two terms in Eq (1). The other form is $-c_n(\partial Q/\partial n)$, where $\partial Q/\partial n$ is the horizontal gradient of Q, and c_n is the wind component normal to Q lines—positive if directed toward higher Q. For example, a west wind 20 m sec^{-1} where the temperature ascendant is southeastward 1° C per 100 km gives advection $-0.7(20$ m sec$^{-1})(1°$ C$/100$ km), as c_n is positive and only seven-tenths the resultant wind speed. There is advective cooling at the rate 0.5° C per hour.

Where it is appropriate to employ geostrophic wind, there are simple methods of deducing horizontal advection by considering the patterns of both contours and Q on isobaric surfaces. Advection is small where either pressure or Q field is flat and large where contours and Q lines are mutually perpendicular and both gradients large. Advection is negative with motion toward higher Q and positive with motion toward lower Q.

Advective temperature and potential temperature changes are analyzed visually in Figure 7.08. At the ground (1000 mb) there is cold advection in the large area from central Canada to the southeastern states. Advection is large in the cold-front zone, especially in its central section. Indeed, cold and warm fronts may be defined as zones of concentrated cold and warm horizontal advection.[14] In southern Texas geostrophic temperature advection is small, since contours parallel isotherms; but the true advection is very large, since actual winds (Fig. 7.03h) have strong components normal to the isotherms. Areas of pronounced warm advection are shown over western Canada, the coast of Labrador, and south of Nova Scotia, the latter of which suggests the possibility of a warm front (as also by the pressure pattern). There is little motion indicated near the west coast of the United States, so, in spite of the large land-

14. An exception is the infrequent case in which, owing mostly to vertical motion, isotherms move horizontally opposite to the advective component of the wind.

sea temperature contrast, horizontal advection is negligible.

At higher levels in the troposphere the coastal and orographic concentrations of isotherms have vanished (e.g., 500 mb), and the "synoptically important" patterns of advection are more clearly defined. Examine the predominant warm advection in the upper right, cold advection in the center of the continent, warm advection over the western part of the continent, and cold advection in the pressure trough at the left.

In most situations temperature advection decreases upward in the troposphere, but a second maximum occurs near the tropopause and in the lower stratosphere, where advection is mostly of sign opposite to advection below (Fig. 7.08a). Areas of strong advection in the lower stratosphere almost coincide geographically with strong advection of opposite sign in the troposphere.

If computed rates of horizontal advection are compared with observed local changes (as determined between successive maps), it is possible to obtain the total effect of dQ/dt and $-w(\partial Q/\partial z)$ in Eq (1). Near the ground $w \rightarrow 0$, and dQ/dt is responsible for most of this residual. For temperature this is due primarily to diabatic (i.e., nonadiabatic) processes; for moisture, to evaporation and condensation. In the free atmosphere, adiabatic vertical motions have a prime role (Eq 3.28[4]); with stable stratification ascent produces local cooling and descent local warming, separate from local changes due to horizontal advection. Hence, where horizontal advective cooling and descent occur together, as they usually do, local cooling is less than advective cooling. Ascent and advective warming have a similar tendency to occur together, and the usual result is less local warming than indicated by horizontal advection.

This relation between observed local change, change due to advection, and change by vertical motion can be used to deduce patterns of vertical motion. Consider the area over the eastern Great Lakes in Figure 7.08c and in the succeeding 500-mb chart, which we suppose is at hand. From pressure and temperature patterns an average value of advective cooling can be computed at certain points in the area. If observed local change shows less cooling than computed by advection, there is indication of subsidence; the average rate of sinking is proportional to this difference and to the stability. This method is particularly applicable to subsidence, since descent is more likely than ascent to be dry adiabatic, but it can be useful even with ascent in saturated areas.

The concept of horizontal advection is important also as regards moisture and, for that matter, cloud layers. Besides, horizontal and vertical variation of advection is a factor in deforming patterns of any quantity.

7.17. Horizontal temperature advection computed from the thermal wind.

—Consider that curve AB in Figure 7.17a is the hodogram of geostrophic wind between levels A and B. Also assume that levels a and b are sufficiently adjacent so that the hodogram segment between them is the same as the straight thermal wind vector $\delta\mathbb{C}_g$. From Eq 7.07(2) and the barometric equation, $\delta\mathbb{C}_g = (980/f)(\partial\Delta Z/\partial n) = (K/f)(-dp/p)(\partial T^*/\partial n)$ for this shallow layer. Horizontal temperature advection is given by $-(c_n)_g(\partial T^*/\partial n)$ from Eq 7.16(1). Therefore,

$$(\partial T^*/\partial t)_{\text{adv}} = -f(c_n)_g\, \delta\mathbb{C}_g/K(-dp/p).$$

It is seen that $(c_n)_g\, \delta\mathbb{C}_g$ is twice the shaded area bounded by the upper and lower wind vectors and the hodogram. For the entire air column between A and B, the advective temperature tendency is

$$(\partial \bar{T}^*/\partial t)_{\text{adv}}$$
$$= -(f/K \ln p_A/p_B)\int_A^B (c_n)_g\, \delta\mathbb{C}_g, \quad (1)$$

which is a constant times area OAB. The sign of advection is evident from the turning of wind with height.

In some conditions the hodogram between two levels may be straight (Fig.

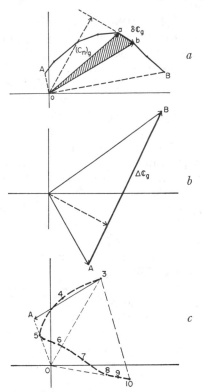

Fig. 7.17.—Computation of horizontal temperature advection from the hodogram.

7.17b); the thermal wind vector is then the same as the hodogram. Eq (1) becomes

$$(\partial \bar{T}^*/\partial t)_{\text{adv}}$$
$$= -(f/K \ln p_A/p_B)[(c_n)_g\, \Delta c_g]. \quad (2)$$

In this form estimation of temperature or thickness advection from a wind sounding is no problem at all.[15] The equation as given is proper for c.g.s. units. If the unit of speed is knots, the unit of f is $10^{-4}\ \text{sec}^{-1}$, the unit

15. Notice that the normal to the thermal vector in Figure 7.17b is the normal component of the winds at both A and B. Thickness advection is obtained by multiplying Eqs (1) and (2) by $K \ln (p_A/p_B)$.

of time 1 hour, and the unit of thickness 1 gpft, then the right side should be multiplied by 0.032. For example, if at 30° latitude $\Delta\mathbb{C}_g = 50$ knots and $c_n = 10$ knots, advective thickness tendency for the layer is $\pm 0.032(0.729)(10)(50) = \pm 12$ gpft/hr.

Frequently the hodogram is curved appreciably, and the scheme illustrated by Figure 7.17*b* is not applicable. Figure 7.17*c* illustrates the discrepancy. By assuming that the hodogram between levels *3* and *10* is straight, strong warm advection (area *0,3,10*) is indicated. However, advection is proportional rather to the area outlined by the wind vectors at levels *3* and *10* and by the hodogram between. In this case it is best to evaluate advection by parts. Since there is cold advection from *3* to *5*, warm advection from *5* to *8*, and no advection from *8* to *10*, we proceed in those three steps, and the net advection for the entire layer is a weighted sum of the three. Symbolically,

$$\text{Adv}_{3,10} = [2\,(\text{Adv}_{3,5}) + 3\,(\text{Adv}_{5,8})$$
$$+ 2\,(\text{Adv}_{8,10})]/7 \,.$$

It is readily seen that, since the uppermost of the three layers shows no advection, and since area *0,5,4,3,0* is larger than *0,5,6,7,8,0*, there is net cold advection for the entire layer.

It is possible to divide layer *3,5* into two smaller layers *3,4* and *4,5* and perform separate computations. But, since the hodogram is most probably curved in those intervals also, we operate with the whole layer *3,5* at once. A straight "thermal vector" is drawn, from point *3* to point *A*, which bounds the same area as does the curve *3,4,5*. The area of triangle *0,3,A* is half the product of the length of this vector and its normal from *0*.

By the approximation of a straight hodogram between two levels, while aware of its shortcomings, it is possible to compute roughly horizontal advection in a pressure layer from its thickness pattern and the topography of either bounding pressure surface. At any geographic point, thickness lines are parallel to the thermal vector for the layer, and thickness gradient is proportional to its magnitude. Further, the geostrophic component normal to the thermal vector for the layer is given inversely by the spacing of contours on either isobaric surface measured along thickness lines. The product of this normal component and the thickness gradient gives an approximate measure of thickness advection. Advection can be visualized in that way from the patterns in Figure 7.06.

Throughout this discussion we limited considerations to geostrophic wind, as distinct from actual wind, because it is the vertical variation of geostrophic wind that is related to horizontal temperature distribution. The discussion applies to actual wind shear if it is sufficiently geostrophic. In most cases, except at low latitudes and near the ground, the results obtained by using actual winds give useful information. In computations for layers which include the ground layer, it is customary to use the gradient-level wind as representative of surface or sea-level geostrophic wind.

7.18. *Relation of hodograph and isobaric analysis.*—In preceding discussions we considered the pressure-wind relation, the thermal wind relation, evaluation of temperature advection from the hodograph, and the relation between geostrophic and actual wind. The twenty-four hodographs in Figure 7.18 afford an opportunity to check some of these and meantime to indicate part of the large amount of information given by a wind sounding. Since space does not permit a thorough discussion of each, the reader is asked to analyze them in detail and to refer to the sets of pressure, temperature, and thickness charts.

The weak westerly winds at all levels at

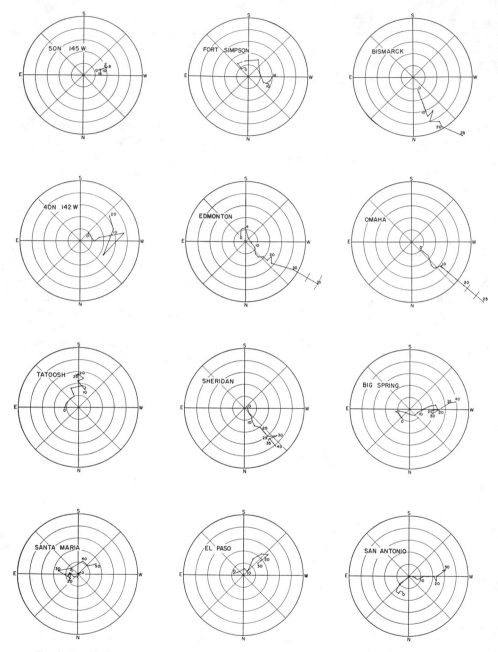

FIG. 7.18.—Hodograms for twenty-four North American stations at 1500 GCT, 1 March 1950

228

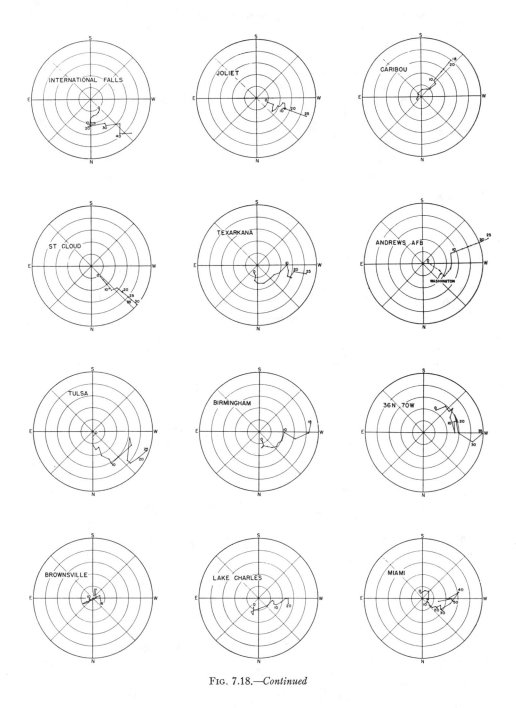

FIG. 7.18.—*Continued*

229

50° N 145° W agree with a position south of but near the center of the LOW and are indicative of a weak thermal field. At Fort Simpson (62° N 121° W) the mean isotherms are oriented northwest-southeast in the first 10,000 feet, with large gradient, and thermal gradients in the layer above are small. At Bismarck there are large temperature gradients at all levels, and isotherms trend NNW-SSE in the lower troposphere and back into more westerly aloft, with negligible advection below and cold advection above. Similar indications are given by Edmonton and, except for high-level cold advection, also by Omaha. Sheridan's lighter winds and lighter thermal winds indicate it is southwest of the main northwesterly current and frontal zone at most levels. Strong thermal wind and weak normal winds at Big Spring suggest a frontal zone oriented west-east lying almost parallel to the winds at most levels. Santa Maria shows generally weak pressure and thermal patterns up to 20,000 feet but strong temperature gradient and warm advection beginning near 30,000 feet. El Paso gives strong evidence of the cold LOW to the west, even though from the surface chart this station might be expected to be influenced mostly by the cold air mass to the northeast. This large temperature gradient in the middle troposphere, indicated by hodograph and synoptic charts, is suggestive of a frontal zone. The hodograph for San Antonio is similar to Big Spring, but, by comparing the two, the proper horizontal curvature of the front over Texas is indicated.

High-level cold advection appears at International Falls, as over Bismarck, but by comparison there is evidence that the center of the northwesterly current and attendant temperature contrast lies nearer Bismarck. High-level cold advection does not appear as far south as St. Cloud and Joliet (compare the four hodographs with

Fig. 7.08*c*). Tulsa, Texarkana, Birmingham, and Lake Charles all indicate post-cold-front cooling, proximity and penetration of the frontal zone, and also horizontal orientation of the front. From more detailed hodographs it might be possible to locate the frontal layer in some of them. Brownsville gives an almost ideal picture of the barotropic tropical air mass. Miami is similar but in high levels shows considerable baroclinity.

That several of the hodographs differ in detail from the analyzed charts is only natural. Geostrophic wind is not always the same as actual wind. Errors in wind observation can be frequent and large. And, also, pressure and temperature charts were prepared independent of the hodographs. In some areas reference to the hodographs would have been a definite help. Nevertheless, a fairly good idea of the pressure and temperature patterns at various levels could have been obtained from Figure 7.18 alone.

7.19. *Some remarks on frontal analysis.*— While attempting not to focus too much attention on this one synoptic example, we might touch on a few common features of frontal analysis given in Figure 7.08. The large-scale frontal pattern is in phase with, but of slightly greater amplitude than, the upper-level flow, with cold fronts in advance of troughs and warm fronts behind (Figs. 7.02*ab*). Occasionally, frontal zones appear continuous with the westerly current at upper levels around large sections of the hemisphere; the system moves along with the wind pattern. Near the ground this frontal wave system most likely appears as a series of cold fronts (Fig. 7.02*a* and Sec. 6.23). Observe in Figure 7.08*e* that there is a distinct cold front over the southeastern United States and some evidence of cold fronts over northwestern Mexico and near the left edge of the map; it is more difficult to define the warm front in the advancing warm air ahead of each.

From the surface upward to 700 mb the temperature patterns lend some suggestion of an "arctic front" extending from the coast of Labrador, south of the Great Lakes, and into northwestern Canada. This zone agrees fairly well geographically with the 500-mb position of the polar front. The eastern part of this concentration is the semipermanent winter land-water contrast in temperature. One might wonder why the southern part of this zone is not merged with the polar front. It is possible that this part is produced and maintained by warming due to subsidence in the cold air south of it, by warming due to vertical transport of heat from the surface south of it, and by sharp southward decrease of northerly winds in that vicinity, or, conceivably, the two zones might have been detached for some time. There is evidence that the first two processes are operating in this case.

There is some indication that the frontal analyses in the three lowest levels near the east coast could have been drawn differently. A more dense network over the Atlantic between 35° and 45° latitude might have indicated a warm front in that area and a frontal wave crest over southern New England at 700 mb and southward at lower levels. With such there would be parallel wave forms in both the polar and the arctic fronts. The analyses as given admittedly reflect the impulse of tradition.

The temperature pattern near the ground is strongly influenced by the properties of the surface. At 850 and 700 mb several of the isolated thermal centers can be ascribed to vertical motions related to terrain. For instance, the warm center in the Brownsville area at 850 mb suggests downslope trajectory from the Mexican Plateau; temperatures are higher than found in oceanic tropical regions at this level. The warm center over California could be due partly to downslope motion along the general terrain and partly to deformation of the isotherms by motion of the cold center from the west. Additional orographic control is seen in the vicinity of Alaska. At higher levels (500 mb) most of these have vanished.

7.20. *Moisture analysis.*—Some patterns of dewpoint temperature and temperature-dewpoint depression are given in Figure 7.20. Figures in parentheses indicate the dewpoint temperature corresponding to relative humidity 20 per cent at those stations for which relative humidity was too low for measurement ("motorboating").

The areas of high moisture are restricted to the warmer regions on each chart, but within these same regions there are also areas of very low humidity. Near the ground the higher dewpoints are found over the tropical maritime surface and in areas where this air is transported. The lower dewpoints over most of North America are typical of continental air in winter (in summer daytime dewpoints over land are frequently higher than over oceans).

An important feature of the surface or 1000-mb chart, other than the general land-sea contrast, is the relatively good similarity between patterns of temperature and moisture (Fig. 7.20f). Upward, this similarity breaks down, and larger ranges are found in temperature-dewpoint depression. The poorer correlation aloft between moisture and temperature leads to complex fields of temperature-dewpoint depression. Areas of moist tropical air at the ground are overlain in places by very dry tropical air (Florida, Cuba, and northeastern Mexico). In other places are found relatively high dewpoints aloft over some of the drier regions at the ground (New Mexico and western Texas). Thus, contrary to the usual behavior of temperature patterns with height, moisture patterns show pronounced tendency to vary, especially in the lowest 5000 feet or so. We

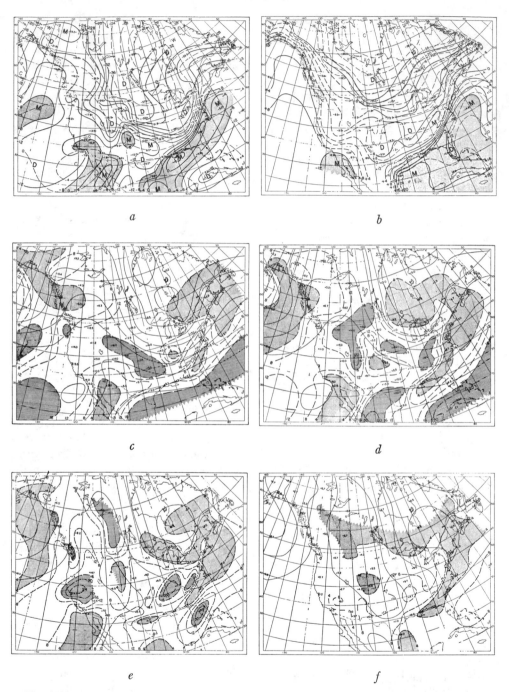

Fig. 7.20.—Isobaric dewpoint temperature patterns (*heavy lines*) and isotherms, 1500 GCT, 1 March 1950. (*a*) 700 mb and (*b*) 1000 mb. Isobaric temperature-dewpoint depression (*dashed lines*) and contours at (*c*) 500 mb; (*d*) 700 mb; (*e*) 850 mb; and (*f*) 1000 mb.

cannot rely on the distribution of moisture or relative humidity at the surface for the patterns aloft except in some areas where rain has been falling.

With necessary caution, we might deduce from upper relative humidity patterns a generalized picture of the field of vertical motion. If moisture decreases upward, areas of high humidities might be explained by ascent and low humidities by descent. On this basis, subsidence is indicated to the rear of the cold front at the surface and ascending motions at and ahead of the surface cold front. In addition, there is indica-

ing by the pressure difference, and dividing by gravity. This may be done conveniently by steps for an entire pT_s sounding on a thermodynamic chart. From computations of total W for several soundings the pattern of W can be drawn for the desired geographic area.

7.21. *Equations of motion for the atmosphere.*—The equations of motion for an air particle of unit mass located at the origin or along the z-axis of a coordinate system so oriented on the earth that x is always directed eastward, y northward, and z upward may be expressed as follows:

$$du/dt - (uv/a)\tan\phi + 2\omega(w\cos\phi - v\sin\phi) = -a(\partial p/\partial x) + F_x ; \quad (1a)$$

$$dv/dt + (u^2/a)\tan\phi + 2\omega u\sin\phi = -a(\partial p/\partial y) + F_y ; \quad (1b)$$

$$dw/dt - 2\omega u\cos\phi = -a(\partial p/\partial z) - g + F_z . \quad (1c)$$

tion of general ascent ahead of cyclonic depressions aloft and descent to the rear. With respect to surface pressure patterns, these areas of general ascent and descent coincide more nearly with the pressure centers. As evidenced by these charts, however, there are significant local departures from the simple correlations between pressure or fronts and the moisture patterns (e.g., the Wyoming-Colorado and Tennessee areas at 850 mb and southern Texas at 700 mb).

Work in hydrometeorology often requires evaluating the amount of precipitable water in a vertical column of air (i.e., the liquid equivalent of the water vapor in that column). The volume dW of liquid water equivalent in an air column of depth dz and unit (1 cm^2) cross-sectional area is $dW = r\rho dz$. Upon substitution from the hydrostatic equation and then integration,

$$\int_{p_1}^{p_2} dW = (1/g) \int_{p_1}^{p_2} r\,(-dp) .$$

$$(r\%_0 \text{ and } p \text{ mb})$$

Thus the depth of precipitable water in a given pressure layer is obtained by evaluating the average mixing ratio in the layer, multiply-

The x- and y-axes at any point are identified by the latitude and longitude through that point.[16] Components u, v, and w are velocity eastward, northward, and upward, respectively. The magnitude of the earth's angular velocity is ω, and distance from the center of the earth to the particle is a.

On the left side of each equation is the component acceleration; du/dt, dv/dt, and dw/dt are the acceleration observed in this reference frame. All other accelerations on the left are due to motion relative to a sphere and to motion on a rotating sphere.[17]

16. We are thus giving the polar coordinate interpretation to u and v. The tan ϕ terms in Eqs (1a) and (1b) are due to this choice of coordinate system (D. Brunt, *Physical and Dynamical Meteorology* [2d ed.; Cambridge: Cambridge University Press, 1939]). This interpretation is desirable for consistency with the meteorological system of directions. For studying local dynamics it may be useful to refer to a Cartesian system fixed on the earth, in which case the tan ϕ terms vanish.

17. Owing to divergence of zeniths, an additional term appears on the left of each equation. These are uw/a, vw/a, and $-(u^2 + v^2)/a$, respectively. They are neglected here because of their small relative contributions.

On the right side of each are the various components of force (per unit mass) along the respective axes. F_x, F_y, and F_z are the components of all extraneous forces, which are usually taken primarily as the viscous (frictional) forces; g is gravity, and $-a(\partial p/\partial x)$, $-a(\partial p/\partial y)$, and $-a(\partial p/\partial z)$ are the components of pressure force. We shall consider individually the effect of each other term on the net apparent acceleration du/dt, dv/dt, and dw/dt.

a) The forces.—The pressure force is along the pressure descendant and gives acceleration in that direction with magnitude equal to the pressure force per unit mass. That is, $du/dt = -a(\partial p/\partial x)$, etc. Gravity acting alone would give downward acceleration of that magnitude.

The accelerating (retarding) effects on the motion of an individual air parcel by drag of the surrounding air is physically evident but difficult to express in a simple manner. The problem is more involved for friction in air flowing over a solid. Discussions of frictional forces are found in certain texts.[18] The viscous forces can be expressed approximately by $F_x = a\mu(\partial^2 u/\partial x^2 + \partial^2 u/\partial y^2 + \partial^2 u/\partial z^2)$, $F_y = a\mu(\partial^2 v/\partial x^2 + \partial^2 v/\partial y^2 + \partial^2 v/\partial z^2)$, and $F_z = a\mu(\partial^2 w/\partial x^2 + \partial^2 w/\partial y^2 + \partial^2 w/\partial z^2)$. After analysis of each term in parentheses for observed distributions of velocity, it is found that for both F_x and F_y the last of the three derivatives ($\partial^2 u/\partial z^2$ and $\partial^2 v/\partial z^2$) is usually dominant; wind varies more rapidly vertically than horizontally. This leads to

$$du/dt = F_x \simeq a\mu(\partial^2 u/\partial z^2)$$

and

$$dv/dt = F_y \simeq a\mu(\partial^2 v/\partial z^2) \, .$$

Also, F_z is extremely small compared to gravity.

The factor μ is the *dynamic viscosity* (ratio of shearing stress to shear), and it is

18. E.g., H. Lamb, *Hydrodynamics* (6th ed.; New York: Dover Publications, 1932).

a function of temperature.[19] The factor $a\mu = \nu$ is *kinematic viscosity*. At standard sea-level pressure and temperature it is about 0.1 cm^2 sec^{-1}; at 200 mb and $-50°$ C, roughly 0.4 cm^2 sec^{-1}.

To examine the relation of F_z to vertical distribution of wind, consider Figure 7.211,

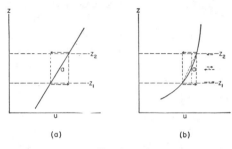

FIG. 7.211

where the heavy curve represents the profile of u and levels z_1 and z_2 are unit distance apart. Similar deductions will hold for F_y in connection with v. In (a) the slower motion at z_1 exerts retarding stress on motion at z_2, and the more rapid motion at z_2 exerts accelerating stress on motion at z_1. Relative motion distorts the fluid layer, but there is no net force acting in the layer, since the vertical variation of u is constant ($\partial^2 u/\partial z^2 = 0$), and stresses at z_1 and z_2 cancel. The motion at the mid-level equals the mean motion of the layer.

In (b) u varies nonlinearly with height ($\partial^2 u/\partial z^2 \neq 0$). If viscous forces are examined in the lower and upper halves of the layer, there is a resultant retarding stress at the mid-level a of the entire layer $z_1 z_2$. The resultant force is proportional to the difference between the mean motion $(u_2 + u_1)/2$ for the unit layer and the motion u_a at the mid-level. Since $\partial^2 u/\partial z^2 < 0$ in the case shown, $du/dt < 0$. This velocity profile is curved as usual for the surface friction layer. The

19. At $0°$ C, $\mu = 1.718 \times 10^{-4}$ gm cm^{-1} sec^{-1} (R. B. Montgomery, "Viscosity and Thermal Conductivity of Air and Diffusivity of Water Vapor in Air," *Journal of Meteorology*, IV [1947], 193–96).

terms F_x and F_y in the equations of motion always can be considered effective there and usually of decreasing significance upward.[20]

b) Accelerations due to rotation of coordinates.—Since the axes in the coordinate system are determined by the local latitude line, meridian, and zenith, the orientation varies on the earth. *Following a moving air particle* in the atmosphere the coordinate system rotates, unless motion is strictly vertical. If the particle follows a meridian the coordinates rotate in yz-plane only, but if there is a component of motion along latitude circles, all three axes rotate (except for motion *along* the equator).

The two component accelerations involving (u/a) tan ϕ result from this change of coordinate direction following a moving particle. Consider a particle moving with west wind at A in Figure 7.212a. If there is no net horizontal force, it follows a horizontal course not along the latitude arc AB but along great-circle arc AC. The particle thus is apparently accelerated southward. Now consider diagram (b) is a cross-section view of the right circular cone tangent to the earth along latitude ϕ. If particle A moves horizontally from west, it requires centripetal acceleration u^2/r to remain along the latitude circle if u is west wind and r the distance from D to A. However, if there is no force to maintain such centripetal acceleration, the particle drifts southward from the latitude circle with apparent acceleration of magnitude u^2/r. That is, $dv/dt = -u^2/r$. The same occurs with east wind at A. Since $r = a/\tan \phi$, $dv/dt = -(u^2/a)$ tan ϕ.

Observe in Figure 7.212(b) that the cone is tangent to the earth *along the rhumb line defined by the wind direction* at point A. The same would be done for winds from other directions. For winds along the equator or

along meridians the rhumb lines are great circles, and there is no such acceleration. For winds of intermediate direction there is apparent acceleration along the latitude arc also; that is, $du/dt = (uv/a)$ tan ϕ. The accelerations $-(u^2/a)$ tan ϕ and (uv/a) tan ϕ are due to departure of the rhumb line from the great circle for the given wind direction.[21]

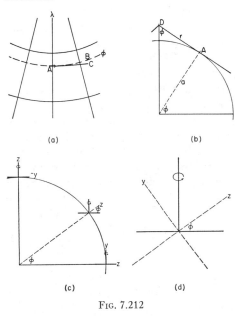

(a)   (b)

(c)   (d)

Fig. 7.212

c) The Coriolis acceleration.—Consider a particle in absence of all forces moving parallel to a rotating plane. Relative to that plane the particle will have speed c_E and an apparent acceleration normal to its relative motion in that plane which gives its path curvature. If ω is the angular velocity of rotation, the magnitude of this apparent acceleration is $2\omega c_E$. If rotation of the plane is positive (counterclockwise), the

20. They are usually neglected in the free atmosphere, but, as $a\mu$ increases upward, they might be significant near pronounced jet streams aloft.

21. More exactly, they are due to the curvature of the rhumb line for the given wind on a map projection for which great circles are straight lines. One can prove that on the polar gnomonic projection the radius of curvature of the latitude circle through point A in Figure 7.212(b) is distance $DA = r$ in that figure.

acceleration appears to the right of the vector relative motion, and, if rotation is negative, the acceleration appears to the left. It is important to remember this deflecting (Coriolis) acceleration, which modifies only direction of relative motion, acts parallel to the rotating plane, and affects only the component of motion parallel to that plane. The same principle is applicable to motions in the atmosphere. The axis of rotation is the earth's axis, and the rotating plane is the equatorial plane.[22]

Figure 7.212(c) shows the axis of rotation (vertical line) and part of the equatorial plane (horizontal line) at the north pole, at the equator, and at latitude ϕ. At the pole Coriolis acceleration occurs only in the xy-plane—the equatorial plane at that latitude—and is independent of w. Since $\omega > 0$, it follows that near the north pole a west-wind component gives southward acceleration observed on the earth, $dv/dt = -2\omega u$; a south-wind component gives eastward acceleration, $du/dt = 2\omega v$.

At the equator the equatorial plane is the xz-plane. Hence, only the u and w components of motion produce accelerations. The vertical acceleration is $dw/dt = 2\omega u$, and latitudinal acceleration is $du/dt = -2\omega w$.

Figure 7.212(d) is a magnified picture of the scheme at latitude ϕ. Velocity w along z has projection $w \cos \phi$ in the equatorial plane outward from the earth's axis. This component of motion gives Coriolis acceleration directed from the plane of the paper, that is, $du/dt = -2\omega w \cos \phi$. Note that dv/dt and dw/dt are independent of w.

The v component has projection $v \sin \phi$ in the equatorial plane directed toward the earth's axis, giving latitudinal acceleration $du/dt = 2\omega v \sin \phi$. Observe that this motion in the equatorial plane has a component along z and therefore gives no Coriolis acceleration along that axis. Finally, consider

22. That is, the plane through the particle in question and normal to the earth's axis.

the acceleration given by motion along the latitude circle. Since this circle lies in the equatorial plane, the u component is the same in that plane. If $u > 0$, acceleration is *outward* from the earth's axis. Acceleration has components along both z and y: $dw/dt = 2\omega u \cos \phi$ and $dv/dt = -2\omega u \sin \phi$. In summary, the component Coriolis accelerations are $du/dt = -2\omega w \cos \phi + 2\omega v \sin \phi$, $dv/dt = -2\omega u \sin \phi$, and $dw/dt = 2\omega u \cos \phi$.

d) Relative magnitudes of the terms in the equations of motion.—The accelerations du/dt, dv/dt, and dw/dt are the principal unknown quantities in the equations of motion. Most of the others can be observed, and for the remainder we know usual ranges of variation, except perhaps for extraneous forces, which we exclude from this discussion.

(i) *Vertical acceleration.*—If we neglect F_z, Eq (1c) becomes

$$dw/dt - 2\omega u \cos \phi = -a(\partial p/\partial z) - g .$$

In our assumption that hydrostatic conditions are satisfied, the left side is negligible in comparison with each of the two terms on the right. The truth of the hydrostatic assumption lies in just how small the net effect of the two terms on the left really is. It is reasonable to assume that in usual conditions dw/dt must be small in comparison with 980 cm sec^{-2}. There is no way of computing dw/dt from data as presently measured, since $\partial p/\partial z$ is obtained by assuming hydrostatic equilibrium.

Now consider the term $2\omega u \cos \phi$. For west wind 10 m sec^{-1} at the equator, upward acceleration is about 0.14 cm sec^{-2}, and ten times that much for a wind 100 m sec^{-1}. This term is ordinarily less than 1/1000 of gravity and is therefore negligible in comparison with gravity. A zonal wind component 100 m sec^{-1} at 45° latitude corresponds to a discrepancy about 1 foot in every thousand in evaluated heights of pressure surfaces.

It follows that with sufficient approximation the terms in the vertical equation of motion due to rotation and spherical shape of the earth are negligible, and

$$dw/dt = -a(\partial p/\partial z) - g. \qquad (2)$$

But in most conditions dw/dt is also negligible.

(ii) *Horizontal accelerations.*—While we are justified in neglecting accelerations in Eq (1c) due to rotation of the earth, we

Table 7.21 shows the values of each acceleration term for given velocity and latitude. Of those listed, the Coriolis acceleration is the largest (except at the equator). Notice also that $2\omega w \cos \phi$ is only about a thousandth of the Coriolis acceleration for vertical velocities between 1 and 10 cm sec^{-1}. If this is a suitable range for w, we can neglect the contribution by $2\omega w \cos \phi$ to the net acceleration except possibly near the equator, where $\sin \phi$ and $\tan \phi$ are zero.

TABLE 7.21

MAGNITUDES OF CERTAIN ACCELERATIONS (CM SEC^{-2}) ON THE EARTH FOR GIVEN VALUES OF WIND SPEED AND LATITUDE

VELOCITY	(uv/a) TAN ϕ		(u^2/a) TAN ϕ		$2\omega v$ SIN ϕ, $2\omega u$ SIN ϕ		$2\omega w$ COS ϕ		$\partial p/\partial x$ OR $\partial p/\partial y$ WITH GEOSTROPHIC BALANCE	
	20°	70°	20°	70°	20°	70°	20°	70°	20°	70²
$u = 10$ m sec^{-1}			0.6 | 4.4 (×10^{-3})		0.5 | 1.4 (×10^{-1})				1 mb/160 km	1 mb/60 km
$v = 10$ m sec^{-1}	0.6 | 4.4 (×10^{-3})				0.5 | 1.4 (×10^{-1})				1 mb/160 km	1 mb/60 km
$u = 100$ m sec^{-1}			0.6 | 4.4 (×10^{-1})		0.5 | 1.4				1 mb/16 km	1 mb/6 km
$v = 100$ m sec^{-1}	0.6 | 4.4 (×10^{-1})				0.5 | 1.4				1 mb/16 km	1 mb/6 km
$w = 1$ cm sec^{-1}							1.4 | 0.5 (×10^{-4})			
$w = 10$ cm sec^{-1}							1.4 | 0.5 (×10^{-3})			

cannot do the same for the other equations. These accelerations are considerably more important in the horizontal equations, owing to the small magnitudes of all horizontal forces compared to gravity.

At the equator the horizontal equations (neglecting friction) reduce to

$$du/dt + 2\omega w = -a(\partial p/\partial x),$$

$$dv/dt = -a(\partial p/\partial y).$$

Since $2\omega w$ is comparatively small in the usual case, air at and near the equator is accelerated horizontally *toward lower pressure* if friction is neglected. That is, $d\mathbb{C}/dt \simeq -a\nabla_h p + \mathbf{F}$.

For winds of order 10 m sec^{-1} the terms with $\tan \phi$ are about 1 per cent of the Coriolis acceleration. With winds 100 m sec^{-1} they are of order 10 per cent of the Coriolis acceleration. At 70° latitude a zonal wind component 100 m sec^{-1} gives $(u^2/a) \tan \phi$ equal to 3/10 the Coriolis acceleration.

The horizontal equations of motion now may be written

$$du/dt - (uv/a)\tan \phi - 2\omega v \sin \phi$$
$$= -a(\partial p/\partial x), \qquad (3a)$$

$$dv/dt + (u^2/a)\tan \phi + 2\omega u \sin \phi$$
$$= -a(\partial p/\partial y), \qquad (3b)$$

if F_x and F_y are not considered. It is seen by comparing these with Eqs 6.18(1a, 1b) that geostrophic wind makes no account for the first two terms in each equation above. Further, geostrophic wind is not defined at the equator (nor does it hold in equatorial regions). It is evident that ageostrophic departures must exist as a rule.

7.22. *Steady motion with curved isobars—the gradient wind.*

The pressure pattern near the surface is mostly a series of HIGHs

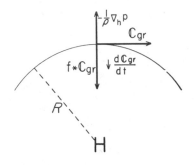

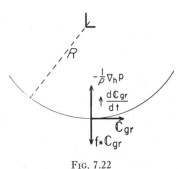

Fig. 7.22

and LOWs, and aloft it assumes more wave-shaped forms of ridges and troughs with fewer centers. In either case air motion must have systematic curvature if it follows the contours. With curved motion a centripetal acceleration exists which is directly proportional to the square of wind speed and inversely proportional to the radius of curva-

ture of the air trajectory (Fig. 7.22). Motion in which the pressure gradient is normal to the wind is known as *gradient motion;* the gradient wind will be designated $\mathbb{C}_{gr}$. This topic is discussed at length in textbooks on dynamic meteorology, and therefore a summary suffices here.[23]

In general, with only pressure forces acting, the vector equation of motion is $-a\nabla_h p + f * \mathbb{C} = d\mathbb{C}/dt$. Geostrophic wind $\mathbb{C}_g$ may be introduced in place of the pressure force:

$$-f * \mathbb{C}_g + f * \mathbb{C} = d\mathbb{C}/dt. \quad (1)$$

The vector $f * \mathbb{C}_g$ is the Coriolis force corresponding to the geostrophic part of the wind. Now assume the air trajectory follows isobaric contours, so that the pressure force is normal to the wind, and the wind is therefore a gradient wind, $\mathbb{C}_{gr}$.

$$-f * \mathbb{C}_g + f * \mathbb{C}_{gr} = d\mathbb{C}_{gr}/dt. \quad (2)$$

Since both gradient and geostrophic winds are parallel to contours (isobars), Eq (2) implies that the *acceleration* of gradient wind is normal to the contours—in other words, perpendicular to the gradient wind itself. This acceleration is therefore a *centripetal* acceleration.

Let R denote the trajectory radius of curvature. If we associate counterclockwise turning with positive curvature $(R > 0)$, the gradient wind equation has the scalar form

$$-fc_g + fc_{gr} = -c_{gr}^2/R,$$

or

$$-980(\partial Z/\partial n)_p + fc_{gr} = -c_{gr}^2/R.$$

The term c_{gr}^2/R is the *cyclostrophic* term;

23. The emphasis placed on the gradient wind equation in the literature might give the impression that this intriguing topic is the ultimate refinement in relating motion to pressure patterns. That it is not is evidenced by the restrictive assumptions; it considers only an idealized case for one of several accelerations present in the atmosphere.

divided by f, it is the difference between geostrophic and gradient wind.

In the northern hemisphere counterclockwise turning corresponds to cyclonic gradient wind and clockwise turning to anticyclonic gradient wind. The preceding equation may be written separately for the two cases:

$$c_{gr} = c_g + c_{gr}^2/f|R| \text{ (anticyclonic, N.H.)} \quad (3a)$$

$$c_{gr} = c_g - c_{gr}^2/f|R| \quad \text{(cyclonic, N.H.)} \quad (3b)$$

Here $|R|$ is the absolute value of the radius of curvature.

From Eq (3a) *gradient wind for anticyclonic curvature is greater than geostrophic wind* for the same pressure gradient. From Eq (3b) *gradient wind for cyclonic curvature is less than geostrophic wind*. In both cases the departure of gradient wind from geostropic is proportional to curvature and the square of wind speed. As the curvature is decreased ($|R|$ increased), the gradient wind approaches geostrophic, and for very large $|R|$ gradient motion is essentially geostrophic.

The departure of actual wind from geostrophic due to curvature is often verified in clearly defined large-scale features of the pressure pattern. Around intense cyclones the wind is much less than geostrophic. Evidence is seen by comparing wind speeds and gradients around the east Canadian LOW with conditions elsewhere in Figure 7.03h. Also, in the upper trough over the eastern United States observed winds are generally subgeostrophic by amounts consistent with the curvature correction. With anticyclones, pressure gradients are weak, and other departures from geostrophic balance (especially at the surface) might completely mask effects of trajectory curvature. It is observed, however, that winds can be supergeostrophic in strong anticyclonic wind currents around upper-level pressure ridges.

When applying the gradient wind correc-

tion in pressure analysis, it is remembered that for given wind the spacing of contours is increased for anticyclonic curvature and decreased for cyclonic curvature relative to geostrophic spacing. There are various graphs for making this correction; Figure O (Appendix) is an example. From this graph one obtains the geostrophic wind for given gradient wind knowing the radius of curvature and latitude or, conversely, the gradient wind for given geostrophic wind, radius of curvature, and latitude. As an example, assume a gradient wind 50 knots at 50° latitude and trajectory radius of curvature 10° latitude. In the middle section of the *Gradient Wind Scale* locate the intersection of the radius of curvature line (vertical) with the line for latitude (slanting). If flow is cyclonic, proceed to the right to the intersection with the vertical line for gradient wind 50 knots. At this point read (from dashed slanting lines) the value of geostrophic wind; c_g is 60 knots. For anticyclonic curvature one reads from the left section a geostrophic wind about 40 knots. Contour spacing then is obtained from a geostrophic scale. The spacing finally determined is consistent for gradient motion.

We now should examine some theoretical aspects of the two gradient wind equations. The quadratic equation for anticyclonic gradient wind can be solved to give

$$c_{gr} = f|R|/2 - \sqrt{[R^2f^2/4}$$
$$-980|R|(\partial Z/\partial n)_p] . \quad \text{(anticyclonic)} \quad (4)$$

It is evident from this that maximum anticyclonic gradient wind is $f|R|/2$. This occurs when

$$(\partial Z/\partial n)_p = f^2|R|/(4 \times 980) . \quad (5)$$

The contour gradient with anticyclonic gradient flow is theoretically limited, and the maximum contour gradient is directly proportional to the radius of curvature. Near anticyclonic centers the pressure field must be weak if balance is maintained, and, with increasing distance from the center, gradients can be larger.

From Eqs (4) and (5) above it follows that *the maximum anticyclonic gradient wind is precisely twice the geostrophic wind for that contour gradient*, or c_{gr} max $= 2c_g$. This is the maximum *equilibrium* wind in anti-cyclonically curved flow; it does not imply that actual winds cannot exceed it.

The quadratic equation for cyclonic gradient wind is solved in a similar manner:

$$c_{gr} = -fR/2 + \sqrt{[R^2f^2/4}$$
$$+980R(\partial Z/\partial n)_p] . \quad \text{(cyclonic)} \quad (6)$$

The quantity beneath the radical is always positive, and its square root is greater than $fR/2$. In this equation there is no restriction on magnitude of pressure gradient to maintain real roots of the quadratic. A useful interpretation is that cyclones can have larger gradients than anticyclones, and such is usually the case. On the pressure charts illustrated observe the asymmetry in distribution of gradients between HIGHs and LOWs at all levels. In fact, the maximum pressure slope is usually somewhat nearer the cyclone center than the center of the adjacent anticyclone. Another verification is that intense tropical storms are observed while analogous anticyclonic circulations are not found.

It is usually assumed in practice that the effect of surface friction decreases rapidly with height, and at levels 500 meters or 2000 feet above ground the wind should attain balance with the pressure field. Such a level can be taken as the *gradient level*. As an aid in sea-level pressure analysis, these winds are often entered on the surface chart. These are called "gradient winds"; more properly, "gradient-level winds." They are not necessarily in gradient balance according to the gradient-wind equations given above.

7.23. *Accelerations with unbalanced forces.* —The principal forces governing motions in the atmosphere are the pressure, the

Coriolis, and the frictional forces. The horizontal acceleration acting on a parcel of air can be considered the unbalanced residual of these forces per unit mass. That is,

$$d\mathbb{C}/dt = f * (\mathbb{C} - \mathbb{C}_g) + \mathbf{F} . \quad (1)$$

$(\mathbb{C} - \mathbb{C}_g)$ is always the vector difference between actual and geostrophic wind. This difference is ageostrophic wind, denoted $\mathbb{C}'$ (Fig. 7.04b). From Eq (1) it follows that the net acceleration is the vector difference

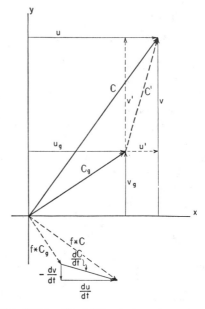

FIG. 7.231.—Graphical relation of acceleration and ageostrophic wind in frictionless motion.

between actual Coriolis force $f * \mathbb{C}$ and that Coriolis force $f * \mathbb{C}_g$ for geostrophic wind, when $\mathbf{F} \approx 0$. (It is $d\mathbb{C}/dt - \mathbf{F}$ when friction is included.) This difference is shown as $d\mathbb{C}/dt$ in the lower part of Figure 7.231. Notice that it is perpendicular to the ageostrophic wind $\mathbb{C}'$ and directed to the right of $\mathbb{C}'$. In absence of friction, *the acceleration is actually the Coriolis force per unit mass for the ageostrophic wind.*

From the illustration above it is evident that, if $\mathbb{C}$ has geostrophic direction but

$c > c_g$, then $\mathbb{C}'$ is along the wind direction, and acceleration is to the right of the wind. Thus, *supergeostrophic winds are accelerated to the right of the geostrophic wind*. Anticyclonically curved flow in gradient conditions is one special case. If $c < c_g$ but both are of same direction, acceleration is to the left of the wind. Therefore, *subgeostrophic winds are accelerated to the left of the geostrophic wind*, and gradient cyclonic flow is one special case of that type.

Acceleration may be given also in terms of the coordinate components of wind. It is seen from further examination of Figure 7.231 that

$$du/dt - F_x = fv' = f(v - v_g) , \quad (1a)$$

and

$$dv/dt - F_y = -fu'$$
$$= -f(u - u_g) . \quad (1b)$$

From the three equations above, the magnitude of acceleration can be determined relative to the ageostrophic wind. Since f is of magnitude 10^{-4} sec^{-1}, acceleration is about 1/10,000 as large in the same system of units as ageostrophic speed in middle latitudes. The change of wind speed by an acceleration acting over a long time may be large. In only 3 hours (10,800 seconds) a constant acceleration gives change of wind speed about equal to the initial ageostrophic wind.

It is of interest to investigate the factors giving accelerations to motion and consequently departures from geostrophic balance, for, if we are to employ the geostrophic and gradient wind equations at all, we should be aware of some conditions with departures. If actual wind is expressed as the sum of geostrophic wind and the ageostrophic component: $\mathbb{C} = \mathbb{C}_g + \mathbb{C}'$; $u = u_g + u'$; $v = v_g + v'$. The total time derivatives of $\mathbb{C}$, u, and v are $d\mathbb{C}/dt = d\mathbb{C}_g/dt + d\mathbb{C}'/dt$; $du/dt = du_g/dt + du'/dt$; $dv/dt = dv_g/dt + dv'/dt$.

It is now seen that acceleration can be examined in two parts, viz., the contribu-tion by geostrophic acceleration and that by ageostrophic acceleration. We shall not investigate the latter. Since geostrophic acceleration can be described adequately in terms of pressure and isallobaric[24] fields, it is worth while finding as much as possible concerning this contribution to total acceleration.

We may transform the geostrophic part of the equations above by use of the differential operator

$$\frac{d}{dt} = \frac{\partial}{\partial t} + \mathbb{C} \cdot \nabla_h + w \frac{\partial}{\partial z}$$

$$= \frac{\partial}{\partial t} + u \frac{\partial}{\partial x} + v \frac{\partial}{\partial y} + w \frac{\partial}{\partial z} .$$

Then, by Eqs (1), (1a), and (1b),

$$d\mathbb{C}/dt - \mathbf{F} = f * \mathbb{C}' = \partial\mathbb{C}_g/\partial t + \mathbb{C} \cdot \nabla_h \mathbb{C}_g$$
$$+ w(\partial\mathbb{C}_g/\partial z) + d\mathbb{C}'/dt - \mathbf{F} , \quad (2)$$

$$du/dt - F_x = fv' = \partial u_g/\partial t + \mathbb{C} \cdot \nabla_h u_g$$
$$+ w(\partial u_g/\partial z) + du'/dt - F_x , \quad (2a)$$

$$dv/dt - F_y = -fu' = \partial v_g/\partial t + \mathbb{C} \cdot \nabla_h v_g$$
$$+ w(\partial v_g/\partial z) + dv'/dt - F_y . \quad (2b)$$

If the last two equations are divided by f and $-f$, respectively,

$$v' = \frac{1}{f}\frac{\partial u_g}{\partial t} + \frac{u}{f}\frac{\partial u_g}{\partial x} + \frac{v}{f}\frac{\partial u_g}{\partial y} + \frac{w}{f}\frac{\partial u_g}{\partial z}$$
$$- \frac{F_x}{f} + \frac{1}{f}\frac{du'}{dt} , \quad (3a)$$

$$u' = -\frac{1}{f}\frac{\partial v_g}{\partial t} - \frac{u}{f}\frac{\partial v_g}{\partial x} - \frac{v}{f}\frac{\partial v_g}{\partial y}$$
$$- \frac{w}{f}\frac{\partial v_g}{\partial z} + \frac{F_y}{f} - \frac{1}{f}\frac{dv'}{dt} . \quad (3b)$$

A similar expression holds for resultant ageostrophic wind $\mathbb{C}'$. Obviously, these are not valid where $f = 0$.

Accelerations $\partial u_g/\partial t$ and $\partial v_g/\partial t$ comprise the *isallobaric acceleration* $\partial\mathbb{C}_g/\partial t$ to geo-

24. Pressure change.

strophic wind, and, when divided by the Coriolis parameter, they give the *isallobaric wind* $\mathbb{C}'_i$, which is that *part* of ageostrophic wind related to isallobaric acceleration. The terms $(u\partial u_g/\partial x + v\partial u_g/\partial y)$ and $(u\partial v_g/\partial x + v\partial v_g/\partial y)$ comprise the (horizontal) *advective acceleration* to geostrophic wind, $\mathbb{C}\cdot\nabla_h\mathbb{C}_g$. The corresponding quantities in the right sides of Eqs (3a) and (3b) give the *advective ageostrophic* wind $\mathbb{C}'_a$. Next, $w(\partial u_g/\partial z)$ and $w(\partial v_g/\partial z)$ are the components of geostrophic acceleration due to vertical motion. As expressed in the last equations, those give the *convective ageostrophic wind* $\mathbb{C}'_c$. Next are the frictional accelerations, which contribute the *antitriptic wind* $\mathbb{C}'_F$. Finally, as seen by Eq (2) the ageostrophic acceleration also contributes to the net ageostrophic wind.

The resultant ageostrophic wind is the vector sum of all contributions:

$$\mathbb{C}' = \mathbb{C}'_i + \mathbb{C}'_a + \mathbb{C}'_c + \mathbb{C}'_F + X ,$$

where X denotes ageostrophic wind related to ageostrophic acceleration. If that term is dropped (not suggesting it is small), we have

$$\mathbb{C}' = \mathbb{C}'_i + \mathbb{C}'_a + \mathbb{C}'_c + \mathbb{C}'_F ; \quad (4)$$

$$u' = u'_i + u'_a + u'_c + u'_F ; \quad (4a)$$

$$v' = v'_i + v'_a + v'_c + v'_F . \quad (4b)$$

It will be remembered henceforth that the resultant ageostrophic wind being described is not the true resultant, since a term is omitted. However, valuable information can be obtained from the simplified form, and the equations we now have are more complete than presently considered in practice.

a) Antitriptic wind.—The most obvious departure from geostrophic balance is that due to friction with the earth's surface. The frictional stress between moving air and the earth first may be viewed as retardation ($\mathbf{F}_g$ in Fig. 7.232) acting opposite to the geostrophic wind. This decreases air motion, thus decreasing the Coriolis force.

There results an acceleration toward lower pressure, as the pressure force is not affected directly, and the air develops a component of motion across contours (isobars) toward lower pressure. In the final adjusted state the wind is less than geostrophic and directed to the left of $\mathbb{C}_g$ (northern hemisphere). In this state the frictional force $\mathbf{F}$, the Coriolis force $f * \mathbb{C}$, and the pressure force are in equilibrium. Observe that

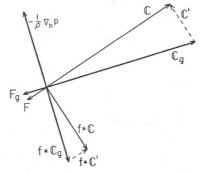

FIG. 7.232.—The effect of friction on geostrophic balance. (Shown here is a state of equilibrium between pressure, Coriolis, and frictional forces.)

$f * \mathbb{C}' = -\mathbf{F}$, which is Eq (2) when $\mathbb{C}'_i + \mathbb{C}'_a + \mathbb{C}'_c = 0$. In this condition $\mathbb{C}'$ and $\mathbb{C}$ are perpendicular, and therefore $c = c_g \cos \delta$, if δ is the angle between $\mathbb{C}$ and the contours (isobars). This departure at the surface is a complicated function of wind distribution, surface roughness, and static stability in the friction layer. The effect of surface friction is shown clearly by comparing the winds in Figures 9.02*ab* with the gradient-level winds in Figure 7.03*h*.

b) Isallobaric wind.—The definition of isallobaric wind found above was first given by Brunt and Douglas,[25] who suggested that with large isallobaric gradients the isallobaric wind could account for the principal ageostrophic flow.

25. D. Brunt and C. K. M. Douglas, "The Modification of Strophic Balance for Changing Pressure Distribution, and Its Effect on Rainfall," *Memoirs of the Royal Meteorological Society*, Vol. III, No. 22 (1928).

Figure 7.233 gives the isobaric topography at an initial time (heavy solid lines), the topography after time δt (heavy dashed lines), and the isopleths (isallohypses, isallobars) of net change. At the right are the initial geostrophic wind $\mathbb{C}_{g0}$, final geostrophic wind $\mathbb{C}_{g1}$, and change $\delta \mathbb{C}_g$ for point A. Vector $\delta \mathbb{C}_g$ parallels isallobars, and its magnitude is proportional to the isallobaric gradient.

The effect of the isallobaric field is given by the first term on the right of Eqs (2). A part of $d\mathbb{C}/dt$ (and a corresponding part of $\mathbb{C}'$) is due to this local variation of geostrophic wind. We denote this contribution by

$$(d\mathbb{C}/dt)_i = f * \mathbb{C}'_i = \partial \mathbb{C}_g/\partial t . \quad (5)$$

For convenience, only the first of Eqs (2) will be used.[26] The quantity $\partial \mathbb{C}_g/\partial t$ is approximated by $\delta \mathbb{C}_g/\delta t$. Accordingly, from Eq (5), $(d\mathbb{C}/dt)_i \delta t = \delta \mathbb{C}_g$. Isallobaric acceleration is *along* $\delta \mathbb{C}_g$, with larger falling tendencies to the left (Fig. 7.233).

Eq (5) shows that an isallobaric acceleration corresponds to isallobaric wind $\mathbb{C}'_i$ directed 90° to the left. Isallobaric wind then must be *directly proportional to and directed along the isallobaric descendant*. This may be shown also by differentiating the geostrophic equation $f * \mathbb{C}_g = 980 \ (\nabla Z)_p$ with respect to time; that is, $f * \partial \mathbb{C}_g/\partial t = 980 \ \nabla(\partial Z/\partial t)_p$. But from Eq (5), $f * (\partial \mathbb{C}_g/\partial t) = f * (f * \mathbb{C}'_i) = -f^2 \mathbb{C}'_i$, or $\mathbb{C}'_i$ is directed 180° from the isallobaric ascendant $\nabla(\partial Z/\partial t)_p$. Then

$$\mathbb{C}'_i = - \ (980/f^2)[\nabla (\partial Z/\partial t)_p] ,$$
or
$$c'_i = (980/f^2)[\partial(\partial Z/\partial t)_p/\partial n] . \quad (6)$$

Direction n is along the isallobaric ascendant. Thus, if isallobars are west-east with

26. For slightly different derivations using coordinate components refer to B. Haurwitz, *Dynamic Meteorology* (New York: McGraw-Hill Book Co., 1941), pp. 155–59, and E. W. Hewson and R. W. Longley, *Meteorology: Theoretical and Applied* (New York: John Wiley & Sons, 1944), pp. 128–30.

larger rises south, isallobaric wind is northward with speed proportional to the magnitude of the isallobaric gradient.

The isallobaric wind speed may be evaluated directly from the tendency field by the scalar form of Eq (6). For example, where $f = 10^{-4} \ \text{sec}^{-1}$ and the tendency gradient is 10 gpft per hour per 100 km (a rather large value), isallobaric wind is about 8 m sec^{-1}.

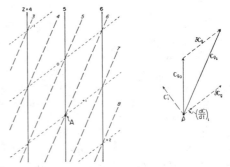

FIG. 7.233.—The isallobaric wind in relation to the isallobaric pattern.

The validity of the isallobaric wind measured from a pattern of pressure change is dependent on how well the field of integrated change during the period indicates the field of instantaneous tendency. In the moving wave pattern in Figure 7.144 the true isallobaric winds would be directed toward the inflection point in the current center in advance of the trough and away from the opposite inflection point. The field of isallobaric wind deduced from the change patterns in the drawing gives quite an erroneous picture. In general, the longer the time interval for pressure change, the greater is the discrepancy between indicated and true isallobaric wind fields.

The degree to which the isallobaric wind $\mathbb{C}'_i$ approximates total ageostrophic wind $\mathbb{C}'$ depends on the contribution by the remaining terms[27] on the right side of Eq (2). It is

27. A more detailed discussion of the relative importance of the isallobaric contribution is given by B. Haurwitz, "On the Relation between the Wind Field and Pressure Changes," *Journal of Meteorology*, III (1946), 95–99.

apparent that even with strong isallobaric wind the ageostrophic wind may be zero or even opposite to the isallobaric wind. Thus *it is improper to attribute existing ageostrophic winds only to the isallobaric pattern until all other factors have been accounted for.* (For gradient winds in an isallobaric field the requirement is that the sum of all terms on the right of Eq [2] equals $d\mathbb{C}_{gr}/dt$.)

c) Advective ageostrophic wind.—As an air parcel moves horizontally, the pressure pattern usually varies along its trajectory. This is true even if the pressure field is steady, for contours are not straight or uniformly curved and not parallel over large distances. Thus, an air parcel which at one instant is in geostrophic (or gradient) balance soon may find itself in a different pressure field. To readjust itself, the parcel is subjected to acceleration and thus develops an ageostrophic component. From this viewpoint alone it is easy to see that geostrophic or gradient wind is rarely precisely fulfilled on synoptic pressure charts.

The effect of horizontal variations in the pressure pattern on accelerations is given by $\mathbb{C} \cdot \nabla_h \mathbb{C}_g$ in Eq (2). We may write $\mathbb{C} \cdot \nabla_h \mathbb{C}_g = c(\partial \mathbb{C}_g/\partial s)$, where s is distance downwind through the point in question, and $\partial \mathbb{C}_g/\partial s$ is the variation of vector geostrophic wind in that direction (Fig. 7.234). If acceleration due to this effect is $(d\mathbb{C}/dt)_a$ and the corresponding ageostrophic wind is $\mathbb{C}'_a$, then

$$(d\mathbb{C}/dt)_a = f * \mathbb{C}'_a = c(\partial \mathbb{C}_g/\partial s) . \quad (7)$$

The advective contribution is illustrated graphically[28] in Figure 7.234. Geostrophic winds are measured from the pressure pattern at points δs apart and equidistant from b. Vector $\delta \mathbb{C}_g$ is the difference in geostrophic winds upstream and downstream from b. Dividing $\delta \mathbb{C}_g$ by δs approximates $\partial \mathbb{C}_g/\partial s$,

28. This can be shown also by superimposing the pressure patterns at two points in a manner similar to Figure 7.233.

which lies along $\delta \mathbb{C}_g$. From Eq (7) the acceleration is along $\delta \mathbb{C}_g$ and directly proportional to wind speed as well as to the magnitude of $\delta \mathbb{C}_g/\delta s$. The ageostrophic wind is 90° to the left of $\delta \mathbb{C}_g$. Notice in this example that the acceleration has a component to the right of the wind, indicating clockwise turning of wind in partial agreement with the contour pattern, and also a component opposite to the wind $\mathbb{C}$, implying retarded motion in agreement with the divergence of contours.

If $\partial \mathbb{C}_g/\partial s$ is normal to the actual wind $\mathbb{C}$, then the ageostrophic wind lies along the

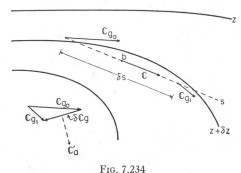

Fig. 7.234

actual wind, the acceleration is centripetal, and the motion is parallel to the contours. This is a statement of the gradient-wind condition for a steady pressure field and horizontal motion. In fact, one can show that $\mathbb{C} \cdot \nabla_h \mathbb{C}$ is the cyclostrophic acceleration in the gradient-wind equation with the same assumptions. By neglecting this advective term in the acceleration, the gradient-wind correction is being neglected.

Figure 7.235 gives simple illustrations of the advective effect on ageostrophic winds. Shown here are the three pressure patterns possible with straight contours, but deductions are also applicable to curved contours. For each case we might assume the wind is geostrophic initially at a. In the first diagram, as the air moves from a to b, it experiences no acceleration due to space variation of the pressure field; this wind remains geostrophic. In the second the contours diverge downwind, the motion is retarded, and ageostrophic wind is to the right. In the last diagram the effect is opposite.

To show the approximate value of ageo-

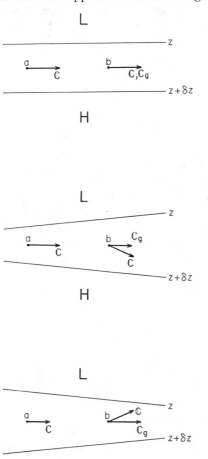

Fig. 7.235.—The departure of actual wind from geostrophic due to changing pressure gradients in the direction of motion.

strophic wind resulting from the advective contribution, average values may be substituted into Eq (7). If $f = 10^{-4}$ sec^{-1}, wind speed is 10 m sec^{-1}, and geostrophic wind from the same direction varies 5 m sec^{-1} per 100 km along the wind, then $c_a' \simeq 5$ m sec^{-1}. This is the same order of magnitude as the isallobaric wind. With stronger winds this effect can be more important than the isallobaric influence.

Deductions made on the basis of Figure 7.235 may be formulated into a set of general rules applicable to most synoptic patterns of pressure and wind. *If air moves into a weaker pressure field, it is deflected to the right of the contours and its motion retarded, and, if air moves into a stronger pressure field, it is deflected to the left of the contours and its motion accelerated,* provided its motion is in reasonable agreement with the pressure pattern initially. These rules exclude contributions by other terms in Eq (2), and they may not be strictly valid in certain critical conditions of curved flow or curved contours.[29]

There is frequent empirical evidence of the general rules given above for advective accelerations to the wind, not so much at the surface, where it is difficult to separate effects of friction, but at levels sufficiently higher, where there are still plentiful wind data. In Figure 7.03g there are several supergradient winds in the south central United States where contours diverge rapidly downstream. Here the winds differ less from those upstream than consistent with the widening of contour channels. Still farther upstream west of Lake Winnipeg, where contours converge downstream, it appears that winds are subgradient in having come from the weak pressure ridge over western Canada.

d) Convective ageostrophic wind.—The discussion above was devoted to varying pressure fields for horizontal air motion only; horizontal accelerations due to vertical displacements from one pressure pattern to another must be considered also. The vertical component of motion is ordinarily very

29. It is possible for the air to be accelerated while moving into a weaker pressure field if the contours vary in curvature sufficiently; in this case, supergradient winds toward lower pressure.

small relative to the horizontal component, but vertical variation of pressure patterns is usually so many times greater than horizontal variation that accelerations resulting from vertical displacements are not always negligible in comparison.

From Eq (2) acceleration $(d\mathbb{C}/dt)_c$ due to vertical displacement may be represented in the same form as Eq (7). Thus,

$$(d\mathbb{C}/dt)_c = f * \mathbb{C}'_c = w(\partial \mathbb{C}_g/\partial z) . \quad (8)$$

Vector $\partial \mathbb{C}_g/\partial z$ is the variation of geostrophic wind with height; it is the thermal wind in a thin layer divided by the depth.

In Figure 7.07 are given the vector geostrophic winds $\mathbb{C}_{g1}$ at lower level z_1 and $\mathbb{C}_{g2}$ at upper level z_2; the difference, $\Delta\mathbb{C}_g$, divided by the layer depth is the approximation to $\partial \mathbb{C}_g/\partial z$, which has the direction of $\Delta\mathbb{C}_g$. For upward motion $w > 0$, and from Eq (8) acceleration $(d\mathbb{C}/dt)_c$ is along $\Delta\mathbb{C}_g$. The related ageostrophic wind is to the left of $\Delta\mathbb{C}_g$. Thus, *ageostrophic wind $\mathbb{C}'_c$ blows across isotherms toward colder air in upward motion; it is directed toward warmer air in downward motion.*

As an example, consider that geostrophic wind is from west at 700 mb and zero at 850 mb. The thermal wind in the layer is then from west. With ascent the air is accelerated eastward and has ageostrophic component from south. For descent the resulting ageostrophic component is from north. If the air rises rapidly, it arrives at 700 mb with actual motion from south of west.

In barotropic regions vertical displacements should give no horizontal accelerations to the velocity. Large horizontal temperature gradients are favorable for development of ageostrophic winds by vertical displacements and can give an appreciable contribution to the total ageostrophic wind. For $f = 10^{-4}$ sec^{-1} and $\partial\mathbb{C}_g/\partial z = 10$ m sec^{-1} per kilometer (horizontal temperature gradient about 3° C/100 km), Eq (8) gives speed for $\mathbb{C}'_c$ 1 m sec^{-1} for each 1 cm sec^{-1} of vertical velocity. That vertical velocity may be considered an average absolute value in large-scale atmospheric motion.[30] The above result for convective ageostrophic wind might appear small in comparison with the others, but there are frequent situations in which vertical velocity and thermal wind are both larger than the values used. In effect, this contribution easily can be as important as those above.

There are several areas in the 850-mb chart for 1 March 1950 through which large vertical velocities might be expected. Along the east coast of the United States north of the Florida peninsula upward motion is evidenced by high humidities aloft and precipitation. Unfortunately, for the same reason wind data are few. Because gradient-level winds are from southwest and the 850-mb geostrophic winds are slightly stronger from west, implying ageostrophic motion northeastward with ascent, 850-mb winds in this area should cross contours to the left. Another region of suspected ascent in the lower atmosphere is the west coast of North America north of 45° latitude. Here the winds in the vicinity of 850 mb are accelerated not just by upward displacement but also by forced flow toward lower pressure induced by terrain.

Over the south central United States strong subsidence is occurring just behind the cold front. Winds at 700 mb are from west or west-northwest with average speeds about 50 knots. The 850-mb chart shows, below this, geostrophic winds from northwest with smaller speeds. In descending to this level, the air is accelerated to southwest (ageostrophic wind from northwest), resulting in northwesterly winds supergeostrophic for that level. In this area at 850 mb

30. H. A. Panofsky, "Large-Scale Vertical Velocity and Divergence," in American Meteorological Society, *Compendium of Meteorology*, pp. 639–46.

the winds are stronger than geostrophic conceivably because of ageostrophic wind by subsidence and by transport horizontally into a field of diverging contours.

Although vertical velocities must terminate at the earth's surface (where it is level), ageostrophic winds due to strong subsidence can be felt even in surface winds if there is sufficient turbulent mixing to carry the effect downward through the friction layer. Accelerations and ageostrophic winds at the surface developed through subsidence should be particularly evident just behind cold fronts, where thermal wind is large and where lack of stability in the friction layer is conducive to turbulent mixing. This could be a major contribution to ageostrophic flow behind the cold front over Texas in Figures 9.02*ab*.

7.24. *Stability for horizontal and vertical displacements.*—In Chapter 3 a discussion of atmospheric stability was given for vertical displacements only. Stability for horizontal and arbitrary displacements deserves similar treatment, but unfortunately little has been done with that phase of the subject. What appears below is a modified version of Van Meigham's summary in the *Compendium of Meteorology* (1951).

Consider a geostrophic current (Fig. 7.24) of invariant wind direction α in a small region of space. To simplify the discussion, the xy-plane is oriented with x-axis downwind, instead of by local grid coordinates on the earth. In this scheme $v_g = \partial v_g/\partial x = \partial v_g/\partial y = \partial v_g/\partial z = 0$, assuming also there is no temperature gradient downwind. Furthermore, if $U_0 = (u_g)_0$ is geostrophic velocity at any point A_0 along x, geostrophic velocity $U_A = (u_g)_A$ at A located a small distance $(\delta y, \delta z)$ transverse to the x-axis is

$$U_A = U_0 + (\partial u_g/\partial y)_0\, \delta y$$
$$+ (\partial u_g/\partial z)_0\, \delta z . \quad (1)$$

In such a coordinate system the equations of motion are

$$du/dt = 2\omega w \cos \phi \sin \alpha + 2\omega v \sin \phi$$
$$- a(\partial p/\partial x) , \quad (2a)$$

$$dv/dt = -2\omega u \sin \phi - 2\omega w \cos \phi$$
$$\times \cos \alpha - a(\partial p/\partial y) , \quad (2b)$$

$$dw/dt = -2\omega u \cos \phi \sin \alpha + 2\omega v$$
$$\times \cos \phi \cos \alpha - a(\partial p/\partial z) - g , \quad (2c)$$

if extraneous forces F_x, F_y, F_z are neglected. In the subsequent discussion we use the

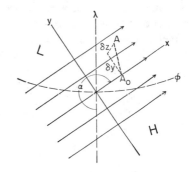

Fig. 7.24

notations: $2\omega \sin \phi = f$, $2\omega \cos \phi \sin \alpha = B$, and $2\omega \cos \phi \cos \alpha = C$.

Now suppose a parcel of unit mass at A_0 is given transverse velocity v_1, w_1 by some temporary impulse; after a small interval dt the parcel occupies position A, which constitutes displacement δy, δz from A_0. At this new position the parcel is assumed to have the pressure of its environment, and its component of acceleration along x is $(du/dt)_A = Bw_1 + fv_1$. Upon integration,

$$u_A = B\delta z + f\delta y + U_0 . \quad (3)$$

This velocity component *relative* to geostrophic flow at A is $u_1 = u_A - U_A$; by substitution of U_0 from Eq (1) into Eq (3),

$$u_1 = (f - \partial u_g/\partial y)\delta y$$
$$+ (B - \partial u_g/\partial z)\delta z . \quad (4)$$

We now can determine the transverse horizontal component dv/dt of the parcel's acceleration at A. From Eqs (2b) and (4)

$$(dv/dt)_A = -f(f - \partial u_g/\partial y)\delta y$$
$$- f(B - \partial u_g/\partial z)\delta z - Cw_1 . \quad (5)$$

Vertical acceleration resulting from displacement of the parcel is

$$(dw/dt)_A = -Bu_A + Cv_1 - c_p\,\theta_0$$
$$\times [\partial(p/1000)^\kappa/\partial z] - g . \quad (6)$$

The next-to-last term on the right is merely the vertical pressure force expressed in potential temperature. We now introduce the assumption that transverse displacement of the parcel is dry adiabatic. Necessity for this assumption[31] is apparent from definition of static stability in Chapter 3.

Vertical acceleration of the *undisturbed* geostrophic motion at A is

$$dw/dt = 0$$
$$= -BU_A - c_p\theta_A \frac{\partial}{\partial z}\left(\frac{p}{1000}\right)^\kappa - g . \quad (7)$$

Upon subtracting Eq (7) from Eq (6) and substituting for u_1 from Eq (4),

$$(dw/dt)_A = -B(f - \partial u_g/\partial y)\,\delta y$$
$$- B(B - \partial u_g/\partial z)\,\delta z + Bv_1$$
$$- c_p(\theta_0 - \theta_A)\frac{\partial}{\partial z}\left(\frac{p}{1000}\right)^\kappa .$$

Now $\theta_0 - \theta_A$ may be expressed in terms of δy and δz by $\theta_0 - \theta_A = -(\partial\theta/\partial y)\delta y - (\partial\theta/\partial z)\delta z$. Also,

$$\frac{\partial}{\partial z}\left(\frac{p}{1000}\right)^\kappa = \frac{1}{c_p\rho\theta}\frac{\partial p}{\partial z} \simeq -\frac{g}{c_p\theta} .$$

31. The assumption of adiabatic displacement should have been made in deriving Eq (5) for a more complete expression for $(dv/dt)_A$, which is obtained from Eq (5) by adding

$$a\,\frac{\partial p}{\partial y}\left(\frac{1}{\theta}\frac{\partial\theta}{\partial y}\,\delta y + \frac{1}{\theta}\frac{\partial\theta}{\partial z}\,\delta z\right)$$

to the right side of the equation, constituting negligible correction for horizontal displacement.

Therefore,

$$(dw/dt)_A = -B(f - \partial u_g/\partial y)\delta y - B(B$$
$$- \partial u_g/\partial z)\delta z + Bv_1 - (g/\theta)$$
$$\times (\partial\theta/\partial y)\delta y - (g/\theta)(\partial\theta/\partial z)\delta z . \quad (8)$$

Eqs (5) and (8) express the transverse horizontal and vertical accelerations resulting from displacement (δy, δz) with perturbation velocity v_1, w_1 of a parcel initially moving with geostrophic wind. The state of stability or instability exists according as the accelerations tend to return the parcel to the reference axis (x-axis) or carry it farther away. For each of these equations there has come to be defined a corresponding type of stability. If the displacement is vertical only, we speak of *static* stability. With displacements in the horizontal only, we have come to associate horizontal or *inertia* stability.

Static stability can be found directly from Eq (8). If $v_1 = \delta y = 0$, and if stability E is defined as the resistance by the atmosphere, per unit distance, to unstable accelerations,

$$E_v = B(B - \partial u_g/\partial z)$$
$$+ (g/\theta)(\partial\theta/\partial z) . \quad (9)$$

This agrees with the previous expression for stability, Eq 3.17(3), except for the term with B. Since $2\omega\cos\theta$ has maximum value $1.46 \times 10^{-4}\ \text{sec}^{-1}$, while $\partial u_g/\partial z$ is usually in the range of magnitude 10^{-3} to 10^{-2} sec^{-1}, and the last term on the right is of the order 10^{-5} to $10^{-4}\ \text{sec}^{-2}$, Eq (9) reduces to

$$E_v \simeq (g/\theta)(\partial\theta/\partial z) . \quad (10)$$

In retrospect, it is evident that, although Eq (10) is the standard definition for stability, it does not express completely the stability along the vertical.

Inertia stability E_h is defined from Eq (5).

$$E_h = -(1/\delta y)(dv/dt)_A \simeq (2\omega\sin\phi)$$
$$\times (2\omega\sin\phi - \partial u_g/\partial y) . \quad (11)$$

As $2\omega \sin \phi$ is positive (northern hemisphere), the atmosphere has inertia stability or instability according as $\partial u_g/\partial y \lessgtr 2\omega \sin \phi$. The sign of inertia stability is dependent only on relative values of the Coriolis parameter and the horizontal shear of geostrophic wind. *There can be inertia instability only when the shear is negative (anticyclonic) and of magnitude exceeding the Coriolis parameter.* In middle latitudes this requires anticyclonic shear about 10^{-4} sec^{-1} or greater. Since that shear seldom is so large, the atmosphere is characterized generally by inertia stability. However, there is one region of the atmosphere in particular where inertia instability frequently exists, namely, immediately to the right of jet streams where anticyclonic shear can be large.

Eq (11) is a valid expression for inertia stability only when the quantity (from Eq [5]) $-f(B - \partial u_g/\partial z)(\delta z/\delta y) - Cw_1/\delta y$ is negligible. This would be the case with no vertical displacements ($\delta z = w_1 = 0$). In regions of strong vertical shear of geostrophic wind the first term above can be large even with small vertical displacements. In such conditions it would have to be considered along with the term retained in Eq (11).

In regions where inertia instability exists, conditions are favorable for horizontal mixing. An air parcel deflected from the geostrophic flow would develop transverse acceleration, carrying it farther away from its original position in the current, and theoretically it could not come to equilibrium with the pressure field until it arrived in a region of stability. All this suggests that regions of strong anticyclonic shear are favorable for horizontal mixing and that inertia instability might be part of the explanation for the weak temperature field to the right of the strong wind current in Figure 7.08*b*.

For the reader interested in reviewing the subject of stability as related to lateral displacements in more detail, we mention a few of the previous works: Solberg,[32] Van Meigham,[33] Palmén,[34] Godson,[35] and Bjerknes.[36]

32. H. Solberg, "Le Mouvement d'inertie de l'atmosphère stable et son rôle dans la théorie des cyclones," *Procès-verbaux de l'assoc. de météor.*, *U.G.G.I.*, *Edimbourg*, 1936, pp. 66–82.

33. J. Van Meigham, "Perturbation d'un courant atmosphérique permanent zonal," *Inst. R. météor. Belg.*, Mem., XVIII (1944), 1–34; "Hydrodynamic Instability," in American Meteorological Society, *Compendium of Meteorology*, pp. 434–53.

34. E. Palmén, "On the Distribution of Temperature and Wind in the Upper Westerlies," *Journal of Meteorology*, Vol. V, No. 1.

35. W. L. Godson, "Generalized Criteria for Dynamic Instability," *Journal of Meteorology*, VII (1950), 268–78; "Synoptic Significance of Dynamic Instability," *ibid.*, pp. 333–42.

36. J. Bjerknes, "Extratropical Cyclones," in American Meteorological Society, *Compendium of Meteorology*, pp. 577–98.

READING REFERENCES

BYERS, H. R. *General Meteorology*, Chap. 16. New York: McGraw-Hill Book Co., 1944.
PETTERSSEN, SVERRE. *Weather Analysis and Forecasting*, pp. 205–21, 378–440. New York: McGraw-Hill Book Co., 1940.

CHAPTER 8

Isentropic Analysis

8.01. *Introduction.*—In air-mass and kine-matic analysis it would be useful to have a set of conservative physical surfaces in the atmosphere. It then would be possible by complete analysis of a surface at successive times to measure, among other things, air motion in three dimensions. While strictly conservative surfaces do not exist, there are surfaces, partially conservative, which satisfy some purposes. The surfaces θ_E and θ_w are the most conservative, but their numerous folds and distortions do not permit a simple analysis. Because atmospheric processes tend to be largely dry adiabatic, at least for short periods, and also because surfaces of θ are more uniform than those of most other thermodynamic properties, isentropic analysis[1] has received the most attention in that respect.

The first outline and practice of synoptic isentropic analysis dates to Sir Napier Shaw.[2] Later, when sufficient aerological data became available, isentropic analysis was given impetus by Rossby and his colleagues at the Massachusetts Institute of Technology.[3] Over a period of a few years, ending about 1945, isentropic data were transmitted in radiosonde reports and used in daily preparation of isentropic charts.

Theory and practice of isentropic analysis are discussed at some length in the

1. "Isentropic analysis" as applied here and in practice refers to analysis at constant potential temperature. Use of the word "isentropic" would be more proper applied to constant θ_E or θ_w.

2. *Manual of Meteorology* (4 vols.; Cambridge: Cambridge University Press, 1926–32), III, 259–66.

"Reading References." This chapter is intended principally as an outline of the subject and to present some pertinent details not stressed sufficiently in the past. For those reasons the following discussion alone might appear too brief and to have an improper distribution of emphasis.

8.02. *Preparation of an analysis.*—The particular isentropic surface(s) selected for analysis depends on the purpose in mind and the region of the atmosphere under study. When used as a daily synoptic chart with emphasis on moisture patterns, the isentropic surface selected for the geographic area should be as low as possible without intersecting the ground. For this purpose Namias suggests the following potential

3. C.-G. Rossby *et al.*, "Isentropic Analysis," *Bulletin of the American Meteorological Society,* XVIII (1937), 201–9; "Aerological Evidence of Large-Scale Mixing in the Atmosphere," *Transactions of the American Geophysical Union, 18th Annual Meeting* (1937), Part I; R. B. Montgomery, "A Suggested Method for Representing Gradient Flow in Isentropic Surfaces," *Bulletin of the American Meteorological Society,* Vol. XVIII, Nos. 6–7 (1937); H. R. Byers, "On the Thermodynamic Interpretation of Isentropic Charts," *Monthly Weather Review,* Vol. LXVI, No. 3 (1938); H. Wexler and J. Namias, "Mean Monthly Isentropic Charts and Their Relation to Departures of Summer Rainfall," *Transactions of the American Geophysical Union, 19th Annual Meeting* (1938), Part I; J. Namias, "Thunderstorm Forecasting with the Aid of Isentropic Charts," *Bulletin of the American Meteorological Society,* Vol. XIX, No. 1 (1938); "The Use of Isentropic Analysis in Short Term Forecasting," *Journal of Aeronautical Sciences,* Vol. VI, No. 7 (1939).

temperatures for isentropic analysis over North America:[4]

Winter	290°–295° K
Spring	295°–300° K
Summer	310°–315° K
Fall	300°–305° K

In any season the surfaces indicated slope upward from near the ground in low latitudes to the vicinity of the tropopause in the coldest areas. For analysis over restricted areas it is advisable to select the most suitable surface independent of the ranges given above. It is usually desirable to use a surface which is smooth, whose topography is representative of adjacent surfaces, and which has the least spurious oscillations, all of which implies that this surface lies within a stable layer of the atmosphere. It is hardly possible to follow this recommendation for analysis in large scale.

Complete analysis of an isentropic surface requires three patterns of lines. These are (1) the *pressure*, which gives also the patterns of temperature and saturation mixing ratio, of density approximately, and of height with still less accuracy; (2) the *moisture content*, drawn for values of isentropic saturation pressure, which gives also the patterns of mixing ratio and equivalent or wet-bulb potential temperature; and (3) an acceleration potential, the "isentropic *stream-function*" ψ, from which geostrophic motion is obtained. Evaluation of pressure P and saturation pressure P_c from soundings was discussed in Section 2.20.

Analyses of the 300° and 310° isentropic surfaces are shown in Figure 8.02; (a) and (b) give the pressure patterns. On the 300° surface the 850 isobar is also the 300° isentrope (about 13° C isotherm) on the 850-mb surface; the 700 isobar is the 300° isentrope (about −2° C isotherm) on the 700-mb sur-

4. J. Namias *et al.*, *Air Mass and Isentropic Analysis* (Milton, Mass., 1940), Chap. 10: "Isentropic Analysis."

face; and so on. Thus, in drawing the pressure pattern of an isentropic surface, any isobaric temperature analyses may be used in addition to the values of isentropic pressure.

With stable conditions, isentropic surfaces slope upward in the direction of cold air. Depressions in the surface (regions of high pressure) are warm centers, and domes are cold centers. The 300° surface in this example intersects the ground in New Mexico and the interior of Mexico; note the absence of data for Albuquerque. This surface intersects the tropopause along the heavy dashed line circling most of the Great Lakes area.

Isobars of P_c are shown in chart (c). The line $P = 415$ at the northern limit is the −40° C isotherm on the surface; that temperature is the present lower limit of moisture measurement. Symbol "M" in place of P_c at some stations indicates relative humidity too low for measurement. Since P_c isobars are lines of r while P isobars are lines of r_s, the difference between P and P_c and the ratio of P_c to P are proportional to (but not exact measures of) relative humidity, and their patterns indicate the field of relative humidity.

The patterns of P and P_c are given together with those of ψ in charts (e) and (f). The three sets of lines, along with the corresponding data and/or wind plotted at each station, usually would appear on the same chart of the surface; different colored lines and shadings are used to make the finished analyses more legible than presented here.

8.03. *Isentropic pressure gradient.*—Variation of pressure along an isentropic surface is a function of the slope of that surface and air density. From the slope formula for θ and the hydrostatic equation, we obtain

$$(\delta p/\delta x)_\theta = \rho g(\partial\theta/\partial x)/(\partial\theta/\partial z) . \quad (1)$$

Thus, for stable conditions, isentropic pres-

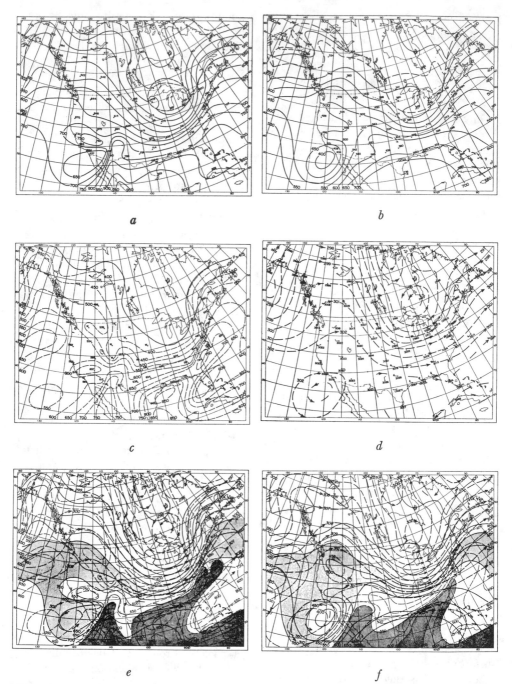

FIG. 8.02.—Pressures, adiabatic saturation pressures, and "streamlines" of the 300° and 310° isentropic surfaces, 1500 GCT, 1 March 1950. (a) Pressure, 300°; (b) pressure, 310°; (c) saturation pressure, 300°; (d) ψ, 300°; (e) composite of a, c, and d; (f) same as e for 310°.

sure gradient is in the same direction as *horizontal* potential temperature gradient, and they are directly proportional. Isentropic pressure gradient varies inversely in magnitude with static stability. With dry adiabatic conditions, isentropic pressure cannot be defined.

Because of the small angle between frontal surfaces and isentropic surfaces, frontal analysis is usually not performed on isentropic charts. In spite of this, the tend-

the thermal wind and inversely proportional to the stability.

The quantity on the right of each equation (except for the factor ρf) is the *shear-stability ratio*.[5] It is a vector along the thermal wind and therefore along the isobars of the isentropic surface. Its magnitude $\theta \left| \partial \mathbb{C}_g / \partial z \right| / (\partial \theta / \partial z)$, when multiplied by ρf is the magnitude of the isentropic pressure gradient. This information, when plotted on the isentropic chart, has been called a "shear-

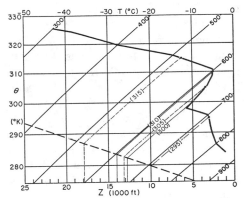 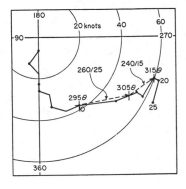

FIG. 8.03.—Computation of the vertical vector change of wind across isentropic surfaces for determining the shear-stability ratio vector.

ency for parallelism between the two can be used in analysis of both.

Isentropic pressure gradient can be evaluated roughly from aerological data at individual observing stations. ρ and $\partial \theta / \partial z$ are obtained from the pT sounding; $\partial \theta / \partial x$ is derived from the change of geostrophic wind with height through the particular isentropic surface. Eq (1) above can be transformed to

$$(\delta p / \delta x)_\theta = \rho f \theta (\partial v_g / \partial z)/(\partial \theta / \partial z) \quad (2)$$

by substitution for $\partial \theta / \partial x$ (actually, $[\partial \theta / \partial x]_p$) from the thermal wind equation. A similar equation,

$$(\delta p / \delta y)_\theta = -\rho f \theta (\partial u_g / \partial z)/(\partial \theta / \partial z), \quad (3)$$

holds for the other coordinate component. Evidently, the isentropic surface slopes upward to the left of the thermal wind vector (for $\partial \theta / \partial z > 0$) at a rate proportional to

stability-ratio vector," directed by the thermal wind and with magnitude equal to the spacing of isobars on the isentropic surface; thus,

$$\delta n = \frac{\delta p}{\rho f} \left(\frac{1}{\theta} \frac{\partial \theta}{\partial z} \Big/ \left| \frac{\partial \mathbb{C}_g}{\partial z} \right| \right). \quad (4)$$

Such information can be of help in drawing the isentropic pressure pattern in the vicinity of each observation.

The shear-stability ratio is determined from the shear of (geostrophic) wind and the variation of θ in unit vertical distance, or from the vertical thickness and vertical change of wind in a unit layer of θ, all near the isentropic surface in question. Figure 8.03 illustrates how this may be done for

5. A. F. Spilhaus, "The Shear Stability Ratio and Its Use in Isentropic Analysis," *Bulletin of the American Meteorological Society*, XXI (1940), 239–48.

the 300° and 310° surfaces. On the thermodynamic chart the heights of insentropic surfaces 5° apart are located from the pressure-height curve; from this, $\partial\theta/\partial z$ is evaluated directly. Heights of the surfaces are located in the hodogram, and the vector change of wind across each desired surface is determined. The change of wind divided by the vertical distance gives the thermal wind shear. Then δn is found by Eq (4).

In Figures 8.02*ab* the vectors at Bismarck, St. Cloud, Omaha, Oklahoma City, San Antonio, Brownsville, and Miami are the vector change of wind vertically in a 10° interval of θ through the isentropic surface. With geostrophic winds, these have the direction of the shear-stability-ratio vector and magnitude directly proportional to that ratio. The directions agree fairly well with the pressure patterns, considering the patterns were drawn independent of them. Pressure gradients appear to be well related with the vector magnitudes, except at Oklahoma City, where the discrepancy could be in the wind observation, in the pT sounding, or in departure from geostrophic winds.

8.04. *Geostrophic wind on isentropic surface.*—By Eq 4.14(1) the horizontal pressure force per unit mass can be expressed

$$-\alpha(\partial p/\partial x) = -\alpha(\partial p/\partial x)_\theta$$
$$+ \alpha(\partial p/\partial z)(\delta z/\delta x)_\theta ,$$

and a similar equation for y. This is simplified to give

$$-\alpha(\partial p/\partial x) = -c_p(\partial T/\partial x)_\theta - g(\partial z/\partial x)_\theta$$
$$= -[\partial(c_p T + 980Z)/\partial x]_\theta .$$

The quantity $c_p T + 980Z$ is the "isentropic stream-function" ψ. It follows that

$$fu_g = -(\partial\psi/\partial y)_\theta , \quad fv_g = (\partial\psi/\partial x)_\theta , \quad (1)$$

and

$$f * \mathbb{C}_g = \nabla_\theta \psi , \qquad fc_g = (\partial\psi/\partial n)_\theta . \quad (2)$$

Numerically, $\psi = (10.04 \quad T° \text{ K} + 98$

$Z_{gpkm}) \times 10^6$ ergs gm^{-1}. In Figure 8.02*d* the the plotted values for ψ are in 10^6, and their isopleths (geostrophic streamlines) are labeled in 10^7 ergs gm^{-1}.

Of the two terms defining ψ, the temperature term is usually dominant in the troposphere. Temperatures usually range between 200 and 300° K, making $10.04T$ vary between about 2000 and 3000. But $98Z$ is at most hardly more than half as large. Predominance by the temperature term is greatest at low levels. Accordingly, errors in evaluating ψ result principally from errors in temperature; an error 1° K in T has about the same effect on ψ as does 100-meter error in Z. Both terms are likely to have their largest relative errors in nearly adiabatic conditions along the vertical.

The constant-pressure geostrophic scale (Appendix, Fig. N) can be applied to ψ on isentropic surfaces just as it is applied to Z on isobaric surfaces. For given wind speed a difference 3×10^6 ergs gm^{-1} in ψ, isentropically, corresponds to nearly 100 gpft in Z, isobarically.

A rather striking feature of the pattern of ψ is its similarity to the isentropic pressure pattern at high elevations in the troposphere and its indifference to the pressure pattern nearer the ground. We recall that through the middle and upper troposphere the wind (streamlines) and thermal wind (isentropic isobars) are nearly parallel, while the angle between them is likely to be large and variable near the ground. Angles between wind and thermal wind are also large in the transition region near and above tropopause, for example, the 310° surface over the Great Lakes and Hudson Bay areas. (Incidentally, because the isentropic surface was so flat here in the vicinity of 300 mb, streamlines for ψ were based on the 300-mb contours.)

8.05. *Deduction of vertical motion from isentropic analysis.*—The most practical uses made of isentropic charts are in analysis and forecasting of vertical motions, and the resulting weather, based on the relation of the ψ, P, and P_c patterns. It is possible to describe the essentials in just a few sentences, but that would leave the way open for misconceptions now prevalent.

From the total time derivative of potential temperature,

$$d\theta/dt = \partial\theta/\partial t + c(\partial\theta/\partial s) + w(\partial\theta/\partial z) ,$$

where $c(\partial\theta/\partial s) = u(\partial\theta/\partial x) + v(\partial\theta/\partial y)$, and s is distance downwind, we obtain for vertical velocity

$$w = (c - c_\theta)(\delta z/\delta s)_\theta$$
$$+ (g/E_v\theta)(d\theta/dt) , \quad (1)$$

where c_θ is horizontal speed of the isentropic surface downwind. With reference to Eq 8.03(1),

$$w = -(a/g)(c - c_\theta)(\delta p/\delta s)_\theta$$
$$+ (g/E_v\theta)(d\theta/dt) . \quad (2)$$

It follows that *for dry adiabatic processes and stationary isentropic surfaces* ($c_\theta = 0$), *there is ascending motion if the wind blows toward lower pressure on the surface and descending motion if it blows toward higher pressure.*

Now let c_θ differ from zero, as it almost always does. The difference $c - c_\theta = c_r$ is horizontal air motion downwind *relative* to that of the isentropic surface, positive if with the wind and negative if opposite to the wind. Thus, from Eq (1)

$$w = c_r(\delta z/\delta s)_\theta = -(ac_r/g)(\delta p/\delta s)_\theta$$

for isentropic processes. In case of no wind, c_r is considered positive and opposite to c_θ, and s is in the direction of c_r. This leads to: *There is ascent where the relative motion* c_r *is toward lower isentropic pressures* (*upslope*), *and descent where* c_r *is toward higher isentropic pressures* (*downslope*), *for a dry adiabatic process.*

The diabatic term in Eq (1) corrects for vertical displacements of isentropic surfaces relative to the air when processes depart from dry adiabatic. For stable conditions ($E_v > 0$), warming by radiation or condensation makes it positive, even though isentropic surfaces move downward relative to the air, to compensate for negative contribution to w by $c_\theta(\delta p/\delta s)_\theta$ resulting from the same process. The opposite is true in diabatic cooling. A simple method for understanding all factors by which w is expressed in Eq (1) is to examine in detail the effects of horizontal motions and displacements of isentropic surfaces in a vertical plane.

The most reliable detection of vertical motion on isentropic charts by Eq (1) is in regions of strong wind; there it is probable that the isentropic surface is displaced horizontally more slowly than the air. From examining consecutive charts illustrated, one finds that such features as the isentropic trough over western Canada, the ridge extending southward from Hudson Bay, and the trough extending northward near the east coast all move slower than the stronger winds across them. Thus the relative motion $c - c_\theta$ is along the wind direction, and, assuming $d\theta/dt = 0$, air ascends the windward slopes of the isentropic surface in areas of strong wind. On this basis one deduces probability of ascent off the Alaska and British Columbia coasts in Figures 8.02, slight descent east of the Rockies, pronounced descent just east of the isentropic ridge over the eastern United States, and ascent east of the isentropic trough line curving from James Bay through Maine to near Cape Hatteras. Ascent is suspected in the area of Lower California and descent west of that isentropic dome. Notice how the moisture pattern seems to conform and how the patterns of saturation decrement ($P - P_c$) given in charts (*a*) and (*b*) of Figure 8.05 show this even better. Reference should be made also to Figure 9.15*c*.

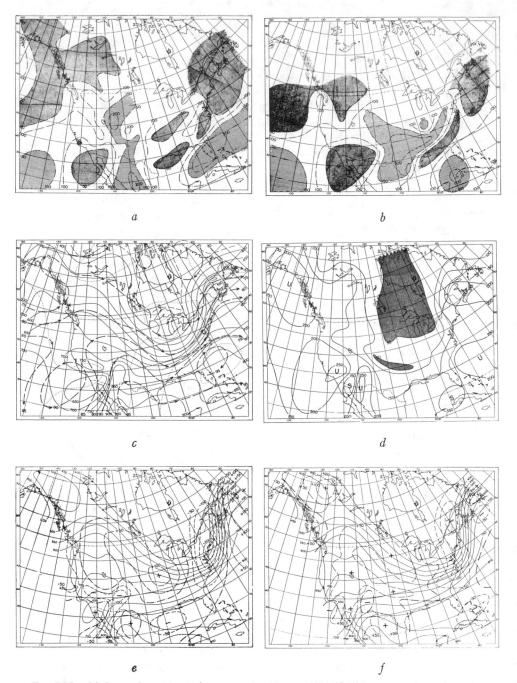

FIG. 8.05.—(a) Saturation pressure decrement, $P - P_c$, on the 300° isentropic surface; (b) $P - P_c$ on the 310° isentropic surface; (c) vertical difference in ψ between the 310° and 300° isentropic surfaces and isobars of the 300° surface; (d) vertical pressure difference between the 300° and 310° surfaces (charts a, b, c, and d for 1500 GCT, 1 March 1950); (e) 12-hour local change in pressure of the 300° surface, with isobars at 0300 GCT, 2 March 1950; (f) 12-hour local change of pressure difference in the 310°/300° layer, with isobars of chart e.

In the lower stratosphere the wind is usually of the same direction as in the troposphere, while isentropic slope downwind is generally the opposite, and therefore vertical motion in the lower stratosphere is likely to be of direction opposite to that found below.

As suggested above, where wind is greater than the horizontal motion of isentropic surfaces, ascent is expected on the windward slopes and descent on the lee slopes of isentropic surfaces. In such places the local change of temperature is less than indicated by horizontal temperature advection, and the difference between the two is a function of the vertical velocity (Sec. 3.28). We thus might expect with strong winds that there is positive correlation between warm advection and ascent and between cold advection and descent.

If it is possible to deduce the direction and relative magnitude of vertical motion, one can theorize on the changes in relative humidity of the air or, in the case of saturated ascent, the probable rates of precipitation of water from the air. Except for diffusion processes, for evaporation from the ground, cloud surfaces, and falling rain, and for condensation of the water vapor, the saturation pressure P_c (mixing ratio and θ_E) is a conservative property on an isentropic surface. Hence, with this condition, upward motion does not affect P_c while P is being lowered and approaching saturation. If the air is lifted beyond saturation, we suppose the excess moisture condenses to form clouds and possibly precipitation.

If $c - c_\theta$ can be determined, it is possible to find the time at which any air parcel having given P and P_c reaches saturation by virtue of ascent. Expressing w in Eq (1) as dz/dt, transforming dz to dp by the hydrostatic equation, and then solving for dt yields

$$dt = \frac{dp}{c_r (\delta p / \delta s)_\theta} + \frac{\rho g^2}{E_v} \frac{d\theta}{\theta} \frac{1}{c_r (\delta p / \delta s)_\theta},$$

where dt is the time interval during which the air parcel changes pressure by amount dp and potential temperature by amount $d\theta$. For a dry adiabatic process $d\theta = 0$. If dp is taken as $P_c - P$, then dt is the time lapsed in reaching saturation. As an example, suppose $P = 700$ mb, $P_c = 650$ mb, $c = 50$ km/hr, $c_\theta = 30$ km/hr, and $(\delta p / \delta s)_\theta = 50$ mb/150 km; dt is then 7.5 hours. The most unreliable element is c_θ, which can be approximated only indirectly. However, even a rough account for c_θ is better than ignoring it altogether.

There is one statement worth emphasizing in concluding this section: *Isentropic motion is considered motion in which the potential temperature of an air particle is conserved (i.e., $d\theta/dt = 0$) and the particle remains on its isentropic surface; it should not be understood as motion necessarily parallel to isentropic surfaces, since the surfaces usually are displaced also.*

8.06. Thermal wind between isentropic surfaces.—The difference in Eq. 8.04(2) on one isentropic surface from the same on an adjacent surface is

$$f * \Delta\mathbb{C}_g = \Delta(\nabla_\theta \psi) = \nabla_\theta(\Delta\psi), \quad (1)$$

where $\Delta\mathbb{C}_g$ is the difference in geostrophic wind and $\Delta\psi$ the difference in ψ, both measured vertically between isentropic surfaces. The pattern of $\Delta\psi$ between the 300° and 310° surfaces is given in Figure 8.05c. According to Eq (1), the vector difference in geostrophic wind through the layer is directly proportional to the gradient of $\Delta\psi$ and is oriented parallel to $\Delta\psi$ lines with higher values of $\Delta\psi$ to the right (northern hemisphere). This difference is the thermal wind vector; it bears the same relation to $\Delta\psi$ between isentropic surfaces as does thermal wind to ΔZ between isobaric surfaces.

The name "relative motion" has been applied misleadingly in the past to $\Delta\mathbb{C}_g$ between adjacent isentropic surfaces; that relative motion is the same as given by the thermal wind between two isobaric surfaces and is not explicitly the motion of air relative to isentropic surfaces, as it has come to be understood. There has been some advocacy for the use of $\Delta\psi$ charts, because, when interpreted along with isentropic topography, the pattern of $\Delta\psi$ gives supposedly the distribution of relative motion and therefore also of vertical motion and weather.[6] Notice in Figure 8.05c that, where thermal wind is toward lower isentropic pressures ("upslope"), the air frequently is nearly saturated and that, where thermal wind is toward greater isentropic pressures ("downslope"), the air is dry. Boundaries between "upslope" and "downslope" thermal winds are indicated by the dotted lines in Figure 8.05c. Some positive correlation is apparent between these areas and the regions of moist and dry air on the isentropic surfaces bounding the layer.

From Eq 8.05(1) an expression for vertical velocity can be found in terms of the cross pattern of thermal wind and isentropic contours. The difference in vertical velocities between top and bottom of the layer can be expressed roughly by

$$w_2 - w_1 \approx |\Delta\mathbb{C}_g| \, (\delta z/\delta s)_\theta, \qquad (2)$$

if $(\delta z/\delta s)_\theta$ is the slope of isentropic surfaces in the direction of $\Delta\mathbb{C}_g$. Eq (2) is conditioned by the following factors being zero or their net contribution negligible: (i) vertical variation of ageostrophic wind; (ii) vertical curvature of isentropic surfaces in the direction of $\Delta\mathbb{C}_g$; (iii) vertical variation of $\mathbb{C}_\theta$, the vector horizontal velocity of an isentropic surface; and (iv) vertical variation of $d\theta/dt$. Occasions when Eq (2) is strictly accurate are probably rare, but it might be reasonable to assume that the

6. A corresponding relation might hold statistically for patterns of isobaric contours with their relative contours—advection.

term on the right is ordinarily more important than the net effect of the ones neglected.

8.07. *Large-scale features of isentropic moisture patterns and their relation to the weather.*—On the basis of the slope of an isentropic surface and the ordinary upward decrease of moisture, the values of P_c are normally largest in tropical regions and in other isentropic depressions. Tongues of high moisture extend from the moist sources in low latitudes into the dry elevated regions of the isentropic surface in higher latitudes. To reciprocate, dry tongues in P_c extend from the cold "dry sources" equatorward and downward along the isentropic surface.

The moist tongues owe their presence to (i) condensation processes, in which saturated air and water droplets can be displaced across higher-valued isentropic surfaces; (ii) vertical and horizontal mixing of heat and moisture, in which the potential temperature and moisture of a given air parcel are modified; (iii) evaporation of water into an isentropic sheet; and (iv) the distribution of air motion, especially its deformation. In areas of intense convection moisture is carried through isentropic surfaces if latent heat is liberated in condensation. The pattern of air motion (exclusive of thermodynamics) would lead to the subsequent pattern assumed by the moisture on the isentropic surfaces. The classic explanation given for moist isentropic tongues is evaporation from the earth's surface in low latitudes, combined with vertical mixing in the friction layer, and then subsequent motion of the moisture isentropically conforming with air movement. Some of the moisture is lost through condensation, precipitation, and isentropic mixing during poleward sojourn of the air.

In summer there is a semipermanent moisture pattern on isentropic surfaces, at least in the vicinity of the United States, which consists of a dry tongue attending an isentropic ridge and predominant subsidence off the west coast, a moist tongue extending from Mexico into the mountainous western states, where isentropic surfaces dip downward due to high temperatures of the elevated terrain, a dry tongue extending southward to the Gulf of Mexico, and the next moist tongue of more variable location somewhere off the Atlantic coast. In winter these moisture waves are considerably more transitory and are as a whole less developed. Convective precipitation is preferred near the axis of a moist tongue. Warm-front cloudiness and precipitation are found near the poleward periphery of a moist tongue, especially with strong motion toward lower isentropic pressures.

In broad scale, moisture tongues appear to take part in characteristic anticyclonic "swirls" or eddies equatorward from the main westerly current, with moist tongues curving eastward into high latitudes and dry tongues curving westward into low latitudes. These moisture eddies are most pronounced in summer, and they have been a subject of interest and speculation in the past. Most explanations assume these eddies describe circulation systems with similar rotation. One theory attributes the eddies to dynamic instability with strong negative shear equatorward from the maximum current on the isentropic surface.[7] Petterssen suggests that the prevalence of well-developed anticyclonic eddies is because axes of anticyclones are nearly perpendicular to isentropic surfaces, while axes of cyclones are more parallel to the surfaces.[8]

The above theories are justifiable and can be the explanation, but there are a few

7. C. L. Pekeris, "Wave-Distribution in a Homogeneous Current," *Transactions of the American Geophysical Union, 19th Annual Meeting* (1938), pp. 163–64.

8. Sverre Petterssen, *Upper-Air Charts and Analyses* (NAVAER, 50-IR-148 [Washington, D.C., 1944]).

remarks worth adding. Moisture eddies do not require actual eddies in air circulation; in fact, numerous clockwise swirls in isentropic moisture patterns are not attended by anticyclonic air circulations. First, considering convection from the surface layer of the atmosphere, the shapes of moist and dry tongues can be controlled or modified by the geographic distribution of factors producing convection. Even if the convection producing the moist tongue initially does not have that pattern, wave patterns of moisture are subsequently distorted into anticyclonic swirls if anticyclonic shear is present, with or without instability. The isentropic surfaces usually chosen slope from light winds near the ground in low latitudes upward into the belt of westerlies. Anticyclonic shear measured along the isentropic surface is in most cases greater than found in any level. The reader might prove graphically how anticyclonic swirls are formed from an initial sinusoidal moisture pattern by advection alone in a region of straight flow with anticyclonic shear. If shear is cyclonic, moisture patterns are distorted into cyclonic swirls.

Where a moisture tongue intersects a current center, its axis should vary in curvature to conform with distribution of shear. Under certain conditions the tongue appears to split near the center of the current on the isentropic surface, one branch curving cyclonically in the cyclonic shear on one side of the current, and the other curving anticyclonically in the anticyclonic shear on the other side of the current. Dry tongues protruding equatorward should have cyclonic curvature prior to intersecting the westerly current center and anticyclonic curvature beyond; if there is a split in the axis of the dry tongue, the split would likely occur poleward of the current center, with one branch cyclonic (eastward) along the poleward edge of the current and the other intersecting the current and thence curving anticyclonically (westward). A moist tongue extending into higher latitudes and splitting near the current center is likely to have an anticyclonic branch following the equatorward side of the current and a cyclonic branch on the other side. Figure 8.02e does not give a good example of swirls and branching tongues in moisture, but even so it does show tendencies in those directions. Note the anticyclonic eddy appearance in the lower right and also the shape of line $P_e = 500$.

8.08. *Pressure difference between isentropic surfaces.*—Figure 8.05d gives the pattern of pressure difference obtained from (a) and (b) in Figure 8.02. This is the distribution of mass between surfaces and approximately the distribution of stability. Charts of isentropic pressure difference have been known as "isentropic weight charts" and more locally as "thick-thin charts."

The area of extreme stability in the south central United States represents the intersection of this isentropic layer with the frontal zone. The relative stability over Arizona and the adjacent part of Mexico lends some evidence for a frontal zone, in addition to large isentropic slope in Figure 8.02a and large temperature gradients in isobaric charts. An area of relative stability between 850 mb and 700 mb, equatorward from the subtropical high-pressure axis and associated with the trade inversion, usually would be indicated by the pattern of pressure difference between appropriate isentropic surfaces. Stability shown in the lower right of Figure 8.05d is likely part of this phenomenon.

8.09. *Dynamic stability.*—As presently defined, *dynamic stability* is the stability of adiabatic (isentropic) motion. The greater significance attached to dynamic stability over inertia stability arises from belief that "turbulent" lateral displacements are likely to be adiabatic, but in the theory the tenuous assumption is made that an adiabatic process requires that air be displaced parallel to isentropic surfaces.

Isentropic shear $-(\partial u_g/\partial y)_\theta$ is related to horizontal shear $-(\partial u_g/\partial y)$ by

$$-(\partial u_g/\partial y)_\theta = -\partial u_g/\partial y + [(\partial \theta/\partial y)/(\partial \theta/\partial z)](\partial u_g/\partial z) , \quad (1)$$

for all conditions when $\partial \theta/\partial z \neq 0$. If we multiply both sides by the Coriolis parameter f and add f^2 to both sides,

$$f[f - (\partial u_g/\partial y)_\theta] = f[f - \partial u_g/\partial y] + f[(\partial \theta/\partial y)/(\partial \theta/\partial z)](\partial u_g/\partial z) . \quad (2)$$

The first term on the right is inertia stability E_h for straight flow (Fig. 7.24). By reference to the thermal wind relation and to static stability, the second term on the right reduces to $-N_{yz}^2/E_v$. If we define dynamic stability E_θ by the term on the left of Eq (2),

$$E_\theta = E_h - N_{yz}^2/E_v$$
$$= (E_h E_v - N_{yz}^2)/E_v . \quad (3)$$

Since in our definition of E_h reference was made to Figure 7.24, the quantity N_{yz} is the baroclinity in the vertical plane normal to the geostrophic current.

As the square of baroclinity is always positive, and since normally $E_v > 0$, Eq (3) shows that *dynamic stability is normally less than inertia stability;* the difference is large in strongly baroclinic conditions and/or low static stability. This relation is given clearly by analysis of Eq (1) or by examining a vertical cross section having analyses of geostrophic wind and potential temperature (Fig. 1.07ab). As one example, in the northern hemisphere south of the polar front in the lower troposphere, horizontal shear is usually negative, but, since $\partial \theta/\partial y < 0$, while $\partial u_g/\partial z > 0$, isentropic shear is greater negative, and thus the condition of instability is more likely to be fulfilled. In summary, with static stability the atmosphere is dynamically unstable if inertially unstable; dynamic instability is probable where inertia stability is small.

READING REFERENCES

BYERS, H. R. *General Meteorology*, pp. 387–405. New York: McGraw-Hill Book Co., 1944.

NAMIAS, J., *et al. Air Mass and Isentropic Analysis*, Chap. 10, "Isentropic Analysis." Milton, Mass.: American Meteorological Society, 1940. (Also appears as Chap. 8 in PETTERSSEN, SVERRE, *Weather Analysis and Forecasting* [New York: McGraw-Hill Book Co., 1940].)

OLIVER, V. J. and M. B. "Construction and Use of Isentropic Charts," in BERRY, F. A., BOLLAY, E., and BEERS, N. R. (eds.), *Handbook of Meteorology*, pp. 848-56. New York: McGraw-Hill Book Co., 1945.

CHAPTER 9

Analysis of the Surface Chart

9.01. *Introduction.*—The primary chart in daily use is the surface synoptic chart. In many areas for lack of aerological data, among other things, this is the only chart employed in weather analysis and forecasting. Because of the completeness and density of data entered on the map, it is really a composite chart not restricted to one level only. From data measured directly at the surface and from visual observations it is possible to obtain quite an elaborate representation of atmospheric state, near the ground from temperature, humidity, and wind, and through deep layers from estimates of clouds, precipitation, and visibility. Within these data and their analyses there are also many clues to atmospheric behavior which are considered in making the forecast.

In view of the usefulness and emphasis on surface analysis in practice, this chapter might appear unreasonably brief in comparison with others. Most concepts employed in surface analysis were discussed before. Avoiding needless repetition here, we shall apply them to that analysis and discuss in detail only those topics not considered sufficiently in preceding chapters and in the "Reading References."

The principal types of analyses currently performed on the surface chart might be classified in an approximate order of emphasis as follows:

(i) Pressure
(ii) Fronts
(iii) Precipitation and other obstructions to visibility
(iv) Clouds and cloud cover
(v) Stability ⎫
(vi) Temperature ⎬ Air-mass analysis
(vii) Moisture ⎭

9.02. *Analysis of sea-level pressure pattern.*—Pressure analysis has long been, and continues to be, the dominant aspect of surface analysis. Pressure analysis is useful because pressure is a simple and uniformly distributed scalar quantity whose pattern is related to the wind field and serves with similar accuracy, but with much convenience, as the objective parameter in forecasting the wind field. The sea-level pressure pattern does not give a true picture of weather and motion at the surface, but, even considering the frictional departure from geostrophic motion at the ground, the pressure pattern remains a worth-while approximation in most areas. When pressure prognosis is secondary and the field of motion is the primary object of sea-level pressure analysis, it appears that our procedure could be more direct, viz., to analyze the wind field rather than only the pressures, realizing that a wind report exists for each pressure report. But, to conform with existing practices, sea-level pressure analysis will be covered with customary emphasis.

As a prelude to discussion of procedure in pressure analysis and the many details involved, it is worth while to examine briefly the pressure pattern over North America for the synoptic situation described previously, viz., 1 March 1950. In Figures 9.02*ab* are the sea-level charts for early morning (1230 GCT = 0630 CST) and mid-

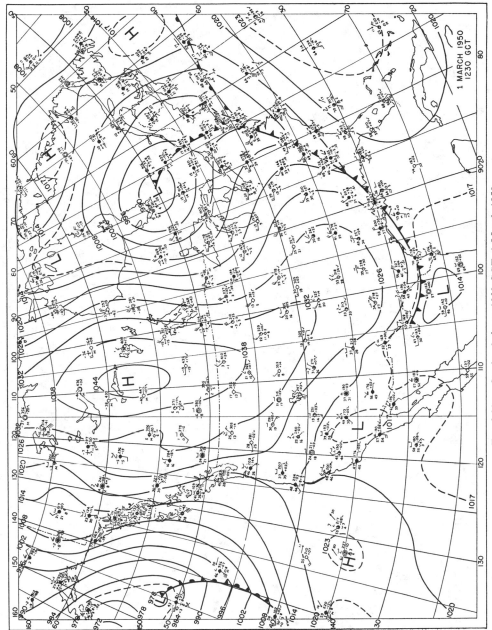

Fig. 9.02a.—Sea-level chart, 1230 GCT, 1 March 1950

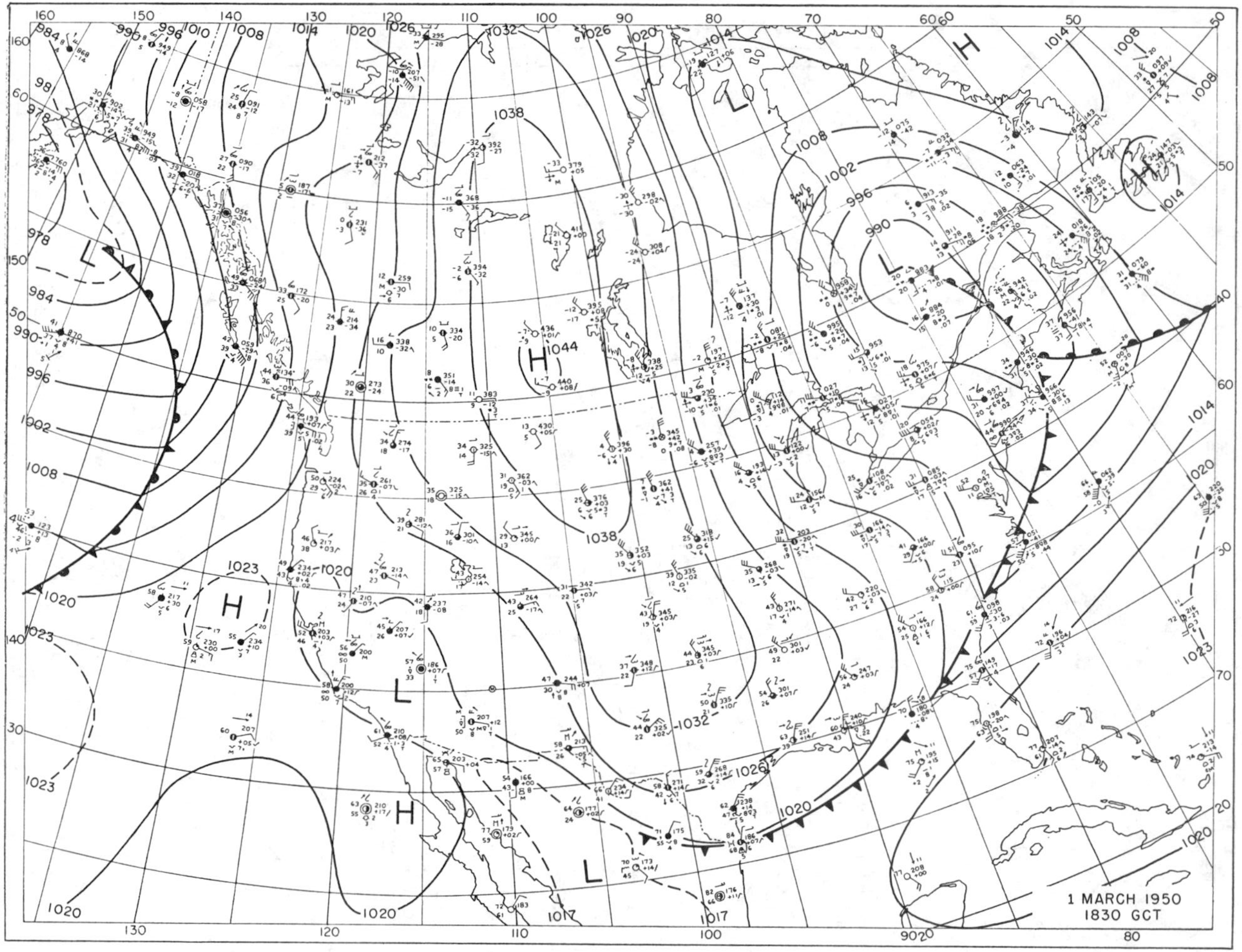

Fig. 9.02b.—Sea-level chart, 1830 GCT, 1 March 1950

day; these times are about 3 hours before and after the charts presented previously. On the surface maps are given the sea-level pressure analysis, the frontal analysis, and some of the data used in drawing them. To obtain a representative sample of synoptic reports for entering into these charts, preference was given first to the radiosonde stations, and then additional stations were chosen to fill in the network or otherwise to clarify some of the more critical features in the pattern. No attempt was made to adjust the analysis according to the reports shown in these illustrations, even though some might have been unrepresentative of prevailing conditions.

The 1230 GCT chart is part of the pattern shown by Figure 7.02*a*. The few local discrepancies between charts are due partly to the different data density used in the analysis, the different frequency of charts, and the differing emphasis and purpose in preparing the charts.

At 1230 GCT the dominant features in the sea-level pressure pattern are the low-pressure complex over the Aleutian area and Gulf of Alaska, the large high-pressure area centered in west-central Canada and extending southeastward into the south-central United States, the deep low-pressure center south of James Bay with its frontal trough extending into the Gulf of Mexico, and the Atlantic subtropical HIGH with a flat ridge continuing northward across the east coast of Canada. These four systems give the appearance of a simple mid-latitude wave pattern consisting of alternating cells of high and low pressure. Of less prominence are the warm Mexican low-pressure area, the small and weak low-pressure center over southern California, and the feeble-looking high-pressure system in the lower left. Although these smaller systems might appear insignificant in the sea-level pressure pattern, they are vital in a complete discussion of the weather. There are several other important features somewhat masked within the broad-scale pattern. Among these are the deep upper-level LOW over northern Hudson Bay whose image is very ill defined at sea level, a new LOW developing over northwestern Canada, and the peculiar pressure pattern near the New England coast. (Compare with the charts in Fig. 7.03.)

Six hours later the pressure pattern is quite similar. The systems have moved generally eastward while largely maintain-

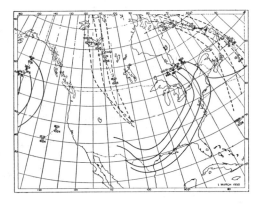

Fig. 9.02*c.*—Continuity chart

ing their identity. Major changes in the period, as well as between previous maps, can be seen from the continuity chart in Figure 9.02*c*, where positions and intensities of the principal systems are entered at 6-hour intervals. Pressure centers are indicated by dots or circles, with the hour and date printed above and the central pressure in millibars below. Besides the general tendency for eastward translation of the entire pattern in the middle latitudes, there are also detectable meridional displacements in some of the systems. *Motions of the systems agree with the winds at upper levels directly over them.* On examining Figures 9.02*abc*, the most obvious conclusion is that the pressure pattern undergoes such a slow process of evolution that events can be followed easily in the existing synoptic schedule.

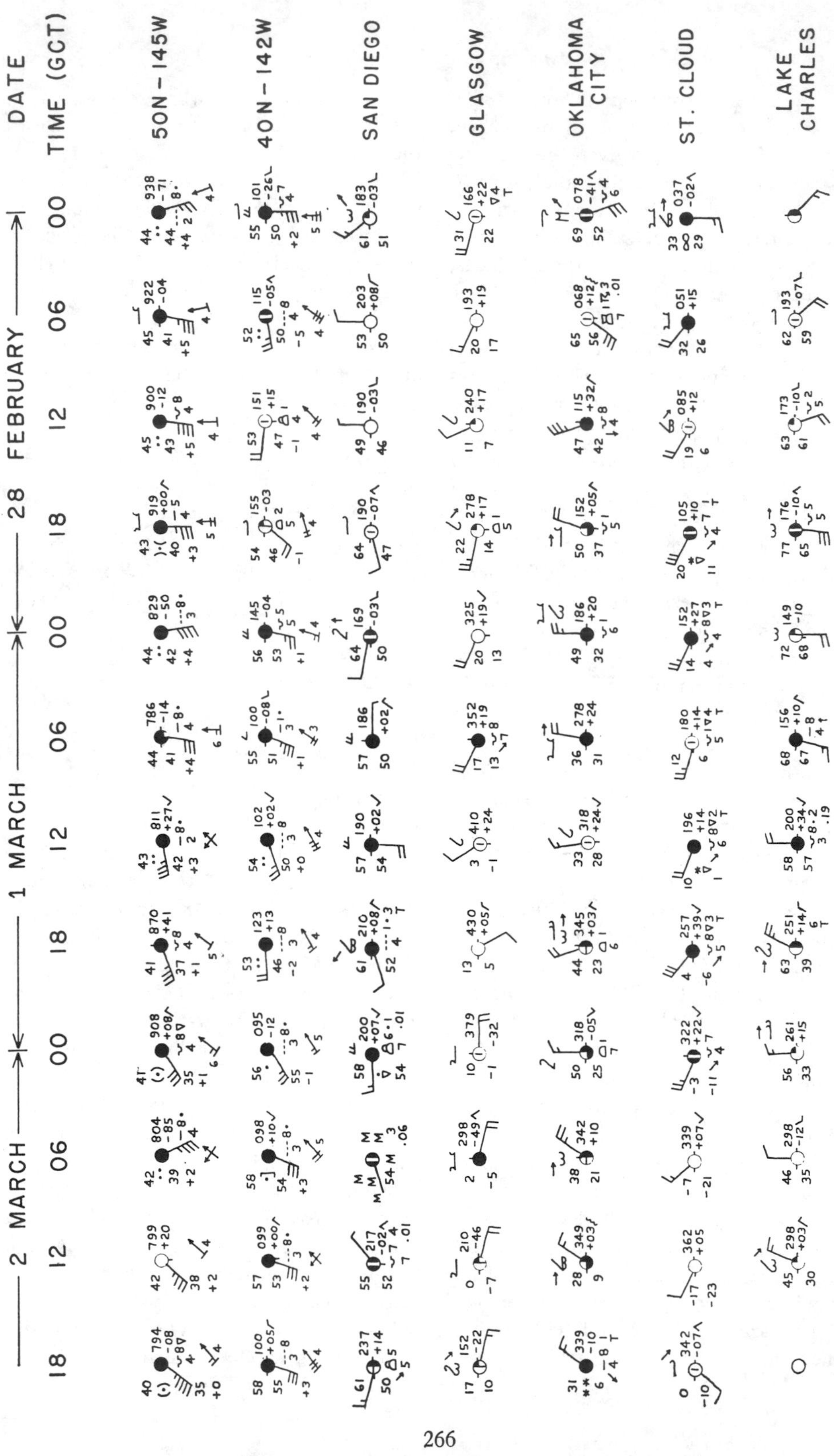

Fig. 9.02d.—Time sections of surface reports for selected stations (year, 1950)

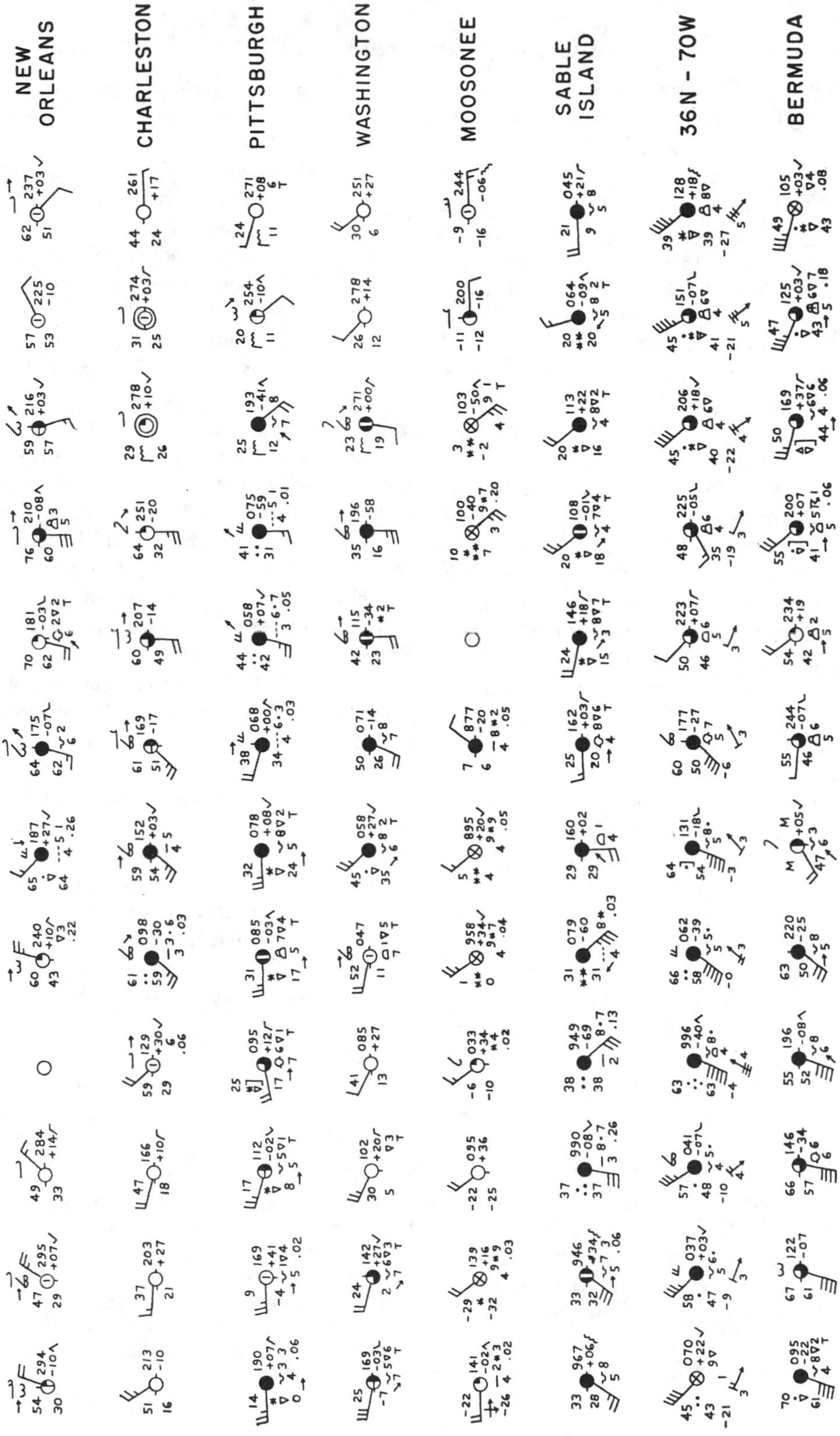

FIG. 9.02d.—Continued

267

Additional features deserving attention in the synoptic charts are the flatness of the pattern near centers and cols, the distribution of curvature about the centers, the sharp discontinuity in pressure and wind fields in some sections of the fronts, the scale of the pressure pattern, and the pressure-wind relation over land and sea.

Over the United States, the adjacent part of Mexico, Alaska, and southern and western Canada there were enough reports to derive the pressure pattern by simple scalar analysis almost entirely without reference to winds and continuity. Over the water areas and northeastern Canada reports were so sparse that help had to be derived from the past sequence of charts (including tendency charts), from greater reliance on the winds, and from analysis of observation sequences at individual stations, as shown in Figure 9.02d. Most of the effort in analyzing the illustrated pressure patterns was devoted to the areas of least dense reports, but, even so, accuracy is still desired in many of those areas.

Because of smoother surface, vector winds over the sea are likely to depart less from geostrophic than over land. Under that assumption it is possible to obtain as reasonable a pressure analysis using a few winds and pressure reports at sea as from denser networks over land. In using ocean surface-wind reports for drawing the pressure pattern, the geostrophic wind scale can be applied if, as a general rule, the winds are corrected by an amount consistent with the frictional departure. This amount varies, but a rough average correction might amount to one unit of Beaufort and 20° in direction. Exceptions are in weak pressure patterns, in the more stable surface air, near rugged coastlines, in certain narrow straits and passages, in localized convection areas, and generally in tropical latitudes. In the latter area the geostrophic approximation is often questionable (it breaks down

at the equator), and winds have some tendency to cross isobars at larger angles.

9.03. *Brief outline of polar-front wave theory.*—The theory of the polar front and its relation to cyclone development was introduced into daily practice by the Norwegian school of meteorologists (hence *Norwegian polar-front theory*), notably through the works of J. Bjerknes, Solberg, and Bergeron.[1] This theory, combined with later emphasis on the large-scale air masses, led by Bergeron in Europe and notably by Willett[2] in the United States, soon became the basis for weather analysis as known today. Some of the earlier ideas have since undergone modification with increasing amounts of aerological data. Complete discussions of frontal theory and of the individual air masses are found in the "Reading References."

As a simple basis for the formation of a frontal wave, first assume a stationary front at the ground which separates polar from tropical air, as in the first diagram of Figure 9.03a. The front lies in a trough of low pressure with easterly winds in the cold air and westerly winds in the warm air; thus the usual cyclonic shear appears at the front. Evidently, this front slopes upward toward the pole, and above the front are westerly winds throughout the troposphere.

It is possible for a disturbance of this uniform pattern to be created at any point

1. J. Bjerknes, *On the Structure of Moving Cyclones* ("Geofysiske Publikationer," Vol. I, No. 2 [Oslo, 1918]); J. Bjerknes and H. Solberg, *Life Cycle of Cyclones and the Polar Front Theory of Atmospheric Circulation* ("Geofysiske Publikationer," Vol. III, No. 1 [Oslo, 1922]); T. Bergeron, *Über die dreidimensional verknüpfende Wetteranalyse* ("Geofysiske Publikationer," Vol. V, No. 6 [Oslo, 1928]), and "Rechtlinien einer dynamischen Klimatologie," *Meteorologische Zeitschrift* (1930).

2. H. C. Willett, *American Air Mass Properties* ("Papers in Physical Oceanography and Meteorology, MIT—Woods Hole," Vol. II, No. 2 [Cambridge, Mass., 1934]).

in the vicinity of the front. This might be initiated by a slight deflection of the flow from instability of either current or simply from superposition of pressure rise or fall near the front. A disturbance results in varying components of motion normal to the front, and a small protuberance or wave is produced on the front. This may be damped out rapidly, as would appear to be

assume convex curvature in the direction of motion, owing to the particular distribution of winds. For the same reason the crest of the wave should coincide with the center of low pressure, assuming the cyclone develops on the front (but there are cases in which the cyclone center is some distance in the cold air from the wave crest).

The new cyclone center is propagated

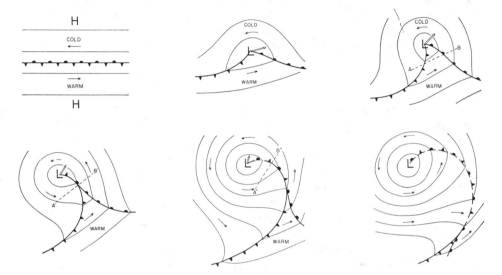

FIG. 9.03a.—Idealized illustration of the life-cycle of a frontal wave cyclone (northern hemisphere)

the usual case, or the wave might be propagated along the front, with little or no further growth for some time. These are *stable frontal waves*. Under certain conditions[3] the wave will be *unstable*, and it continues to grow into a circulation of dimensions observed in the surface pressure patterns.

Development of an unstable frontal wave is depicted schematically in Figure 9.03a. The disturbed flow makes the front move as a cold front west of the wave crest and as a warm front in the east. Near the new cyclone center both cold and warm fronts

3. See Sverre Petterssen's discussion on cyclone waves (*Weather Analysis and Forecasting* [New York: McGraw-Hill Book Co., 1940], pp. 313–24).

(steered) by the warm air above the front, as partially explained by local changes in mass due to the motions of the warm and cold fronts. If pressures and winds at upper levels are not changed in the process, retreat of the cold wedge ahead of the cyclone contributes pressure falls at the surface, and advance of the cold wedge to the rear contributes pressure rises at the surface. As a consequence, the pressure center would move with the warm air. Departures result from changes in the pressure pattern aloft.

It is useful to examine the surface pressure about the wave cyclone, especially in the warm sector. In spite of proximity to the cyclone center, the isobars are abnormally straight in this region, with curvature increasing cyclonically toward the center. In advance of the cold front, some cyclonic curvature is found due equatorward from the cyclone center.

As a rule, a large fraction of the tropical air is displaced upward above the warm front, and that front moves more slowly than the winds in the warm sector. The cold front remains the leading edge of the cold-air current; it should move with the gradient-level flow normal to the front in the cold air. The wave increases in amplitude, and the warm sector narrows with time. During this stage (third diagram) the cyclonic circulation intensifies, and the cyclone is usually deepening.

At still a later stage of development (fourth diagram) the cold front overtakes the warm front, resulting in an *occluded front* (or occlusion), and the frontal wave is said to be occluded. This is ordinarily the time of maximum intensification of the cyclone.[4] The front occludes further, and meantime the cyclonic circulation becomes larger and more symmetric. In the final stage shown the frontal system is disintegrating or *frontolyzing*. This old occluded front has little significance beyond a slight wind shift at the surface. The circulation is now dominated by a nearly circular cyclonic vortex imbedded in the cold air. At this period the cyclone is drifting quite slowly, usually with a poleward component, but it might just as well begin moving westward as continue eastward.

In Figure 9.03*b* are cross sections illustrating relative positions of the fronts along the dashed lines indicated in the previous figure. In the first the warm sector at the ground is wide but is narrowing by differential motions of the cold and warm fronts. The second cross section is near the tip of

4. One of Bergeron's rules in analysis: A frontal wave is occluded if the pressure difference between the cyclone center and the last closed isobar is 15 mb or greater.

the warm sector at the ground, that is, near the point of occlusion at the surface. A short time later the cold front makes contact with the warm front in this plane. The remaining cross sections show two possible results. In both cases the warm sector is lifted from the surface, and the coldest wedge of

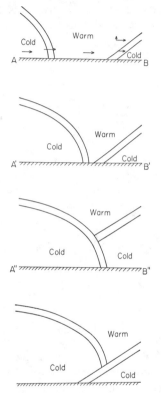

Fig. 9.03*b*.—Schematic cross sections through various stages of the frontal wave in Figure 9.03*a*.

air hugs the ground. If the air behind the cold front is colder than the air ahead of the warm front, then the warm front is lifted too, and the occluded front has the character of a cold front. This is the *cold-front occlusion* (Fig. 9.04); this type of occlusion process was assumed in preparing Figure 9.03*a*. The warm front appears only as an *upper warm front*. The line of contact between the warm front and the surface of the cold front in three dimensions is designated as the position of the upper warm front. It is behind the surface occlusion. This line also marks the lowest elevation of the lifted warm sector; it reaches the ground at the point of occlusion between the surface cold and warm fronts. It is also the line connecting the point of occlusion in the frontal wave at successive levels in the atmosphere.

In the last drawing of Figure 9.03*b* it is assumed that the retreating cold wedge is colder than the advancing air behind the cold front. In this case the cold front and the air behind it are pushed up the slope of the warm front, and the occluded portion of the system appears as a weak warm front. This is the *warm-front occlusion*. Attending this occlusion is an *upper cold front* which is found ahead of the occlusion at the surface (Fig. 9.04).

It appears that most occlusions are of the cold type. The fresh cold air moving southward behind the cyclone is usually more dense than the returning cold air ahead of the cyclone which has been warmed during its sojourn in lower latitudes. However, depending on the season, there are several localities in which warm-type occlusions are preferred. Over western parts of continents in winter the polar maritime air to the rear of the wave cyclone may be considerably milder than the polar air in the low levels cooled by radiation over the continental interior. A large fraction of the frontal cyclones developing just east of the Rocky Mountains in winter tend to be of this variety, since the low-level air overlying the central part of the continent can be much colder than air of Pacific origin which has descended the mountains from west or northwest.

The upper warm front associated with an occluded frontal wave is of little consequence in the surface weather. In fact, there is usually insufficient evidence for locating this feature of the occluded system. The

upper cold front is frequently detectable in the surface weather, in the surface pressure pattern, and therefore also in the field of barometric tendencies.

Although the frontal-wave theory described above agrees with observed events in many respects and forms a primary reference in weather analysis, there are many questions unanswered. For example, it was assumed at the outset that a front at the surface preceded formation of the cyclone. Indeed, there is evidence that formation or intensification of the front (*frontogenesis*) and development of the cyclone often occur simultaneously and also that frontogenesis at the surface often follows cyclogenesis. Until occlusion, at the surface there is usually a frontogenetic process operating in advance of the wave cyclone and a frontolytic process to the rear, so that the maximum intensity of the front is found near the peak of the warm sector, and the surface front is rather ill defined as a general rule at large distances from the cyclone center. (Examine Figs. 7.08 and 9.02*ab*.)

9.04. *Remarks on the structure of occluded fronts.*—There are many problems in analysis of occluded fronts in three dimensions. For the present we shall work from the simple models illustrated in cross section in Figures 9.03*b* and 9.04. In the latter figure are the idealized patterns of potential temperature with each type of occlusion structure. The schemes given here are based on similar illustrations by Byers.[5]

With the cold-type occlusion the warm front separates the tropical air of the warm sector from the modified polar air below. The occluded portion of the frontal system separates the modified polar air from the fresh polar air to the rear. The cold front marks the greatest contrast in temperature; there the warmest and coldest air masses

5. H. R. Byers, *General Meteorology* (New York: McGraw-Hill Book Co., 1944), pp. 324–25.

are adjacent. The most intense portion of this frontal complex therefore should be the cold front, and it would be the most easily identifiable in temperature distribution in vertical soundings and over level charts. In

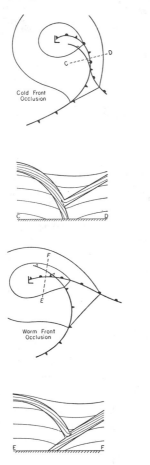

Fig. 9.04.—Idealized patterns of cold-front and warm-front occlusions with cross sections of potential temperature.

contrast, the principal part of the warm-front occlusion is the warm front. It includes all the air-mass contrast of the cold and occluded sections combined.

The warmest air in the cross sections is near the occluded front. On level charts there is usually a pronounced bulge in the

isotherms extending from the warm sector toward the cyclone center along the occluded front. In the retreating cold wedge the air near the occluded front has had trajectory from the lowest latitudes. Behind the occluded front there is also a temperature gradient from the front outward into the advancing wedge of cold air.

In preceding chapters it was shown that fronts are usually found with pressure troughs at the ground. Certainly, a major consideration in the technique of locating fronts on the surface chart is properly analyzing the pressure pattern near frontal troughs as an objective means for placing the front. This is applicable to both warm and cold fronts at the surface, and, by stretching the principle, we ordinarily are led to locate the surface occlusion within a pressure trough also. In Figure 9.03a the occluded front was placed in the pressure trough merely for consistency with existing models, and the same was done in the analysis of Figures 9.02ab. In fact, the occluded front was drawn all the way into the center of low pressure. However, there is insufficient reason for doing this.[6]

We may refer to the cross sections in Figure 9.04 to deduce in a qualitative way that the surface occluded front need not coincide with the trough in the pressure pattern. Suppose these cross sections are taken along an isobaric contour of the uppermost pressure in the drawings. This may or may not be along the lines CD or EF, but that does not matter. In the plane of the cross section the distribution of surface pressure then is a function only of temperature distribution. The lowest surface pressure will be found below the warmest vertical sounding, which is where the warm sector is

6. For an interesting discussion on this subject see K. R. Postma, "The Formation and Development of Occluding Cyclones: A Study of Surface-Weather Maps," *Staatsdrukkerij-en Uitgeverijbedrijf* (The Hague, 1948).

nearest the ground (i.e., at the position of the upper warm front and upper cold front, provided the two cold masses are not very different in temperature near the ground). Note that in the drawing at the top the surface pressure trough would be slightly to the rear of the surface occlusion. With warm-front occlusion the surface pressure trough would be ahead of the surface occlusion. As surface winds are controlled by the pressure distribution, the primary wind shift need not coincide with the surface occlusion either.

9.05. *The pattern of fronts on 1 March 1950.*—Aloft (Figs. 7.06 and 7.08) the isotherm concentration extends across Nova Scotia toward the cyclone near James Bay, thence southwestward through the Great Lakes region, curving gradually through the central states, and then directed through western Canada. The surface frontal wave is shown occluded with considerably more amplitude. It was shown previously that the frontal region over the western United States was broad horizontally. To avoid erroneous interpretations, it is found best in entering fronts into the chart to forget about such weak regions of the front and restrict attention only to the portion with steeper slope and distinct wind shifts attending it at the surface.

In Figure 9.02*a* the warm front off the east coast has had no previous history at the surface, and it is entered here as warm frontogenesis. The occluded front is placed within the surface pressure trough, according to standard practice, but there is some evidence that it might be east of the indicated position. The cold-front section of this system is very sharply defined in the pressure and wind patterns as far south and west as southern Texas. Beyond this it is next to impossible to locate a surface front with any meaning.

The frontal system entering the chart over the Pacific Ocean is shown as an occluded front, but such a choice is open to some question. In practice the analysis of surface fronts is usually drawn on the basis of air-mass distribution *at the surface*, even though it can be very misleading three-dimensionally. Here we find the polar air on both sides of the front; the surface air to the east apparently never penetrated far enough south to attain tropical characteristics before turning back northward ahead of the pressure trough. In the thickness pattern for the (700/1000)-mb layer (Fig. 7.06*d*) the gradient immediately ahead of the front is not as would be required of an occluded system. Instead, the warm front (if any) appears to be near the continental coast, and the modified polar air apparently is a thin film near the surface. The analysis at a level just a short distance above the sea might show an open frontal wave with perhaps only a small part of the system occluded. In such case the surface front as given in the analysis should be a cold front.

The succeeding chart (Fig. 9.02*b*) shows essentially the same pattern of fronts with allowance for translation with the winds. The system in the east is now indicated as a cold-front occlusion agreeing with the model in Figure 9.03*a*. But such representation is debatable again. Note from the extreme eastern end of Figure 6.13*d* that no substantial evidence appears for entering a frontal zone between the polar air at the surface and the air of more tropical nature aloft. As very often the case near the center of vigorous depressions, it is difficult to define clearly all three sections of an occluded front in cross section.

9.06. *Outline on air-mass analysis.*—As basically defined, fronts are distinct zones between broad-scale air masses of different temperature. The development of air-mass analysis was thus a logical parallel in the early application of frontal theory to daily weather analysis. Air masses have additional characteristic features in moisture, stability, probability of and sensitivity to modification, predominant vertical motions (to some extent), and therefore also certain types of weather. Attempts to relate as many of these features as possible into a simple scheme for describing air-mass characteristics and phenomena led to air-mass classification systems by which an air mass could be "labeled." The basic system, and the most universal, is the one by Bergeron which we follow below. Because of the difficulty in distinguishing between air masses at upper levels, except generally the difference between polar and tropical, and also because the lower and upper portions of a given air column can have had distinctly different trajectories, the discussion pertains to the lower levels only, say, below 700 mb.

In keeping with the intended objective of merely outlining the long descriptive subject of air-mass analysis, we will avoid repeating much of the material covered in Chapter 3 or readily accessible among the "Reading References." There is no intention of minimizing the value of air-mass analysis, in spite of favorable veering of approach during recent years from thorough analysis of the surface chart and several soundings to thorough analysis of the surface chart and charts for several other reference levels, but it has its serious defects as the mode of frontal analysis and analysis of weather processes beyond the surface layer.

a) Classification by source region.—From our discussion in Chapter 3 concerning the modification of air columns by the underlying surface, it is apparent that air sojourning over large areas of the globe, each having nearly uniform surface characteristics, will result in low-level air masses of about that lateral extent whose properties conform with the character of the underlying surface in temperature and moisture. Areas capable of thus modifying air masses of significant lateral extent are known as air-mass *source regions.* There are numerous types of source regions, different in degree; but, to devise a classification as simple as possible, it is necessary to restrict the types to a small number taking in the major source regions only. On basis of temperature, the distinction is between cold and warm. About half the area of the globe, generally equatorward from 30° latitude, has comparatively homogeneous surface temperatures (excluding the usually small differences between continent and ocean); these are the *tropical* (*T*) source regions. Although less homogeneous in surface temperature, the higher latitudes contain the *polar* (*P*) source regions, which may be subdivided into polar and *arctic* (*A*).

The basic thermal classification of air masses given above agrees roughly with mean circulation patterns over each hemisphere (and also with daily flow patterns aloft). The belt of westerlies corresponds to the average belt of the polar front, and the areas of weaker flow to either side correspond to the more barotropic tropical and polar-air source regions.

Air brought into the tropical source region is modified from below by addition of heat and moisture and, as evidenced particularly in the eastern ends of subtropical anticyclones, modified at upper levels by predominant subsidence, thus producing a well-mixed surface layer capped above by the dry "trade inversion." Tropical air from this laterally homogeneous pool, when injected into higher latitudes in the area between an anticyclone and the adjacent cyclone to the west, is subject to predominant horizontal convergence near the ground and ascent which carries the moisture to higher levels, most pronounced in the western flank of the poleward current (as revealed by moist tongues on isentropic charts). With exception of this and the organized convection areas in the tropics, the pool of tropical air is characterized by dryness[7] above about 2 km. A mean tropical sounding for the Caribbean area appears in Table 13.05.

Polar (arctic) air, on the other hand, is formed by radiation cooling, most rapidly from below. In migration toward lower latitudes the polar air is heated and obtains moisture from the surface and at the same time undergoes net subsidence except in the area immediately adjacent to the cyclone. It is important to recognize that much of the low-level polar air in middle latitudes partakes in rather short meridional oscillations corresponding with the traveling cyclones and anticyclones. This intermediate air mass, which is quite heterogeneous in temperature and moisture, usually has not had a wholly direct trajectory from the poleward source region; hence, the distinction in winter between polar and arctic air[8] or between modified polar and true polar air.[9] However, since except near coastlines there is usually no concentration in temperature gradient separating the polar and arctic air (i.e., no arctic front) such as found with the polar front, distinction between polar and arctic air is often pointless.

The necessity for exchange of air from one major source region to the other requires that the polar front be discontinuous.[10] The break in the otherwise continuous frontal wave occurs at the surface generally between the equatorward-most part of the wave and the succeeding wave crest to the west, which suggests a reason why frontal waves at the surface are characterized as a rule by long cold fronts and relatively short warm fronts.

A secondary classification of air-mass source regions distinguishes between land and water surfaces, since the distinction is imparted to the air masses themselves. The designators are continental (c) and maritime (m), usually prefixed to the thermal designators, thus: cT, mT, cP, mP, cA, mA. The cT source regions comprise the larger tropical and subtropical land masses in summer; in the northern hemisphere these are Mexico and the southwestern United States, the Sahara Desert area, and interior southwestern Asia. This air is hot and of lower than average moisture content for the latitude. mT air is moisture-laden and has temperatures in the source region agreeing with the uniform surface temperatures of the tropical oceans; mT air in the surface layer is the most homogeneous of air masses in temperature and moisture. Most tropical air masses affecting the United States are of maritime origin (mT).

Both cP and cA air masses are formed by modification over continents through cooling by radiation. cP may be obtained directly from cA by southward land trajectory. As expected, mP air is usually more moist than cP and is somewhat warmer in winter and cooler in summer than cP for the same latitude. mP is found over the oceans in middle and high latitudes and over windward regions of continents in those latitudes; in winter it may be transported great distances inland before attaining continental characteristics. Most mP air masses invade the United States from the Pacific Ocean, and these are usually modified (also by the mountainous topography) before reaching the Mississippi River. Such air reaching the south central states on a direct trajectory is usually very dry. Occasionally, surges of mP air cover the northeastern states and eastern Canada from the

7. Often called "superior" (S) air. Because of its dryness and source from above, the tropical air above subsiding (and subsided) polar air masses in the middle latitudes is often called S air also.

8. Byers, *op. cit.*

9. H. C. Willett, *Descriptive Meteorology* (New York: Academic Press, 1944).

10. Further discussion is given by E. Palmén, "The Aerology of Extratropical Disturbances," in American Meteorological Society, *Compendium of Meteorology* (Boston, Mass., 1951), pp. 599–620.

Atlantic during periods of persistent easterly flow.

Arctic air in its source region is continental. In outbreaks of arctic air over oceans the air is designated mA, and then only temporarily, since it rapidly attains the characteristics of mP air.

b) Classification by type of modification from below.—Subclassification of air masses is applied to distinguish between warming and cooling from below; that is, whether the air is warmer (w) or colder (k) than the surface. This distinction provides the most direct means of describing by label the weather phenomena in the surface layer. The suffixes w and k also imply differences in stability of the surface layer (not including stability at higher levels). A w air mass is being cooled from below or is moving over a colder surface, and the weather is featured by steady surface winds of large ageostrophic departure in speed and direction, fog or low clouds of stratified nature, precipitation of the drizzle type, subnormal visibility, and large diurnal temperature range in the absence of clouds. A k air mass is being warmed from below and has features of relative instability in the low levels: gusty winds of subnormal ageostrophic departure, clouds of the turbulent or convective varieties, precipitation of a showery type (including snow flurries and some intermittent rain), generally good visibility, and subnormal diurnal temperature range. These features, and the k or w characteristic, should be considered apart from the common effects of diurnal variation in low-level stability.

By applying these two suffixes to the designations already listed, the following array is obtained: cTw, cTk, mTw, mTk, cPw, cPk, mPw, mPk, cAw, cAk, mAw, and mAk. This list requires some clarification, as at least one is inconsistent, and often there is no clear distinction between some of them.

cT air would result from modification of mP, cP, or even perhaps mT by passage over a hot land surface; the air mass would be k during formation but w in moving out of the source region. mT is predominately w in moving poleward over land or water in winter and over water in summer; in passing over continents in summer, it is likely to be heated and moistened from below (as seen in surface temperature and dewpoint patterns inland from coasts). In general, the P air masses are k in moving equatorward (also for A) and w in moving poleward. The designator mAw is not applicable, since arctic air moving over open water is an unstable and rapid transformation from cA or cAk to mAk and mPk. cAw is generally not applicable either, except conceivably in the source region.

Seasonal averages of certain of the physical characteristics (T, r, θ_E) for various air masses affecting the United States were published in tabular form by Showalter.[11] Those tables are of use as a guide, in spite of the short period of data on which they were based. A different statistical approach to the thermal classification of air masses was suggested recently by McIntyre.[12] Air-mass classification at upper levels must be based almost wholly on potential temperature (tropical or polar); the moisture aloft is not well related to the horizontal trajectory of the air.

9.07. Thermal patterns and fronts.—The surface cold front is clearly indicated by the (700/1000)-mb thickness pattern in Figure 7.06d. Schematic illustrations of thickness

11. A. K. Showalter, "Further Studies of American Air Masses," *Monthly Weather Review*, July, 1939. Cf. F. A. Berry, E. Bollay, and N. R. Beers (eds.), *Handbook of Meteorology* (New York: McGraw-Hill Book Co., 1945), pp. 608–20.

12. D. P. McIntyre, "On the Air-Mass Temperature Distribution in the Middle and High Troposphere in Winter," *Journal of Meteorology*, Vol. VII, No. 2 (1950).

patterns in the vicinity of occluded waves are shown in Figure 9.16. With respect to ideal cold and warm fronts, the discontinuity in thickness gradient occurs at the surface position of the front, and the strong gradient extends into the cold air to the vicinity of the cold boundary of the frontal zone at the upper level. Because the cold air is normally quite baroclinic in the lower levels, the cold boundary of the frontal zone for the layer is usually difficult to detect in the thickness pattern. Also, a few of the thickness lines may intersect the surface front due to the usual meridional variation of temperature. Excluding that, the thickness lines can be viewed roughly as contours of the frontal surface, since the slope of a front is directly proportional to the horizontal temperature gradient in the frontal zone.

The (700/1000)-mb thickness pattern can be an objective aid in properly identifying and locating fronts on the surface chart. There are times when fronts are poorly indicated by the pressure and temperature patterns at the surface, yet comparison of tentative pressure analyses at 1000 and 700 mb yields the thermal pattern representative of a deep layer and may show that a real front exists.[13] Such often would be the case with fronts whose topographies resemble Figures 6.23*bcdf*. Regardless of whether we like to call these "surface" or "upper" fronts, they are significant in the analysis.

The use of thickness analysis in connection with surface analysis raises a question of basic philosophy. In drawing fronts on the surface chart, the primary purpose is to indicate by simple scheme the horizontal temperature pattern through a significant depth of the atmosphere. The thickness pattern does that directly, and does it more accurately than can possibly be done by subjective use of a limited number of models. Besides, thickness analysis provides ap-

13. See Figure 9.192.

proximately for three techniques all at once—horizontal frontal analysis, air-mass analysis, and frontal contour analysis.[14]

A related objective tool for locating and following frontal systems, as well as for drawing proper upper-flow patterns about them (over oceans especially), is the pattern of departure of temperature from normal (Fig. 9.07).[15] Some important features of this pattern are the similarity in shape to

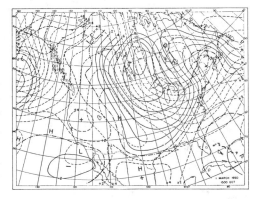

Fig. 9.07.—5000-foot isobars (*dashed*) and departure of the temperature pattern in the layer between sea level and 10,000 feet from the normal.

frontal systems, the large gradients in the cold air next to surface fronts, and the large departures occurring over land. Notice the suggestion of occlusion structure in the pattern over Maine and the Maritime Provinces of Canada. In low latitudes daily departures from normal are usually very small, as partially indicated in the chart. The advantage of this type of analysis arises from the greater conservativeness of temperature departure as an air-mass property, in value and pattern, as compared with temperature itself.

The difference in time between surface

14. A. M. Crocker, W. L. Godson, and C. M. Penner, "Frontal Contour Charts," *Journal of Meteorology*, IV. (1947), 95–99.

15. The patterns indicated can be taken as the 850-mb topography and the departure of thickness in the (700/1000)-mb layer from normal.

and upper-level charts in North America is a handicap to co-ordinated analysis. The thermal pattern obtained between the surface and upper charts will be in some error in the more rapidly developing and rapidly moving systems. The discrepancy can be determined from the shapes of the upper and lower pressure patterns and their relative displacement during the time difference. For example, the usual surface pressure trough in middle latitudes appears too warm if the surface chart is the later and too cold if it is the earlier.

9.08. *Analysis of the pressure-tendency field.*—The pressure-tendency field is analyzed by drawing isallobars. Isallobars for intervals 1 mb per 3 hours from tendencies in surface synoptic reports are shown in Figure 9.08*b*. In Figure 9.08*c* are 12-hour pressure changes.

It is seen that the isallobaric pattern is of

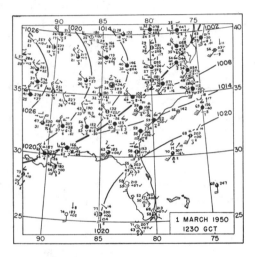

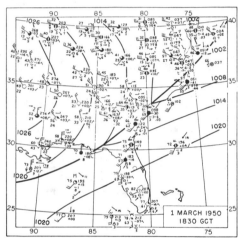

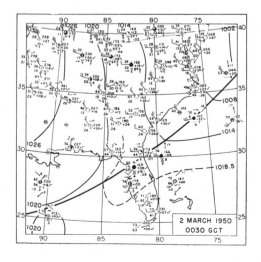

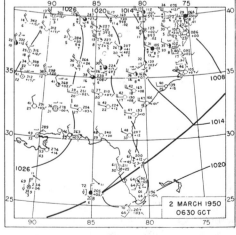

FIG. 9.08*a*

smaller scale than the pressure pattern, which is especially evident for 3-hour pressure changes. As the length of the time interval is increased, isallobaric systems be-

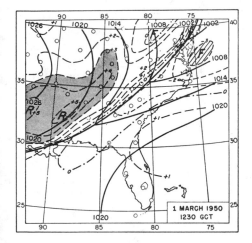

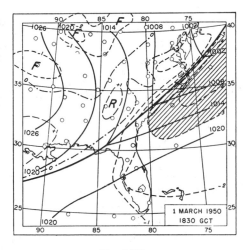

FIG. 9.08*b*

come more clearly defined in area and attach themselves more definitely to the individual moving pressure systems. For general forecasting purposes the longer-period pressure changes are more reliable,[16] as the

16. Forecast centers of the U.S. Weather Bureau presently place the most emphasis on the 12-hour pressure-change chart.

local and temporary irregularities in pressure tendency are smoothed out to a large extent, and also their patterns can be drawn more accurately. Nevertheless, some use is found for 3-hour isallobars in analyzing conditions of more synoptic nature.

There are several features of interest in the 3-hour isallobaric patterns shown in Figure 9.08*b*. Ahead of the moving frontal trough is a band of falling pressure and to the rear a band of rising pressure, both

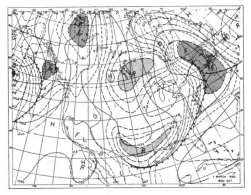

FIG. 9.08*c*.—12-hour pressure change at sea level and the pressure pattern at the end of the period.

paralleling the trough line. This is to be expected of any moving trough not deforming radically, whether or not it contains a front, and it is more pronounced the sharper the trough in the isobars. The isallobaric gradient behind the trough line is very large in this case and is a reflection of the speed of the pressure pattern as well as the large local shift of pressure gradient as the trough moves by. Because of the discontinuity in the pressure pattern at the front, the magnitude and gradient of the isallobars just behind it are not representative of instantaneous pressure tendency or of its distribution.

Textbooks in the past have favored drawing isallobars discontinuously at fronts, because a front supposedly marks a local discontinuity in pressure tendency. However, isallobars are drawn not for instantaneous pressure tendencies but for changes over finite periods of time. Since the 3-hour pressure change represents an *integral* of the barogram, the discontinuity appears in the 3-hour isallobars as one of first order.[17] Hence, isallobars are kinked, and more likely just bent, but not discontinuous at the front.

9.09. *The semidiurnal pressure wave.*— Upon closer examination of the 3-hour isallobaric patterns in Figure 9.08*b*, we find significant differences in the values of pressure rise and pressure fall from one chart to the next. In the first chart of the sequence there appears a net average pressure rise for the whole chart. In the next chart the average tendency is negative. There is no detectable area of 3-mb pressure rise, but there is a large area of 3-mb pressure fall. Over the succeeding chart (0030 GCT) the average change again is positive, as in the first one. The chart for 0630 GCT showed no pronounced leaning one way or the other, but pressure rises behind the front were smaller and pressure falls ahead of the front larger than in the preceding chart. The sequence gives the impression of an independent pressure oscillation superimposed on the pressure changes due to motion of the pressure pattern. This oscillation is the *semidiurnal pressure wave*, of a tidal nature, which follows the sun and whose amplitude is an inverse function of latitude.

The semidiurnal wave[18] is illustrated schematically in Figure 9.09. It can be determined for any station from long-term averages of hourly surface-pressure readings. The primary maximum in the wave occurs at about 1000 local time, the second-

17. To be more realistic, a second-order discontinuity (shown by a dip in the barogram).

18. For a more detailed discussion refer to G. R. Jenkins, "Diurnal Variation of Meteorological Elements," in Berry, Bollay, and Beers (eds.), *Handbook of Meteorology*, pp. 746–49.

ary maximum near 2200, the primary minimum near 1600, and the secondary minimum near 0400. These times vary by about ± 1 hour seasonally and from one geographic area to another, and it might be expected they also vary with particular meteorological conditions. For stations outside the tropics the amplitudes are greater in summer than in winter. Near the equator the amplitude of this wave (half the range from minimum to maximum) is of the order 2 mb.

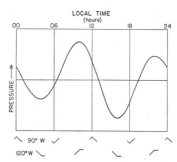

Fig. 9.09.—Schematic profile of the semidiurnal pressure wave.

It decreases with latitude, and in polar regions it is hardly perceptible at all. In low latitudes the pressure changes due to this wave ordinarily overshadow the changes due to motion and development of the pressure pattern. Even in middle latitudes the effect is large enough to make the 3-hour, the 6-hour, and to some extent the 12-hour local pressure change misleading; and in most uses made of pressure tendencies this semidiurnal change should be subtracted from the observed change. For this purpose one may use such charts and graphs as given in U.S. Weather Bureau Technical Report No. 1, 10-*Year Normals of Pressure Tendencies and Hourly Station Pressures for the United States* (1943).

The influence of this pressure wave is clearly evidenced in the reported barometric tendency symbols in low latitudes, especially with the ordinary flat pressure fields there. Characteristic tendency profiles due to this semidiurnal wave for the four synoptic times are shown for the 90th and 120th meridians in the lower part of Figure 9.09. Over the southeastern United States at 1230 GCT this wave would result in rising or check ($\checkmark$) tendencies in the absence of other pronounced effects. Notice in the corresponding chart of Figure 9.08*a* the predominance of such tendencies away from the immediate influence of the frontal trough. The effect is shown even just in advance of the trough, where advance of the trough should give steady pressure falls, but, owing to superposition of the semidiurnal wave, many of the reports have the form $\diagdown\,\cdot$.

Upper-level data are still too infrequent to verify the vertical extent and nature of semidiurnal pressure variations aloft. However, there is a detectable *diurnal* variation of pressure aloft[19] which represents the excess of diurnal variation of temperature in the layer below over the surface diurnal pressure variation.

9.10. *Isallobaric field and changes in the pressure pattern.*—The material on deepening, intensification, and displacement of pressure patterns presented in Section 7.14 is equally applicable to pressure at constant levels. The formulas are converted for that use by changing the variable Z to p.

9.11. *Correction of ships' pressure tendencies.*—The 3-hour pressure changes reported by ships at sea include the local

pressure change (as reported by land stations) plus the effect of the ship's movement across isobars. In order to interpret properly the report from a ship, it is necessary to remove the effect of the ship's motion from the given pressure change.[20]

The pressure tendency dp/dt observed aboard the *moving* ship is

$$dp/dt = \partial p/\partial t + c_s(\partial p/\partial s) \, .$$

The first term on the right is the desired local tendency; c_s is the speed of the ship; and $\partial p/\partial s$ is the horizontal variation of pressure in the direction toward which the ship is moving. By solving for the local tendency, $\partial p/\partial t = dp/dt - c_s(\partial p/\partial s)$. Further, if c_s is given in knots, if unit distance δs is 1° of latitude (60 nautical miles), and if the unit of time is 3 hours, this equation reduces numerically to

$$\partial p/\partial t = dp/dt - (c_s/20)(\delta p/1°\phi) \, . \quad (1)$$

Thus, for a ship moving with speed 20 knots, the correction to the reported tendency is 1 mb for each mb/°ϕ variation of pressure along the path of the ship. Obviously, if the ship is stationary, if the pressure pattern is flat, or if the ship moves along an isobar, the tendency observed aboard ship is the local tendency.

When the pressure pattern is reasonably uniform, the ship's tendency can be corrected easily by measuring the pressure variation along the course of the ship, using the current pressure analysis and the reported speed of the ship. The problem is not so simple if a sharp frontal discontinuity has passed the ship (or has been passed by the ship) in the last 3 hours, since then the present distribution of pressure in the vicinity is not representative of $\partial p/\partial s$ during the 3-hour period.

19. S. Teweles, "The Tentative Normal Diurnal Height Change of the 700-Millibar Surface over the United States and Adjacent Areas," *Monthly Weather Review*, Vol. LXXVII (1949); Oliver R. Wulf, Mary W. Hodge, and Stanley J. Obloy, *The Meteorological Conditions in the Upper Reaches of the Radiosonde Flights over the United States: Diurnal Effects in Temperature and Pressure* ("Miscellaneous Reports, Department of Meteorology, University of Chicago," No. 21 [Chicago: University of Chicago Press, 1946]).

20. Nomograms for correcting ships' tendencies are given by B. E. Olson, *Bulletin of the American Meteorological Society*, June, 1946.

9.12. *Barometric tendencies at fronts.*— For properly locating surface fronts in the chart a good deal of emphasis is placed on reported symbolic barometric tendencies. This element of the report is useful for placing a front the proper distance past the station after the approximate position of the

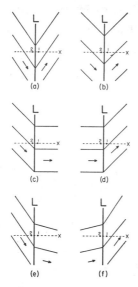

Fig. 9.12.—Various types of pressure patterns about surface fronts (idealized).

front has been determined from air-mass distribution and from larger-scale indications in the pressure and wind fields.

Moving fronts of significant intensity are found at the surface in discontinuities in the pressure pattern; and evidently, as a front passes by a fixed observing point, there is a corresponding change in the local pressure tendency shown as a change in the slope of the barogram. The magnitude of the shift in the pressure tendency is a function of the speed of the trough, the change of pressure gradient at the front, and such other influences as the semidiurnal pressure wave, the rate of deformation within the pressure pattern, and local disturbing effects of convective or instability phenomena.

The relation of the pressure pattern to the

change of local pressure tendency at a front can be deduced ideally by referring to Figure 9.12. Points *1* and *2* indicate successive locations of the same observing station with respect to the moving system. In the first two examples, (*a*) and (*b*), the trough is symmetric about the front. If the trough conserves its shape and if its speed is constant, then pressure is falling at time *1* and rising at the same rate at time *2*. This results in the check tendency. It is apparent that the sharpness of the change is directly proportional to the speed of the trough. Also, for given trough speed the tendencies at *1* and *2*, and also the sharpness of the tendency discontinuity, is less in (*b*) than in (*a*). The other examples are left for analysis by the reader.

The effect of deepening on the tendencies ahead and behind the front can be visualized in each illustration by shifting the pattern of isobars along the front. This may change the sign of tendencies at either or both of times *1* and *2*, but the change of tendency (the angle made in the barogram) is not altered if there is no deformation in the pattern, that is, if deepening is the same at *1* and *2*. The possibility of deepening is always expected, and therefore the magnitude of the *change of tendency* is more valid than the sign of the tendency to either side of the front.

Because of their peculiar location in advance of moving cyclones, warm fronts almost invariably have pressure falls at the surface both ahead and behind, and the frontal passage is marked by change from large falls to smaller falls. Cold-front troughs can be undergoing deepening, filling, or no change at all. Near a deepening cyclone or where the cold front has moved ahead of the primary trough, pressure falls can be found on both sides of the cold front. At greater distances from the cyclone there may be pressure rises on both sides of the front, as seen over southern Texas and northern Mexico in Figures 9.02*ab;* this general pressure rise is the effect of eastward displacement of the pressure ridge at the surface and aloft. At some intermediate location along the front (e.g., near 35° latitude) there is no deepening, and the signs of the tendencies in the two air masses are consistent with the geometry and motion of the trough.

The combined effects of deepening and displacement of the frontal trough on the change in *observed* pressure tendency can be incorporated into a single expression. The Petterssen formula, Eq 7.14(5), may be rewritten

$$(\partial p/\partial t)_2 - (\partial p/\partial t)_1$$
$$= U_x[(\partial p/\partial x)_1 - (\partial p/\partial x)_2] , \quad (1)$$

in which the subscripts denote the times before and after passage of the front. Except in high latitudes the semidiurnal pressure variation also must be considered when comparing tendencies over an interval of hours. This contribution may be added to give

$$(\partial p/\partial t)_2 - (\partial p/\partial t)_1$$
$$= U_x[(\partial p/\partial x)_1 - (\partial p/\partial x)_2]$$
$$+ [(\partial p/\partial t)_2 - (\partial p/\partial t)_1]_{sd} . \quad (2)$$

The last term is the change of semidiurnal pressure tendency during the period. In low latitudes this may be large, easily of magnitude 2 mb per 3 hours, and with weak slow-moving frontal troughs it may completely overshadow the change of tendency due to movement of the front. At certain times during the day (near 0400 and 1600 local time) this pressure variation accentuates the frontal discontinuity in the local pressure tendency. At other times (near 1000 and 2200 local time) it subtracts from the frontal discontinuity, and its relative effect may be so large as to give actual 3-hour tendency of the form ∧ as the front passes by the station.

Besides the semidiurnal pressure wave there are many other factors operating which may make the local barometric tendency misleading for locating fronts on weather maps. As one example, the small-scale convection cells revealed as showers and thunderstorms have isallobaric fields of their own.

From the equations above, it follows that the change in barometric tendency at a front is a direct reflection of the discontinuity in the pressure pattern. It was shown earlier (Sec. 6.21) that the kink in isobars is a function of the slope of the front and the discontinuity in temperature gradient. Thus cold fronts usually have sharp kinks in the barogram. In contrast, warm fronts have small slopes near the ground, giving isobars a more rounded shape across the surface position of the front, and therefore less distinct transition in the barogram as the front goes by. Merely through force of habit, we are led to accentuate the kinks in isobars in drawing the pressure analysis at warm fronts especially, and an inconsistency arises between local tendencies and the geometry of the moving pressure pattern.

9.13. *Discussion of isallobaric charts.*— The series of charts in Figures 9.08*ab* cover the region where the front was most clearly defined within the sea-level pressure pattern. The isallobars were drawn from complete sets of synoptic data.

Observe the irregularity of these short-period isallobaric patterns outside the neighborhood of the frontal trough. In the vicinity of intense moving features of the pressure pattern the 3-hour isallobaric field is well defined and meaningful. Elsewhere, smaller-scale disturbances have greater relative effect, and the isallobaric pattern is in such confused state that analysis and use are quite subjective.

In Figure 9.08*b* the isallobaric pattern even near the front appears to have some cellular structure, and isallobars just behind the front tend to be of wavy form. There is little doubt that many more actual irregularities were smoothed out in the process of analysis. Such waves in the isallobars might suggest that the front does not move with uniform sweep over its entire extent (even though we permit drawing them such), but its motion is subject to accelerations varying with the locality and might correspond to small stable waves along the front.

In Figure 9.08c there are several other features of importance to either the motion or the development of the pressure pattern. For example, the pressure falls east of the Pacific frontal trough and pressure rises to the west indicate eastward progression. The isallobaric minimum over Great Bear Lake is a case of strong local deepening in a region with initially straight isobars. It is not associated with movement of a trough or cyclone in the surface pressure pattern. The continued deepening soon gave a new LOW which eventually moved southeastward and became the principal cyclone over North America.

The minimum of pressure rise near the Great Lakes is an interesting feature in view of strong cold advection in the low levels. This indicates large pressure falls aloft. An interesting feature in the east is the presence of the primary isallobaric minimum a large distance southeast of the cyclone center. On succeeding charts the original cyclone continued its motion eastward, but a new cyclone developed southwest of Newfoundland and soon became the dominant one in that area.

The cyclonic disturbance affecting southern California is hardly apparent in the pattern of pressure change. Although this system is moving and is of vital importance in the weather, the attendant isallobaric field must be weak, because pressure gradients about it are also weak.

9.14. *Remarks on analysis of pressure and fronts.*—We have given attention to the use of pressure, pressure tendency, and wind in surface frontal analysis but little attention here to the use of air-mass analysis for that purpose. The thermal element was mentioned in Sections 6.22, 6.23, 7.06, 7.19, 9.06, and 9.07, among others. In using the surface temperature pattern for locating fronts, some caution is necessary to remove effects of stability and diurnal variation over land, cloud cover, local modifications by the surface, and the geographically controlled temperature contrasts. As a rule, the dewpoint temperature is the more reliable air-mass property for frontal analysis at the surface. Variations in low-level stability and weather types are also objective parameters for that purpose.

A problem arises in drawing the sea-level pressure pattern near fronts in elevated areas. The pressure analysis refers to sea level, but the frontal analysis is for the surface, which may be several thousand feet above. Here the ordinary methods and models for pressure analysis are defeated, even though the surface front can be located from other elements of the reports.

Excluding local influences on the pressure pattern and the difficulties resulting from reduction to sea level, analysis of the pressure pattern is no problem where the data network is dense. In other parts of the chart, however, the data are so few and far between that other considerations surmount in importance. Here the plotted synoptic data can be used to fix the pattern usually after it has been determined by other means. It is from this viewpoint that the concepts of continuity and thorough "single-station analysis" are of greatest value in obtaining a complete analysis of the synoptic chart.

Time sections of surface reports, as shown in Figure 9.02d, can be maintained for ship and island stations, for critical isolated stations over the continent, and even for moving ships at sea. The potentialities of such a procedure are numerous but still not fully explored in routine analysis.

Examine the sequence for the stationary ship at 50° N 145° W in the Pacific Ocean. In addition to the ordinary synoptic data for land stations, the reports of air-water temperature difference (plotted below the dewpoint temperature) and sea swell are included. Throughout February 28 and the early part of March 1 this station reported strong southerly winds, air temperatures several degrees warmer than the sea surface, generally falling pressure, and overcast skies with precipitation. The conclusion is that a trough of low pressure, or a cyclone center with strong pressure gradient, was approaching the ship from west. At 0930 GCT on March 1 (not shown), the ship reported southerly winds and pressure falling slowly. However, at 1230 GCT the wind is from almost due west, and the pressure has begun to rise. Evidently, the trough must have passed the station between 0930 GCT and 1230 GCT. From the check tendency of pressure it might be inferred that the trough was not smoothly rounded but had the shape of a frontal discontinuity. On this basis the front should be placed slightly east of the position of the ship at 1230, if there is a front at all. Note that the following observations for 1830 and 0030 substantiate this shift in wind, and therefore indications by the 1230 report cannot be ignored as due to an insignificant small-scale disturbance in the pressure field or to errors in observation or transmission. It is seen also that, though the surface wind never attains a northerly component behind the trough, the lower temperatures and dewpoints might indicate that the isobars are probably curved cyclonically and that the air actually has a trajectory from a higher latitude. This evidence was used as part of the basis for drawing the cyclone center shown in Figure 9.02a.

From this same set of reports it is possible to obtain further information for drawing the pressure analysis on the 1230 and subsequent charts. The strength of the winds indicate that pressure gradients are large. The southerly winds in advance of the frontal trough imply isobars with south-southwesterly orientation, and the winds behind the trough show that isobars are oriented about west-east. Since the winds are not of very different speeds, it follows that the pressure trough is oriented northwest-southeast near the station. There is evidence that the attendant cyclonic center probably is moving with a northward component, since winds remain westerly behind the trough while pressure rises rapidly. The backing of wind into more southerly direction on March 2 would indicate the approach of another pressure trough; this one is shown south of the Aleutians in Figure 7.02a. Sea swell moves out from the region of wave generation, and accurate reports are of aid in determining the wind field in a wide area.[21]

The reader should analyze the entire sequence of reports for the stationary ship at 40° N 142° W. There it will be found that the most likely time of frontal passage consistent with the other ship is just prior to 1230 on March 1. Between the first two reports of the sequence there was another sharp shift in the wind and pressure pattern, indicating the probable passage of a front. This one is not given in Figure 9.02a (but it was shown dissipating near the continental coast in Figure 7.02a). It is seen from these sequences that the drop in temperature with the front is very small compared to that found with the cold front over the continent. In general, the surface temperature change through fronts is small over the oceans because of the relative invariance of water sur-

21. See "Reading References."

face temperatures. Over the oceans, fronts are marked more clearly as contrasts in low-level stability, or in upper-level temperature, rather than as contrasts in surface air temperature except in areas of large sea-surface temperature gradient.

At San Diego is found a good example of a depressionary cloud sequence that would be associated with a polar-front wave. However, the surface frontal picture in this area is vague, if the surface charts are examined carefully, and all we might say with any certainty is that in the period March 1–2 a weak trough of low pressure passed by the station. This trough is identified with the closed cold cyclone on the upper-level charts for March 1.

The sequence for Glasgow, Montana, shows pressure rising with northwest winds through 1230 March 1, indicating eastward motion of a pressure ridge toward the station. Between 1230 and the following 0030 report the pressure reaches a maximum, and the wind changes from northwest to east. The flat pressure ridge passes during this period. The particular shift of wind implies that the high-pressure center passes north of the station.

At Oklahoma City it is evident that the cold front passed between 0630 and 1230 on February 28 and marked a sharp transition from mT to cPk at the surface. The rise in temperature behind the cold front from 1230 to 1830 is a reflection of the diurnal temperature variation within the polar air mass. If this change is subtracted, large cooling would be found in the cold mass throughout the day and would indicate progression of the cold front. At St. Cloud the front passed early in the sequence of reports without any large air-mass change. The cold front passed Lake Charles and New Orleans between 0630 and 1230 on March 1 but probably nearer to 1230. Here again the front separates real mT air from cPk at the surface. At Charleston the frontal

passage occurs late on March 1, at Pittsburgh between 0030 and 0630 March 1, and at Washington just prior to 1230 March 1. Incidentally, notice that rapid clearing followed the cold front in the south, while at the more northerly stations (St. Cloud, Pittsburgh) the cloudiness and precipitation persisted for a long time after frontal passage.

The sequence of reports for Moosonee indicates that the cyclone center passed south of the station, moving with a component eastward, and was due south of the station at some time during the 12 hours prior to 0630 March 1. Considering the rapid changes of wind direction and usual speeds of pressure patterns, the indication is that the cyclone center passed near the station. The northwesterly winds and pressure rises later might indicate the cyclone was moving toward north of east.

At Sable Island the northwesterly winds early in the period backed gradually into west and then into south. Thus, a small pressure ridge passed early on March 1, and the associated high-pressure center must have been located at a latitude south of Sable Island. Later, the wind increases from southeast, and pressures fall rapidly. This would agree with northeastward motion of the warm front trough shown in Figures 9.02ab and might indicate that the ridge which passed the station earlier was part of a small closed anticyclone which moved northeastward. With time, the wind veers slowly into south and then southwest, pressure begins to rise, and precipitation ceases. The fact that temperatures never get much above freezing indicates that the surface warm front does not pass the station. It is logical to conclude that the occluded front trough passed near 0630 March 2, while the larger trough associated with the cyclone over southeastern Canada lags some distance behind.

The ship at 36° N 70° W in the Atlantic

Ocean shows northwest winds at the beginning which also back through west and are very strong through March 1. The pressure ridge evidently passed between February 28 and March 1 and was followed by a surge of tropical air to the rear. The cold front passed early on March 2 with extremely well-marked wind shift and air-mass discontinuity. (This station lies in the region of largest sea-surface temperature gradient attending the Gulf Stream Current, and frontal passages can give sharp temperature changes here compared to other parts of the ocean.)

Bermuda is some distance east and south of the ship. Therefore, the same events should take place some time later. An important clue concerning the pressure pattern in this area can be found by comparing the reports for the two stations at 1230 and 1830 March 1. Notice that wind directions are consistent, indicating that directions of isobars are about the same in the vicinity of both stations. However, the temperatures and dewpoints are higher at the ship, which is located at higher latitude. From this, one might conclude that air reaching the ship has had more southerly trajectory than air reaching Bermuda, which should be reflected in the synoptic pressure pattern.

The strong northwesterly winds with low temperatures and unstable weather on February 28 at the three Atlantic stations are indicative of a large and intense cyclone situated some distance east. From the pressure rises at the stations it appears that the cyclone is drifting eastward. Figure 7.02*a* shows that it is near 40° W in the Atlantic Ocean at 1230 March 1, and it still has strong northwest winds to the rear. The instability clouds and precipitation result from rapid motion of very cold continental air across sea-surface isotherms toward higher temperature.

In the above discussion we barely touched on the amount of information that can be had from observation sequences at individual stations for deriving the current pressure analysis. Without need for elaborating, it is obvious that a study of events at a few critically located stations will give much of what is required in the pressure and frontal patterns. In analysis over oceans and remote land areas, this will be found not just a supplemental aid but actually a primary basis for efficient analysis.

9.15. *Representing the visible weather.*[22]—Our methods of analysis and representation of the physical variables are quite modern compared to the methods used for the weather information contained in surface synoptic reports. The usual procedure in "analysis" of condensation phenomena and obstructions to visibility is merely one of making those plotted elements more visible from a distance, that is, application of green shading over a report of continuous precipitation, shower symbols over shower reports, yellow shading over fog reports, etc., with little or no attempt to indicate *distributions* of phenomena. Also, present methods do not include any means of representing patterns of cloud types and cloudiness on the charts.

Though it is not entirely practical in routine analysis where time and effort impose critical limitations, the outline given can be of some value in learning types and distributions of weather in relation to other patterns on the chart, and some parts might be recommended for routine analysis. The outline is not wholly original; it reflects a good deal of the Bergeron philosophy.

a) Representing precipitation.—Since we take the liberty to interpolate and extrapolate patterns of any of the physical variables between stations and between synoptic times, it appears reasonable to proceed

22. Conventions for indicating weather on the surface chart are outlined in the current weather service manuals.

likewise with patterns of weather on standard charts. *Areas* of continuous or intermittent rain (snow) or drizzle and *areas* of shower activity should be designated such by accounting not only for the few plotted stations reporting this weather but also for the specific areas giving direct and indirect evidence.

Because of difficulties in properly encoding precipitation, many individuals reports may not be representative of the general area or of a longer interval of time. As a result, a report of showers may be surrounded by reports of continuous rain or snow, one of continuous rain by intermittent rain, one of drizzle by light rain, and so on. In view of this, it is logical in the usual large-scale analysis to smooth out those prominent inconsistencies to the extent of difficulties in observation and local variation. In more detailed local analysis it can be advisable to distinguish between reports.

Evidence of existence and extent of precipitation areas can be obtained from reports of past as well as present weather. Past-weather reports are useful especially in shower areas, since showers are usually a scattered phenomenon for which chances of precipitation currently at the station are small. Frequently the only evidence of showers is the report of cumulonimbus. If the weather is associated with a line phenomenon, it should be indicated as such. Over spaces with no data, it is useful to indicate the expected weather if there is a fair amount of certainty for the physical cause.

Owing to difficulties in printing, no distinction was made in Figures 9.15*ab* between present and past weather, between showers and other types of precipitation, or between intermittent and continuous precipitation.[23] It is possible that the two rain areas in the extreme west could have been con-

23. The jagged edges in the hatching indicate that the boundary is unknown.

nected, although it is likely that the topography of southern Alaska could produce a separate rain area or one of more intense rain. In chart *b* the rain area in the southwestern United States contains inconsistent reports which do not necessarily reflect on the competence of observers. A suitable way of representing this entire precipitation area is by intermittent rain. Over the Great Lakes and the area to the west in *b*, the standard form of indication is snow showers, even though they may be due primarily to low-level turbulence. Near and ahead of the large cyclone is the area of "continuous" precipitation. Near the cold front the prevailing type is more difficult to determine, but there is evidence of predominant showers at the surface front[24] and more steady forms ahead.

b) Representing fog and other restrictions to visibility.—As in the case of precipitation, properly indicating fog, smoke, and dust should be based on knowledge of the cause. Fog formed by upslope motion is due to adiabatic cooling of moist air and would be found on windward slopes near steep coast lines and in river valleys and frequently over the eastern slopes of the Rockies, particularly with easterly winds from a rain area. In each case the patterns of surface temperature and motion and the source and pattern of moisture will indicate the approximate area of fog. The pattern of valley fogs conforms to the shape of the valley. Fogs formed by motion of warm moist air over a cold surface usually have a large area which can be defined from the patterns of air motion and moisture and the temperature distribution of the underlying surface; over water the pattern of fog will agree with the pattern of surface-air temperature advection. Coastal fog, resulting from differ-

24. The *present* precipitation area may extend only a few miles behind the surface position of the front; note that the precipitation area is for the last 3 hours.

ences between land and sea, generally parallels the coastline.

Smoke spreads downwind from its source (industrial areas, forest fires), and dust is lifted by turbulence over dry exposed surfaces. In both cases it is possible to locate areas with fair approximation from geography and the nature of the motion.

c) Cloud analysis.—On the whole, cloud analysis is not emphasized sufficiently, which is probably one reason why there is so much to be desired in cloud forecasting. Frequently no attention is given to this plotted information during surface analysis.

The major groups of cloud types have areas which usually can be defined easily

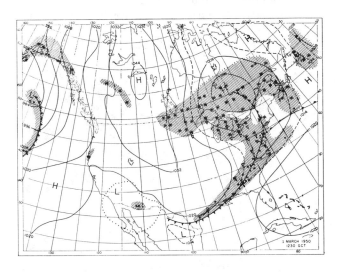

a

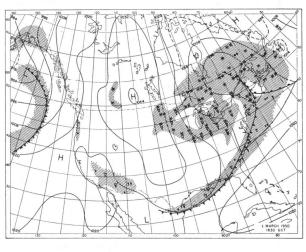

b

Figs. 9.15*ab*.—Areas in which precipitation occurred during the 3 hours preceding the time indicated; (*a*) 1230 GCT, (*b*) 1830 GCT.

on the synoptic chart (Fig. 9.15c). The middle clouds of stratiform types cover large areas in and ahead of moving cyclones, and they are associated with the areas of more or less continuous rain. Cirrostratus decks normally precede and coexist with those areas of middle cloud. Low clouds are dependent more on the nature of the underlying surface than on large-scale dynamics. Cumulonimbus is an indication of strong

with cloud areas in relation to other patterns on the surface chart and upper-level charts. It should be possible to devise a scheme of analysis, using either lines or shading, to distinguish also between the various cloud types.

There should be some relation between cloud and horizontal patterns of relative humidity, bearing in mind, however, that sharp vertical moisture gradients can exist

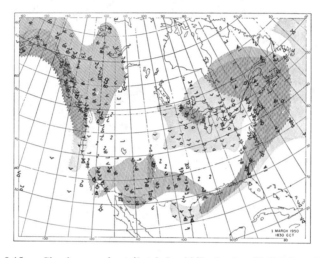

Fig. 9.15c.—Cloud-cover chart (*hatched*, middle clouds; *stippled*, low clouds)

convection and instability through deeper layers.

Figure 9.15c shows a simple and incomplete form of cloud analysis. Areas of more than 5/10 cover are indicated by hatching (middle cloud) or by stippling (low cloud). Where the sky was overcast with low clouds, it was not possible to determine the middle cloudiness (northern Great Lakes). Over Pacific Ocean areas the reports were too sparse to give a unique analysis, but a fair analysis could have been made even so. The cloud symbols given were taken from individual station reports.

Cloud analysis of this type, or even more detailed, would be useful in short-period forecasting and in familiarizing one's self

and that horizontal moisture patterns can vary widely with height. On comparing Figure 9.15c (and also 9.15ab) with Figures 7.20cdef, a good deal of similarity is seen if proper levels are considered. The middle-cloud areas agree in general with high relative humidity at 700 mb (and 600 mb) and, in some places where the clouds may be higher, better with 500 mb. Notice in particular the correspondence in the extreme eastern and southern states. The pattern of low clouds agrees in most places with the 850-mb pattern, although the clouds may be totally above or below 850 mb in some areas. The surface relative-humidity pattern usually also will be closely related to low-cloud distribution and also to fog.

9.16. *Distribution of weather about frontal waves.*—The pattern of weather around frontal waves has some common features, but many are likely to vary with the individual wave, the stage of development, and the character of the underlying surface. We shall consider the weather about an idealized wave[25] (Fig. 9.03*a*), and then point out some of the more important discrepancies. Reference can be made to Figures 12.06*ab*.

a) Warm front.—The classical picture of the warm front contains a wedge-shaped cloud mass bounded below by the warm-front surface. The leading portion of the cloud mass is cirrus, followed by cirrostratus, altostratus, and finally nimbostratus in the deepest portion of the cloud mass, which ends near the surface position of the front. The upper boundary of the clouds is the top of the high-cloud layer. The cirrus at the leading edge may extend several hundred miles in advance, the altostratus up to 500 miles in advance, and the rain area normally 100–300 miles in advance of the surface front. The presence of low clouds in the cold air is conditioned by the initial humidity of the air mass and by evaporation from falling rain. Normally, the original low clouds become more stratified locally with approach of the front; they are likely to be fragmented with strong winds.

As quite frequently the case, the cloud mass in the warm air is well layered, but, where rain is falling, the mass is more continuous and may extend to the ground. Frequently there are "hard cores" of cumulonimbus extending through several cloud layers, especially near the cyclone center. It often appears that the high and middle cloud decks are based several thousand feet above the frontal surface, as moisture in the warm air is variable and stratified. Though

in many cases in vertical plane both frontal location and moisture analysis are questionable, there is reason to believe that it is the exception rather than the rule when frontal cloud masses are bounded below by the frontal surface. The graphic description of cloud patterns given by Conover and Wollaston illustrates this point excellently.[26] When the tropical air is relatively dry, there may be no precipitation and only small amounts of attenuated middle cloud, if any.

With the typical warm front the precipitation ceases slightly ahead of the surface position of the front, and the main cloud mass might end there too. This separation is related directly to the lifting condensation level for the surface tropical air and inversely to the slope of the front. When the warm front has topography similar to Figures 6.23*bcf*, the main cloud mass and precipitation may be separated by several hundred miles from the surface position of the front.

Whether the precipitation is rain or snow depends on the temperature at which the precipitation is formed (the amount of supercooling varies).[27] If the ground is below freezing, rain will freeze on contact ("freezing rain"). Over mid-latitude continents in winter, in favorable conditions the precipitation area ahead of the warm front will be characterized by almost parallel bands in the order: snow, freezing rain, rain.

The spreading of warm-front weather on the cold side of a stationary or quasi-stationary front is good evidence of incipient wave formation, in addition to the field of barometric tendency. The area of cloud and precipitation normally increases until the

25. A more complete discussion is given by Schereschewsky, in Berry, Bollay, and Beers (eds.), *Handbook of Meteorology*, Sec. XI.

26. J. H. Conover and S. H. Wollaston, "Cloud Systems of a Winter Cyclone," *Journal of Meteorology*, VI (1949), 249–60.

27. The region of supercooling of cloud droplets, which is the area of greatest danger for aircraft icing, lies mostly between the 0° C and −20° C isotherms (Byers, *op. cit.*, chap. 22).

cyclone reaches its maximum rate of development; and, although the area might lengthen along the front, it narrows appreciably in the later stages. With occlusion of the wave, formation of a closed cyclonic circulation aloft is typical, and part of the warm-air cloud mass (or moist air) can be carried westward over the poleward side of the surface cyclone. In fact, in decaying stages of the frontal wave the weather might be restricted only to that area of the system.

The weather ahead of the surface warm front is due mostly to ascent of the moist warm air. There are several ways of viewing this. From the classical approach, "the warm moist air glides up the frontal surface" with little or no ascent of the retreating cold air. That distribution of vertical motion does not account for the usually strong horizontal convergence of mass near the ground in advance of the cyclone, but it is perhaps the easiest way of describing the warm-front cloud mass. Warm-front weather also agrees with other methods of relating weather to patterns on certain charts. The warm front is the ideal correlation of weather with the rate of horizontal temperature advection. It is also one of the better illustrations of the relation of weather to isentropic relative motion. The forward edge of the area in which the altostratus cloud mass is being produced should be closely associated with the line connecting points of tangency between streamlines and isobars on proper isentropic surfaces (or contours and isotherms on proper isobaric surfaces). The forward edge of the clouds can be some distance downstream by virtue of horizontal transport.

A useful relation between the location of warm-front weather and the upper flow pattern will be seen by comparing Figures 9.15 with Figures 7.03*de*. Notice that the middle clouds and associated rain area occur on the downstream side of the upper troughs. In the southwestern states the upper flow pattern gives better indication of the type of weather than does the surface pressure pattern. Thermal patterns representative for the troposphere are almost as good indicators as the flow patterns in the upper troposphere.

b) *Cold front.*—In most simple view the cold front is considered "a line of showers or thunderstorms with check (barometric) tendencies and sharp wind shift." This has led at times to accommodating any such lines of weather by cold fronts on the surface-weather map. It has also given misleading impressions to the public and to flying personnel[28] and is the source of belief that a cold front can be *seen* from the air. Very often the main line of weather is some distance ahead of the surface cold front and still farther ahead at upper levels, or there may be no visible weather phenomena at all.

Since normally the surface cold front lies in quite a sharp pressure trough, the frictional departure of low-level winds gives a maximum of horizontal convergence of air at the front. There is then preference for ascending motions above the surface front, at least through the lowest few thousand feet. If the moisture in the warm surface air is sufficient, a line of convective clouds, showers, or even thunderstorms can develop.

There are two general types of cold fronts worth examining in some detail (Fig. 9.16). The two sets of patterns given represent the surface (1000-mb) analysis, the contours of an upper pressure surface (e.g., 700 mb), and the relative contours of the layer between. Notice that only slight difference is shown in the two patterns for the surface, but the difference in weather can be large. In *(a)* the cold front is intense and has relatively constant intensity over most of its

---

28. One might also be led to believe falsely that advance of cirrus clouds marks reliably the approach of a warm front and rain.

extent, as seen from the gradients of thickness. In (*b*) the intensity of the cold front for the layer decreases equatorward, owing either to decrease of horizontal temperature gradient in the frontal zone at any level or to decrease in slope of the front, but most

front is not decreasing rapidly with time. The band of cloudiness and precipitation is more narrow than at warm fronts, as attested by the different frontal slopes. This cold front is likely to retard; it is a slow mover compared to (*b*).

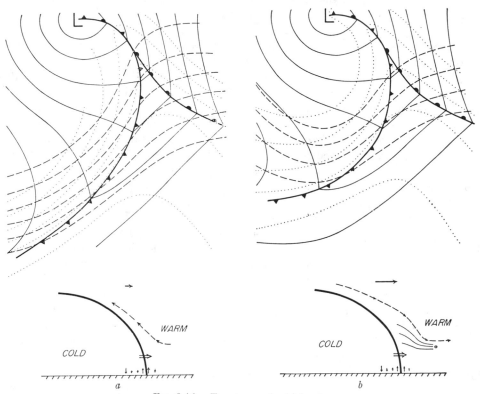

FIG. 9.16.—Two types of cold fronts

probably the latter. In the central section of the cold front the low-level geostrophic wind normal to the front is the same in each case, but the different distributions of thickness give a stronger normal component of wind at the upper level in (*b*) than in (*a*). It is possible for the surface front in (*a*) to be moving faster than the normal component at the upper level,[29] thus indicating relative upslope motion of the warm air along the frontal surface and producing warm-front weather behind the surface position of the front, all provided the slope of the

In case (*b*) the vertical increase of wind normal to the surface front gives downslope motion of the upper warm air, supergeostrophic motion of the front in the low

29. This could be true if the low-level winds in the cold air are supergeostrophic or if the upper winds are subgeostrophic, as would be expected with cyclonic curvature. Supergeostrophic surface winds could be due to one or a combination of several factors existing: the advective ageostrophic wind (diverging contours downstream), isallobaric wind (largest rises behind the front), and convective ageostrophic wind (sinking and spreading of the cold air).

levels, and greater low-level turbulence behind it. The weather, if any, occurs at or ahead of the front, often along a "squall line" or *instability line*[30] lying nearly parallel to the front and up to a few hundred miles in advance of the surface front. Instability lines appear to form near the surface front and preferably below the region of strong upper winds. They move out ahead of the front, sometimes for great distances in the warm sector, and then dissipate. At times they extend across the warm front, and often a series of lines is present in the warm sector. This squall line is characterized by vigorous convection, sharp shifts in wind at the surface, and sharp changes in the barogram.[31] The general area appears to be the favorite birthplace of tornadoes in North America.[32]

A common feature in the warm air ahead of this cold front is a stable layer characterized in most localities by dryness above; its formation agrees with the differential vertical motion above the surface position of the front. With a gradient of horizontal mass convergence at the ground, the stable layer likely can have the tilt indicated, which could be a mechanism for pressure jumps and convection in Tepper's explanation of squall lines. The dry air above the inversion does not entirely prohibit the de-

velopment of strong convection and precipitation. Potential instability can be great if the surface air is very warm and humid, and instability can be released by the surface convergence, among other things.

c) Warm sector.—The weather in the warm sector, except near the cyclone center and instability lines, is of air-mass type and is largely dependent on the nature of the surface and history of the air. There may be significant ascent in the warm sector near deepening cyclone centers, and the weather is then hardly distinct from that with warm front.

With the poleward flow in the warm sector there is preference for stabilization of the surface layer, which is most prominent in regions of large temperature gradient of the underlying surface (coastlines, snowlines, concentrations of sea-surface isotherms). Fog is common with light winds, moist surface, and large surface temperature gradient in the direction of motion. With stronger winds stratus cloud, perhaps with drizzle, is the usual thing. Over dry land, especially in summer, diurnal variations of stability can be large, and there may be some convective cloudiness or perhaps no clouds at all. In general, the weather improves toward higher pressure in the warm sector.

d) Advancing cold-air mass.—To the rear of the cold front and cyclone the advection of cold air maintains relative instability of at least the surface layer, but the usual subsidence in this region has a stabilizing effect above. If the cold air can absorb moisture from the surface, the cloudiness is ordinarily of the cumulus and stratocumulus varieties, the amount and depth depending on availability of moisture and on stability. There are characteristic differences between land and sea; the conservative sea-surface temperatures and the ready source of moisture favor greater convection and wider areas of cloudiness. With slowly moving

30. J. R. Fulks, "The Instability Line," in American Meteorological Society, *Compendium of Meteorology*, pp. 647–52; I. W. Brunk, "The Pressure Pulsation of 11 April 1944," *Journal of Meteorology*, VI (1949), 181–87.

31. M. Tepper, "A Proposed Mechanism of Squall Lines: The Pressure Jump Line," *Journal of Meteorology*, VII (1950), 21–29; J. C. Freeman, "Map Analysis in the Vicinity of a Pressure Jump," *Bulletin of the American Meteorological Society*, XXXI (1950), 324–25.

32. E. J. Fawbush and R. C. Miller, "The Tornado Situation of 17 March 1951," *Bulletin of the American Meteorological Society*, XXXIV (1953), 139–45; E. M. Brooks, "Tornadoes and Related Phenomena," in American Meteorological Society, *Compendium of Meteorology*, pp. 673–80.

flow patterns, the subsidence is less; and, with moderate to strong winds over large sea-surface temperature gradients, strong convection is the rule (this is the case over the western North Atlantic on 1 March 1950).

In the immediate vicinity of the surface cyclone, and also just to the rear, persistent low cloudiness and turbulence showers are favored by the cold advection, by the supply of moisture from the ground recently wet by the warm-front rain area, and by virtue of the air having passed through the warm-front rain. Low-level "convergence lines" radiating from the cyclone can provide for zones of locally more intense showers. The contours of the terrain will modify the patterns in low-level weather, in this region as in any other. A cyclone centered over Iowa, for example, would be expected to have a larger than normal area to the northwest with low cloudiness, perhaps with light precipitation (and even with fog), and clearer weather to the south and southwest. A cyclone centered over the northeastern states has more persistent weather on the windward slopes of the Appalachians than on the lee slopes (compare Pittsburgh and Washington in Fig. 9.02*d*).

9.17. *Nonfrontal weather.*—Much of the cloudiness and precipitation occurs some distance from surface fronts. Some of this is warm frontal in character (the southwestern United States in Fig. 9.15*b*) and produced by the same large-scale ascent, even though the temperature gradients may or may not be large enough to apply the name "front." Other weather may be produced through ascent resulting from more or less local convergence areas or zones at the surface (equally by divergence aloft); some such cases are considered in Chapters 10 and 13. The weather resulting from strong flow of moist air impinging on mountain ranges is evident. Clouds and precipitation

may be formed by vertical mixing of moist air in almost any part of the pressure pattern but least in areas of light winds and preferably with moist air and upslope motion.

While the principal cause of cloudiness and precipitation is the ascent of moist air, we should not ignore altogether the effect of radiation cooling. A mass of moist air traveling horizontally from the tropics into high latitudes in winter may be cooled by several degrees, enough to permit condensation and perhaps light precipitation, by net loss of heat to the surface and to space. Air moving equatorward would experience the opposite effect. Thus, with similar conditions, more cloudiness should be found in poleward than in equatorward air currents through most of the year.

The tropics are a broad area where most of the weather is nonfrontal. Tropical storms, easterly waves, and convergence lines have more or less characteristic patterns of weather, yet the atmosphere is virtually barotropic. The analysis of this weather must take into account directly the fields of divergence, as distinct from pressure and frontal analysis, and also the features of the underlying surface. Similar analysis is applicable to air-mass weather as a whole.

9.18. *Fronts in mountainous terrain.*— Surface frontal analysis in mountains is a difficult problem, both for location and for proper indication in the analysis. Difficulties with the sea-level pressure pattern were already commented on briefly, but there are many other ways in which rugged terrain affects our standard modes of analysis. While the subject is a serious and detailed one, we can barely touch on it in hopes of stimulating more thought.

Slopes and rugged terrain impose certain retarding and accelerating effects on air motion, trap shallow masses of cold air in basins and valleys, produce variations in radiation, set up local circulations and eddies, and, with given large-scale flow patterns, establish vertical circulations which modify the distribution of weather. These problems are due to orography, but many difficulties in proper analysis stem directly from our viewpoints. First, instead of viewing fronts and surface contours as lines on a map, there is something to gain in viewing them as relief (i.e., in three dimensions). Another misleading viewpoint we must consider for analysis over mountains, plains, and oceans concerns the use of surface-air temperatures as the primary basis for identifying fronts. The point to be stressed is that our objective is meteorological analysis as contrasted with micrometeorological analysis; we should not let the surface temperatures, whose patterns may be representative of the lowest hundred-foot layer only, dictate an analysis which represents the thermal pattern for a deep layer and broad area, the large-scale weather pattern, and the evolution of important systems. If we like to indicate the surface temperature pattern specifically, the most efficient approach is drawing isotherms. If it is desired to perform small-scale friction-layer analysis on the common synoptic chart, we should develop appropriate methods and models probably much different from the polar-front theory of large-scale phenomena.

For a basic discussion of the influence of mountain ranges on fronts, the reader is referred first to Petterssen's book.[33] To avoid too much repetition, our discussion will concentrate on other details. Figure 9.18*a* shows, in an admittedly simplified way, how a surface *cold* front might be distorted by mountains. The surface front is retarded on the windward sides of peaks due both to the blocking of motion and to the fact that the slope of the terrain is opposite

33. *Op. cit.*, pp. 298–302.

to the slope of the front. The front is free to move through the passes and is accelerated there by the cold air forced around the peaks, perhaps in partial disrespect for the pressure pattern, which we know not. The cold air funneling through the passes, often in tremendous force, spreads down the valleys with winds generally paralleling the valley. It is safe to assume that some peaks and their lee slopes can become isolated for a while by the cold air, and a station situated on a lee slope might well experience

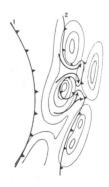

Fig. 9.18*a*.—Surface cold front distorted by mountains (previous and present positions of front).

frontal passage from the direction opposite to the average motion of the front. Detailed analysis of the type shown is not being advocated for routine work, but it is useful in pointing out how adjacent station reports can differ.

The weather is expected to be extremely varied, considering the varied previous conditions and the large upward and downward motions. After the front has traversed the mountains, with apparent acceleration and development downward on the lee side, it would tend to regain its normal shape, but the distortions might be evident for some distance beyond. At levels above the mountain peaks, the front may not be affected significantly in shape or movement during passage over the mountains. Shallow cold-air masses can be blocked altogether by the mountain ranges.

A warm front crossing mountains is affected quite differently. For one thing, the similarity of frontal and terrain slopes contributes to acceleration of the surface front on the windward sides of mountains in excess of the actual air motion. Warm fronts lying across the Appalachians are frequently seen to have a position more advanced in the mountains than on both sides. Other effects are left for thought by the reader; some primary considerations will be the small slope of the front near the ground and the prevailing stable conditions about the front.

The series of cross sections in Figure 9.18*b* is intended for nothing more than to illustrate frontal passages over mountain valleys and basins in wintertime. The drawings are a good deal simplified, mostly next to the ground, but they do show why frontal analysis from surface data consisting mostly of reports from valley stations can be so mystifying and often misleading.

During winter the bottoms of mountain valleys and basins of middle and high latitudes are storage areas for dense air accumulated from a previous cold air invasion, by radiation and downslope sinking of the air ("mountain winds"), or both. This pool of cold air trapped between the ranges is difficult to dislodge because of its density, the stability at its upper boundary, and protection by the ranges. Above, the air and the weather may glide over, in the sense that the atmosphere glides over a body of water. The stagnant cold air is removed by strong winds, by motion of cold air above it, or by replacement with a cold outbreak along the valley.

The first drawing shows initial conditions and the approach of a warm front. The chances are that the warm front will glide along the valley inversion with no distinct temperature change at the surface and, as

usual for warm fronts, without pronounced effect on the pressure and wind fields on the surface synoptic chart, although the pattern of barometric tendency might be some indi-

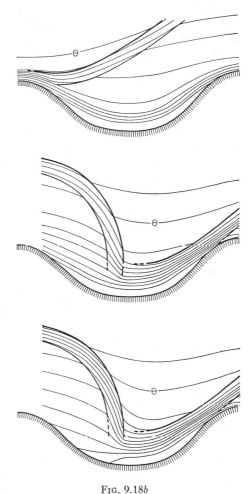

FIG. 9.18*b*

cation. The most prominent evidence of the front from the ground is the warm-front weather system, and it is likely to be attenuated or otherwise modified by virtue of passage over at least one mountain range.

The warm front leaves behind it a stronger valley inversion (second drawing); if there was precipitation with the front, the chances for valley fog and stratus are now better. An appropriate designation on the surface chart is "upper warm front." With passage of a cold front some time later, the cold film of air in the valley may or may not be affected significantly, depending on the strength of the flow and the temperature difference between the new cold mass and the air at the valley floor. As shown in the second drawing, the fresh cold air may be relatively mild, and the basin air thus can be relatively undisturbed. The passage of the front above would affect the pressure and wind fields, the local barometric tendency, all more distinct than with the warm front, and perhaps the weather. An appropriate designation on the surface chart is "upper cold front."

If the fresh cold air moving across is somewhat colder (third drawing), there can be different events observed locally. The pressure and wind fields are better defined, as also the isallobaric field, and more intense weather is expected. Here the surface temperatures are affected also, but the chances are good that surface warming is experienced at the cold front by more intense vertical mixing and by motion down the slope of the valley. This would be called a "cold front," either surface or upper, but too often it is called something else—"warm-type occlusion"—from surface evidence only. If the invading air mass is still colder, falling temperatures might be observed at the surface, in which case there is no difficulty in identification.

Figure 9.18b gives a fair indication of frequent occurrences in the western United States in winter. The prevailing movement of air masses is from the Pacific Ocean. The cold basin air may be mP air modified by cooling from below, or it may be the remains of a cP (or cA) mass occasionally breaking through the mountains from north or east. The fresh cold-air mass in the drawings is mP air which is ordinarily warmer in potential temperature. Thus, many frontal

systems moving over that region are difficult to detect in several elements of the surface observation except clouds and weather, to which as a rule too little attention is given during analysis, and confusion frequently rules. Because of all this, there is likely to be found from examining surface analyses too few warm fronts, an abundance of occlusions (of the warm type especially), and warm-front-type cloud and rain areas far in advance of cold fronts and occlusions.

Since the surface layer does present such a problem in proper frontal analysis, it is worth while suggesting that some effort be given in training to frontal analysis in layers or at levels above pronounced surface control (e.g., 850 and 700 mb). Then, if necessary, the surface analysis can be made, using that as a guide.

While Figure 9.18b is at hand, it is well to point out that preferred conditions for the *foehn* or *chinook* are indicated by the first drawing, if the right half of the drawing is ignored. It then represents a common winter condition east of the Rockies—cold surface layer of (arctic) air overlain by milder polar air of more recent Pacific origin. Passage of the warm front will intensify the temperature inversion, resulting in large contrast of potential temperature through it. If strong westerly winds attend or follow the warm front, there is rapid removal of the cold air away from the mountains. The surface air thus removed from the lee slope has to be replaced partially or totally by the tropical ("superior") air of the upper warm sector, accounting for a rapid local increase of temperature near the ground.

9.19. *Difficulties in frontal analysis over cold surfaces.*—The stabilizing influence of continents in winter and of cold ocean areas in summer presents problems in proper frontal analysis from surface data only. In middle latitudes the air aloft moves freely and rapidly from westerly directions, as do the frontal waves. At the ground, however, the flow is more cellular and sluggish; air moving equatorward behind one frontal wave will likely be drawn back poleward in advance of the next wave before reaching the tropical oceans. Particularly if moving over a cold surface, the returning polar air may be very shallow, stable, and intermediate in temperature between fresh polar and tropical air. The structure illustrated by Figure 9.191(*a*) is likely to result.

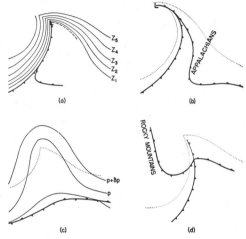

FIG. 9.191.—Schematic illustrations of frontal systems giving difficulty in surface analysis.

On the basis of air-mass analysis at the surface only, the indicated frontal analysis might be proper. The occlusion separates fresh polar air to the rear from the warmed polar air ahead; true tropical air might be found at the surface in the lower part of the diagram. In the surface-pressure and wind fields there may be evidence of a trough, about where the dashed line is given, and ahead of it perhaps a fairly prominent temperature gradient with warm-front weather.

The remaining lines in the drawing are contours of the frontal surface; the zero contour is at the given surface cold and warm fronts. Above the friction layer is found an open wave consistent with all ele-

ments in the surface reports except temperature (and dewpoint). There are at least two possible ways of indicating the surface analysis: as given, with an additional (surface or upper) warm front at the advanced position; or an open wave, using the advanced position for the warm front (shown surface or upper) and, if necessary, entering some distinguishing line to separate the two *surface* air masses on the right. The second alternative is perhaps the less deceptive and gives emphasis to the more important fronts.

Figure 9.192 illustrates a similar situation more concretely. The surface chart was traced from an official analysis; only a fraction of the reports could be entered on the chart shown. Note the surface warming occurring with the passage of the front, which gives reason for entering the front occluded. Notice also the large middle-cloud area extending southeastward from the cyclone just in advance of a detectable pressure trough with its attendant isallobaric gradient—all as with a warm front.

The 850-mb chart[34] lends all the desired evidence for an open frontal wave, whose cold front accounts for the surface-"occluded" front and warm front for the pressure, isallobaric, and cloud patterns at the surface. A cross section through this frontal wave evidently has components similar to drawings (*c*) and (*d*) in Figure 6.23. The (700/1000)-mb thickness pattern (drawn using the above surface chart and 700-mb chart 3 hours earlier) verifies this frontal wave, and it also indicates that the sea-level HIGH centered over western Colorado is fictitious. The illustrated pT soundings indicate the extreme shallowness of the "friction-layer cold-air mass" beneath the warm sector.

Figure 9.191(*b*) conveys a similar idea, but,

34. This chart will not be completely valid for the purpose when the thin layer of cold air extends to 850 mb or higher (see Grand Junction sounding).

since that pattern is so characteristic of the eastern states, a separate illustration is devoted to it. The mountains are an effective barrier to the cold air circulating from the Atlantic, and the surface warm front is thus retarded east of the mountains. At 850 mb the front (dashed line) may not be affected at all. The type of weather found east of the near the coast or out over the Gulf of Mexico, and the cold air in this case is shallow for some distance inland. The dotted line is the front at 850 mb, and the front is steep at levels above. A pressure trough at the surface, with perhaps a closed cyclone, is found beneath the upper frontal wave.

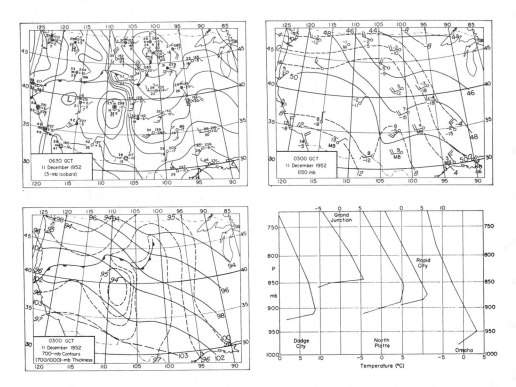

FIG. 9.192.—Situation of 11 December 1952 at 0630 GCT (surface) and 0300 GCT (upper levels). Contours and isotherms at 850 mb, and contours and (700/1000)-mb thickness lines on the 700-mb chart.

mountains is left for consideration by the reader. Incidentally, this is a favorable structure for formation of the peculiar "Cape Hatteras secondary" cyclone.[35]

Figure 9.191(c) shows schematically a structure found occasionally in the south central states. The surface front shown is

35. J. E. Miller, "Cyclogenesis in the Atlantic Coastal Region of the United States," *Journal of Meteorology*, III (1946), 31–44.

Figure 9.191(d) shows a structure quite common for the northern plains states in winter. A frontal wave at upper levels (dotted line), which might appear as a surface occlusion for reasons aforementioned, moves over the wedge of arctic air extending southward east of the mountains and initiates a frontal wave upon it. Constructing this pattern in three dimensions makes an interesting problem.

While Figure 9.191 gives a very limited number of the interesting situations that arise, it is hoped this is enough to provoke the interest to examine and visualize such things in three dimensions. During surface analysis of such areas, reference should be made to temperature patterns at levels above immediate surface control (e.g., 850 or 700 mb). The 850-mb temperatures plotted on the surface chart are of direct help in the surface frontal analysis.

9.20. *Secondary fronts and frontal paraphernalia.*—We saw in the previous section that secondary warm fronts can appear, far ahead of the boundary between polar and tropical air at the surface, by virtue of varying slope in the front. The similar thing can appear in the fresh cold air as an upper cold front reflected as a secondary cold front at the surface. Given a rather uniform distribution of pressure above the cold dome in Figure 6.23b, the wave shape in the frontal surface should be reflected in a troughlike shape of pressure surfaces near the ground below the upper front. If this line is normal to surface isobars, it appears as a regular trough; but, if parallel to isobars, it is either the trough axis of a col or more likely the troughlike axis *tt'* in Figure 4.104. Such features at the ground, combined with frictional effects on the motion, can lead to accumulation of isotherms locally. It is proper to recognize this structure, but whether or not to call it a surface front is again subject to debate and depends on the case.

If the leading bulge of the cold air spreads and subsides far ahead of the main dome, the secondary zone formed near the ground will probably merge by frontogenesis with the upper front and become the more significant. Indicating this in the surface chart is certainly appropriate. However, in most cases multiple slopes in the cold dome are temporary, and caution is needed in the continuity of analysis.

The structure in Figure 6.23b can be expected on occasion with the type of system shown in the lower part of Figure 9.04. The surface cold front, moving with the winds, can move far ahead of the same front at upper levels where the flow may be parallel to the isotherms. If a secondary steepening of the front occurs, it will lie across the surface isobars; and chances are that the air blows *through* the reflected pressure trough at the surface. Indicating a secondary surface front could be questioned in this case.

The scheme in Figure 6.23b takes place more prominently in meridional view across the belt of westerlies. A cold outbreak may be cut from its supply of cold air in high latitudes.[36] In moving equatorward, this air spreads and sinks, and meantime the portion of the front in high latitudes redevelops in the upper levels and favors formation of an "arctic-front" extension near the ground. The conditions preceding Figure 9.191(a), Figure 6.02b, and the "cut-off" cold dome off California on 1 March 1950 can be viewed as having occurred in this manner. The name "secondary front" is usually not applied to the portion redeveloping in high latitudes. In fact, it becomes the dominant front immediately. If the shallow cold air moves over warmer surface, it may be modified rapidly, and the initial cold front disappears. In summertime over continents, the remains of the initial cold front might be found merely as a dewpoint contrast—a "dewpoint front."

A common secondary cold front at the surface is indicated in Figure 9.191(d). This one is a pre-existing temperature gradient, blocked by the mountains, which is drawn southward behind the moving cyclone. Others of this nature can occur behind deep cyclones at the low-level zone of separation

36. Also to be interpreted as the westerly current remaining in high latitudes.

between the air carried around from the eastern side of the cyclone and the fresh cold air subsequently drawn into the circulation from higher latitudes.

Strong flow of cold continental air across coasts, or soon across zones of large sea-surface temperature gradient such as the Gulf Stream Current, maintains a similar air-temperature gradient in the cold air, which may overshadow the gradient at the cold front moving out ahead. The stationary or quasi-stationary temperature gradient is apt to be pronounced at upper levels also, owing to strong convection. A detectable surface trough and line or band of showers would be expected with the temperature gradient, but to call this a "front" is questionable[37] when the air is gushing through. Such phenomena are common in winter along the east and northwest coasts of North America and less frequently near the coast of the Gulf of Mexico. In the return circulation these are likely to behave as warm fronts.

37. But we do need terminology for these things.

Finally, a few words should be said about "back-bent," "bent-back," or "broken-back" occlusions, however they may be known, which without climatic change have come to adorn fewer weather situations. These are the oldest part of an occlusion believed bent back by the flow in the rear of the cyclone, by migration of the cyclone along the occlusion, or by development of a new cyclone nearer the surface warm sector. A detailed cross-section analysis through at least one of them would be in order. Preferred regions of secondary cyclonic development at or near the surface point of occlusion are strategically located south of mountain barriers in the northern hemisphere. Some are the southern tip of Greenland, the northern Gulf of Alaska (in summer), the western Mediterranean, the southern tip of Norway, and even southern British Columbia. Associated upper-level sequences in these and other cases are branching of the southwesterly flow and increase in poleward temperature gradient from the new surface cyclone.

READING REFERENCES

AMERICAN METEOROLOGICAL SOCIETY. *Compendium of Meteorology*, "Marine Meteorology," pp. 1057–1100. Boston: The Society, 1951.

BERRY, F. A., BOLLAY, E., and BEERS, N. R. (eds.). *Handbook of Meteorology*, pp. 573–674, 703–35, 746–95, 882–926, and 1030–56, and Sec. XI (clouds). New York: McGraw-Hill Book Co., 1945.

BYERS, H. R. *General Meteorology*, Chaps. 6, 8, 9–15, and 17–24. New York: McGraw-Hill Book Co., 1944.

HEWSON, E. W., and LONGLEY, R. W. *Meteorology: Theoretical and Applied*, pp. 105–22, 271–89, 343–79, 396–403, 408–27. New York: John Wiley & Sons, 1944.

PETTERSSEN, SVERRE. *Weather Analysis and Forecasting*, pp. 1–49, 138–204, 274–350, 441–90. New York: McGraw-Hill Book Co., 1940.

SVERDRUP, H. U. *Oceanography for Meteorologists*, pp. 133–46, 223–35. New York: Prentice-Hall, Inc., 1942.

TANNEHILL, I. R. *Preparation and Use of Weather Maps at Sea*. U.S. Weather Bureau Circular R. Washington, D.C.: Government Printing Office, 1949.

UNITED STATES NAVY HYDROGRAPHIC OFFICE. *Wind Waves at Sea, Breakers and Surf*. H.O. Pub. No. 602. Washington, D.C., 1947.

———. *Techniques for Forecasting Wind Waves and Swell*. H.O. Pub. No. 604. Washington, D.C., 1951.

CHAPTER 10

Kinematic Analysis

10.01. *Introduction.*—Owing to the relation between wind and pressure, conventional methods of weather analysis hinge largely on use of pressure. But pressure analysis is for most purposes not an end in itself; it is merely an *expedient* for obtaining the large-scale pattern of motion. Moreover, our use of the pressure field in describing or explaining weather is mostly association —sometimes good, sometimes poor, depending on areal dimensions and accuracy desired. In low latitudes particularly, these methods of analysis and forecasting break down because the pressure pattern is comparatively ill defined and wind is less geostrophic. It is necessary there to place greater emphasis on direct wind analysis.

Most methods of drawing the wind field have existed for some time, but they remained dormant during emphasis on pressure and frontal analysis. Application of hydrodynamic methods to analysis of atmospheric motion is seen in early works by Sandström[1] and V. Bjerknes,[2] the latter of which is a classic in meteorology and serves as the principal source for the first part of this chapter. Some of the techniques were applied to large-scale analysis by Werenskiold.[3] Theoretical and practical aspects of atmospheric hydrodynamics were later discussed more fully in *Physikalische Hydrodynamik* by V. Bjerknes, J. Bjerknes, H. Solberg, and T. Bergeron.[4] Petterssen later elaborated on the frontogenetic properties of the wind field.[5] By necessity, motions in the tropics have been studied almost exclusively by direct wind analysis, especially during and since the last war.[6] Still more recently, analysis of the upper wind field has been investigated by Gustafson, partly through whose efforts some of the methods are applied on a daily basis by analysis centers.[7]

Air motion is a function of space and time and can be analyzed by means mostly simi-

4. Berlin: Julius Springer, 1933.

5. Sverre Petterssen, *Contribution to the Theory of Frontogenesis* ("Geofysiske Publikationer," Vol. XI, No. 6 [Oslo, 1935]); see also his *Weather Analysis and Forecasting* (New York: McGraw-Hill Book Co., 1940).

6. [E. M. Kindle], *An Application of Kinematic Analysis to Tropical Meteorology* (Air Weather Service Technical Report 105-51 [Washington, D.C., May, 1945]); Civilian Staff, Institute of Tropical Meteorology, Rio Piedras, Puerto Rico, "Tropical Synoptic Meteorology," in F. A. Berry, E. Bollay, and N. R. Beers (eds.), *Handbook of Meteorology* (New York: McGraw-Hill Book Co., 1945), pp. 763–803; H. Riehl, "On the Formation of Typhoons," *Journal of Meteorology*, Vol. V (1948); C. E. Palmer, *The Tropical Pacific Project, 14th Report* (Los Angeles: University of California, October, 1951), and "Tropical Meteorology," in American Meteorological Society, *Compendium of Meteorology* (Boston, Mass., 1951), pp. 859–80.

7. A. F. Gustafson, *The Upper-Level Winds Project, Final Report* (Los Angeles: University of California, June, 1949); see also "High-Level Isotach Analysis," in *Air Weather Service Manual 105-26* (Washington, D.C.: Headquarters, Air Weather Service, 1951).

1. J. W. Sandström, "Über die Bewegung der Flussigkeiten," *Annalen der Hydrographie*, Vol. XXXVII (1909).

2. V. Bjerknes, Th. Hesselberg, and O. Devik, *Dynamic Meteorology and Hydrography*, Part II: *Kinematics* (Washington, D.C.: Carnegie Institution, 1911).

3. W. Werenskiold, *Mean Monthly Air Transport over the North Pacific Ocean* ("Geofysiske Publikationer," Vol. II, No. 9 [Oslo, 1922]).

lar to the physical variables. The vertical component of motion is not measured; it has to be inferred either from horizontal motions or indirectly by reference to weather patterns or certain changes in atmospheric structure. The wind field is analyzed by drawing *streamlines* and *isotachs*, which give, respectively, the course of flow and its speed. An alternate method for analyzing wind directions is by isogon curves or *isogons*, that is, isopleths of wind direction. Those topics occupy the first parts of this chapter. Then considerable space will be devoted to the horizontal derivatives of motion, which consist of shear and dilatation (stretching) and give by various combinations the properties *vorticity* (or rotation), *divergence*, and *deformation* in the velocity field.

That knowledge of the nature of the particular variable is necessary in the process of analysis becomes evident at once in dealing with the wind. Near the surface, motion is modified by friction and by physiographic peculiarities. Local circulations due to convection or orography may extend to upper levels and cause discrepancies between local winds and the flow pattern of the surroundings. Also, knowledge of observation methods is a prerequisite.

To obtain a complete analysis of wind at high levels, the data often must be supplemented by reference to the pressure pattern and by necessary emphasis on time and space continuity, since on the whole the amount of wind data in the upper troposphere is still insufficient to bring out the desired amount of detail in the motion. In this regard, one might be aided by the geostrophic equation and by approximating accelerations.

10.02. *Analysis of wind speed.*—The wind can be treated as a continuous function of horizontal space everywhere except at sharp fronts, certain *shear lines*, and, of

course, also at the ground. Therefore, excluding these, isotachs can be drawn from plotted wind speeds by the principles of scalar analysis. Where discontinuities exist, analysis to each side is carried out

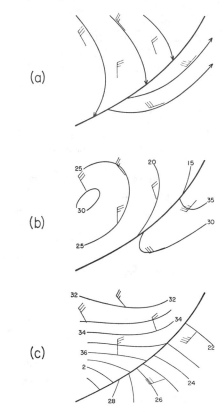

Fig. 10.021.—Schematic illustration of (*a*) streamline, (*b*) isotach, and (*c*) isogon patterns at an ideal frontal discontinuity in the wind field.

separately, and isotachs end (or start) at the discontinuity (Fig. 10.021). Singular isotachs for a value greater or less than the wind speed on both sides should not be drawn. However, singular points are found at centers of maximum wind speed and at calms. On the surface chart the number of singular points varies diurnally over land, and it may be necessary to smooth out some of them. As in other forms of analysis, it is useful in drawing isotachs first to locate tentatively the singular lines (wind-shift lines), singular points, and axes of isotach systems. These determine the general shape of the entire pattern.

Figure 10.022 shows a simplified isotach (and streamline) analysis at 500 mb based on actual data and other considerations. In Figure 10.031(*b*) are shown iostachs (in knots) at 4000 feet. This level was selected high enough to minimize the surface influence and low enough to benefit by a dense

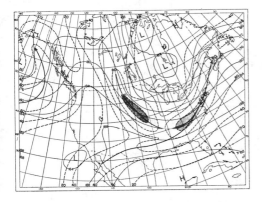

Fig. 10.022.—Isotachs in knots and pressure contours at 500 mb, 1500 GCT, 1 March 1950.

network of winds. Isotachs were not drawn near the front or in the warm air, because reports were insufficient in that part of the pattern. Charts (*e*) and (*f*) give isotachs of west wind and south wind, respectively. The data were obtained graphically from wind reports and supplemented by computations from the isotach-streamline pattern of (*b*). Such analyses of the velocity components facilitate work with horizontal derivatives.

10.03. *Drawing streamlines.*—The field of horizontal flow is entirely represented by two sets of curves, one indicating flow intensity or wind speed, described by isotachs, and the other set showing direction or course of motion. The latter may be analyzed by drawing isogons of wind direc-

tion, but the more direct method is by streamlines, or "lines of flow." *Streamlines are a series of curves tangent to the wind throughout an instantaneous flow pattern.*

A streamline analysis of Figure 10.031(*a*) is shown in (*b*). In Figure 10.032(*a*) is the streamline pattern for the surface chart, and the next chart shows the same streamline pattern and the sea-level pressure pattern. Notice, incidentally, that there are wide departures between flow at the surface and the sea-level pressure pattern, even over water and relatively smooth terrain. From either of the streamline analyses the pattern of motion can be interpreted immediately and indicates certain peculiarities not easily discernible from pressure patterns. Streamlines in the free air (Fig. 10.031[*b*]) are more simple and in better agreement with pressure than at the ground. On closer examination of the surface chart, one sees it is about impossible to account for all reports, especially the lighter winds. To a degree dependent on the purpose in mind, it is necessary to select and make maximum use of consistent reports from knowledge of local peculiarities at each station. At the surface there are usually enough reports over the continent to permit some selection, but usually not aloft.

Wind directions plotted on the map as shafts or vectors are a set of line segments giving the slopes of streamlines in the horizontal plane. Drawing streamlines is thus merely a process of integrating by sight the curves whose slopes are reported in the winds. The number of streamlines to be drawn is a matter of choice.

In Figure 10.032(*a*) it is seen that, in looking downstream, some streamlines branch and others join. This is permissible for practical convenience if, in branching or joining, the streamlines are drawn so that at the point of contact they have the same direction. A finite wind must have a single direction, and therefore outside a calm the

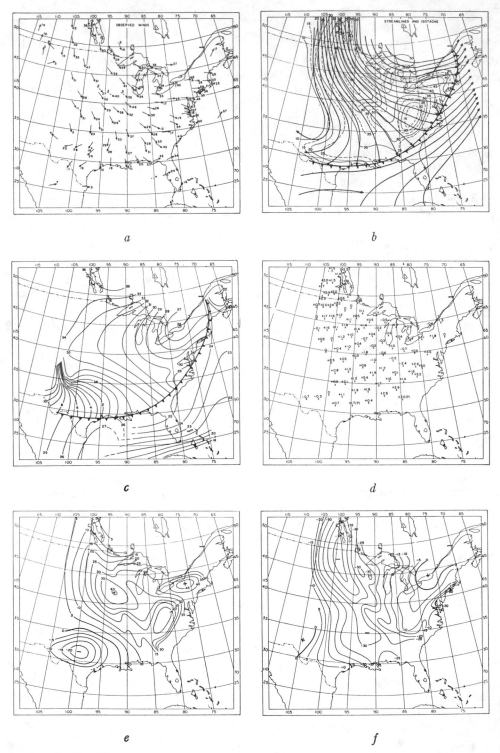

FIG. 10.031.—Wind analyses for 4000 feet, 1500 GCT, 1 March 1950. (a) Winds in knots; (b) isotachs and streamlines; (c) isogons; (d) horizontal divergence measured from b in units 10^{-5} sec^{-1}; (e) west wind; (f) south wind; (g) $\partial u/\partial x$; (h) $\partial v/\partial y$; (i) $\partial u/\partial x + \partial v/\partial y$; (j) $\partial v/\partial x$; (k) $-\partial u/\partial y$; and (l) $\partial v/\partial x - \partial u/\partial y$. In charts g through l units are knots per degree latitude; to convert to 10^{-5} sec^{-1}, multiply indicated values by 0.46.

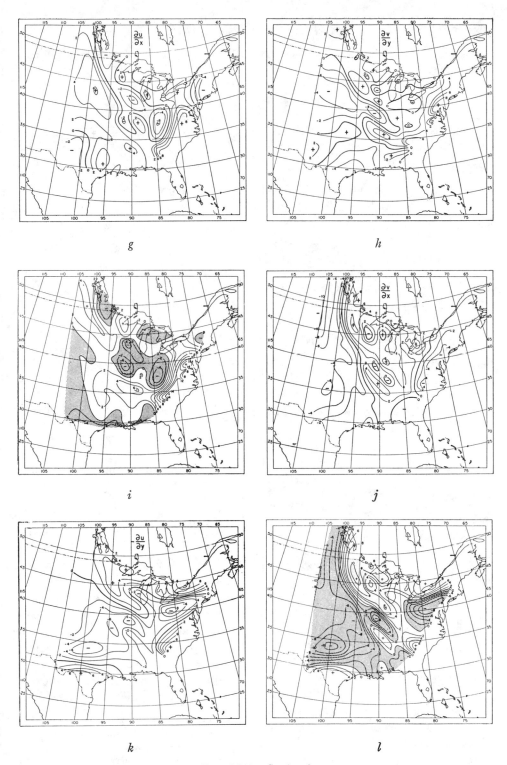

$$g \qquad h$$

$$i \qquad j$$

$$k \qquad l$$

FIG. 10.031.—*Continued*

307

streamlines cannot intersect or meet at finite angles.

Streamlines may begin or terminate at lines along which the wind is discontinuous. On the surface chart the illustrated cold front is accompanied by sharp wind shift, and thus the streamlines in the cold air were terminated at the front. There is in-

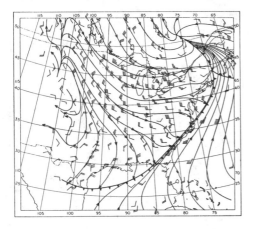

Fig. 10.032a.—Surface streamlines at 1830 GCT, 1 March 1950; winds in Beaufort scale.

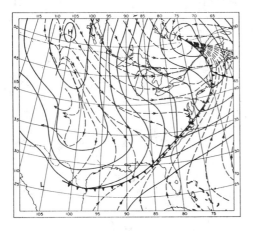

Fig. 10.032b.—Streamlines of chart *a* and sea-level isobars.

sufficient evidence that the warm and oc-cluded fronts in the chart are attended by such distinct shift, and the streamlines were drawn continuously with only maximum

curvature there. It is logical to end or begin horizontal streamlines also where they reach the earth.

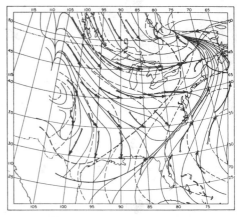

Fig. 10.032c.—Various air trajectories during the 12-hour period from 0630 (*dashed streamlines*) to 1830 GCT (*continuous streamlines*), 1 March 1950.

10.04. *Singular lines and points in the streamline pattern.*—Not only may stream-lines fork or join but they may also spread outward from, or converge into, a singular streamline asymptotically in infinite num-bers. The singular streamline is a line of divergence or a line of convergence of streamlines.[8] An example is found in Figure 10.032(*a*). In general, these singular lines are located within a relative minimum of wind speed.

A calm is a singular point for both speed and direction. It can be void of any stream-lines, or it can be the intersection or focus of several streamlines, but it cannot be repre-sented by any single one (Fig. 10.04). The most simple singular point in streamlines is the *center* of a cyclonic or anticyclonic circu-lation, about which streamlines are closed. Toward the center, wind speeds diminish to

8. It is well to remember that a convergence line (or zone) in streamlines is not necessarily a line of maximum horizontal velocity convergence, and conversely; similarly for divergence.

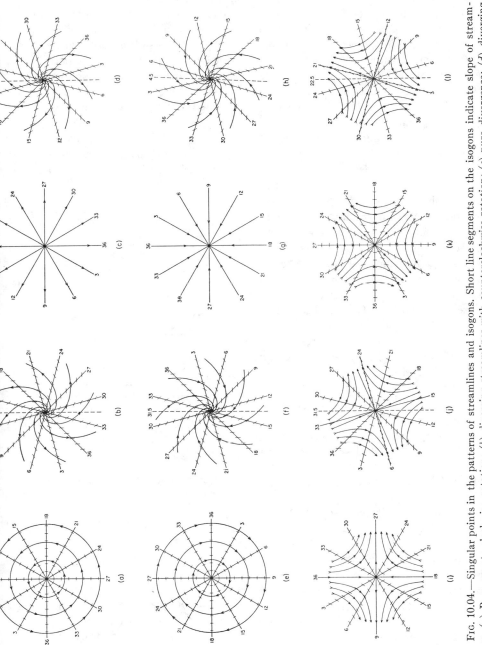

FIG. 10.04.—Singular points in the patterns of streamlines and isogons. Short line segments on the isogons indicate slope of streamlines. (a) Pure counterclockwise rotation; (b) diverging streamlines with counterclockwise rotation; (c) pure divergence; (d) diverging streamlines with clockwise rotation; (e) pure clockwise rotation; (f) converging streamlines with clockwise rotation; (g) pure convergence; (h) converging streamlines with counterclockwise rotation; (i, j, k, l) hyperbolic singular points.

zero, and streamlines reduce from closed curves to a point.

The other singular points also represent calms, but instead of no streamlines there are two or more streamlines meeting at the point. One type is the so-called *neutral point*, illustrated by the bottom row in Figure 10.04, and corresponds to the col of the pressure pattern. The simple neutral point is shown as a point of divergence of streamlines along one axis and of convergence of streamlines along another, the net effect being no wind in the center.

A third general type of singular point in the streamlines is the *point of divergence* (or convergence). The flow radiates from or into this point, usually with rotation superimposed so that streamlines appear as spirals. In all cases the wind is zero at the center and increases outward. Although not always justified, we might associate (*d*) in Figure 10.04 with HIGHs and (*h*) with LOWs in the northern hemisphere, and (*b*) and (*f*) with HIGHs and LOWs, respectively, in the southern hemisphere. A cyclonic spiral can be seen in Figure 10.032*a*.

In drawing the streamlines, it will be found helpful first to identify and examine the possible singular points, singular lines, and discontinuities, even though there is difficulty in finding *exact* locations. It is also advisable to concentrate first on the central regions of data and proceed outward.

10.05. *Drawing isogon curves.*—Analysis of wind direction by isogon curves is a possible aid in drawing streamlines. Isogons are drawn by scalar analysis of wind direction α. In the *meteorological system* of directions winds are reported to 36 points of the compass (tens of degrees) beginning from north and increasing in *clockwise* sense. An east wind has direction 9, south wind 18, west wind 27, and north wind 36. Although this order of directions is contrary to the positive sense of rotation, we will be consistent with the data and base all directions on that system.

The isogon analysis of (*a*) in Figure 10.031 is shown in (*c*). At the front the wind was assumed to shift abruptly in direction. In the northwestern part of the pattern isogons are widely separated—almost straight air flow. Where isogons are closely spaced, the flow is greatly curved, and the sense of curvature is given by the order of isogons in the direction of flow. If we designate radius of curvature R_s, positive counterclockwise, then from Figure 10.05 $R_s = -\delta s/\delta \alpha$, if δa

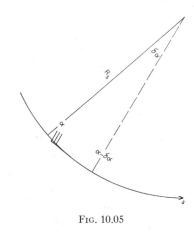

FIG. 10.05

is the change of wind direction in radians over distance δs downstream. If streamline curvature is K_s, then

$$K_s = 1/R_s = -\partial \alpha/\partial s, \quad (\alpha \text{ radians}), \quad (1)$$

or

$$K_s = 1/R_s = -(\pi/180)(\partial \alpha/\partial s), \ (\alpha \text{ degrees}).$$

If curvature is counterclockwise ($K_s > 0$), wind direction decreases downstream, and wind *backs* downstream.[9] For the converse, wind *veers* downstream. It is evident that perfectly straight flow is attended either by absence of isogonal curves or by parallelism of isogons with streamlines, the first case also

_____

9. This terminology is appropriate for the northern hemisphere.

indicating that streamlines are parallel and the second that streamlines are divergent or convergent.

a) Singular points in isogons.—All isogons meet at a singular point in the streamline pattern (Fig. 10.04). Isogons in (*a*) give a set of streamlines for a simple positive circulation. If this system of isogons is rotated counterclockwise 45°, the resulting pattern of streamlines is the set of divergent counterclockwise spirals in (*b*). Further turning of the isogon system 45° gives the third pattern of the sequence, and through successive steps the remainder of the first eight patterns is obtained. All these patterns have isogons increasing in clockwise sense about the singular point.

To show how unstable and infrequent some of the eight patterns really are, examine the consequence of rotating the systems of isogons in both directions through infinitesimal angles. Patterns (*a*), (*c*), (*e*), and (*g*) are merely critical cases of the spiral or "whirl" streamline patterns. It is thus expected that *spiral streamlines are the predominant type of pattern about singular points with clockwise order of isogons;* the others seldom are found in nature.

Drawings (*i*) through (*l*) have isogons increasing in counterclockwise sense about the origin. These are neutral points (cols) in the streamline pattern. Rotation of the isogon system through any angle, clockwise or counterclockwise, does not distort the pattern. This implies that the neutral point is stable and is therefore a common type of singularity. (Note that, in the process of rotating the isogons through a finite displacement about the neutral point, the streamline pattern is rotated by precisely half that amount.)

In Figure 10.04 the isogons are shown as straight radii spaced equally about the origin. Actual cases are not so simple; isogons are usually curved and irregularly spaced about the singular point. Spirals are

therefore seldom logarithmic, neutral points seldom rectangular hyperbolae, and closed streamlines seldom concentric circles.

b) Additional properties of isogon patterns.—The singular points described by intersections of isogons are singular points both for streamlines and for isogons. There are additional singularities in the isogon pattern not revealed as such in the streamlines. These are centers of maximum or minimum wind direction and lie within closed isogon curves (eastern Great Lakes in Fig. 10.031*c*), and they indicate inflection in curvature of that streamline with amplitude greater or less than adjacent streamlines. A center of maximum direction attends inflection from clockwise to counterclockwise curvature downstream, and a center of minimum direction attends the opposite inflection.

c) Use of isogon patterns in streamline analysis.—Drawing streamlines is an operation quite different from ordinary scalar analysis, where one locates in the data a finite number of curves with predetermined values. Theoretically, these few scalar lines have definite locations on the map, and the task is one of finding them. In streamline analysis there is an infinite number of lines to choose from, and, in drawing any one of them, the analyst is integrating the equation of that curve from indications of its slope at scattered points. For most purposes the number of wind-direction reports is usually sufficient for a first approximation to the true streamline pattern, and these reports can be supplemented by indications from analyses of pressure. Where greater accuracy and detail are required, it is often advisable to draw first the pattern of isogons, which can be done with some accuracy, and then use the isogons to derive the streamlines.

Once isogons have been drawn on the chart, the next step is to enter at convenient intervals on each isogon a series of short segments whose direction with respect to the earth is the same as the value of that isogon,[10] as indicated in Figure 10.04. These segments serve as additional wind-direction data, greatly augmenting the observations. With a dense network of segments and observations, the task of integrating the streamlines is simplified, and a more reliable streamline pattern is thus possible.

10.06. *Horizontal air trajectories determined from streamlines.*—The horizontal streamline analysis is a static picture of air movement. It is drawn for instantaneous direction of motion for all particles in the field. In contrast, a *trajectory* or path is a curve tracing successive locations of an *individual air parcel*. Since a streamline represents air motion in space for a fixed time, while a trajectory is a function of air motion in space and time, the two are not synonymous; in most cases air trajectories cannot be inferred reliably from a streamline pattern.

Estimation of air trajectories, both past and future, is essential in countless ways to weather analysis and forecasting. The trajectory taken by an air parcel or an entire air mass determines its modifications in heat, moisture, and impurities, especially in the layer at the surface. In diagnosing present weather, the question of recent history of the air as well as differing histories is always important. In forecasting, a first consideration for determining future air properties over an area is the origin and path of the air mass. (Also, the radius of curvature of the air trajectory is required to determine gradient wind from geostrophic wind.)

The task of computing and estimating horizontal air trajectories with desired precision is a difficult one, and only a brief survey of the problem can be afforded here.

10. This operation was first suggested by Sandström (V. Bjerknes *et al.*, *op. cit.*).

First, we might examine the basic relation between streamlines and trajectories, since this principle underlies all techniques for constructing trajectories either mentally or graphically from synoptic information.

As streamlines are always tangent to the motion, the streamline pattern must give the instantaneous path for all particles in the field. If the air moves horizontally, the instantaneous path of any parcel also will be given by the streamlines on successively later charts. Hence, if air motion is known at sufficiently small time intervals, the trajectory of a given parcel of air can be found by integrating the curve always tangent to a streamline and whose arc displacement is consistent with wind speeds. With only this simple approach, the 12-hour surface trajectories shown in Figure 10.032c were obtained from three streamline charts 6 hours apart.

By superposition of one chart on the other, it is possible to trace the locus of each parcel, using a variable speed determined from the two networks of winds. The trajectory starts tangent to the streamline on the earlier chart, its direction at any later time is determined by interpolation between the two given streamline patterns, and, finally, at the end of the period the trajectory is tangent to a streamline for that time. Where streamlines are steady, the trajectory follows a streamline, and the task is merely one of displacing the parcel a required distance. In regions of rapid local turning of wind it is difficult to estimate trajectory curvature, but in such areas the wind is frequently light and thus minimizes the error of position.

Where wind data are numerous, it is unnecessary to draw the entire pattern of streamlines (and isotachs) to determine desired parcel trajectories. After a little practice one finds it sufficient to sketch lightly just a few streamlines in the neighborhood of the parcel to serve as guides in drawing the trajectory. The amount of preparatory work should vary with local change in the flow pattern and with the density of wind data in space and time.

A more objective technique for computing trajectories from a series of streamline charts involves first approximation to the acceleration in the interval between charts. Assume that we desire to determine the path taken by parcel A in Figure 10.061. The streamline through A on the first chart is AB. Displacement of A through the interval δt between charts with its present speed takes it to B. On the subsequent chart the streamline through location A is the dashed line. Displacement of A through the same time interval with the speed on the second chart takes it to C. Finally, the midpoint between B and C (point D) is assumed to approximate the position of the parcel on the second chart.

This approach may be modified by displacing the parcel in steps for intervals $\frac{1}{2}\delta t$, that is, 3-hour displacements with charts 6 hours apart. On the first chart, parcel A is displaced along the streamline from A to E, which is midway between A and B in Figure 10.061. Point E is then transferred to the next chart, and another computation locates a point corresponding to D. On this same chart the parcel is displaced through another interval $\delta t/2$; the resulting point is transposed to a third chart, and the process is carried on. Once all points are located, the trajectory can be drawn by fitting a smooth curve to the irregular course so obtained. The accuracy of this method and others depends on the uniformity of the patterns of streamlines and isotachs. The reliability of an entire trajectory so computed diminishes with time from the initial point.

To develop ability for deducing horizontal air trajectories visually from a series of wind or pressure charts, the relation between streamlines and trajectories is ex-

amined further. Consider that wind direction a, which is also the trajectory direction at a given time, is a property of the moving air parcel and a function of time t and horizontal downstream distance s. Thus,

$$da/dt = \partial a/\partial t + (\partial a/\partial s)(ds/dt)$$
$$= \partial a/\partial t + c(\partial a/\partial s) . \quad (1)$$

Since a is direction, da/dt is change in direction *following the air parcel* or trajectory. In Eq (1) $\partial a/\partial t$ is the *local turning* of wind, c is wind speed (speed of parcel), and $\partial a/\partial s$ is variation of direction along the streamlines (curvature of streamlines).

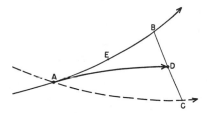

FIG. 10.061.—Graphical computation of a trajectory by first approximation from initial and final wind patterns.

The individual derivative da/dt also may be written $(da/ds)(ds/dt)$, and Eq (1) becomes $\partial a/\partial t = c(da/ds - \partial a/\partial s)$. From Eq 10.05(1), $-\partial a/\partial s = K_s$. The curvature of the trajectory is evidently $K_t = -da/ds$. By substitution into the last equation, we get Blaton's formula:[11]

$$\partial a/\partial t = c(K_s - K_t)$$
$$= c(1/R_s - 1/R_t) . \quad (2)$$

This shows that in steady-state motion curvatures of streamlines and trajectories are identical. Also, where winds at fixed points turn rapidly with time, the departure between streamline and trajectory curvature can be great; the departure is directly proportional to $\partial a/\partial t$ and in-

11. J. Blaton, "Zur Kinematik nichtstationärer Luftströmungen," *Bull. Soc. Geophys. Varsovie* (Warsaw) (1938).

versely proportional to wind speed. Where wind is veering locally, $K_t < K_s$; the counterclockwise curvature of the trajectory is less than that of the streamline. The converse holds if winds are backing locally. For large values of $\partial a/\partial t$ it is easily possible for K_s and K_t to be of opposite sign.

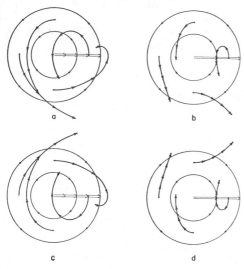

FIG. 10.062.—Future trajectories of certain particles with respect to the initial streamline pattern. The circulation systems are moving slower than the winds in (*a*) and (*c*) and faster than the winds in (*b*) and (*d*).

It is useful to investigate the trajectories about moving cyclonic and anticyclonic wind systems (Figs. 10.062 and 10.063). If the angle between the wind and the motion of the system is ψ, and the speed of the system is U_x, then[12]

$$K_t = K_s[1 - (U_x/c) \cos \psi]. \qquad (3)$$

Here U_x/c is always positive, and $\cos \psi$ varies between 1 and -1. In the illustrations the streamlines are the concentric circles, trajectories are the heavy curves, and the wind system moves from left to right. The arrowheads are successive positions of air parcels and the center of the

12. For derivation of this formula refer to Petterssen, *Weather Analysis and Forecasting.*

wind system. All trajectories were drawn assuming constant wind speed throughout each system.

Diagram (*a*) shows a cyclone (northern hemisphere) moving slower than the winds. Eq (3) shows that all trajectories in this case have counterclockwise curvature. Where streamlines are perpendicular to the displacement of the system, $K_t = K_s$. Maximum curvature in the trajectories occurs directly to the left of the center, and minimum counterclockwise curvature in trajectories occurs where winds have the same direction as the displacement of the center. If we now examine the corresponding slow-moving clockwise circulation in (*c*), we find again that, since $1 > U_x/c > 0$, the trajectories are curved in the sense of the

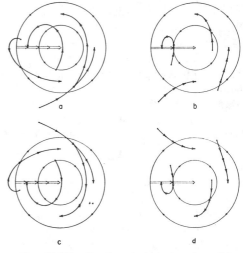

FIG. 10.063.—Previous trajectories of particles with respect to the same patterns shown in Figure 10.062.

streamlines. Along the path of the center $K_t = K_s$. To the right of the system trajectories have greater clockwise curvature than streamlines.

Now imagine a wind system moving with the speed of the winds. Along the path of the center $K_t = K_s$, but at points directly to the left of the counterclockwise center or directly to the right of the clockwise center $K_t = 2K_s$. Where the wind is the same direction as the motion of the system, $K_t = 0$.

If $U_x/c > 1$, for either type of circulation, there is a sector in the neighborhood of $\psi = 0$ in which trajectories and streamlines are of opposite curvature. The angular width of that sector as well as the departure of K_t from K_s varies directly with U_x/c. Here alone is sufficient evidence that streamlines on a given synoptic chart should not be interpreted generally as actual air trajectories, particularly at the surface.

The trajectories shown were based on assumptions of uniform wind speed, concentric circular streamlines, and no acceleration of the system. They may be adjusted qualitatively for any departure from those assumptions. Furthermore, the trajectories are applicable to any curved flow patterns; for example, in troughs and ridges.

10.07. *Horizontal trajectories from pressure charts.*—Since at present streamline charts are not drawn on a daily basis, it is desirable to compute or visualize trajectories directly from series of pressure charts. Furthermore, in many regions and at many levels where knowledge of trajectories is valuable, wind data are insufficient to give a unique wind analysis without major reference to the pressure field. In the free atmosphere the problem is not too difficult, since isobars (contours) ordinarily serve the purpose of streamlines in middle and high latitudes. Wind speed can be determined from reports or approximated from geostrophic or gradient wind. Where there is evidence that winds depart from the pressure pattern significantly, this might be taken into account.

The problem is considerably more difficult near the surface. In absence of sufficient empirical evidence, it is ordinarily advisable to assume a certain frictional departure of the wind from the pressure pattern. One might assume, for example, $\frac{3}{4}$ the geostrophic wind speed and 30° angle with sea-level isobars reasonable for use over land and less departure over oceans. The procedure illustrated by Figure 10.061 may suffice in determining the trajectories from successive sea-level pressure charts. That pressure pattern should be used for this purpose only over rather smooth terrain and where it is likely to be consistent with gradient-level flow.

10.08. *Three-dimensional trajectories on conservative surfaces.*—Displacements or trajectories computed on level charts in the free atmosphere are not parcel trajectories in the true sense unless motions are strictly horizontal. Vertical motions and vertical wind shear may make the actual parcel trajectory depart widely from that computed only from winds at one level. In barotropic regions, even where vertical motions are pronounced, the displacement computed at any level is the horizontal projection of the three-dimensional displacement, but it gives nothing concerning the trajectory's vertical component. On the other hand, in strongly baroclinic regions displacements computed on horizontal charts are misleading in terms of the horizontal projection of a trajectory if vertical motions exist.

It is necessary to consider three-dimensional trajectories. Excluding possible use of tracers to determine only the *past* trajectory of an air parcel, one solution to the problem is the use of wind analysis on a conservative surface. By computing displacements from the given wind fields, it is possible to obtain not just the horizontal (geographic) projection of the trajectory but the vertical displacement as well. From wind analyses on isentropic charts trajectories can be determined by the procedure for level surfaces. If wind reports are insufficient, geostrophic or gradient wind may be used. In any case, the resulting trajectory gives the geographic displacement of an air parcel. To determine its vertical displacement in pressure, isentropic pressures at the initial and final positions are compared.

An indirect method for obtaining three-dimensional trajectories is theoretically possible, without reference to fields of motion, by using three conservative air properties. The first defines substantial surfaces dividing the atmosphere into layers of air always bounded by the same surfaces. By choosing the interval between surfaces small, the layers may be treated as thin sheets or individual surfaces. Surfaces of another property intersect the first to form substantial tubes, strips, or lines. A third set of surfaces intersects to define parcels of air as cuboids, thin quadrilaterals, or points. If all three properties are strictly conservative, an air parcel is located by its same numerical values of each property from one chart to the next in time, and its trajectory can be drawn without knowledge of motion. In most conditions potential temperature can serve as the first conservative property, mixing ratio as the second, and for the third Starr and Neiburger[13] suggested possible use of *potential vorticity*. This property, however, is difficult to measure with required accuracy and gives no immediate promise as the practical answer.

10.09. *Horizontal derivatives of the velocity field.*—The first-order horizontal derivatives of motion are expressed in Cartesian components by $\partial u/\partial x$, $\partial u/\partial y$, $\partial v/\partial x$, and $\partial v/\partial y$; each is illustrated schematically in Figure 10.09a. These are scalar quantities with dimensions *time*$^{-1}$. The basic unit in *c.g.s.* is *sec*$^{-1}$, but in some instances they are expressed in unit *hr*$^{-1}$. In the lower atmosphere dilatation and shear of wind are

13. V. P. Starr and M. Neiburger, "Potential Vorticity as a Conservative Property," *Journal of Marine Research*, Vol. III, No. 3 (1940).

ordinarily[14] 10^{-5} to 10^{-4} sec^{-1}. These values vary somewhat with wind speed; large values of the derivatives are found more often with strong winds, particularly in shear. Also, the possible magnitudes vary inversely with the scale. In small-scale phenomena, such as thunderstorms or eddies due to surface obstacles, the wind may vary horizontally by 10^{-3} sec^{-1} or more. Over distances corresponding to the dimensions of the primary waves in the upper westerlies, for example, these derivatives may not be any larger than 10^{-7} to 10^{-6} sec^{-1}.

10.10. *Linear properties of motion.*—The distribution of horizontal velocity around a fixed point in space may be expressed by Taylor expansions of the x and y components of velocity about the point, which serves as the origin of the coordinate system. At any point in the neighborhood of the origin

$$u = u_0 + (\partial u/\partial x)_0 x + (\partial u/\partial y)_0 y + \ldots ,$$
$$v = v_0 + (\partial v/\partial x)_0 x + (\partial v/\partial y)_0 y + \ldots .$$

The subscripts "0" denote properties of motion at the origin. We are justified in neglecting higher-order derivatives if we consider only the (*linear*) distribution of velocity *in the vicinity of the origin* of the system, that is, at the point in question. The fact that fields of motion are actually nonlinear does not invalidate use of these expansions unless the distances x and y are large.

The four space derivatives of velocity in the xy-plane may be combined as sums or differences into the following possible expressions involving both x and y:

$$\partial u/\partial x + \partial v/\partial y = 2b = \text{div } \mathbb{C} ,$$
$$\partial v/\partial x - \partial u/\partial y = 2c = \text{rot } \mathbb{C} ,$$
$$\partial u/\partial x - \partial v/\partial y = 2a = \text{def } \mathbb{C} ,$$
$$\partial v/\partial x + \partial u/\partial y = 2a' = \text{def}' \mathbb{C} .$$

The quantity $2b$ defines horizontal *diver-*

14. I.e., in common "synoptic" scale.

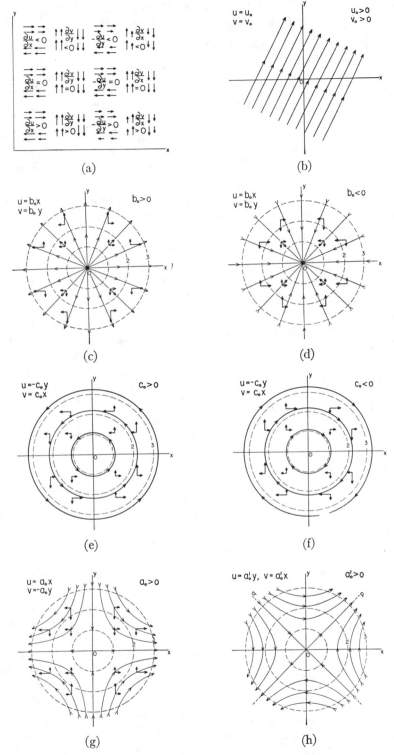

FIG. 10.09.—Elementary properties of motion in the vicinity of a point in the wind field; (b) pure translation; (c) divergence; (d) convergence; (e) positive vorticity; (f) negative vorticity; (g) deformation (stretching); and (h) deformation (shearing).

gence, 2c *vorticity* (more explicitly, the vertical component of relative vorticity), and 2a and 2a' horizontal *deformation*. From the above one finds $\partial u/\partial x = b + a$, $\partial u/\partial y = a' - c$, $\partial v/\partial x = a' + c$, and $\partial v/\partial y = b - a$. Upon substitution into the Taylor expansions,

$$u = u_0 + b_0 x - c_0 y + a_0 x + a_0' y + \ldots,$$

$$\tag{1}$$

$$v = v_0 + b_0 y + c_0 x - a_0 y + a_0' x + \ldots,$$

which shows that motion relative to the origin is a composite of velocity divergence, vorticity, and deformation at the origin. Using these equations, we investigate the character of horizontal motion in the vicinity of the origin by considering each basic property separately.[15]

(i) $u = u_0$, $v = v_0$.—In absence of net divergence, vorticity, and deformation, the velocity in the neighborhood is the velocity at the origin. Such a distribution of velocity, pure *translation*, is illustrated by Figure 10.09b. Streamlines are straight and have directions dependent only on the velocity at the origin. In case velocity is zero at the origin, there are no streamlines at all (i.e., no singular point of calm).

(ii) $u = b_0 x$, $v = b_0 y$.—With only the terms involving b_0 in the right of Eq (1) the motion is pure *divergence*. Streamlines are straight radii emanating from the origin in infinite numbers. Isotachs are concentric equally spaced circles centered at the origin and increasing outward. If $b_0 > 0$, streamlines diverge from the origin (Fig. 10.09c). This is positive velocity divergence and may be interpreted as horizontal spreading of the air centered at the origin. For $b_0 < 0$, streamlines converge into the origin. This is negative divergence, or simply *convergence*, and implies horizontal shrinking of the air.

15. We are maintaining the subscript "0" to indicate *constant* value of a property (i.e., strictly linear motion).

A feature of divergence is its complete independence of the orientation of coordinates. Another feature, though quite unrelated, is the apparent source or sink at the origin if the atmosphere is a continuous and virtually incompressible medium. This implies vertical shrinking or stretching of air.

(iii) $u = -c_0 y$, $v = c_0 x$.—For this type of motion streamlines are circles about the origin. When $c_0 > 0$, u has sign opposite to y and v the same as x, giving counterclockwise rotation (Fig. 10.09e). When $c_0 < 0$, there is clockwise rotation. In the northern hemisphere the positive rotation shown here is known as *cyclonic vorticity;* the opposite rotation, *anticyclonic vorticity*. It is seen by inspection that vorticity also is independent of the orientation of coordinates.

(iv) $u = a_0 x$, $v = -a_0 y$.—For motion of this kind streamlines are rectangular hyperbolae (Fig. 10.09g). When $a_0 > 0$, the sign of u is the same as x and the sign of v opposite to y. If $a_0 < 0$, the pattern of streamlines is the same, but the flow is reversed.

This hyperbolic pattern of motion is pure *stretching deformation*. A deformation field has two characteristic axes—one toward which streamlines converge asymptotically and the other from which streamlines diverge asymptotically. The axes themselves are streamlines meeting in a singular point at the origin. For $a_0 > 0$, x is the *axis of dilatation* and y the *axis of contraction*. If $a_0 < 0$, the definitions are reversed. In a pure deformation field the fluid parcel is stretched along the axis of dilatation and compressed by the same amount along the axis of contraction, so there is no net divergence, but there is change in shape of the parcel. Observe from the definition of 2a that deformation consists of the difference between the dilation components $\partial u/\partial x$ and $\partial v/\partial y$, while divergence is their sum.

A peculiar aspect of both types of deformation is dependence on the orientation of coordinate axes. Notice that, by rotating the coordinates in the diagram, the streamline pattern remains fixed but changes with respect to the rotating coordinates. This peculiarity introduces the principal difficulty in properly analyzing and interpreting deformation.

(v) $u = a_0'y$, $v = a_0'x$.—In this case streamlines are again rectangular hyperbolae, but for $a_0' > 0$ the axes of symmetry are 45° to the left of those in the deformation field $a_0 > 0$. Deformation $2a'$ will be called *shearing deformation;* it distorts fluid elements through difference in shear. Notice how it is related to vorticity in a manner similar to the relation of stretching deformation to divergence.

The two types of deformation can be combined into a *resultant* deformation (def$_r$ $\mathbb{C}$) by appropriate choice of coordinates. These adjusted coordinates, known as the *principal axes*, have to be found for each point at which deformation is determined.

In summary, to describe completely the motion about a point, the following information is required: (1) wind speed and direction; (2) scalar value of divergence; (3) scalar value of vorticity; and (4) scalar value of resultant deformation and direction of the principal axes. The importance of wind speed and direction is evident in terms of advection of air properties. In following sections we investigate each of the derivative kinematic properties. Each plays a vital role in the mechanics and dynamics of the atmosphere. Divergence is a dominant factor in vertical motion, clouds and precipitation, and pressure change. Vorticity is related to the field of force. Deformation is a primary factor in the mechanics of frontogenesis. However, it must be emphasized that the actual flow about any point is almost invariably a combination of two or more of these basic properties; it is usually a composite of all of them in varying propor-

tions. The presence of any one is difficult to detect unless the entire flow is decomposed into the independent patterns in Figure 10.09. One finds by experience that in many conditions divergence is larger in comparatively straight and parallel flow than in the vicinity of most singular divergence points in the streamlines. In the case of vorticity, there is evidence that magnitudes are larger near strong wind currents of relatively little curvature than in cyclonic and anticyclonic centers. Rates of deformation also are large with strong wind shear—even larger than found in the ordinary neutral points or cols.

10.11. *Divergence of wind.*—Horizontal divergence of velocity is defined by

$$\operatorname{div} \mathbb{C} \equiv (\partial u/\partial x + \partial v/\partial y) = \nabla_h \cdot \mathbb{C} . \quad (1)$$

Divergence in three dimensions is $\partial u/\partial x + \partial v/\partial y + \partial w/\partial z$, w being vertical velocity defined positive upward.

The physical significance of horizontal divergence may be seen by examining the variation with time of the two axes of an infinitesimal fluid cross (Fig. 10.11) whose

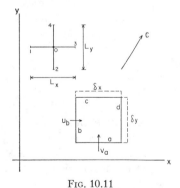

Fig. 10.11

center moves with the wind. Suppose at a given time the x and y arms have lengths δx and δy both equal to L. It is seen that the difference $u_3 - u_1 = \delta u$ gives the change in length per unit time of the x arm. Thus, $dL_x/dt = u_3 - u_1$. Dividing both sides by $L_x = \delta x$ gives $(1/L_x)(dL_x/dt) = \delta u/\delta x \approx$

$\partial u/\partial x$. Hence, $\partial u/\partial x$ is merely the percentage change in length per unit time of a fluid filament parallel to x. By considering v_4 and v_2, a similar relation is found for $\partial v/\partial y$ in terms of the stretching of L_y. Adding the two, we may write

$$[(1/L)(d/dt)](L_x + L_y) \simeq \partial u/\partial x + \partial v/\partial y \ .$$

This becomes an equality as L approaches zero. Horizontal divergence is the percentage rate of change of the sum of the horizontal dimensions of a fluid particle with time. Three-dimensional (volume) divergence includes a third dimension L_z.

To obtain another view of horizontal divergence, consider that the plane area of horizontal dimensions δz and δy, and area A, in Figure 10.11 is a fluid parcel in a region of linear wind distribution. The average wind normal to the left edge is u_b, and the average wind normal to the lower edge is v_a. Now, the wind normal to side d is $u_d = u_b + (\partial u/\partial x)\delta x$, and similarly $v_c = v_a + (\partial v/\partial y)\delta y$. Increase in the x dimension of the element per unit time is evidently $u_d - u_b = (\partial u/\partial x)\delta x$, and increase in the y dimension is $v_c - v_a = (\partial v/\partial y)\delta y$. The increase in area in unit time dt due to stretching along x is $(\partial u/\partial x)\delta x\delta y$, and along y is $(\partial v/\partial y)\delta x\delta y$. Therefore, areal expansion is $dA/dt = (\partial u/\partial x + \partial v/\partial y)\delta x\delta y$. Division by $A = \delta x\delta y$ gives

$$(1/A)(dA/dt) = \partial u/\partial x + \partial v/\partial y \ . \quad (2)$$

This shows that *horizontal divergence is percentage areal expansion per unit time* of the moving air.[16] Had we considered a volume element in Figure 10.11, divergence in three dimensions would be

$$(1/V)(dV/dt) = (\partial u/\partial x + \partial v/\partial y + \partial w/\partial z) \ , \quad (3)$$

16. Note that divergence was not defined in terms of initial and final areas of the particle, that is, the difference in the product of its horizontal dimensions. There is no discrepancy if areas are infinitesimal, but it can be significant if the area has dimensions corresponding to the usual distance between observing stations.

where V is the volume of the element considered. Space divergence is percentage volume expansion per unit time.

Convergence is negative divergence. When both components $\partial u/\partial x$ and $\partial v/\partial y$ are zero, or of equal magnitude but opposite sign, horizontal velocity is *nondivergent*. It is found empirically that $\partial u/\partial x$ and $\partial v/\partial y$ are *usually* of opposite sign;[17] that is, in most cases stretching along one axis is compensated by shrinking along the other, and there is attendant deformation. This implies for the average case that horizontal divergence is only the residual of two factors contributing oppositely, and divergence is of smaller magnitude than the derivatives themselves.

There is even more physical justification for suspecting that the change in volume of a fluid represents a small residual of opposite contributions by horizontal divergence and vertical stretching $\partial w/\partial z$. With horizontal divergence a parcel of air most likely shrinks vertically, and with horizontal convergence it stretches vertically. To show this more concretely, we differentiate the gas equation for adiabatic motions and substitute into Eq (3):

$$-(0.71/p)(dp/dt) = \text{div } \mathbb{C} + \partial w/\partial z \ . \quad (4)$$

Near the surface we might assume that motions are nearly horizontal (except over steeply sloping terrain) and that therefore dp/dt experienced by a moving air parcel is due to local pressure changes and to ageostrophic motion. To establish an approximate upper limit, assume the pressure change on the moving parcel is 5 mb hr^{-1}. For $p = 1000$ mb, this gives 10^{-6} sec^{-1} for the left side of Eq (4). It follows that the two terms on the right usually are of opposite sign, since horizontal divergence is itself greater than 10^{-6} sec^{-1} in that scale. As it is

17. Tabulations of like and unlike signs in the computations performed for Figure 10.031(g) and (h) showed this was true in about two-thirds of the cases.

only a fractional residual of contributions by horizontal and vertical divergence, the left side of Eq (4) may be neglected as a rough approximation for many purposes in explaining the weather. Thus

$$\partial w/\partial z \approx - \operatorname{div} \mathbb{C} = -(\partial u/\partial x$$
$$+ \partial v/\partial y) . \quad (5)$$

10.12. *Usual relations between horizontal and vertical divergence.*—On the basis of the last equation, we deduce that horizontal divergence is associated with vertical shrinking ($\partial w/\partial z < 0$) and horizontal convergence with vertical stretching ($\partial w/\partial z > 0$) of air columns. That equation states nothing explicitly about the relation between horizontal divergence and *direction* of vertical motion.

Eq 10.11(5) is illustrated by the several unrelated diagrams in Figure 10.12. Vertical velocities are indicated by direction and length of the vertical vectors. Horizontal divergence is a two-dimensional property but is implied by the horizontal vectors. Notice that in all cases of horizontal convergence there is vertical divergence, and conversely for horizontal divergence. In each diagram there is deformation in the vertical plane. In general, with either horizontal convergence or divergence the vertical velocities may be upward, downward, or in few cases zero.

The earth is a boundary to air motions, and, although tangential motions can vary, the component of motion normal to the surface must cease at the ground. Thus, with a level surface vertical motion at the ground is zero, and only with strong winds over steep terrain can the vertical velocity be appreciable. From Eq 10.11(5) we conclude that *horizontal convergence at the surface is attended by both upward motion and vertical stretching in air columns and horizontal divergence by downward motion and shrinking of the air columns.*

In the discussion of stability in Chapter 3,

it was seen that in a stable atmosphere vertical stretching is destabilizing and vertical shrinking stabilizing. Divergence in the flow is an important agent in determining vertical distribution of temperature and moisture. Near the ground, where the relation between divergence and vertical motion is more definite, patterns of horizontal divergence also are associated with certain

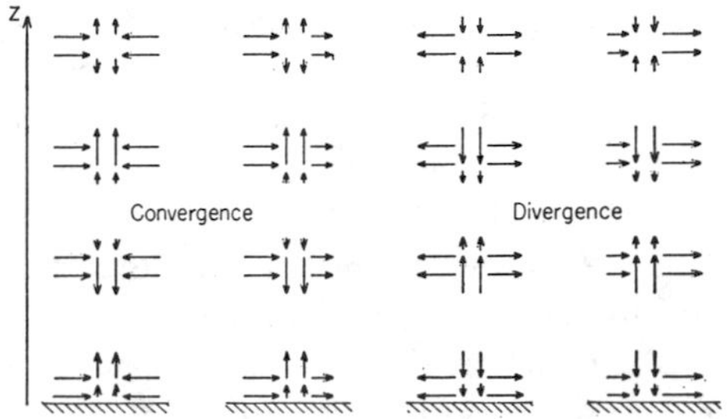

Fig. 10.12.   Cross-section schemes of horizontal divergence and convergence with attendant types of vertical stretching. The lowest row of drawings is for the layer next to the ground; other schemes do not necessarily have preferred locations in the vertical.

patterns of weather—convergence with upward transport of moisture, adiabatic cooling, condensation, clouds, and precipitation; divergence with general drying of the lower atmosphere. In the main the direction of vertical motion found just above the surface extends through most of the troposphere, and therefore the predominant sense of vertical motion in the troposphere in many cases can be inferred from the sign of divergence near the ground.

10.13. *Measurement of horizontal divergence by finite differences.*—Horizontal divergence may be approximated by $\delta u/\delta x + \delta v/\delta y$, which can be adapted to measurement of divergence from horizontal analyses of u and v. The method is based on the scheme of the fluid cross illustrated in Figure 10.11. With a convenient interval of distance for δx and δy (e.g., 1° or 2° of latitude), $\delta u/\delta x$ and $\delta v/\delta y$, and thus horizontal divergence, can be evaluated for any point or grid of points over the map.[18]

This operation was used in obtaining Figure 10.031*i*. From diagram (*e*), $\delta u/\delta x$ was evaluated at each 1° intersection of latitude and longitude, using distance interval 2° of latitude. Its pattern is shown in (*g*). A similar process for $\delta v/\delta y$ gave (*h*). The sum of the two is horizontal divergence (*i*), expressed in knots per degree of latitude. Multiplying the given values by approximately half gives divergence in 10^{-5} sec^{-1}.

The pattern in (*i*) is admittedly not a very accurate picture of the divergence field at that level, but it is as reliable as could be expected from the data and from known methods of analysis. The process of graphical differentiation can be done with all the required precision, but the results are most sensitive to the initial wind analysis. Small areas of the final pattern are doubtful, because all steps in analysis were based on what was considered questionable data in the first place (e.g., just south of Lake Superior and the edge of the analysis behind the cold front). Nevertheless, the analysis shown gives at least the general pattern of divergence and its approximate magnitudes.

On comparing Figures 10.031(*b*) and (*i*), we find no clear-cut relation between divergence of streamlines and velocity divergence. Throughout most of the chart streamlines spread outward downstream, yet superimposed are scattered centers of divergence

and convergence due to rapid variations in wind speed along the streamlines. Here is an excellent example of the precaution to take in deducing horizontal divergence from streamlines.

Since in this case streamlines spread and curve very uniformly over most of the chart, the centers of velocity divergence and convergence are closely related to downstream gradients of wind speed. Along a line from Iowa toward the Carolinas are two sharp decreases in speed separated by a region of more uniform wind in the vicinity of western Kentucky. Notice how they are reflected in the pattern of divergence. Over and just east of the Appalachians rapid downstream increase in speed gives large positive divergence. In a band near the east and south coasts there appears to be prevailing convergence attending the cold front, and slightly upstream is a broad band of divergence paralleling the front also.

Over the northern plains states, to the rear of the main wind maximum, the two contributing effects—downstream increase in speed (stretching) and diverging streamlines—are additive. Farther south over western Kansas and Texas there is predominant horizontal convergence. Notice also that the humidities aloft are higher over this region than to the east.

If we assume that this pattern of divergence is representative of the lower atmosphere, we might postulate from this the distribution of vertical motions, using Eq 10.11(5). With due consideration for other factors such as moisture content, evaporation from the surface, orography, and surface instability, good agreement is found between the divergence pattern and the distribution of clouds and weather in Figures 9.15*abc*.

18. If the coordinate system is identified with the grid lines on the earth, with *y* northward and *x* eastward, horizontal divergence is $\delta u/\delta x + \delta v/\delta y - (v/a)\tan\phi$, where *a* is the earth's radius, ϕ the latitude, and *v* the south-wind component. The correction is due to the convergence of meridians and amounts to $-(1.57 \times 10^{-9}$ cm$^{-1})v \tan\phi$. It is important only at high latitudes and for strong meridional component of wind. For instance, a north wind 10 m sec^{-1} at 45° latitude makes the correction only 1.57×10^{-6} sec^{-1}. It is permissible to neglect this term in most conditions except near the pole.

10.14. *Measurement of horizontal divergence from kinematic directions.*—Divergence is a point property of the fluid, although we usually extend the definition to include areas or volumes, and we are free to select a suitable orientation of coordinates for measuring divergence. By placing the origin of coordinates at the point in question with one axis along the streamline, we can measure directly from streamline and isotach patterns the stretching along the wind direction and the stretching along the normal to the wind direction, the sum of

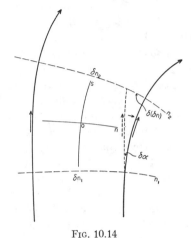

Fɪɢ. 10.14

which is horizontal divergence. Thus at least one step is saved in the operations, since analysis of u and v is not required. These new directions, oriented relative to the streamlines, are called natural or *kinematic* directions.

Figure 10.14 shows this system in relation to the streamlines. The origin is placed at the point for measurement; the s-axis ("streamline axis") is directed downstream, and the n-axis ("normal axis") is perpendicular and to the *right* of the wind. The ns system has here the same sense as xy. Evidently, $\partial u/\partial x = \partial u/\partial n$ and $\partial v/\partial y = \partial v/\partial s$ at the origin.

By placing y tangent to the streamline at o, it is apparent that $\partial v/\partial y = \partial c/\partial s$. This part of the divergence is the *stretching* term; it is evaluated from change of wind speed over finite distance along the streamline.

The stretching contribution to horizontal divergence is readily estimated by visual inspection of the streamline and isotach patterns.

The second part, $\partial u/\partial x$, is the *spreading* term, seen as variation in the normal wind component from left to right across the wind vector at o. Consider that the area bounded by the two streamlines and by the normals n_1 and n_2 is the greatly magnified version of an element of area centered at o. The streamlines are δn_1 apart at n_1, δn_2 apart at n_2, and δn apart across the origin, where the distance from n_1 to n_2 is δs. The difference in directions of the two streamlines is δa. The wind speed c is the speed at o. From this it can be shown that $\partial u/\partial x = \partial u/\partial n = c(\partial a/\partial n)$. Therefore,

$$\text{div } \mathbb{C} = \partial c/\partial s + c(\partial a/\partial n) . \qquad (1)$$

The second term is found by the product of the wind speed at the origin and the change of streamline direction (in radians) per unit distance normal to the streamlines. For angular measurements in degrees, the result for that term should be multiplied by $\pi/180 = 0.0175$ radian per degree. Measurement of $\partial a/\partial n$ is facilitated if the isogon analysis is already given. One then obtains the difference in isogon values at both ends of finite distance δn and multiplies by $\pi/180\ \delta n$.

In Eq (1) the sign of the spreading term depends only on the sign of $\partial a/\partial n$. By the manner we have defined a and n, there is contribution to *horizontal convergence if the streamline channel narrows downstream and divergence if the channel broadens downstream.* The stretching term gives *convergence if wind speed decreases downstream and divergence if wind speed increases downstream.* In comparison with the definition of divergence in Cartesian coordinates, this one is more convenient for visually estimating divergence from elementary analyses of the wind field.

From Figure 10.14 the angle $\delta\alpha$ can be expressed $\delta\alpha = \delta(\delta n)/\delta s$, where $\delta(\delta n)$ is the widening of the channel in distance δs downstream (i.e., $\delta[\delta n] = \delta n_2 - \delta n_1$). By redevelopment of Eq (1),

$$\text{div } \mathbb{C} = \frac{\partial c}{\partial s} + \frac{c}{\delta n}\frac{\delta(\delta n)}{\delta s}. \qquad (2)$$

The last term is now more easily revealed as the product of wind speed and the rate of channel widening downstream, or the "spreading of flow" downstream.

In most cases Eq (2) is found more convenient than Eq (1) for both numerical evaluations and visual estimations of divergence. The term $\partial c/\partial s$ is obtained in the same manner for either formula, but the spreading term is handled differently. With a set of dividers, two points are located within the channel a selected distance δs apart and equidistant from the origin. The dividers are then reset for the width of the current at the upstream point, and without change of setting they are placed across the current at the position downstream with one point of the dividers placed on a bounding streamline. The quantity $\delta(\delta n)$ is the excess of channel width downstream relative to that upstream. This excess obtained, the contribution by spreading is found upon multiplication by $c/\delta n\delta s$.

By this method of measurement, divergence in Figure 10.031*b* was evaluated directly at a number of points in the area. Raw results appear in (*d*). The principal source of error in this procedure lies in measurement of $\delta(\delta n)$. This is usually a small quantity in comparison with δn and δs, which were taken 1°–3° of latitude; and, since divergence is directly proportional to $\delta(\delta n)$, a large percentage error is therefore possible in the results. However, in spite of difficulties, there is significant relation between the results in (*d*) and (*i*).

From Figure 10.032*a* we may draw several deductions concerning divergence

and vertical motion which are not possible from the pressure pattern only. In the warm air ahead of the cold front, streamline channels narrow rapidly downstream. Although wind speed increases downstream and there is positive contribution to divergence by the stretching term, the narrowing of the channels overcompensates to give prevailing convergence in a broad band extending from the Gulf of Mexico northeastward in advance of the front. This band broadens over the maritime provinces of Canada and then covers the cyclone area centered southeast of Hudson Bay. Except for small scattered areas, there appears to be surface convergence everywhere within a radius at least 5° of latitude from the cyclone center. Notice how this and the convergence attending the cold front coincide with the patterns of clouds and weather in Figures 9.15*abc*. Much of the weather in advance of the warm front can be attributed directly to this. The cloudiness and precipitation in advance of the cold front are well borne out by convergence in the surface winds. Indeed, convergence is a common feature of that part of the flow pattern and provides for the principal precipitation in the warm sector of frontal waves.

In following the streamlines from Manitoba across the northern Great Lakes, one might identify a zone of convergence extending eastward from just south of Lake Winnipeg. Another region of convergence appears east of the Great Lakes about over New York State. Both areas are characterized by predominance of snow showers. This precipitation area extends to the windward of the Appalachians, where orographic lifting of the moist and turbulent air is sufficient to produce precipitation without needing convergence in surface flow.

The northwestern part of the map is dominated by diverging streamlines with wind speeds increasing downstream. This is the only large area on the map with clear skies, and in fact it is the only large area where horizontal divergence is large and unquestionable. Farther south, over extreme western Kansas and eastern Colorado, it appears that surface convergence prevails, even though this area is featured by anticyclonically curved isobars. This convergence is consistent with cloudiness and moisture patterns aloft.

10.15. *Planimetric evaluation of divergence.*—In preceding sections it was shown how horizontal divergence could be evaluated by various means using two or more types of analyses consisting of isotachs, streamlines, isogons, or u and v components of wind. While these methods depend on prior preparation of a number of analyses, there are means for computing divergence directly from the wind reports. These methods involve measurement of areas.

Eq 10.11(2) shows that divergence can be obtained by measuring the percentage increase in area per unit time indicated by actual winds. We may extend this concept to include finite areas outlined by wind-observing stations on the understanding that for such large areas the divergence measured is only a rough *average* over the area. Let points a, b, and c in Figure 10.15(a) represent three stations whose wind reports are the vectors. The area outlined by the three stations is a constant A_0 which can be measured on the map by geometry or planimeter. Now suppose the three parcels are displaced through finite time each in a direction and distance conforming with its wind report. Any convenient interval of time δt may be used (e.g., 1, 2, 3, or more hours), but the same interval must be used for the entire computation. The most convenient time interval to use is determined by the scale of the map and the wind speeds. The locations of the three air parcels at the end of δt are then entered on the map (a', b', c').

As a rule, the final area is bounded by curved or wavy lines, since the wind field is nonlinear. In order to carry out the compu-

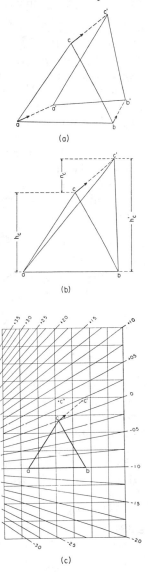

Fɪɢ. 10.15

tation, we must assume that the variation of wind between stations is linear or nearly so, in order that the three final positions may be connected by straight lines to outline another triangular area A_1. After A_1 has been measured, it is substituted along with other known quantities into the formula

$$(A_1 - A_0)/A_0 \delta t \approx \text{div } \mathbb{C}. \qquad (1)$$

Divergence is obtained in a unit corresponding to that for δt. Since the area appears in both numerator and denominator, for ordinary displacements on common meteorological map projections, the scale factor (or, rather, the square of the scale factor) for the map is practically canceled. Therefore, it is necessary to determine only *map areas* instead of corresponding areas on the earth. As an example, if the interval used in displacing the parcels is 3 hours (10,800 seconds), if the initial area is 5 cm² on the map and the final area 7.2 cm², div $\mathbb{C} = +4.1 \times 10^{-5}$ sec⁻¹.

If long time intervals are used, one frequently finds that in cases of convergence the triangle becomes inverted. This can be visualized by assuming the wind at a is much stronger than the others. If a long time interval is used, at some time during displacement a' would lie on $b'c'$, giving $A_1 = 0$, and at a later time in the displacement a' would lie beyond $b'c'$, giving $A_1 < 0$.

Bellamy devised a technique for evaluating horizontal divergence based on the principles above, but which uses a nomogram overlay to give results directly.[19] The divergence of area abc in diagram (b) can be expressed as the sum of the partial divergence contributed by each wind. Symbolically, $\text{div}_{abc} = \text{div}_a + \text{div}_b + \text{div}_c$. The contribution by any wind, for example, c, is found by displacing that point in the same manner as before while holding the other two vertices fixed. This gives a second area abc', and

$$\text{div}_c = \frac{1}{\delta t} \left(\frac{a\,b\,c' - a\,b\,c}{a\,b\,c} \right).$$

19. J. C. Bellamy, "Objective Calculations of Divergence, Vertical Velocity, and Vorticity," *Bulletin of the American Meteorological Society*, Vol. XXX, No. 2 (1949).

According to Figure 10.15(b), the initial area abc is $(ab)h_c$, and the final area abc' is $(ab)h_c'$. Therefore,

$$\text{div}_c = \frac{1}{\delta t} \left(\frac{h_c' - h_c}{h_c} \right) = \frac{1}{\delta t} \left(\frac{n_c}{h_c} \right).$$

Similar expressions hold for the remaining two winds, and

$$\text{div } \mathbb{C} \approx \text{div}_{abc} = (n_a/h_a + n_b/h_b$$
$$+ n_c/h_c)/\delta t. \qquad (2)$$

The ratios are evaluated by use of the nomogram in (c). It consists of a rectangular grid and series of slanting lines radiating from a point off the chart. The -1.0 line is horizontal, and all other slant lines are drawn with constant spacing along each vertical line. Labels on the radiating lines are values of the ratio n/h. To evaluate n_c/h_c, the -1.0 line is placed along the base ab of the triangle, and the chart is shifted so that the 0 line traverses point c. A point c'' is located at the intersection of the horizontal line through c' and the vertical line through c. The value of the radiating line through c'' is the value of n_c/h_c. The ratios n_a/h_a and n_b/h_b are evaluated similarly.

10.16. *Measurement of lateral divergence between conservative surfaces.*—Consider the

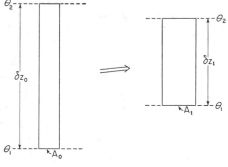

FIG. 10.16

air column at time $t = 0$ in the left of Figure 10.16. It is bounded by two conservative unit surfaces θ_1 and θ_2; it has horizontal cross-section area A_0, vertical depth δz_0, volume $A_0 \delta z_0$, and mass $\rho_0 A_0 \delta z_0$. At later time t the *same column* of air has the dimensions shown at the right: depth δz_t, area A_t, volume $A_t \delta z_t$, and mass $\rho_t A_t \delta z_t$. Through conservation of mass, $\rho_t A_t \delta z_t = \rho_0 A_0 \delta z_0$. Substitution from the hydrostatic equation gives $A_t \delta p_t = A_0 \delta p_0$, where δp is the pressure difference between surfaces. From this, $A_t/A_0 = \delta p_0/\delta p_t$. If 1 is subtracted from both sides, the result divided by δt, and the left side then compared with Eq 10.11(2), it follows that

$$\text{div } \mathbb{C} \approx (\delta p_0/\delta p_t - 1)/\delta t . \quad (1)$$

Evidently, lateral divergence can be measured by comparing the *pressure depth* of an individual column of air at the beginning and end of its displacement between two conservative surfaces.

The pressure difference between two isentropic surfaces is given for any geographic location by the "isentropic weight" chart (Fig. 8.05d). If a column between the two surfaces is traced by its trajectory, then the average lateral divergence experienced by the column during the time interval is found by substituting for δp_0 in Eq (1) the pressure difference at its initial position and time and for δp_t the pressure difference at its final position and time. For example, if its initial depth is 100 mb and its depth 12 hours later is 50 mb, then

$$\overline{\text{div } \mathbb{C}} \approx (100/50 - 1)/43{,}200 \text{ sec}$$
$$= 2.32 \times 10^{-5} \text{ sec}^{-1} .$$

This provides a practical though qualitative approach for estimating divergence in the free atmosphere. The principal difficulty in quantitative measurement is in tracing the air column (i.e., the problem of trajectories). Another source of doubt is introduced with deep layers in baroclinic conditions; there can be large differential in the displacement of the top and bottom of the

layer. This possible error could be reduced by employing thin isentropic layers, but then the relative error in measuring δp is increased. Another source of error is non-conservativeness of θ.

**10.17. *Streamlines for constant horizontal transport.*—Consider the element in Figure 10.17a outlined by two streamlines and two normals to them. If c_2 is the wind speed at

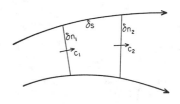

Fig. 10.17a

the downstream face of the horizontal element, the increase in area of the element per unit time due to c_2 is $c_2 \delta n_2$. Similarly, the decrease in area due to c_1 is $c_1 \delta n_1$. Thus $dA/dt = (c_2 \delta n_2 - c_1 \delta n_1)$. Dividing this by the area of the element gives the horizontal divergence, Eq 10.11(2).

For *nondivergent* motion the above expression shows that $c_2 \delta n_2$ must equal $c_1 \delta n_1$, or that $c\delta n = $ constant. In other words, if flow is nondivergent in the horizontal, the spacing δn of streamlines varies inversely with wind speed. This is the condition for *solenoidal* motion in a horizontal plane.[20] Evidently, if there is no divergence, streamlines can never touch or intersect. Reasoning conversely, if streamlines do touch, there is indication of horizontal velocity divergence.

Suppose for the moment that horizontal motion is nondivergent. With an arbitrary value for the constant in the last equation, all streamlines on the chart then could be drawn with spacing everywhere inversely

20. For a discussion of solenoidal motion see V. Bjerknes *et al.*, *Dynamic Meteorology and Hydrography*, Part II: *Kinematics*.

proportional to wind speed. A scale similar to the one in Figure 5.05a and based on a convenient value for the constant could be prepared which gives map distance δn for any wind speed. This streamline pattern would indicate not only wind direction but also wind speed at any point in the field. The relation of wind to this streamline pattern would be analogous to the relation of geo-

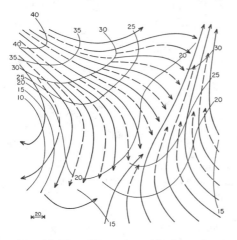

FIG. 10.17b.—Illustration of horizontal transport lines with isotachs.

strophic wind to pressure pattern (latitude variations excluded).

In actual conditions the flow is usually divergent. The condition $c\delta n =$ constant is not fulfilled, and continuous streamlines spaced inversely with the wind speed cannot be drawn. However, we may use this relation to draw certain "streamlines" which have all the properties of ordinary streamlines except they are discontinuous even where the motion itself is continuous. These we shall distinguish as *transport lines*. It becomes evident that, where new transport lines originate in the pattern, there must be horizontal divergence in the flow, and, conversely, where transport lines disappear, there must be convergence.

Figure 10.17b gives an example of transport lines. The isotachs are superfluous but

are included in the drawing for this discussion. The first step in drawing the pattern was selecting a convenient distance interval δn. This distance for 20-knot wind is shown in the lower left. A central streamline in each current then was drawn carefully as a basis for locating adjacent transport lines. Then, from a preliminary sketch of isotachs, the 30-knot line was chosen along which to space transport lines from the base streamlines. With δn two-thirds the length for 20 knots, the intersections of transport lines were marked off along the 30 isotach. Finally, transport lines were extended outward through these points and parallel to the winds. Alternate ones were dashed to indicate values of δn between 1 and 1/2 the spacing required by the winds.

In the right side of the chart a number of transport lines terminate in the pattern, that is, the flow is converging horizontally. The rate of convergence would be, in a sense, inversely proportional to the length of the dashed transport lines. It is now evident that this analysis fulfils the purpose of streamlines and to a lesser extent isotachs, but it also gives a better visual indication of divergence than does any single form of analysis discussed previously.

The method outlined above is only one objective approach in drawing transport lines. It can and should be modified after some practice, since the simple analysis obtained can be misleading. Notice the relative number of lines beginning or ending at the 30-knot isotach in Figure 10.17b. This resulted merely from the choice of a particular isotach in the initial steps, and it might give the false impression that divergence occurs principally along this line. The false indication can be removed by shifting a few transport lines a short distance relative to the initial streamline and then adjusting the pattern accordingly. Also, the use of prior isogon analysis is as useful here as in ordinary streamline analysis.

10.18. *Divergence of geostrophic wind.—* By differentiating isobarically the geostrophic equations for west wind $u_g = -(980/f)(\partial Z/\partial y)_p$ and for south wind $v_g = (980/f)(\partial Z/\partial x)_p$ with respect to x and y, respectively, and then adding,

$$\text{div } \mathbb{C}_g = (\partial u_g/\partial x + \partial v_g/\partial y)$$
$$= -(v_g/a) \cot \phi, \quad (1)$$

where a is the earth's radius. The term with $\cot \phi$ is due to meridional variation of the Coriolis parameter.[21]

Eq (1) shows that geostrophic wind is divergent *only if* it has a meridional component, that is, if isobaric contours intersect latitude circles. Divergence of geostrophic wind may be illustrated by a system of mutually parallel contours oriented across latitude circles; geostrophic wind is strongest in the equatorward part of the contour channel. However, this contribution is negligible in ordinary scale. At 45° latitude, $v_g = 10$ m sec^{-1} gives for horizontal divergence of geostrophic wind 1.57×10^{-6} sec^{-1}. It becomes apparent that *the primary contribution to horizontal divergence must be found elsewhere than in the pressure distribution.* This fact is readily seen by reference to (c) and (d) in Figure 10.09. Except for variation of latitude, no pressure pattern can balance the types of flow indicated.

10.19. *Divergence of ageostrophic wind.—* By horizontal differentiation of $\mathbb{C} = \mathbb{C}_g + \mathbb{C}'$,

$$\text{div } \mathbb{C} = \text{div } \mathbb{C}_g + \text{div } \mathbb{C}'$$
$$= -(v_g/a) \cot \phi + \text{div } \mathbb{C}'. \quad (1)$$

Since we neglect divergence of geostrophic wind, it is obvious that *horizontal divergence*

21. Note that Eq (1) was obtained by *isobaric* differentiation and that that equation is properly for isobaric instead of horizontal divergence of geostrophic wind. The difference is a bothersome solenoid term of magnitude usually less than 10^{-6} sec^{-1}.

lies in the divergence of ageostrophic wind. This means that patterns of vertical motion and weather are controlled primarily by departure of motion from the pressure pattern.

For convenience, ageostrophic wind can be considered the vector sum of known departures from geostrophic wind, as shown in Section 7.23. It follows that

$$\text{div } \mathbb{C}' = \text{div } \mathbb{C}'_F + \text{div } \mathbb{C}'_i + \text{div } \mathbb{C}'_a$$
$$+ \text{div } \mathbb{C}'_c + \dots .$$

In the following each contribution is considered separately.

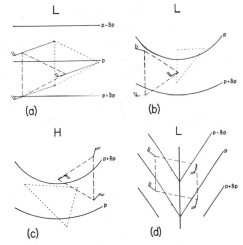

FIG. 10.191.—Horizontal divergence with friction and curved isobars.

a) *Divergence of antitriptic wind.—* With reference to Figure 7.232, it is deduced that, with invariant geostrophic wind and constant influence of friction, ageostrophic winds due to friction will be of constant speed and direction over the region. In this case, as shown by the "area method" in Figure 10.191(a), antitriptic wind is nondivergent. By introducing a gradient in frictional influence or one in geostrophic wind, it is evident that divergence results.

The other drawings are given to indicate how divergence is produced by action of friction in common pressure patterns. In all cases the angle between the wind and isobars is 20° or less—a conservative value for the surface. With cyclonic curvature and friction (b), the winds converge (spirally) toward the center, and for any given speeds convergence is indicated. With anticyclonic curvature and friction (c), the motion is divergent. The last diagram is left for analysis by the reader. It might be assumed that the discontinuity is a front moving with the actual winds behind it. In addition to certain divergence, there is indicated deformation with axis of dilatation fairly parallel to the trough line.

The reader is now asked to analyze visually the probable distribution of vertical motion from Figures 9.02ab based only on this simplified relation between horizontal divergence and action of surface friction on curved and discontinuous pressure patterns. Comparison with the patterns of cloudiness and moisture (aloft) is then in order.

To obtain an idea of the large-scale horizontal divergence maintained by friction with curved isobars, assume that the wind crosses isobars with angle 30°. The wind component normal to isobars is then half the total wind speed. For isobars of radius 500 km and actual winds 10 m sec^{-1}, the corresponding horizontal divergence is 2×10^{-5} sec^{-1} in magnitude. The value varies directly with wind speed and inversely with radius of curvature. It is thus possible to obtain very large values of divergence merely through effect of friction.

Besides the relation of divergence to isobar curvature with friction, one can also examine effects of varying friction and even varying geostrophic wind speed. For example, if friction increases in the direction of motion, such as might be the case in air motion from ocean to land, there should result horizontal convergence, and conversely for air moving from the opposite direction. In addition, for given angular deviation from the isobars, there is expected con-

vergence to the left and divergence to the right of the strongest geostrophic winds (northern hemisphere).

b) *Divergence of isallobaric wind.*—From Eq 7.23(6) we obtain for the divergence of $\mathbb{C}'_i$

$$\operatorname{div} \mathbb{C}'_i = \frac{1}{\rho f^2} \operatorname{div}_h \left(-\nabla_h \frac{\partial p}{\partial t} \right)$$
$$= \frac{980}{f^2} \operatorname{div}_h \left[-\nabla \left(\frac{\partial Z}{\partial t} \right)_p \right].$$

Divergence of isallobaric wind is directly proportional to, and of the same sign as, divergence of the isallobaric descendant. Thus, owing to isallobaric influence alone, there should be *horizontal convergence in areas of large pressure fall* and *horizontal divergence in areas of large pressure rise*. Cloudiness and precipitation should be favored in advance of moving cyclones and troughs at the surface, and clearing behind them, if only the isallobaric effect is considered. The contribution of isallobaric wind to horizontal divergence can be appreciable where isallobars are closely spaced and greatly curved. For example, isallobaric winds only 1 m sec^{-1} blowing across an isallobar 200 km in radius give divergence 1×10^{-5} sec^{-1} in magnitude.

c) *Divergence due to varying pressure gradients downwind.*[22]—If neighboring isobars are straight and parallel, there should be no divergence of this ageostrophic wind. However, if next to a channel of parallel isobars the adjacent isobar channels diverge or converge downstream, a pattern of horizontal divergence exists (Fig. 10.192). *Where isobars diverge more rapidly to the right or where they converge more rapidly to the left, there is positive horizontal divergence.* For the converse there is horizontal convergence, all considering this one effect.

Figure 10.193 illustrates the distribution

22. The discussion is a great deal simplified—in the manner that the gradient wind equation is simplified.

of divergence due to variation of gradient wind in a sinusoidal isobar channel.[23] Consider the wave pattern stationary and the winds gradient. For the westerly current, ageostrophic wind is opposite to geostrophic in the trough and in the direction of geostrophic in the ridge. If motion is parallel to isobars (contours), this results in horizontal divergence between the trough and the next ridge downstream, and convergence between the ridge and the next trough. The same interpretation is valid for an easterly

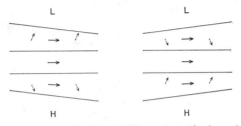

Fig. 10.192.—Schematic illustration of horizontal divergence due to varying isobar spacing in the horizontal. Heavy vectors represent the geostrophic wind component; dashed vectors, the ageostrophic component.

sinusoidal current. It is not necessary that isobars be parallel, since geostrophic and gradient winds are inversely proportional to channel width, and the relative transport at the trough and ridge is what matters. Even if some allowance is made for lag in adjustment of gradient wind to the pressure pattern and for progression of the wave pattern itself, a similar distribution of divergence and convergence is found if the wave moves slower than the winds. For discussions of divergence of gradient wind for various other conditions the reader is referred to appropriate textbooks in meteorology.[24]

The magnitude of divergence in the wave

23. J. Bjerknes, "Theorie der aussertropischen Zyklonenbildung," *Meteorologische Zeitschrift,* Vol. LIV (1937).

24. In particular, J. Holmboe, G. E. Forsythe, and W. Gustin, *Dynamic Meteorology* (New York: John Wiley & Sons, 1945), and Petterssen, *Weather Analysis and Forecasting.*

patterns in Figure 10.193 can be large where variation of gradient wind speed is large. For geostrophic wind 20 m sec^{-1}, $\phi = 50°$, $R = 1000$ km $\simeq 9° \phi$ at the trough and ridge, and wave length $L = 2000$ km, the variation in gradient wind from trough to ridge is about 10 m sec^{-1}, giving magnitude 1×10^{-5} sec^{-1} for average divergence over half the wave length. The value of divergence increases in magnitude with decreasing latitude, decreasing wave length, increasing amplitude, or increasing geostrophic wind. It is thus possible to obtain values larger than 1×10^{-5} sec^{-1} due only to cyclostrophic contribution.

d) Divergence due to vertical motions in baroclinic regions.—As shown previously, ageostrophic wind $\mathbb{C}'_c$ is directly proportional to the product of horizontal temperature gradient and vertical velocity, and for ascent it blows toward colder air. Horizontal divergence of this component is therefore a function of horizontal variation in vertical velocity as well as horizontal variation in temperature gradient. Figure 10.194 illustrates effects of both variants in a field of straight isotherms.

Diagrams (*a*) and (*b*) show the effect of varying horizontal temperature gradient with constant upward or downward motions. *For ascent convergence occurs to the cold side of the strongest temperature gradient and divergence to the warm side; the opposite holds for descent.* The remaining diagrams show the effect of varying vertical motions. There is *horizontal convergence to the cold side of the maximum upward velocities and to the warm side of the maximum downward velocities.* Horizontal divergence is found on the opposite side. In actual situations effects of varying temperature gradient and varying vertical motion are combined.

Although this contribution to horizontal divergence has been obscured in the past by emphasis on surface friction, isallobaric wind, and cyclostrophic correction, the effect of vertical motions is not entirely negligible in comparison, even if it is more difficult to deal with. To show what magnitudes of divergence can be obtained in ordinary conditions, we may take an example based on diagram (a) or (b). If vertical velocity is constant and x is along isotherms with y toward cold air,

$$\operatorname{div}\mathbb{C}'_c = \frac{w}{f}\frac{\partial}{\partial y}\left(\frac{\partial u_g}{\partial z}\right).$$

Since in the illustrations $\partial u_g/\partial z = -(g/$

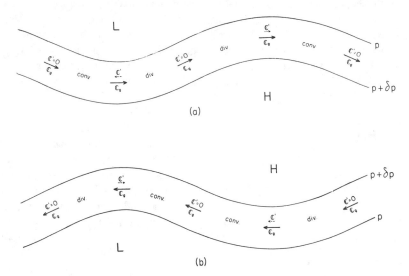

Fig. 10.193.—Horizontal divergence with gradient wind conditions in a channel of parallel sinusoidal isobars: (a) with westerly winds and (b) with easterly winds.

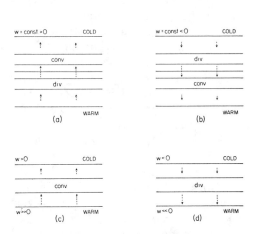

Fig. 10.194.—Horizontal divergence due to vertical motions in a baroclinic region with straight isotherms; dashed vectors are the ageostrophic winds.

$fT)(\partial T/\partial y)_p$, then

$$\operatorname{div}\mathbb{C}'_c \simeq - (3.6 \times 10^8 \text{ cm deg}^{-1})$$

$$\times w \frac{\partial}{\partial y}\left(\frac{\partial T}{\partial y}\right)_p,$$

for $g = 980$ cm sec^{-2}, $f = 10^{-4}$ sec^{-1}, and $T = 270°$ K. If over a distance 100 km the temperature gradient varies from 3° C/100 km to 1° C/100 km, div $\mathbb{C}'_c \simeq (0.72 \times 10^{-5}$ cm$^{-1})w$, showing that with these conditions a vertical velocity only 1.4 cm sec^{-1} is required to give horizontal divergence of magnitude 1×10^{-5} sec^{-1}. Since vertical velocities and variations in temperature gradient larger than these do frequently occur, this divergence can be considerable.

10.20. *Notes on vertical distribution of horizontal divergence.*—For determining the patterns of vertical motion and weather, the vertical variation of horizontal divergence is more important than divergence at a certain level. This subject has been discussed widely in the literature. Summaries of ideas—past and present—can be found, together with recent findings, among the studies by the New York School.[25]

The problem may be approached deductively from empirical evidence on pressure changes and horizontal mass divergence near the ground. The local derivative $\partial\rho/\partial t$ is related to horizontal mass divergence through the *equation of continuity*,

$$\partial\rho/\partial t = -\,(\partial\rho u/\partial x + \partial\rho v/\partial y$$
$$+\,\partial\rho w/\partial z) = -\,\mathrm{div}\,\rho\mathbb{C} - \partial\rho w/\partial z\,. \quad (1)$$

By substitution into Eq 3.12(2), the local pressure tendency at the surface ($Z = 0$) is

$$\partial p_0/\partial t = -\,980\int_0^\infty (\mathrm{div}\,\rho\mathbb{C})\,dZ. \quad (2)$$

The integral of $\partial\rho w/\partial z$ vanishes, since $w = 0$ at the (level) ground and $\rho = 0$ at the top of the atmosphere. Eq (2) states that with hydrostatic equilibrium *the pressure tendency at the ground is given by the integral of horizontal mass convergence above that point.*

Now consider a vertical column fixed geographically and extending from the ground to the top of the atmosphere (Fig. 10.20a). In the lower part of this column the motion is permitted to diverge or converge horizontally. However, suppose for the moment that above a level Z net horizontal mass divergence is zero. Thus the integral from Z to ∞ is zero, and the pressure tendency at

25. J. E. Miller, *Studies of Large-Scale Vertical Motions of the Atmosphere* ("Meteorological Papers, New York University," No. 1 [New York, 1948]); R. G. Fleagle, "The Fields of Temperature, Pressure, and Three-dimensional Motion in Selected Weather Situations," *Journal of Meteorology*, Vol. IV (1947), and "Quantitative Analysis of Factors Influencing Pressure Change," *ibid.*, Vol. V (1948); H. A. Panofsky, "Large-Scale Vertical Velocity and Divergence," in American Meteorological Society, *Compendium of Meteorology*, pp. 639–46.

the ground is due only to mass convergence below Z.

As a first and only simplifying assumption,[26] merely for present purposes, let us say that horizontal advection of density is zero; then $\mathrm{div}\,\rho\mathbb{C} = \rho\,\mathrm{div}\,\mathbb{C}$, and

$$\partial p_0/\partial t = -\,980\int_0^Z (\rho\,\mathrm{div}\,\mathbb{C})\,dZ$$
$$\simeq -\,980\,\overline{(\rho\,\mathrm{div}\,\mathbb{C})}\,Z\,.$$

The mean values of density and divergence are with respect to the depth from the

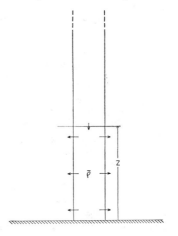

Fɪɢ. 10.20a

ground to level Z. If, for example, Z is 4 km above ground, $\bar\rho = 1 \times 10^{-3}$ gm cm^{-3}, p is in millibars, velocity divergence is in units sec^{-1}, and δt is 1 hour, $\delta p_0 = -(14.1 \times 10^5$ mb sec) $\overline{\mathrm{div}\,\mathbb{C}}$.

For positive horizontal divergence 1×10^{-5} through the lower portion of the column the surface pressure falls 14.1 mb hr^{-1}. Obviously, if horizontal divergence is 1×10^{-4}, the surface pressure changes 141 mb hr^{-1}. Such a value for divergence is not unreasonable in the scale and time over which pressure tendency can be measured. Furthermore, 4 km represents only a fraction of the mass of the atmosphere; and, since in the above formula the pressure tendency is proportional to the depth of the

26. A perfectly valid assumption in the tropics.

layer considered, a constant value of horizontal mass divergence throughout the troposphere would result in excessive pressure changes at the ground, even if divergence is lowered to such small values as 10^{-6} sec^{-1}. On the other hand, surface pres-

It appears therefore, in comparing ordinary values of pressure tendency and horizontal divergence, that the divergence observed in the lower troposphere must diminish with height and give way to divergence of opposite sign aloft.[27] Owing to the rela-

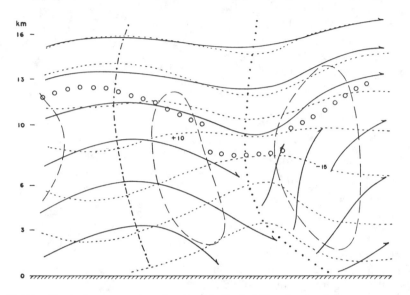

FIG. 10.20b.—Vertical cross section showing idealized fields of potential temperature, pressure change, west-east and vertical components of velocity, and ridge and trough lines. Full lines indicate the streamlines projected onto the vertical plane, and dashed lines indicate isallobars for ± 6 mb (12 hr)$^{-1}$. The vertical velocity is exaggerated 200 times the horizontal velocity. (After Fleagle, 1947.)

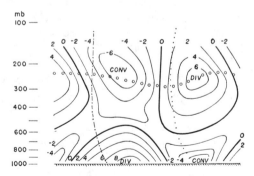

FIG. 10.20c.—Vertical cross section showing average space distribution of specific horizontal mass divergence (ρ^{-1} div$_h$ $\rho\mathbb{C}$) in units 10^{-6} sec^{-1}. (After Fleagle, 1948.)

tive incompressibility of air in nature and its freedom to expand vertically in response to horizontal divergence, there are related distributions of vertical motion. As a general rule, *horizontal divergence near the ground is overlain by horizontal convergence and connected with it by predominant descending motions in the troposphere; horizontal convergence near the ground is overlain by horizontal divergence and connected with it by predominant ascending motions.* The above principle has been known and discussed for some time in connection with observed weather and surface pressure patterns. Only

sure tendency 1 mb hr^{-1} gives 7×10^{-7} sec^{-1} for horizontal divergence.

27. For a favorable argument consult R. C. Sutcliffe, "Cyclonic and Anticyclonic Development," *QJRMS*, Vol. LXV (1939).

recently has it become possible with adequate upper-level data to substantiate it quantitatively.

The weight of evidence, both theoretical and empirical, suggests that in the main there are two "levels" of maximum absolute divergence—one at the ground and the other near the tropopause; the lower and upper divergence fields are of opposite sign and of comparable magnitude. Between these, maximum vertical velocities occur near a so-called "level or surface of nondivergence," which is not level at all and can intersect both the ground and the tropopause (Fig. 10.20c). It seems more practical to refer to a *level of minimum absolute divergence*, located on the average almost midway between the surface and the tropopause. In middle latitudes this would be near the 600-mb level. This figure is nearly the same determined theoretically by Charney[28] (580 mb) and empirically by Fleagle.[29]

Figure 10.20c gives a pattern of horizontal *mass* divergence per unit mass in vertical cross section. Horizontal velocity divergence is actually the difference between the given pattern and the distribution of horizontal advection of density per unit mass, $(1/\rho)\mathbb{C}\cdot\nabla_h\rho$. However, from the same paper, the advective contribution is almost an order of magnitude less than the values in Figure 10.20c, implying that the pattern of horizontal velocity divergence is nearly the same as illustrated.

If horizontal divergence and convergence are a maximum near the tropopause, then this region is also characterized by maximum vertical shrinking and stretching, indicated by corresponding large changes in stability. (Maximum horizontal conver-

gence might be found below the tropopause, and maximum horizontal divergence above it, to agree with the vertical distribution of stability across it.) But the upper maximum of absolute horizontal divergence as a rule does not coincide with the location of minimum vertical velocities. From evidence cited by Fleagle and by Palmén and Nagler,[30] it appears that minimum vertical motions are found in the lower stratosphere a short distance above the region of largest horizontal divergence and convergence. Near the tropopause itself the large-scale vertical motions are predominantly of the same sign as in the middle troposphere, but magnitudes are smaller.

Figure 10.20b shows the idealized distribution of motion in a west-east cross section through what may be considered an eastward-moving wave in the westerlies. Pressure troughs and ridges are the tilted axes, and the tropopause is the series of circles. In the lower troposphere general ascent occurs with low pressure and descent with high pressure. Vertical motions in the middle and upper troposphere are descent from ridge to trough and ascent from trough to ridge, facing downstream. With respect to the typical wave in the easterlies, general ascending motions occur to the rear (east) of the pressure trough and descending motions in advance (west) of the trough.[31] Figures 10.20bc now should be analyzed on the basis of the previous discussion on divergence of ageostrophic wind.

10.21. *The pressure tendency equation.*— The detailed causes of local pressure change

28. J. Charney, "The Dynamics of Long Waves in a Baroclinic Westerly Current," *Journal of Meteorology*, Vol. IV (1947).

29. "The Fields of Temperature, Pressure, and Three-dimensional Motion in Selected Weather Situations," *op. cit.*

30. E. Palmén and K. M. Nagler, "The Formation and Structure of a Large-Scale Disturbance in the Westerlies," *Journal of Meteorology*, Vol. VI, No. 4 (1949).

31. H. Riehl, *Waves in the Easterlies and the Polar Front in the Tropics* ("Department of Meteorology, University of Chicago, Miscellaneous Reports," No. 17 [Chicago: University of Chicago Press, 1943]).

at the ground are age-old topics of conjec-
ture, study, and debate, approached from
various viewpoints, and still a matter of
controversy. We do know that pressure
tendency is the integral of mass convergence
above the point, but the relative importance
of the three factors—(i) horizontal velocity
convergence, (ii) horizontal density advection, and (iii) vertical transport—is the
point of doubt. Only a brief outline of the
problem is given here; extensive discussions
are to be found in the literature.[32]

At any height z the pressure tendency is

$$\partial p_z / \partial t = - g \int_z^\infty (\rho \operatorname{div} \mathbb{C} + \mathbb{C} \cdot \nabla_h \rho$$
$$+ \partial \rho w / \partial z) \, dz \,, \quad (1)$$

or, considering $\rho w = 0$ at the top of the
atmosphere,

$$\partial p_z / \partial t = - g \int_z^\infty (\rho \operatorname{div} \mathbb{C}$$
$$+ \mathbb{C} \cdot \nabla_h \rho) \, dz + g (\rho w)_z . \quad (2)$$

The last term on the right is the vertical
mass transport through $z;$ it vanishes when
z is at the (level) ground. The surface pressure tendency can be considered as due only
to integral effects of horizontal velocity
divergence and horizontal density advection; that is,

$$\partial p_0 / \partial t = - g \int_0^\infty (\rho \operatorname{div} \mathbb{C}) \, dz$$
$$- g \int_0^\infty (\mathbb{C} \cdot \nabla_h \rho) \, dz .$$

a) Horizontal velocity divergence.—The
net effect of divergence on pressure change
can be expressed $-g(\overline{\rho \operatorname{div} \mathbb{C}})$, the average
being for the entire air column above the
level. This net effect is usually the small
difference between large contributions of
opposite sign near the ground and the tropopause. It is roughly an order of magnitude
less than the contribution in each of the

regions; its sign is given by the larger of the
two, which may differ with the case.

In this connection it is useful to examine
the apparent paradox of divergence and
pressure change at the ground. With the
ordinary middle-latitude cyclone the largest
pressure falls (slightly in advance) are associated with low-level *convergence* due to
friction, due to meridional variation of the
Coriolis parameter, and due to variation in
gradient wind (according to Petterssen),[33]
and with *convergence* of isallobaric wind (and
perhaps of convective ageostrophic wind)
through deep layers, all nicely verified by
wind analysis about the cyclone (Figs.
10.032*ab*). The fact that pressure does fall
must be due to (i) much greater *divergence*
aloft or (ii) greater net contribution by
warm advection, or both combined. Existence of divergence aloft is indicated by
prevailing upward motion, but whether
divergence there is effectively greater than
low-level convergence remains a question. A
similar type of analysis is appropriate for
moving anticyclones.

b) Horizontal advection.—To visualize the
effect of density (temperature) advection on
the pressure tendency at a level below, consider that a certain volume of air is removed
from the air column and replaced horizontally by air of different density. The contribution by advection is large with strong wind
components across concentrations of isotherms. Cold advection contributes to pressure rise, and warm advection to pressure
fall, which is the usual correlation observed
with frontal advection in low levels. The primary regions of advection are the lower
troposphere and the lower stratosphere
(temperature charts in Chap. 7), but for the
same rate of temperature advection the effect is greater in low levels. In some places,
such as behind surface cold fronts, the two
regions contribute oppositely; but else-

---

32. See the references by J. M. Austin, "Mechanism of Pressure Change," in American Meteorological Society, *Compendium of Meteorology*, pp. 630–38.

33. *Weather Analysis and Forecasting*, pp. 233–
37.

where compensation may not be so evident.

For ordinary conditions the value of $(1/\rho)\mathbb{C}\cdot\nabla_h\rho$ is near 1×10^{-6} sec^{-1}. This is apparently about the same order of magnitude as the *net* contribution by horizontal divergence. Thus, in some instances, the advective effect might be dominant in the pressure tendency, while in others the divergence effect prevails. The latter must be the case in the more barotropic conditions.

While local pressure tendencies with common mid-latitude wave systems are generally well correlated with density advection in at least the lower troposphere, the same could hardly be said for systems in the tropics, where thermal gradients are small. But there is one prominent case in middle latitudes which defies the usual agreement in sign of pressure tendency and advection. An example is shown over the north central United States on 1 March 1950. Notice the area of negative and small pressure tendencies in the region of strong cold advection west of the Great Lakes. After examining the patterns of advection in the lower stratosphere, of divergence near the ground, and of weather, one might conclude that strong high-level positive divergence is the dominant factor here. However the pressure tendency may contradict the pattern of advection, that example is in agreement with a useful rule in forecasting: Renewed cold advection in the upper (pressure) trough leads to deepening of the cyclone at the surface.

10.22. *The vertical component of relative vorticity.*—By definition, relative vorticity of the (horizontal) wind is

$$\zeta \equiv \partial v/\partial x - \partial u/\partial y .\qquad(1)$$

Physically, it is the curl of the vector wind, defined positive for counterclockwise rotation. Thus

$$\nabla\times\mathbb{C}=\mathbf{k}(\partial v/\partial x-\partial u/\partial y) .$$

In speaking of relative vorticity of the wind, the "vertical component of relative vorticity" is implied. The vorticity of the wind field is qualified as "relative" to distinguish it from the earth's own vorticity. There should be no cause for ambiguity, and we henceforth refer to relative vorticity merely as *vorticity*.

Interpretation of vorticity in terms of rotation can be seen by examining the rotation of a unit fluid cross (Fig. 10.22) due to shear in the velocity field. Consider the

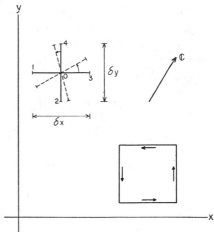

Fig. 10.22

center o of this cross is moving with wind velocity $\mathbb{C}$. The y component of speed at extremity *3* relative to that at o is $v_3 - v_0 = \frac{1}{2}(\partial v/\partial x)\delta x$, if δx is the total length of the arm from *1* to *3*. This relative speed also can be seen as the tangential velocity of rotation about o, and therefore the relative speed $v_3 - v_0$ is related to an angular velocity ω_x about o: $\frac{1}{2}\omega_x\delta x = v_3 - v_0$. By combining the two equations, $\omega_x = \partial v/\partial x$. The same procedure for the y arm of the cross gives $\omega_y = -\partial u/\partial y$. Then

$$\zeta = \omega_x + \omega_y = 2\bar{\omega} .$$

Vorticity is thus twice the average angular velocity of a fluid parcel in the moving medium.

The reader might prove that vorticity is the circulation per unit area of fluid represented by the small square in Figure 10.22:

$$\zeta = \delta C/\delta A = (1/A)\oint (u\,dx + v\,dy) .$$

10.23. *Measurement of vorticity by finite differences.*—Vorticity can be approximated in finite differences by $\zeta \simeq \delta v/\delta x - \delta u/\delta y$,[34] and thus evaluated for any point from analyses of u and v. The procedure is similar to that employed for horizontal divergence.

The quantities $\partial v/\partial x$ and $-\partial u/\partial y$ evaluated by the above method are shown in (j) and (k) of Figure 10.031. Vorticity is given in (l). Difficulties and errors inherent in this analysis are about the same as for horizontal divergence. A number of small and irregularly distributed centers could have been avoided by using larger distances but thereby minimizing to some extent the magnitudes of vorticity.

10.24. *Measurement of vorticity from kinematic directions.*—The kinematic directions fixed relative to the motion at a point also can be used for evaluating vorticity. Vorticity expressed in these coordinates not only facilitates actual measurement but also provides a more simple way of visualizing it in any field of motion.

Consider the ns system centered in the streamline pattern at o in Figure 10.24. At o, $\partial v/\partial x = \partial c/\partial n$, and $-\partial u/\partial y$ is a function of the rate of turning of the flow through o. The proper expression for $-\partial u/\partial y$ in these coordinates is $-c(\partial a/\partial s)$,

34. If y is identified with a meridian on the earth (y northward), vorticity is actually

$$\zeta = \partial v/\partial x - \partial u/\partial y + (u/a)\tan\phi .$$

The correction is negligible in comparison with $\partial v/\partial x - \partial u/\partial y$ except at very high latitudes or with strong zonal wind components. In low and middle latitudes it is in most cases negligible even with strong zonal winds, since, as u increases in magnitude, the chances are that $\partial v/\partial x - \partial u/\partial y$ also increases proportionately due to increased shear.

as will be found from analysis of Figure 10.24. Thus,

$$\zeta = \partial c/\partial n - c(\partial a/\partial s) . \tag{1}$$

Vorticity is now expressed as the sum of a *shear term* $\partial c/\partial n$ and a *curvature term* $-c(\partial a/\partial s)$. Cyclonic shear and backing of wind downstream both contribute to cyclonic vorticity; anticyclonic shear and veering of the wind downstream both contribute to anticyclonic vorticity.

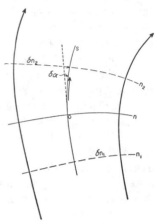

FIG. 10.24

Eq (1) may be used for evaluating vorticity. The term $\partial c/\partial n$ is the difference in wind speed δc from left to right, across the stream channel, divided by the width δn of the channel. The other term is found by taking the wind speed at o and the angular turning of the streamline (in radians) per unit distance through o. If isogons are drawn, δa is the difference in isogon values at n_2 and n_1 along s, converted from degrees to radians, and δs the distance from n_1 to n_2 also along s.

The second term on the right of Eq (1) can be expressed in terms of the spreading of the normals to the streamlines:

$$\zeta = \frac{\partial c}{\partial n} - \frac{c}{\delta s}\frac{\partial(\delta s)}{\partial n} . \tag{2}$$

The quantity δs is determined as before. The term $\partial(\delta s)/\partial n$ is evaluated by subtracting the distance between n_1 and n_2 along the left streamline from that distance along the right streamline, and then dividing by the width of the stream channel δn across the origin.

On comparing Eq (1) above with Eq 10.05(1), one finds

$$\zeta = \partial c/\partial n + c/R_s = \partial c/\partial n + cK_s. \quad (3)$$

This expression is the most convenient for visually estimating vorticity, but it is usually not so practical as Eq (1) for actual measurement.

Simple interpretations of vorticity follow immediately from either definition above. By the shear term only, *cyclonic shear contributes to cyclonic vorticity and anticyclonic shear to anticyclonic vorticity.* From the curvature term, *cyclonic curvature of flow contributes to cyclonic vorticity and anticyclonic curvature to anticyclonic vorticity;* the magnitude of this contribution is a function of wind speed as well as streamline curvature. Precisely at the center of a circulation system the streamline curvature is infinite, but, since wind is zero, vorticity is indeterminate. Further, *vorticity along the axis of an isotach maximum is dependent only on the curvature term, and vorticity in straight flow is dependent only on the shear term.* To see these relations more clearly, compare charts (*b*) and (*l*) in Figure 10.031.

10.25. *Planimetric evaluation of vorticity.* —Take a vector $\mathbb{C}''$ of the same magnitude but 90° to the right of the wind vector $\mathbb{C}$ in Figure 10.25. The coordinate components of this second vector are u'' and v''. Then $u = -v''$, $v = u''$. If the first is differentiated with respect to y and the second with respect to x: $\partial u/\partial y = -\partial v''/\partial y$ and $\partial v/\partial x = \partial u''/\partial x$. Subtraction of the first from the second gives $\partial v/\partial x - \partial u/\partial y = \partial u''/\partial x + \partial v''/\partial y$, or

$$\zeta = \text{div } \mathbb{C}''. \quad (1)$$

Hence, *vorticity is horizontal divergence of the vector winds rotated 90° to the right.*

The above relation provides a simple method for evaluating vorticity from a set of three wind reports. After first rotating the wind vectors 90° to the right, one follows the same procedure used in measuring divergence by either of the methods illustrated in Figure 10.15. Incidentally, all methods for evaluating divergence and vorticity are related in this way!

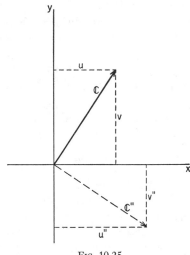

Fig. 10.25

10.26. *Distribution of vorticity in simple current systems.*—Figure 10.271 gives approximate patterns of actual vorticity in the indicated flow patterns, and those will be discussed shortly. Here we consider only circular flow patterns.

For the counterclockwise system in Figure 10.26, suppose that, to approach actual conditions, the strongest winds are found midway between the two most closely spaced streamlines. Vorticity is then positive from the center outward to beyond the current axis. The circular zero vorticity line is where anticyclonic shear equals in magnitude c/R_s. The maximum cyclonic vorticity is somewhere between the current axis and the center of the system. Similar reasoning applied to the clockwise circulation gives the zero vorticity line outside the current axis and negative vorticity over central portions of the system.

With respect to the counterclockwise system, one can conceive of such a distribution of wind speeds that vorticity is everywhere zero (except exactly at the center). This would require that, at any point, anticyclonic shear be equal in magnitude to c/R_s and that wind speeds decrease outward from the center. In such a critical condition $\partial c/\partial R + c/R = 0$, $R(\partial c/\partial R) + c = 0$, or $\partial(cR)/\partial R = 0$. Therefore, $cR = $ constant.

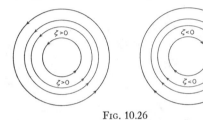

FIG. 10.26

This is the "VR-vortex" frequently applied in discussions of tropical storms and tornadoes. It is "vorticitiless" (i.e., irrotational), even though the flow is circular. It is also possible to define an anticyclonic circulation without vorticity.

Another critical type of circulation system is *solid rotation*, in which vorticity is everywhere constant and finite. The wind along any streamline expressed as tangential velocity about the center of curvature is $c = \omega R$. The shear term for vorticity is then $\partial c/\partial R = \omega$, and the curvature term is $c/R = \omega$. The vorticity is $\zeta = 2\omega$; such a flow pattern is illustrated by Figures 10.09(*e*) and (*f*). In fact, there is solid rotation ′in any homogeneous vorticity field, whether flow is circular or not. The vorticity of the earth is an example of solid rotation and equals twice the angular velocity about its axis ($\zeta_e = 2\omega = 1.458 \times 10^{-4}$ sec^{-1}).

10.27. Geostrophic vorticity.—By differentiating the geostrophic equations for v_g and u_g (at constant pressure) with respect to x and y, respectively, and then subtracting the second result from the first,

$$\zeta_g \equiv \frac{\partial v_g}{\partial x} - \frac{\partial u_g}{\partial y} = \frac{980}{f}\left(\frac{\partial^2 Z}{\partial x^2} + \frac{\partial^2 Z}{\partial y^2}\right)_p$$
$$-\frac{1}{f}\left(v_g\frac{\partial f}{\partial x} - u_g\frac{\partial f}{\partial y}\right), \quad (1)$$

where ζ_g is geostrophic vorticity; it has the relation to geostrophic motion that ζ does to actual motion.

The derivatives $(\partial^2 Z/\partial x^2)_p$ and $(\partial^2 Z/\partial y^2)_p$ reflect the *curvature of the pressure surface*. They are both positive if the surface curves concave upward ("depression") along both coordinate axes, and they are both negative if it curves convex upward ("dome"). It should be evident that *geostrophic vorticity is a function of the shape of an isobaric surface in three dimensions;* it lies in the geometric pattern of isobaric contours.

This term in Eq (1) above may be written $(980/f)(\nabla^2 Z)_p$ and evaluated by use of five grid points on the isobaric surface, in the manner described in Section 5.06:

$$(980/f)(\nabla^2 Z)_p = (980/f)(Z_x + Z_{-x}$$
$$+ Z_y + Z_{-y} - 4Z_0)/(\delta s)^2 .$$

The quantities Z are heights (c.g.s. units) of the surface at the five points. The result is positive (cyclonic geostrophic vorticity) or negative according as the average height at the four outer points is greater or less than at the origin.

The last term in Eq (1) is merely a correction to the preceding one and accounts for variation of Coriolis parameter with latitude. With ′y northward, $\partial f/\partial x = 0$, and Eq (1) becomes

$$\zeta_g = (980/f)(\nabla^2 Z)_p + (u_g/a) \cot \phi . \quad (2)$$

As found by experience, the first term on the right is ordinarily of magnitude 10^{-5} to 10^{-4} sec^{-1}. At $\phi = 45°$, for $u_g = 10$ m sec^{-1} the second term equals 1.57×10^{-6} sec^{-1} and is really negligible in comparison with the first. For $u_g = 100$ m sec^{-1} the value is 1.57×10^{-5} sec^{-1}, which is also quite small considering the much stronger geostrophic shear found with such winds. Therefore, in many cases it is practical to neglect this contribution to geostrophic vorticity and assume[35]

$$\zeta_g \simeq (980/f)(\nabla^2 Z)_p = (980/f)(\partial^2 Z/$$
$$\partial x^2 + \partial^2 Z/\partial y^2)_p . \quad (3)$$

In comparison with actual vorticity, ζ, the usefulness of ζ_g stems from the ease with which it is determined from the pressure pattern even in areas where wind reports are too scarce for ζ to be obtained with similar accuracy. Since we consider motions largely geostrophic, geostrophic vorticity is a valuable substitute for actual vorticity.[36] The general distribution of ζ_g approximates actual patterns of ζ, particularly above the level of surface friction. Possible discrepancies in values and patterns of ζ_g and ζ are yet to be fully investigated. However, it is worth while stressing at this point that, while horizontal divergence is determined by ageostrophic motion, this does not hold in the case of vorticity.

Figure 10.271 gives several idealized pressure patterns with corresponding distributions of geostrophic vorticity. These were prepared in a plane coordinate system using constant Coriolis parameter 0.98×10^{-4} sec^{-1}. Thus, with $k_1 = 980/f = 10^7$ sec, geopotential height Z of the surface is

35. The reader can prove that geostrophic vorticity on an isentropic surface is

$$\zeta_g \simeq (\nabla^2 \psi)_\theta/f = (\partial^2 \psi/\partial x^2 + \partial^2 \psi/\partial y^2)_\theta/f$$

if $\psi = c_p T + 980 Z$.

36. This substitution is made (with little reservation) in the forecasting aids developed by Sutcliffe (*QJRMS*, LXXIII [1947], 370, and LXXVI [1950], 189) and in the numerical forecasting technique by the Princeton group (Charney, *Journal of Meteorology*, Vol. VI [1949]).

a simple stream function, and $u_g = -k_1(\partial Z/\partial y)$, $v_g = k_1(\partial Z/\partial x)$, and $\zeta_g = k_1(\partial^2 Z/\partial x^2 + \partial^2 Z/\partial y^2)$.

Diagram (a) shows the distribution of ζ_g about a straight shearing current with maximum geostrophic velocity 40 m sec^{-1}. An interesting feature is the large gradients of vorticity with only the small shear visible from the pressure pattern. This reveals how extremely sensitive geostrophic vorticity is to details in the pressure pattern and, indirectly, how sensitive actual vorticity is to the field of motion.

The second gives ζ_g in a sinusoidal wave with constant zonal geostrophic speed $(\partial u_g/\partial y = 0)$. The maximum geostrophic speeds (32 m sec^{-1}) are at the inflection of contours, and the minimum speeds in the trough and ridge. Maximum vorticity coincides with the pressure trough, minimum with the pressure ridge, and the zero line connects inflection points of adjacent contours. Absolute values of ζ_g are proportional to the zonal speed. Observe that the largest gradients of vorticity are found at the minimum absolute contour curvature—at the inflections—and the smallest where contour curvature is greatest. These gradients are directly proportional to zonal speed and wave amplitude and inversely proportional to wave length.

The third diagram (c) represents a composite of the preceding two. The u component of geostrophic wind is 40 m sec^{-1} on Z_0 and varies linearly along y on both sides. The maximum geostrophic speed is 62.5 m sec^{-1}. Although central geostrophic speeds might be extreme for most levels outside the tropopause region, the *distribution* of geostrophic vorticity can be considered typical of the common wave in the westerlies.

Diagram (d) shows a pattern of geostrophic vorticity in a region where uniform zonal motion $(\partial u_g/\partial y = 0)$ spreads downstream. Owing to the required variations in

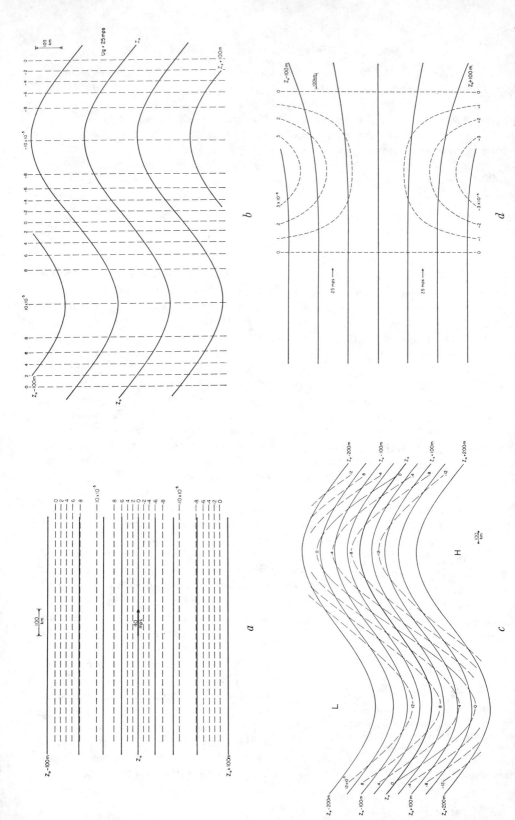

FIG. 10.271.—Geostrophic vorticity (10^{-5} sec^{-1}) for given isobaric topographies; Coriolis parameter constant with value 0.98×10^{-4} sec^{-1}. (In d a line of $\zeta_g = 0$ coincides with contour Z_0).

contour curvature, there is positive geo-strophic vorticity to the left and negative to the right of the straight contour.

Figure 10.272 shows geostrophic vorticity in frequently occurring small-scale features of the upper pressure pattern. Diagram (*a*) gives appropriate patterns found above a young frontal wave at the surface; (*b*)

10.28. *Absolute vorticity.*—The earth has vector angular velocity $\omega = 7.29 \times 10^{-5}$ sec^{-1} directed upward at the north pole, directed parallel to sea level at the equator, and at intermediate latitudes makes an angle with the local sea level equal to the latitude of that point. The projection of this angular velocity vector (of magnitude

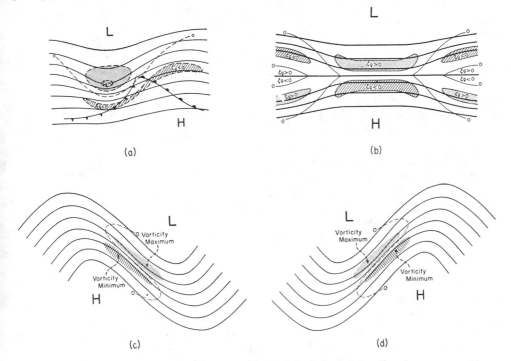

Fig. 10.272.—Isobaric contours and geostrophic vorticity in certain small-scale pressure patterns aloft.

shows a version of the pressure and vorticity patterns about a pronounced concentration of wind or small "jet maximum." Diagrams (*c*) and (*d*) show the distribution in an asymmetric wave having strongest winds near the inflections. These indicate that extremes of geostrophic vorticity are not always associated with troughs and ridges of the pressure field. Here the contribution by horizontal shear is far greater than the curvature effect, and there is decrease of vorticity from inflection to trough on the left side of the current.

ω) upon the local vertical is $\omega \sin \phi$. The vorticity ζ_e of the earth is twice its own angular velocity; therefore, at any latitude, the component of earth vorticity with respect to the local vertical is $\zeta_e = 2\omega \sin \phi = f$.

Absolute vorticity about a vertical axis at any point is the sum of the relative vorticity ζ and the local earth vorticity ζ_e. Thus, *absolute vorticity is the sum of relative vorticity and the Coriolis parameter*. By analogy, geostrophic absolute vorticity is the sum of geostrophic vorticity and the Coriolis parameter. At the equator absolute vorticity is identical with relative vorticity. The pattern of absolute vorticity (actual or geostrophic) is obtained by adding the simple pattern of f to the pattern of ζ or ζ_g.

The Coriolis parameter varies from zero at the equator to 1.46×10^{-4} sec^{-1} at the north pole. Absolute vorticity therefore is usually positive in the northern hemisphere, except near the equator, where its sign is governed by relative vorticity. It is negative only when the magnitude of anticyclonic relative vorticity exceeds the Coriolis parameter. The probability for negative absolute vorticity is great in sharp anticyclonic bends of strong wind currents or on the right side of a jet stream (northern hemisphere). Because of the strong winds necessary, this can occur almost exclusively in the upper troposphere and tropopause region above strongly baroclinic conditions and more locally in tropical storms and tornadoes. Negative absolute vorticity is usually taken as a criterion for *inertia instability;* examine Eq 7.24(11).

10.29. *Factors modifying vorticity.*—Consider the horizontal fluid disk bounded by the solid circle in the upper left of Figure 10.29. Assume that this element lies in linear motion; its vorticity is then solid rotation. The angular velocity of the disk with respect to the earth is $\frac{1}{2}\zeta$, and the component of the earth's angular velocity about the vertical through the element is $\frac{1}{2}f$; thus the absolute angular velocity of the element is $\frac{1}{2}(f + \zeta)$. If r is the radius of the disk, conservation of absolute angular momentum (per unit mass) requires that $(f + \zeta)r^2 = $ constant. Or, if $A = \pi r^2$,

$$(f + \zeta)A = \text{const}. \qquad (1)$$

This is one of the forms of the vorticity theorem: *The product of absolute vorticity and horizontal area of a fluid is conserved.*

This principle can be examined more

thoroughly in its differential form. From Eq (1)

$$\frac{1}{f + \zeta} \frac{d(f + \zeta)}{dt} + \frac{1}{A}\frac{dA}{dt} = 0.$$

By substitution from Eq 10.11(2),

$$d(f + \zeta)/dt = -(f + \zeta)\,\text{div}\,\mathbb{C}. \qquad (2)$$

Expansion gives

$$d\zeta/dt = -\,df/dt - \zeta\,\text{div}\,\mathbb{C} - f\,\text{div}\,\mathbb{C}. \qquad (3)$$

The relative vorticity of a fluid parcel thus varies with (*a*) the change in latitude; (*b*) the product of relative vorticity and hori-

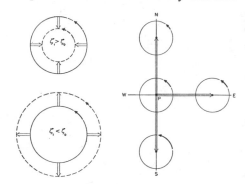

Fɪɢ. 10.29

zontal divergence; and (*c*) the Coriolis effect on horizontal divergence.

a) Changes in latitude.—Consider the horizontal fluid disk centered at P in the right-hand portion of Figure 10.29. If the disk is displaced northward in the northern hemisphere without change in area, Eq (1) shows that, as f increases, ζ must decrease. If the air is displaced southward, its relative vorticity increases to compensate for the decrease of the earth's vorticity about the vertical. Eastward and westward displacements contribute nothing to the change of relative vorticity.

The time derivative of the Coriolis parameter is

$$df/dt = (2\,\omega\cos\phi)d\phi/dt$$

$$= (2\,\omega\cos\phi)(v/a),$$

if a is the earth's radius and v the south-wind component. Following Rossby, $(2\omega \cos \phi)/a = \beta = \partial f/\partial y$, and therefore

$$d\zeta/dt = - \beta v - \zeta \operatorname{div} \mathbb{C} - f \operatorname{div} \mathbb{C}. \quad (4)$$

The factor β is the northward variation of the Coriolis parameter; it is zero at the pole and $2.289 \times 10^{-13} \sec^{-1} \mathrm{cm}^{-1}$ at the equator. For intermediate latitudes see Table 10.29.

With average value $\beta = 1.6 \times 10^{-13} \sec^{-1} \mathrm{cm}^{-1}$ in middle latitudes and meridional wind 10 m $\sec^{-1}$, ζ changes at the rate

increases the numerical value of vorticity but does not alter its sign. By the lower diagram, cyclonic and anticyclonic vorticities are both made smaller by positive horizontal divergence. In a manner of speaking, divergence "dilutes" rotation and convergence "concentrates" rotation.

c) *The Coriolis effect.*—Horizontal divergence is an ageostrophic property of motion. It is obvious that the divergence fields in Figures 10.09(c) and (d) cannot be in equilibrium with a pressure pattern,

TABLE 10.29

Values of $\beta=(2\omega \cos \phi)/a$ in Units 10^{-13} Sec^{-1} Cm^{-1}
at Various Latitudes

ϕ	β	ϕ	β	ϕ	β
0°.	2.289	30°.	1.982	60°.	1.144
5.	2.280	35.	1.875	65.	0.967
10.	2.254	40.	1.753	70.	0.783
15.	2.211	45.	1.618	75.	0.592
20.	2.151	50.	1.471	80.	0.397
25.	2.074	55.	1.313	85.	0.199
				90.	0.000

$1.6 \times 10^{-10} \sec^{-2}$ or $1.5 \times 10^{-5} \sec^{-1}$ per day. Thus, in one day the relative vorticity of moving air may easily halve or double due only to displacement across latitudes.

b) *Effect of divergence on existing vorticity.*—The second term on the right of Eq (3) is a purely kinematic consequence of the motion. Unlike the other terms, it is independent of the field of force and neither creates nor destroys vorticity; it only *modifies* the scalar value. In average conditions this term is easily as large as the first. For example, with vorticity and divergence both $2 \times 10^{-5} \sec^{-1}$, change of vorticity is $4 \times 10^{-10} \sec^{-2}$ or $3.5 \times 10^{-5} \sec^{-1}$ per day.

The change of vorticity due to $- \zeta \operatorname{div} \mathbb{C}$ may be examined from the conservation of angular momentum using the left-hand drawings of Figure 10.29. In the upper one the fluid element is compressed horizontally, reducing its radius and thereby increasing its angular velocity. Convergence

except possibly at the equator. Since divergent motions are not balanced by pressure forces, there must be net acceleration to the right of the motion in the northern hemisphere. If Coriolis acceleration is added to the straight outward flow in Figure 10.09(c), clockwise rotation follows. If the same is done to (d), counterclockwise rotation follows. The patterns of motion take the forms of clockwise and counterclockwise spirals, respectively. Evidently, in the northern hemisphere divergence is a source for anticyclonic vorticity and convergence for cyclonic vorticity.

Except at low latitudes, where f is small, and except in certain local circulations where ζ may be very large numerically, this term is larger than the other terms on the right of Eq (3). Consequently, this factor is usually the principal one in modifying vorticity. In most discussions of this equation the two terms containing divergence are combined as the product of absolute vorticity and divergence: $-(f + \zeta)$ div $\mathbb{C}$. As absolute vorticity is nearly always positive, the effect of horizontal divergence is to decrease cyclonic vorticity (increase anticyclonic vorticity), and the effect of convergence is the opposite.

When $\zeta = -f$, the last two terms of Eq (3) cancel, and the change in vorticity then is a function only of meridional motion. However, this condition is only temporary, since, as soon as vorticity is changed by meridional displacement, $f + \zeta$ becomes different from zero, and divergence then becomes effective.

To determine the limitations inherent in Eq (3), it is useful to refer to the equations of motion, from which the following dynamic relations are obtained considering horizontal acceleration as a function of x, y, z, and t:

$$d\zeta/dt = \nabla \times (d\mathbb{C}/dt) - \zeta \text{ div } \mathbb{C} - [(\partial w/\partial x)(\partial v/\partial z) - (\partial w/\partial y)(\partial u/\partial z)] , \quad (5a)$$

$$d(\text{div } \mathbb{C})/dt = \text{div } (d\mathbb{C}/dt) - \tfrac{1}{2}(\text{def } \mathbb{C})^2 - \tfrac{1}{2}(\text{div } \mathbb{C})^2 + \tfrac{1}{2}\zeta^2 - \tfrac{1}{2}(\text{def}' \mathbb{C})^2$$
$$- [(\partial w/\partial x)(\partial u/\partial z) + (\partial w/\partial y)(\partial v/\partial z)] , \quad (5b)$$

$$d(\text{def } \mathbb{C})/dt = \text{def } (d\mathbb{C}/dt) - (\text{div } \mathbb{C})(\text{def } \mathbb{C}) - [(\partial w/\partial x)(\partial u/\partial z)$$
$$- (\partial w/\partial y)(\partial v/\partial z)] , \quad (5c)$$

$$d(\text{def}' \mathbb{C})/dt = \text{def}' (d\mathbb{C}/dt) - (\text{div } \mathbb{C})(\text{def}' \mathbb{C}) - [(\partial w/\partial x)(\partial v/\partial z)$$
$$+ (\partial w/\partial y)(\partial u/\partial z)] . \quad (5d)$$

We are concerned only with Eq (5a) now; its expansion based on Eq 7.22(1) is

$$d\zeta/dt = -f \text{ div } \mathbb{C} - df/dt + N_{xy}$$
$$- \zeta \text{ div } \mathbb{C} - [(\partial w/\partial x)(\partial v/\partial z)$$
$$- (\partial w/\partial y)(\partial u/\partial z)] . \quad (6)$$

The term N_{xy} is the solenoid density in the horizontal plane, and the last term on the right is the effect of "tilting horizontal planes of air" due to horizontal gradients of vertical motion. The curl of the horizontal viscous force has been omitted.

On comparing Eqs (3) and (6), there are two terms not accounted initially. Both may be considered negligible in the more barotropic regions. The one involving gradients of velocity is zero for strictly horizontal displacements; it is difficult to account for in most conditions, since, although vertical wind shear might be determined, horizontal gradients of vertical motion are unknown. Because of this, it is usually not even considered. However, except at the ground and at the "level" of maximum wind, this term can be easily the same magnitude or larger than any of the others in strongly baroclinic regions. If vertical wind shear is 10 m sec^{-1} per kilometer and vertical velocities vary 1 cm sec^{-1} per 100 km, for example, this term can be of magnitude 10^{-9} sec^{-2}, which is equivalent to f div ζ for divergence 10^{-5} sec^{-1} in middle latitudes.

While that term in Eq (6) together with friction is usually neglected in discussions of vorticity, the solenoid term is overemphasized. For an approximate magnitude of solenoid density in a horizontal plane, consider horizontal pressure gradient 3 mb per 100 km near sea level and temperature variation 1° C per 100 km along the isobars; $N_{xy} = 8.6 \times 10^{-11}$ sec^{-2}. Thus, for these rough average values the change of vorticity due to horizontal solenoids is many times smaller than other contributing factors; it vanishes if reference is made to isobaric instead of level surfaces.

10.30. *Potential absolute vorticity.*—Consider the column of air represented in cross section on the left side of Figure 10.16. It is bounded above and below by conservative surfaces a pressure difference Δp apart. If the horizontal area of this column is A, then the mass of the column is $M = (A/g)\Delta p$. As the layer stretches and shrinks vertically, Δp and A both may vary, but the mass is conserved. Thus, with gravity constant, $A(\Delta p) = $ constant, and by Eq 10.29(1)

$$(f + \zeta)/\Delta p = \text{const.}, \tag{1}$$

or, in terms of initial conditions, $(f + \zeta)/\Delta p = (f_0 + \zeta_0)/\Delta p_0$.

This form of the vorticity theorem was developed and used rather extensively by Rossby.[37] The quantity $(f + \zeta)/\Delta p$ is the general expression for *potential (absolute) vorticity*. Applied to air columns between substantial isentropic surfaces, Δp is the vertical pressure difference through an arbitrary interval of potential temperature or the pressure depth of a layer of unit potential temperature difference.

Eq (1) is the same principle as Eq 10.29(1), since the depth of an air column and its horizontal area are related by continuity. Nevertheless, Eq (1) above shows explicitly that *vertical stretching of air columns increases vorticity* and that *vertical shrinking decreases vorticity*. If $f_0 = 10^{-4}$ sec^{-1} and $\zeta_0 = 0$, relative vorticity must increase 5×10^{-5} sec^{-1} if the air column stretches to twice its initial depth or must decrease by that amount if the air shrinks to half its initial depth. This provides a useful method for studying large- and small-scale divergence, and vertical shrinking of air masses, in terms of absolute vorticity.

10.31. *Constant absolute vorticity (CAV) trajectories.*—While in search for methods of describing and forecasting atmospheric flow patterns in hemispheric scale, Rossby

formulated the concept of CAV trajectories.[37a] These are the fictitious horizontal paths traced by individual air parcels as each maintains its initial absolute vorticity. This assumes that changes in relative vorticity are due only to variations in latitude; in such idealized process the effects of horizontal divergence, friction, and vertical motion are negligible either individually or in sum. The stringency of the assumptions in this development quite naturally limits its application on a quantitative basis, but even the partial success obtained for forecasting primary fields of motion over periods of from 1 to 3 days suggests that large-scale flow patterns at least *tend* to adjust themselves downstream to the steady state corresponding to CAV trajectories. Even this is helpful.

For a thorough treatment of the topic the reader is referred to Rossby's original discussions[38] and to the report by Fultz.[39] Subsequent refinements in theory and application, results of experimental tests, and also graphical and tabular aids for computing the trajectories have been provided.[40] Renewed interest in the subject has led to Wobus' mechanical contrivance[41] (the "wiggle-wagon") for computing trajectories of greater accuracy directly on the working chart. In the subsequent discussion we only review briefly the highlights of the theory and application of the concept.

37. C.-G. Rossby, "Planetary Flow Patterns in the Atmosphere," *QJRMS*, Vol. LXVI, Suppl. 68 (1940).

37a. *Ibid.*

38. *Ibid.*; C.-G. Rossby *et al.*, "Forecasting of Flow Patterns in the Free Atmosphere by a Trajectory Method," in V. P. Starr, *Basic Principles of Weather Forecasting* (New York: Harper & Bros., 1942), Appendix.

39. Dave Fultz, *Upper-Air Trajectories and Weather Forecasting* ("Department of Meteorology, University of Chicago, Miscellaneous Reports," No. 19 [Chicago: University of Chicago Press, 1945]).

40. J. C. Bellamy, "Slide Rule for Constant Vorticity Trajectories," appended to Fultz, *op. cit.*

41. *First Progress Report, Task 19, U.S. Navy Bureau of Aeronautics* (Project AROWA, January, 1953).

If an air parcel maintains CAV, then

$$f + \zeta = f + \partial c/\partial n + c/R_s = \text{const.}$$

along its path. The trajectory of the parcel is always tangent to its streamline; in steady state the trajectory coincides with the streamline $(R_t = R_s = R)$, and speed is constant. Further, in a broad uniform current, or in the *axis* of a well-developed current, horizontal shear $\partial c/\partial n$ is zero. With these provisions, $f + c/R = f_0 + c/R_0$.

As a first approximation the Coriolis parameter is represented by a linear func-

origin of coordinates in Figure 10.31. As the air moves to higher latitudes (curve *a*), f increases, and, to maintain CAV, the relative vorticity of the parcel decreases. Since we assume that changes in vorticity are manifest only as changes in trajectory curvature, its path curves clockwise. As the path turns, the poleward component decreases, until eventually the current is from due west. As the parcel continues on, y decreases, and clockwise curvature decreases. When the parcel reaches its initial latitude, the trajectory is straight once more. It

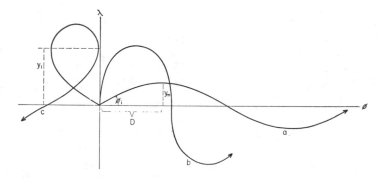

FIG. 10.31.—Selected constant absolute vorticity trajectories

tion of distance northward from the latitude of f_0, that is,

$$f = f_0 + (\partial f/\partial y)y = f_0 + \beta y .$$

If β is taken constant for the range of latitude embraced by the trajectory, then $c/R = -\beta y + c/R_0$. If we take only those parcels situated initially in straight flow or at inflection points in curved flow, then $R_0 = \infty$, and

$$c/R = -\beta y, \quad \text{or} \quad Ry = -c/\beta . \quad (1)$$

Initial points having finite streamline curvature also can be used if corrections are made by successive approximations (as from the slide rule developed by Bellamy).

Consider a wind from southwest[42] satisfying the prescribed initial conditions at the

42. Northern hemisphere.

develops increasing counterclockwise curvature toward lower latitudes. This trajectory describes a periodic sinusoidal oscillation about the initial latitude. If the given wind is from south (curve *b*), the trajectory assumes a similar distribution of curvature; but for the same wind speed the trajectory has greater amplitude and shorter wave length. In accord with the general rule for all trajectories, as the wind speed is increased, the wave length and amplitude of the trajectory are also increased proportionately.

The CAV trajectory for an initial wind from west should coincide (nearly) with a latitude circle. A given wind from due north describes a CAV trajectory (in the northern hemisphere) the inverted image of curve *b*, and one from northwest gives a trajectory the inverse of *a*. Winds from easterly directions describe CAV trajectories quite differently. For example, curve *c*, drawn for initial wind from 120°, gives closed loops north and then south of the initial latitude, and the trajectory progresses westward. For the critical wind direction about 130° (southeast) or 50° (northeast) the trajectory forms a figure of eight.

The information required for computing a CAV trajectory are wind speed *c*, wind direction ϕ_i (reckoned from the latitude circle), and latitude ϕ. From these parameters it is possible to evaluate wave amplitude y_m, quarter wave length *D*, and quarter period τ. Once these points for the trajectory are entered on the chart, the trajectory is drawn freehand, and the position and wind direction for any desired time then can be found. For an easterly trajectory forming loops, there are provisions for finding distance y_1 from the initial latitude to the latitude at which the trajectory has meridional direction.

The assumptions underlying CAV trajectories are so numerous it is surprising the results are of any value at all. Absolute vorticities of individual air parcels are not strictly conserved, but quite likely divergence patterns are of such small horizontal dimensions compared to dimensions of the trajectory that the net effect over the entire length is small. Another factor which can invalidate CAV trajectories is vertical motion, which changes the levels of individual air parcels and carries them into different velocity and vorticity. This difficulty is partially minimized by dealing with deep-seated and rather extensive air currents (of relatively invariant direction with height) in which vertical mixing is permitted, instead of referring only to certain levels. If vorticity and speed are approximately linear functions of height, the most representative trajectory for a column of air is found at the intermediate level of average

vorticity and speed. From experience with both 700-mb and 500-mb levels it appears that in general the most reliable trajectories are obtained somewhere between—perhaps 600 mb.

The most perplexing problem by-passed is the assumption that relative vorticity is reflected as trajectory curvature. Initial points can be selected with minimum horizontal shear, but there is no assurance the selected parcels or columns remain in the center of the current. Once vorticity resolves itself from curvature to shear, the actual CAV trajectory deviates from the computed trajectory. There is no way of anticipating the sense or magnitude of this departure; some degree of error has to be allowed in using the results.

10.32. *Application of CAV trajectories to large-scale flow patterns and weather.*—The concept of CAV trajectories has been applied with some success to forecasting primary fields of motion in the middle troposphere for periods up to about 3 days. The methods in use are concerned primarily with the CAV paths taken by centers of well-developed *current systems*, particularly the stronger ones of recent origin or intensification.

A commonly observed sequence in the upper flow pattern is illustrated roughly by Figure 10.32*a*. Suppose the initial state is represented by straight isobaric contours with westerly winds. Then, for some reason, a large pressure fall (cyclogenesis) occurs in the upstream portion of the pattern, giving rise to a strong southwesterly current just east of the large drop in pressure. The common sequence of events downstream is portrayed—although by no means explained—by a CAV path for this new current. While progressing toward higher latitudes, the current turns clockwise about an area of pressure rise, and it eventually returns as a northwesterly current separating the pressure rises from incipient pressure falls farther east. This is viewed as a downstream adjustment of the wave pattern to the initial pronounced disturbance. The process has been described by Yeh, considering energy dispersion downstream from a source coinciding in location with the initial pressure fall.[43] The sequence is frequently verified from observation of daily upper-air flow patterns.[44] Similar processes can be visualized for newly formed intense currents from other directions.

Daily use of CAV trajectories is made by the Extended Forecast Section, U.S. Weath-

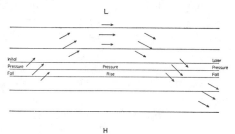

FIG. 10.32a

er Bureau, in forecasting 5-day-mean pressure patterns.[45] From experience with the technique and a number of statistical and synoptic investigations on departures arising from orographic influences and preferred distributions of horizontal divergence, certain qualitative corrections are made to improve the reliability of computed trajectories.

CAV trajectories also may be used as a

43. T. C. Yeh, "On Energy Dispersion in the Atmosphere," *Journal of Meteorology*, Vol. VI (1949).

44. Department of Meteorology, University of Chicago, "On the General Circulation of the Atmosphere in Middle Latitudes," *Bulletin of the American Meteorological Society*, Vol. XXVIII (1947); G. P. Cressman, "On the Forecasting of Long Waves in the Upper Westerlies," *Journal of Meteorology*, Vol. V (1948).

45. J. Namias, *Extended Forecasting by Mean Circulation Methods* (Washington, D.C.: U.S. Weather Bureau, 1947).

basis for describing the distribution of weather in certain large-scale flow patterns. Consider once more a broad current of rather uniform velocity, or the axis of an intense current; in either case relative vorticity is indicated by c/R_s. If the shear term does not change along trajectories, departures of *actual* trajectories from CAV are the result of processes changing absolute vorticity, the principal of which is horizontal divergence. Near the ground the effect of friction might be significant also, but we might exclude specific reference to the surface layer.

Suppose either a or b in Figure 10.31 is the CAV trajectory for such a current. If the actual path is in close agreement with CAV, indications are that the motion is approximately nondivergent. The magnitude and sense of departure are indicators of magnitude and sign of divergence. In the northern hemisphere, *if broad air currents take paths to the left of representative CAV trajectories, there is indication of horizontal convergence; if they take paths to the right of representative CAV trajectories, there is indication of horizontal divergence.* This rule is applicable to instantaneous flow patterns only if the pattern is reasonably steady.

As corollaries to the above rule the following statements are in order (for northern hemisphere): *Large masses of air moving southward along straight or anticyclonically curved trajectories are undergoing horizontal divergence and vertical shrinking; only if they curve very much cyclonically is there indication of horizontal convergence and vertical stretching. Large masses of air moving northward along straight or cyclonically curved trajectories are undergoing horizontal convergence and vertical stretching; only if they curve very much anticyclonically is there indication of horizontal divergence and vertical shrinking.* These rules may be explained also with reference to potential vorticity. The conclusions might be extended to include the sign of prevailing vertical motion through the troposphere, and hence the attendant types of weather, if attention is restricted to certain levels or layers in the troposphere.

A classic example of the relation of horizontal divergence and weather with the trajectories taken by air masses in the lower troposphere is offered by Figure 10.32b. Cold outbreaks to the rear of traveling cyclones in middle latitudes spread southward and eastward. The western portions take

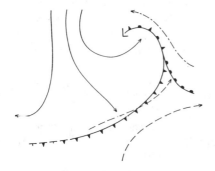

FIG. 10.32b.—Typical trajectories taken by the warm and cold air masses in the lower levels relative to a final surface frontal pattern.

trajectories with predominant anticyclonic curvature, horizontal divergence, and sinking. The portion of the cold mass nearer the cyclone center almost invariably follows a cyclonic course, especially near intensifying cyclones. This is the portion of the fresh cold air most likely possessing horizontal convergence, vertical stretching, net ascending motions, persistent cloudiness, and instability precipitation.

The cold air in advance of the cyclone usually will have undergone horizontal divergence during its previous sojourn in lower latitudes, but, as the intensifying cyclone approaches, this air is carried into the circulation with increasing cyclonic curvature and later may curve even more sharply to the rear of the center. This precyclone convergence can be directly responsible for

much of the precipitation occurring with the warm front.

The warm air in moving northward also shows certain preferred trajectories in relation to the cyclone center. The warm air nearest the center may have started moving northward with anticyclonic curvature while under domination of high pressure. With approach and broadening of cyclonic circulation, the northwest portion of this air is drawn more cyclonically. Warm air at greater distances from the cyclone assumes trajectories more anticyclonic and shows less likelihood for cloudiness and precipitation.

10.33. *Dynamic factors in horizontal divergence.*—It is seen from Eqs 10.29(5) that individual change of a linear property of velocity is the sum of the corresponding property of the acceleration field and certain kinematic products. Excluding the last term, we found the change of vorticity dominated by the curl of acceleration, except in equatorial regions, where $f \simeq 0$. It is at least suggestive that in each of the other equations the acceleration field is a major contribution in ordinary scale.

For the individual change of divergence, Eq 10.29(5b), we will describe only the dynamic effect,[46] that is, acceleration divergence. This is the field of force (per unit mass) *causing* the fluid to spread horizontally. Its effect can be seen graphically in applying the area method for measuring divergence to three winds, identical initially, each with different acceleration during interval δt.

By differentiating the equations of motion in the form given by Eqs 7.21(1), di-

46. Analysis of the other terms and their contribution in special conditions is left for the reader. These have been outlined by J. C. Bellamy (Institute of Tropical Meteorology, Rio Piedras [Puerto Rico], 1943) (mimeographed); L. Sherman, "On the Scalar-Vorticity and Horizontal-Divergence Equations," *Journal of Meteorology*, Vol. IX, No. 5 (1952).

vergence of accelerations, with only pressure and Coriolis forces acting, is

$$\partial \dot{u}/\partial x + \partial \dot{v}/\partial y = f\zeta - f\zeta_g - (u$$
$$- u_g)\partial f/\partial y + (v - v_g)\partial f/\partial x , \quad (1)$$

where $\dot{u} = du/dt$ and $\dot{v} = dv/dt$. If y is northward, then $\partial f/\partial x = 0$ and $\partial f/\partial y = \beta$, and, if $u - u_g$ is u', then

$$d(\text{div } \mathbb{C})/dt \approx f(\zeta - \zeta_g) - \beta u' . \quad (2)$$

For the west (or east) ageostrophic wind we might consider values from 1 to 10 m sec^{-1}. For $\beta = 1.6 \times 10^{-13}$ sec^{-1} cm^{-1}, $-\beta u'$ is 1.6×10^{-11} to 1.6×10^{-10} sec^{-2}. In con-

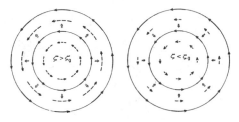

FIG. 10.33.—Illustration of divergence and convergence of accelerations about a point.

trast, $f(\zeta - \zeta_g)$ is usually of larger magnitude (10^{-9} sec^{-2} for $f = 10^{-4}$ sec^{-1} and ageostrophic vorticity 1×10^{-5} sec^{-1}). Therefore, except in latitudes where f is very small, and except with extreme zonal ageostrophic wind, the first term on the right of Eq (2) dominates.

Acceleration divergence due to Coriolis effect on ageostrophic vorticity ($\zeta' = \zeta - \zeta_g$) is illustrated graphically by Figure 10.33. The dashed vectors represent motion about the origin in uniform vorticity ζ. The solid vectors represent geostrophic motion about the same point in a region of uniform[47] geostrophic vorticity ζ_g. Hence, in these fields, ζ' is also constant. In the pattern at the left motion is supergeostrophic, giving rise to net accelerations everywhere to the right of the ageostrophic components and therefore

47. There is no loss of generality in our use of *uniform* fields, which is simply to facilitate the discussion.

outward from the origin. The same is found for $\zeta > \zeta_g < 0$. Thus, *if actual vorticity is greater than geostrophic, accelerations are directed outward, and motion is forced toward divergence.* If motion is already divergent, divergence should be increased, whereas if motion is converging, accelerations tend to decrease convergence.

In the diagram at the right ζ and ζ_g are both positive, but $\zeta < \zeta_g$, and the net acceleration is to the left of the motion. With clockwise ageostrophic circulation, accelerations are inward. The same is found for anticyclonic vorticity with $\zeta < \zeta_g$. Hence, *if actual vorticity is less than geostrophic, air is forced toward convergence through action of the accelerations.*

From the above it is evident that, for given latitude, where vorticity is nearly geostrophic there is little tendency for velocity divergence of either sign to be changed. Where vorticity differs greatly from geostrophic, acceleration divergence is large and acts to change velocity divergence. In this respect, departure of actual from geostrophic vorticity is vitally important to the dynamics of weather, even though it might be only a small fraction of the vorticity.

Geostrophic vorticity is a geometric property of the pressure field and not a property of the air parcels within it. The pattern of ζ_g is propagated not with air motion but with the pressure field instead, and it undergoes individual changes in structure only as the pressure field changes. Now the air has motion relative to the geostrophic vorticity pattern; almost everywhere winds differ from the propagation velocity of the pressure field. In particular, in the center of certain wind currents aloft, air can be moving ten times as fast as the pressure pattern. In lighter winds the pressure pattern may be moving faster than the air. Consequently, *air generally moves through the geostrophic vorticity pattern.*

Consider an air parcel in Figure 10.271*b* situated initially along the line of maximum ζ_g. Suppose at its initial position $\zeta_0 \approx \zeta_{g_0}$. As the parcel moves to the right relative to the pressure pattern, it finds itself in successively lower values of ζ_g. Since at a later time its ζ exceeds ζ_g, $\zeta' > 0$, and the parcel is subjected to acceleration divergence which should continue as long as the parcel is crossing ζ_g lines toward lower values. Meantime, as the parcel diverges it decreases its ζ (if $f + \zeta > 0$), according to Eq 10.29(3), and thus the motion "absorbs" vorticity from its ambient pressure field. *The parcel is forced to diverge during the process of adjustment between pressure field and motion.*

Now consider a fast-moving air parcel upstream from the ζ_g maximum. As it crosses vorticity lines toward higher values, its vorticity is subgeostrophic, and accelerations are convergent. The tendency for dynamically enforced convergence of the parcel should be effective at least until it reaches maximum ζ_g. During this travel the parcel also increases its ζ through convergence, and its vorticity thus trends toward the vorticity of its new pressure field.

A detailed analysis of the expression $d\zeta'/dt = d\zeta/dt - d\zeta_g/dt$ using Eq 10.29(3) reveals that, following an air parcel, (i) motion relative to the pressure pattern provides a mechanism for actively maintaining divergence and (ii) there are preferred similar *trends* of ζ and ζ_g (so long as $f + \zeta > 0$), so that, even if ζ and ζ_g may differ, $d\zeta/dt \rightarrow d\zeta_g/dt$. Some lag might be expected in their rates. Only for qualitative inferences concerning horizontal divergence patterns, we might permit the rough equality, $d\zeta'/dt \approx 0$, and Eq 10.29(2) becomes

$$\text{div } \mathbb{C} \approx -[(c - \gamma)(\partial \zeta_g / \partial s)$$
$$+ \beta v]/(f + \zeta) . \quad (3)$$

Here c is wind speed, s is distance down-stream, and γ is the speed of the ζ_g pattern in that direction, which can be determined by a path method or any other method applicable for displacement of patterns. It is the speed of the pressure pattern corrected, if necessary, for intensification; $(c - \gamma)$ is merely air motion *relative* to the ζ_g (pressure) pattern. Eq (3) also may be expressed in the form

$$\text{div } \mathbb{C} \approx -\frac{(c - \gamma)}{f + \zeta} \frac{\partial}{\partial s}(f + \zeta_g) . \quad (4)$$

With additional approximation, ζ may be replaced by ζ_g. In Eq (4), $(c - \gamma)$ is air motion relative to the pattern of geostrophic absolute vorticity.[48]

The effect of βv is to give convergence in poleward flow and divergence in equatorward flow. For *meridional* wind component 30 m sec^{-1} and $f + \zeta \approx 10^{-4}$ sec^{-1}, βv gives divergence of magnitude about 5×10^{-6} sec^{-1} in middle latitudes. In comparison, the relative motion term gives divergence of 10^{-5} sec^{-1} for relative speed 10 m sec^{-1} and downstream vorticity gradient 1×10^{-5} sec^{-1} per 100 km. The latter effect apparently dominates except in rather straight and strong currents directed meridionally.

Let us consider practical applications of the above analysis with reference to Figure 10.271. Since we can cover only a few cases here, the reader is encouraged to examine more of the patterns occurring daily. In (*b*) maximum values of $\partial \zeta_g / \partial s$ occur at inflections of curvature in pressure contours. If $c > \gamma$, the greatest probability for velocity divergence and convergence occurs near those inflection points—divergence ahead of the trough and convergence to the rear. This agrees with conclusions given independently by the gradient wind in Figure

48. The above equations are not restricted only to horizontal displacements. They are applicable to three-dimensional motion if c is regarded as total air speed, γ the speed of ζ_g(or $f + \zeta_g$) surfaces in the direction of air motion, and $\partial/\partial s$ the variation in that direction. This helps to explain Sec. 10.19(*d*).

10.193. In (c) preferred locations for divergence and convergence should be similar, but, owing to strong shearing motion, $(c - \gamma)(\partial \zeta_g / \partial s)$ varies with respect to the center of the current. It has maximum magnitudes on the current axis; outward from this axis it decreases and perhaps even reverses in sign. Here is evidence that *largest divergence and convergence in an upper-level wave are expected near inflections of contour curvature and within the current center.*

The straight current in Figure 10.271a has large gradients of ζ_g, but, since motion is apparently parallel to ζ_g lines, there is no indication of enforced divergence anywhere. However, only slight perturbation of the current can set air in motion across vorticity lines and thus provide the mechanism for strong horizontal divergence.

A related vorticity field is illustrated by Figure 10.26. Interesting deductions follow when we superimpose either circular system upon a straight uniform basic current.[49] In this composite field the circular ζ_g pattern is effectively unchanged. However, the (geostrophic) winds then have components *across* vorticity lines which give divergence downwind from the ζ_g maximum and upwind from its minimum, and conversely for location of convergence, if the basic current is stronger than the speed of the system. When interpreted in terms of the relation of divergence to local pressure changes, this mechanism suggests some explanation for the "steering" of perturbations by a basic current, including propagation of tropical storms.

In Figure 10.271(d) air moving from the upper left part of the pattern is subjected to horizontal convergence as it enters the area of positive ζ_g, and the opposite in the lower part of the pattern. The general distribution of ζ_g—positive on the left and negative on the right of the straight contour—is the only type possible for a straight

49. See Figure 11.02.

uniform current spreading downstream without discontinuities in geostrophic wind. From this it follows that in diverging contours there should exist horizontal divergence on the right and convergence on the left of the straightest flow, provided the air moves faster than the pressure pattern (compare with Fig. 10.192).

In Figure 10.272(b) the rapid variation in shear from the straight to the diverging contours overcompensates the required increase in curvature and gives downstream gradient of ζ_g opposite to Figure 10.271d and thus opposite locations of preferred divergence and convergence in the spreading contours. From comparing the two drawings and considering the numerous possible variations of Figure 10.272(b), it is found that distribution of horizontal divergence in areas of spreading (or converging) contours varies with the case. *Divergence should depend on relative motion and the pattern of geostrophic vorticity and is not determined uniquely by divergence of contours.*

The foregoing discussions offer a physical explanation for preferred divergence fields in relation to geometry of the pressure pattern. It is made apparent that details in the structure of the pressure pattern have important bearing on distribution of weather and also on evolution in the pressure pattern itself. The latter relation has been used from a deductive approach by Sutcliffe in an attempt to associate future developments in the pressure field with the vorticity in pressure patterns at the surface and aloft and also in the temperature distribution between.[50] Sutcliffe bases his work on the vorticity theorem, Eq 10.29(3), in which the substitution of geostrophic for actual vorticity is made; that is, $d\zeta_g / dt = -(f +$

50. R. C. Sutcliffe, "A Contribution to the Problem of Development," *QJRMS*, LXXIII (1947), 370–83; R. C. Sutcliffe and A. G. Forsdyke, "The Theory and Use of Upper Air Thickness Patterns in Forecasting," *QJRMS*, LXXVI (1950), 189–217.

ζ_g) div $\mathbb{C} - \beta v$. This merely assumes $d\zeta'/dt = 0$.

From the pressure and vorticity patterns in Figures 10.271, there are further practical though qualitative rules which follow immediately. First, it is seen that air situated in a geostrophic vorticity maximum is *potentially divergent*, since, regardless of the direction it takes outward, the chances are that ζ' will be positive, and the air is subject to divergence. In a similar way, air located in a minimum of geostrophic vorticity is *potentially convergent*. Where air moves more rapidly than do features of the pressure field, *the greatest tendency for horizontal divergence should exist downstream from highest positive geostrophic vorticity, and the greatest tendency for horizontal convergence should exist downstream from highest negative geostrophic vorticity*. The opposite holds for pressure patterns, or regions of pressure patterns, which move faster than the air.

We might assume that such wave patterns as in diagrams (*b*) and (*c*) are characteristic of the middle and upper troposphere for middle and high latitudes. They are usually propagated with the same speed at all levels up to the lower stratosphere, and we may then say that γ is approximately independent of height. Since wind speeds in the baroclinic westerlies increase upward to the tropopause, $\partial c/\partial z > 0$, and $\partial \zeta_g/\partial s$ most likely also increases in magnitude with height up to that level. Therefore, excluding the lower troposphere where the pressure pattern is usually quite different, owing to both factors $(c - \gamma)(\partial \zeta_g/\partial s)$ *should have its maximum absolute values near the tropopause*. The reality of this statement is Figure 10.20c. At lower levels this term is apt to be smaller with the same sign; nearer the ground it is likely to have sign opposite to that above.

If the geometry of pressure patterns at upper levels does exert such a control on horizontal divergence, then it also will show its influence on vertical motions and isallobaric fields. Predominant ascent and pressure fall are found downstream from maximum vorticity and predominant descent and pressure rise downstream from minimum vorticity in the pressure patterns of the upper troposphere, in areas where winds are stronger than the propagation velocity of pressure patterns.[51]

Since pressure patterns in the upper troposphere almost parallel the temperature distribution in the troposphere below, similar reasoning is applicable to a representative thermal pattern in the troposphere. Sutcliffe and Forsdyke discuss this adaptation with respect to the relative topography between 1000 and 500 mb.[52] The "thermal geostrophic vorticity" pattern and thermal wind in the relative topography can be viewed in most cases as indicative of vorticity distribution and wind velocity in the upper troposphere. The reader is encouraged to analyze the situation for 1 March 1950 from pressure, temperature, and weather charts with these principles in mind.

10.34. *Horizontal deformation.*—Horizontal *stretching deformation* is

$$\text{def } \mathbb{C} = \partial u/\partial x - \partial v/\partial y = 2a . \quad (1)$$

Its geometric interpretation is seen by the change in shape of a fluid element in Figure 10.09g, or by the types of analysis used in Figure 10.11. Its magnitude is the differential stretching along the two coordinate axes, and its sign indicates the direction in which the fluid element is elongated. In the special case $\partial u/\partial x = \partial v/\partial y$, the motion is *nondeformative* by stretching, but there is divergence.

51. See L. G. Starrett, "The Relation of Precipitation Patterns in North America to Certain Types of Jet Streams at the 300-mb Level," *Journal of Meteorology*, VI (1949), 347–52.

52. *Op. cit.*

Horizontal *shearing deformation*,

$$\text{def}' \, \mathbb{C} = \partial v / \partial x + \partial u / \partial y = 2a' , \quad (2)$$

is the difference in rates of horizontal shearing along the two coordinate axes. Distortion in shape due to shearing is visualized readily from differing angular velocities of

gives individually the desired total rate and sense of deformation in the motion, unless the other is zero. To derive full value from this concept, the two types are combined to give the *resultant direction* of elongation and *net rate* of deformation. This is done by rotating the coordinate axes through such

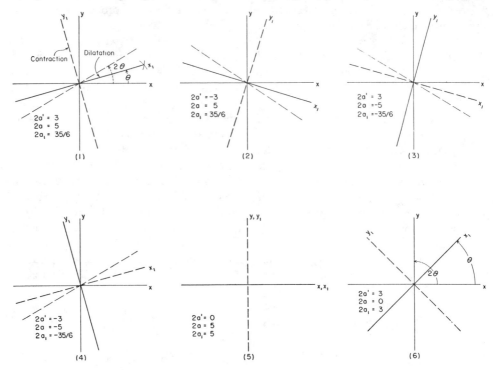

Fig. 10.34a.—Resultant axes of deformation for given values of shearing deformation (2a') and stretching deformation (2a). The axis of dilatation is the heavy continuous line; the axis of contraction, the heavy dashed line.

the two arms of the fluid cross in Figure 10.22. If $\omega_x = \omega_y$ (i.e., $\partial v / \partial x = -\partial u / \partial y$), there is rotation but no change in shape, and the motion is nondeformative by shearing. For positive-shearing deformation the axis of dilatation is 45° counterclockwise from the x-axis, and for negative-shearing it is 45° clockwise from x. Regardless of the signs or rates of deformation, the axes of dilatation in both types of deformation are always 45° apart.

Neither deformation described above

an angle that in the new system one type of deformation vanishes.

If we select resultant deformation $\text{def}_r \, \mathbb{C}$ as of the stretching type (i.e., eliminate shearing deformation), then[53]

$$\tan 2\theta = (\partial v / \partial x + \partial u / \partial y) / (\partial u / \partial x \\ - \partial v / \partial y), \quad (3)$$

53. For this derivation see W. J. Saucier, "Horizontal Deformation in Atmospheric Motion," *Transactions of the American Geophysical Union*, Vol. XXXIV, No. 5 (1953).

if θ is the angle between the principal axes and the initial xy coordinates. If both deformations are of same sign, the coordinates are rotated counterclockwise through angle θ.

The magnitude of resultant deformation is[54]

$$\text{def}_r\ \mathbb{C} = (\partial u/\partial x - \partial v/\partial y)\ \sec 2\theta$$
$$= (\partial v/\partial x + \partial u/\partial y)\ \csc 2\theta . \quad (4)$$

This equation is not valid for $\theta = 0$ or $\theta = \pm 45°$, but in those cases it is not needed.

The procedure in determining resultant deformation is: (1) measure stretching and shearing deformation in the original coordinates; (2) from Eq (3) find θ; (3) locate

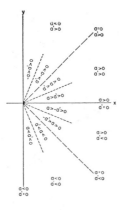

Fig. 10.34b.—Directions or sectors in which the axis of dilatation lies for the various relative values of stretching (a) and shearing (a') deformation.

the principal axes $\theta°$ from initial axes; (4) determine def$_r$ $\mathbb{C}$ from Eq (4); (5) if def$_r$ $\mathbb{C} > 0$, the principal x-axis is that of dilatation, but, if def$_r$ $\mathbb{C} < 0$, the axis of dilatation is the principal y-axis; and (6) the abso-

54. From the trigonometric identity sec $2\theta = \sqrt{(\tan^2 2\theta + 1)}$, and substituting for $\tan 2\theta$ from Eq (3) into Eq (4), it follows that

$$|\text{def}_r\ \mathbb{C}| = \sqrt{[(\partial u/\partial x - \partial v/\partial y)^2 + (\partial v/\partial x}$$
$$+ \partial u/\partial y)^2]} .$$

lute value of deformation and direction of dilatation are assigned to the point in question. The information in (6) gives a complete description of horizontal deformation about that point.

This operation is clarified by an example (Fig. 10.34a). Suppose from the first step

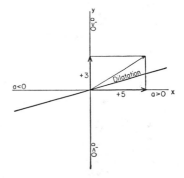

Fig. 10.34c.—Graphically obtaining resultant rate and direction of deformation.

$(\partial u/\partial x - \partial v/\partial y) = 5$ and $(\partial v/\partial x + \partial u/\partial y) = 3$ in units 10^{-5} sec^{-1}. Then $\tan 2\theta = 3/5$, $2\theta = 31°$, and $\theta = 15\frac{1}{2}°$. As sec $2\theta \simeq 7/6$, def$_r$ $\mathbb{C}$ is $+35/6 \times 10^{-5}$ sec^{-1}, and the principal x-axis is that of dilatation.

The resultant axis of dilatation and rate of deformation can be determined graphically and more readily by the scheme in Figure 10.34c. Stretching deformation is plotted to scale as a vector along x, positive along $x > 0$, and negative along $x < 0$, and shearing deformation is entered similarly along y. The length of the vector resultant gives the desired (absolute) rate of deformation,[55] and the axis of dilatation is the bisector of the angle between the positive x-axis and this vector resultant.

10.35. *Measurement of deformation by finite differences.*—Stretching deformation can be determined from field distributions of west wind (u) and south wind (v) by

$$\text{def}\ \mathbb{C} \simeq \delta u/\delta x - \delta v/\delta y - (v/a)\ \tan \phi ,$$

55. See n. 54.

and shearing deformation by

$$\text{def}' \; \mathbb{C} \simeq \delta v/\delta x + \delta u/\delta y + (u/a)\tan\phi \, .$$

The correction terms accounting for convergence of meridians are negligible in most instances except near the pole. In subsequent discussions they are omitted altogether.

The pattern of $\partial v/\partial y$ subtracted from that of $\partial u/\partial x$ in Figure 10.031 gives the stretching deformation in Figure 10.35a relative to the latitude-longitude grid of the earth, and $-\partial u/\partial y$ subtracted from $\partial v/\partial x$ gives the shearing deformation in Figure 10.35b. From these two sets the resultant rate and direction of deformation were found at each intersection of unit meridians and parallels by use of Eqs 10.34(3) and (4). The final analysis of the deformation field is Figure 10.35c. Heavy lines are isogons for

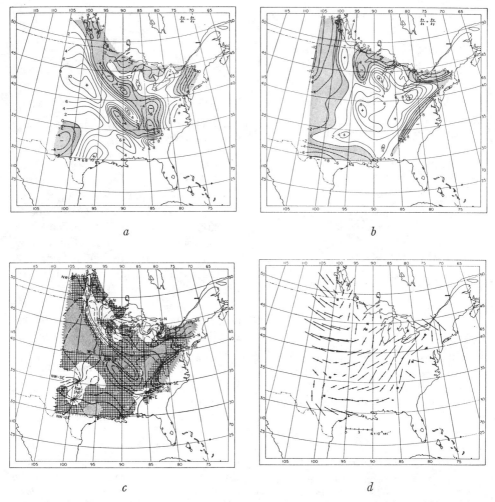

a

b

c

d

FIG. 10.35.—Field of deformation at 4000 feet, 1500 GCT, 1 March 1950. (a) $\partial u/\partial x - \partial v/\partial y$; ($b$) $\partial v/\partial x + \partial u/\partial y$; ($c$) resultant direction and rate of deformation; (d) vector representation of (c); (a) and (b) are in units knots per degree latitude; in (c) crosshatching is 2 to 4×10^{-5} sec^{-1}, stippling $>4 \times 10^{-5}$ sec^{-1},

dilatation, for intervals 15°, and the shading indicates rate of deformation. Where deformation is zero, direction of dilatation is indeterminate and is shown by intersections of isogons in Figure 10.35c. Notice by comparison with Figure 10.031b that these singularities bear no simple relation to wind speed.

Figure 10.35d gives a more legible but discontinuous representation of the same field of deformation. Dilatation is indicated by shafts at selected points over the map; the length of the shaft is drawn proportional to the local rate of deformation. An inter-

10.36. *Measurement of deformation from kinematic directions.*—From our previous discussions on transition from Cartesian coordinates to kinematic directions with reference to divergence and vorticity, it is seen that

$$\partial u/\partial x - \partial v/\partial y = c(\partial a/\partial n) - \partial c/\partial s,$$

and

$$\partial v/\partial x + \partial u/\partial y = \partial c/\partial n + c(\partial a/\partial s)$$
$$= \partial c/\partial n - cK_s.$$

The two deformations can be measured from streamline-isotach patterns by means simi-

TABLE 10.36

DIRECTION OF DILATATION RELATIVE TO THE STREAMLINE (OR WIND) DIRECTION DUE TO INDIVIDUAL COMPONENTS OF DEFORMATION

Longitudinal stretching ($\partial c/\partial s > 0$)	:	along streamlines
Longitudinal shrinking ($\partial c/\partial s < 0$)	:	normal to streamlines
Diverging streamlines ($\partial a/\partial n > 0$)	:	normal to streamlines
Converging streamlines ($\partial a/\partial n < 0$)	:	along streamlines
Cyclonic shear ($\partial c/\partial n > 0$)	:	45° to the right of streamlines
Anticyclonic shear ($\partial c/\partial n < 0$)	:	45° to the left of streamlines
Cyclonic curvature ($K_s > 0$)	:	45° to the left of streamlines
Anticyclonic curvature ($K_s < 0$)	:	45° to the right of streamlines

esting feature of this chart, when compared with Figure 10.031b, is the large average angle between the wind and the axis of dilatation and also the large rates of deformation occurring in apparently uniform flow. This shows definitely that deformation is not to be associated only with neutral points or cols, where, in fact, deformation is usually quite small with the light winds.

On comparing the rates of deformation with the rates of divergence given in Figure 10.031i, it is found that deformation is generally somewhat larger in magnitude than divergence. For the case illustrated, the average magnitude of divergence was 2.2, while for deformation it was 5.4 in units 10^{-5} sec^{-1}. In the majority of cases the magnitudes of $\partial u/\partial x$ and $\partial v/\partial y$ are subtractive for divergence but additive for deformation.

lar to those employed for divergence and vorticity. The angle of rotation θ (*measured from the streamline*) follows by substitution into Eq 10.34(3). Thus

$$\tan 2\theta = (cK_s - \partial c/\partial n)/[\partial c/\partial s$$
$$- c(\partial a/\partial n)], \quad (1)$$
and

$$\mathrm{def}_r \mathbb{C} = [\partial c/\partial s - c(\partial a/\partial n)] \sec 2\theta$$
$$= (cK_s - \partial c/\partial n)\csc 2\theta. \quad (2)$$

Positive deformation from Eq (2) implies that the axis of dilatation makes angle θ with the streamline.

From Eqs (1) and (2) the effect of each factor in determining direction of dilatation is summarized in Table 10.36. In most conditions several of the individual components are effective, and the resultant direction is found by giving the proper weight to each component direction.

With the table as a guide, it is now possible to interpret some of the more prominent features of Figure 10.35d with respect to Figure 10.031b. In the extreme northwestern part of the pattern along the current axis and upstream from the strongest winds there is little shear, curvature, or streamline divergence, but rapid longitudinal stretching places dilatation very nearly along the streamlines. Near 40° latitude and 90° longitude there is rapid longitudinal shrinking and some streamline divergence, while shear and curvature appear to be small, and therefore contraction along the streamlines. On the right flank of this wind maximum are several places where stretching deformation seems small, but relatively strong anticyclonic shear and cyclonic curvature combine to place dilatation about 45° to the left of the wind. Observe that the opposite is the case on the left side of the wind maximum where cyclonic shear predominates.

**10.37. *Planimetric evaluation of deformation.*—It is possible to determine deformation from three wind reports. Each of the three winds is rotated to vectors $\mathbb{C}''$ bearing the following relation to the actual winds: $u'' = u$, $v'' = -v$. The orientation of initial coordinates is arbitrary, but the same coordinates are used for the entire process for each set of three winds. The divergence of the new vectors is $\partial u''/\partial x + \partial v''/\partial y = \partial u/\partial x - \partial v/\partial y$. The percentage rate of areal expansion given by the new vectors is the stretching deformation for the actual winds.

A second set of vectors $\mathbb{C}''$ is now chosen, satisfying the following relation to the actual winds: $u'' = v$, $v'' = u$. Their divergence is $\partial u''/\partial x + \partial v''/\partial y = \partial v/\partial x + \partial u/\partial y$, which is the shearing deformation for the three given winds. After both types of deformation are obtained, the total rate of deformation and resultant axis of dilata-

tion are found by methods described previously.

**10.38. *Geostrophic deformation.*—A convenient method for measuring and visualizing horizontal deformation is provided by the geostrophic approximation. Stretching

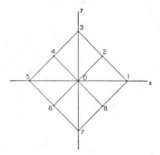

Fig. 10.381.—Grid for geostrophic deformation

and shearing geostrophic deformations (at constant pressure) are

$$\frac{\partial u_g}{\partial x} - \frac{\partial v_g}{\partial y} = -2\frac{980}{f}\left(\frac{\partial^2 Z}{\partial x \partial y}\right)_p ,$$

$$-\frac{1}{f}\left(u_g\frac{\partial f}{\partial x} - v_g\frac{\partial f}{\partial y}\right),$$

$$\frac{\partial v_g}{\partial x} + \frac{\partial u_g}{\partial y} = \frac{980}{f}\left(\frac{\partial^2 Z}{\partial x^2} - \frac{\partial^2 Z}{\partial y^2}\right)_p \tag{1}$$

$$-\frac{1}{f}\left(v_g\frac{\partial f}{\partial x} + u_g\frac{\partial f}{\partial y}\right).$$

If we neglect variation of f, then, following Eq 10.34(3),

$$\tan 2\theta \simeq -[(\partial^2 Z/\partial x^2 - \partial^2 Z/\partial y^2)/ \\ 2(\partial^2 Z/\partial x \partial y)]_p , \tag{2}$$

and, from Eq 10.34(4),

$$\text{def}_r\,\mathbb{C}_g \simeq -2(980/f)(\partial^2 Z/\partial x \partial y)_p \sec 2\theta \\ \simeq 980/f)(\partial^2 Z/\partial x^2 \\ - \partial^2 Z/\partial y^2)_p \csc 2\theta . \tag{3}$$

The approximate geostrophic deformation can be evaluated conveniently by use of a grid system such as illustrated in Figure 10.381. If the diagonal distance across the smaller squares is δs, then

$$\tan 2\theta \simeq -\tfrac{1}{2}(Z_1 - Z_3 + Z_5 - Z_7)/(Z_2 - Z_4 + Z_6 - Z_8) .$$

Eq (3) becomes

$$\text{def}_r\, \mathbb{C}_g \simeq -2\,\frac{980}{f\,(\delta s)^2}\,(Z_2 - Z_4 + Z_6 - Z_8)\sec 2\theta$$

$$\simeq \frac{980}{f\,(\delta s)^2}\,(Z_1 - Z_3 + Z_5 - Z_7)\csc 2\theta .$$

It is found that

$$|\,\text{def}_r\, \mathbb{C}_g\,| \simeq 980\sqrt{[(Z_1 - Z_3 + Z_5 - Z_7)^2 + 4(Z_2 - Z_4 + Z_6 - Z_8)^2]}/f(\delta s)^2 .$$

Deformation so determined applies to the central point about which data are taken, that is, point *0* in Figure 10.381. Angle θ is measured from diagonal *5,0,1*. The grid system either can be fixed on the map or

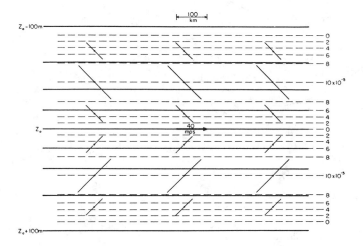

a

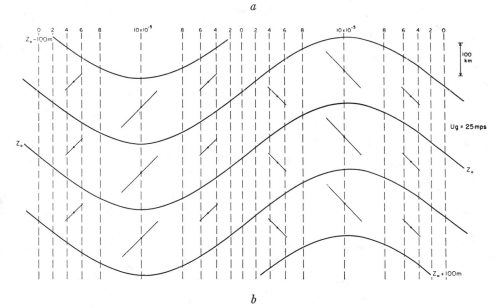

b

Fig. 10.382.—Deformation in idealized pressure patterns

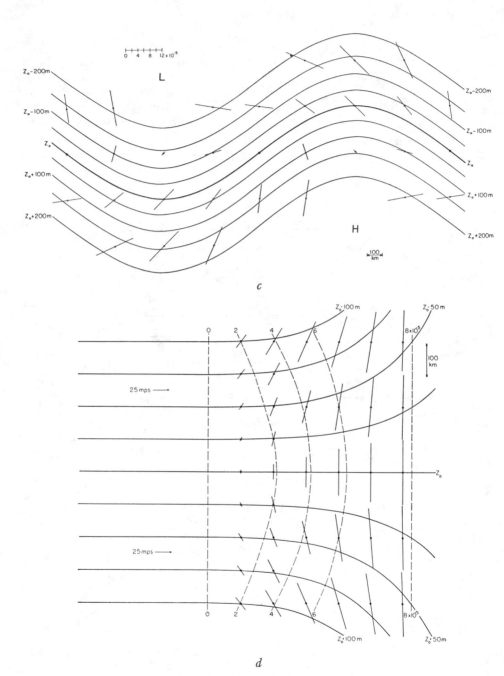

c

d

FIG. 10.382.—*Continued*

shifted relative to the map to any desired location and direction.[56]

On the basis of Eqs (2) and (3), deformation was determined in each of the pressure patterns shown in Figure 10.271 to derive Figure 10.382. Deformation in (a) is due only to the shear, and rate of deformation is the same as the absolute value of relative vorticity. In (b) the rate of deformation is again the same as the absolute value of vorticity, since $\partial^2 Z/\partial y^2$ and $\partial^2 Z/\partial x \partial y$ are both zero. Here deformation corresponds only to curvature of the flow. Pattern (c) shows deformation in a more general type of pressure pattern. All parts of the pattern should be analyzed by Table 10.36.

The last pattern brings out clearly the effect of diverging streamlines on direction of dilatation. Along contour Z_0 geostrophic flow is straight but spreads downstream, and dilatation is normal to the winds. If motion was from right to left, dilatation would be *along* the contour Z_0. By combining the results obtained from Figures 10.382ad, one can deduce the pattern of deformation about a local velocity maximum or "isotach center" in the wind field, which in geostrophic sense is a constriction in the pattern of pressure contours.

10.39. *Accumulation of horizontal gradients.*—Let $\partial Q/\partial n$ be the gradient magnitude of any physical quantity Q in space; n is normal to the surfaces and toward higher values of Q. The expression $d(\partial Q/\partial n)/dt$ represents the concentration of moving Q surfaces with time. This total time derivative we shall call *accumulation* of gradient, denoted by $A(Q)$. This subject has been treated by Petterssen[57] and by others as *frontogenesis*, but we make a distinction, because in generation of fronts accumulation of gradient is necessary but insufficient.

Consider that Q is a function of three dimensions and time. If we now differentiate its individual time derivative dQ/dt along n and rearrange terms,

$$A(Q) = -\frac{\partial u}{\partial n}\frac{\partial Q}{\partial x} - \frac{\partial v}{\partial n}\frac{\partial Q}{\partial y}$$
$$- \frac{\partial w}{\partial n}\frac{\partial Q}{\partial z} + \frac{\partial}{\partial n}\left(\frac{dQ}{dt}\right), \quad (1)$$

which is the general expression for accumulation in space.

If we restrict attention to accumulation of Q lines in a horizontal plane, we take n along the horizontal ascendant. Also, $u\,\partial Q/\partial x + v\,\partial Q/\partial y = c_n\,\partial Q/\partial n$, where c_n is the wind normal to the Q lines and directed toward higher Q. Accordingly, Eq (1) becomes

$$A_h(Q) = -\frac{\partial c_n}{\partial n}\frac{\partial Q}{\partial n} - \frac{\partial w}{\partial n}\frac{\partial Q}{\partial z}$$
$$+ \frac{\partial}{\partial n}\left(\frac{dQ}{dt}\right), \quad (2)$$

which is the complete equation for accumulation of gradients in horizontal plane. A similar expression, applicable to formation of stable layers and large vertical moisture gradients, is found for accumulation in vertical plane.[58] The terms on the right of Eq (2), all of which can be the same order of magnitude, are, respectively, (i) differential horizontal advection, $A_1 = -(\partial c_n/\partial n)(\partial Q/\partial n)$; (ii) differential vertical advection, $A_2 = -(\partial w/\partial n)(\partial Q/\partial z)$; and (iii) differential "nonconservative" term, $A_3 = (\partial/\partial n)(dQ/dt)$. Thus, $A_h(Q) = A_1 + A_2 + A_3$. The first two terms treat Q as conservative; individual changes in Q are incorporated in the last one.

(i) *Differential horizontal advection* (A_1). —If $\partial c_n/\partial n > 0$ (i.e., horizontal stretching

56. A nomograph for evaluating geostrophic deformation from this grid was developed by W. P. Elliott and H. A. Brown, *Scientific Report No. 3* (Contract AF 19 [604]-559 [GRD, AFCRC] [College Station: Department of Oceanography, Texas A. and M. College, November, 1953]).

57. *Weather Analysis and Forecasting.*

58. Notice the correspondence of the factors in Eq (2) to the list of processes forming stable layers.

normal to Q lines), lines are spread apart with time ($A_1 < 0$), and, if $\partial c_n/\partial n < 0$ (shrinking normal to Q lines), they are brought closer together with time ($A_1 > 0$). The conditions are illustrated by the three examples in the upper row of drawings in Figure 10.391, in which the wind normal to the Q lines is indicated by short vectors. Observe that the character of the *total wind* is of no concern. The only effective part of

where convergent. Observe that this effect is independent of the orientation of Q lines. One may show that $\partial c_n/\partial n$ in a convergence field is equal to half the rate of convergence, and therefore, owing to this effect alone,

$$A_1 = -\tfrac{1}{2}(\partial Q/\partial n) \text{ div } \mathbb{C}.$$

Diagram (*b*) illustrates the effect of horizontal divergence—spreading the lines. Evidently, *convergence accumulates property*

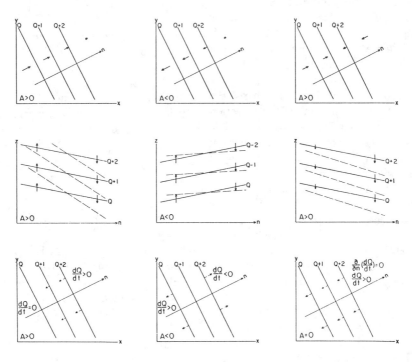

FIG. 10.391

the motion is the component *normal* to the Q lines, and accumulation of the gradient lies only in the distribution of the normal components.

Since horizontal shrinking is involved, horizontal convergence and deformation must play a part. Figure 10.392 illustrates the effect of the three first-order properties of motion on packing or spreading of Q lines with time. The convergence pattern (*a*) brings Q lines closer together, as the normal components (dashed vectors) are every-

lines, and divergence spreads them apart. It is also seen that convergence (divergence) does not alter the orientation of Q lines.

The pure rotation pattern in (*c*) gives quite different results. Since $\partial c_n/\partial n = 0$, *accumulation of property lines by vorticity is zero.* The lines are rotated, but the *magnitude* of horizontal gradient is unchanged. In (*d*) the lines are 45° from the axis of dilatation in a deformation field; lines are rotated, while the gradient magnitude is unchanged. Accumulation of Q lines for angles less than

45° is shown by (*e*). When the lines are parallel to the axis of dilatation, $-\partial c_n/\partial n = \frac{1}{2}|\mathrm{def}_r\ \mathbb{C}|$, and then $A_1 = \frac{1}{2}(\partial Q/\partial n)\cdot|\mathrm{def}_r\ \mathbb{C}|$. In the last drawing, Q lines are initially more than 45° from the axis of dilatation. Because the normal components of motion are divergent, the lines are being spread apart.

Petterssen[59] shows that the accumulation of property lines by deformation can be expressed by

$$A_1 = \tfrac{1}{2}(\partial Q/\partial n)\cdot|\mathrm{def}_r\ \mathbb{C}|\ \cos 2\beta\ ,$$

where β is the angle between Q lines and the axis of dilatation. If we combine effects of both horizontal convergence and deformation on accumulation of property lines in the plane, then

$$A_1 = \tfrac{1}{2}(\partial Q/\partial n)[|\mathrm{def}_r\ \mathbb{C}|\ \cos 2\beta$$
$$- \mathrm{div}\ \mathbb{C}]\ . \quad (3)$$

This expression is the contribution by the first term on the right of Eq (2). It shows explicitly that *the rate of increase in horizontal gradient of a quantity by horizontal motion is the product of its gradient and the sum effects of convergence and of the contraction by deformation.* Horizontal convergence is always accumulative, so long as a horizontal gradient of the property exists, and horizontal deformation is accumulative only if the axis of dilatation is less than 45° from the property lines. Although deformation is usually several times as large as convergence, both deformation and convergence must be considered important to the process.

Interesting and important conclusions concerning horizontal accumulation by deformation can be drawn from analysis of the idealized patterns in Figures 10.382. In (*a*), if property lines are less than 45° to the left of the wind, those lines are accumulated to the right of the current center and spread apart to its left, and the opposite results if the lines are less than 45° to the right of the

59. *Weather Analysis and Forecasting.*

wind. In (*b*) accumulation by deformation can be most rapid in the trough and ridge if property lines are favorably oriented. If the pattern of the property in question is wave-shaped also, most rapid accumulation could be expected when this pattern lags about 45° behind the pressure pattern. In

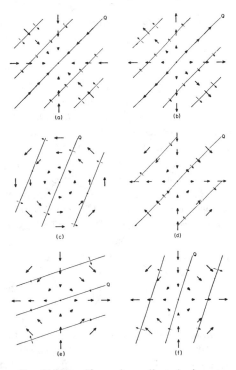

FIG. 10.392.—Change in gradient of a (conservative) quantity Q by the various kinematic properties.

(*c*) possible rates of accumulation by deformation vary even more; the most rapid accumulation is possible to the right of the current center in the trough and to the left of the current center in the ridge.[60]

(ii) *Differential vertical advection* (A_2).— The contribution by $-(\partial w/\partial n)(\partial Q/\partial z)$ is shown schematically by the middle row in Figure 10.391. The cross-section planes are normal to Q lines in the horizontal, but results are applicable also in arbitrary vertical plane. In the first example $\partial Q/\partial z > 0$, and, as indicated by the distribution of vertical

60. Examine Figures 7.08.

velocities, $\partial w/\partial n < 0$. This results in horizontal accumulation of Q lines; it is seen as the decreased horizontal spacing of Q lines due to rotation of their surfaces toward vertical. The next drawing shows decrease in horizontal gradient if surfaces are rotated toward horizontal. The vertical gradient of Q varies in the opposite manner.

(iii) *Differential individual changes* (A_3). —While the first two terms on the right of Eq (2) describe the entire kinematic process of horizontal accumulation, the last term accounts for varying individual changes in the property. This latter effect is illustrated in the bottom row of Figure 10.391 and results from motion of Q lines relative to air parcels. To simplify discussion, suppose the air is stationary. In the first drawing $dQ/dt = 0$ on the left, and line Q remains fixed. If in the region to the right the air for some reason increases in Q, the consecutive Q lines are shifted to the left—toward lower values of Q—and at some later time a line of value $Q + 3$ or $Q + 4$ occupies the former position of $Q + 2$. This obviously gives crowding of the lines, and $A_3 > 0$. The increase of gradient is easily visualized with reference to the temperature when the heating of the air is greatest at the higher temperatures. If we now permit translation of the air (i.e., wind), the same effect is seen relative to the moving air.

10.40. *Increase in horizontal gradients of certain air properties*.—The quantity Q may be identified with any air property—T, θ, θ_E, r, etc. Effects of differential advection in the horizontal and vertical are the same in all cases, but A_3 depends on the nature of the property. The effect of various thermodynamic processes on horizontal gradients of each property can be analyzed in detail.

If potential temperature increases upward, its horizontal gradient is increased by convergent horizontal advection and also by vertical advection if the warm air sinks more rapidly than the cold air (or if the cold air ascends more rapidly than the warm air). This distribution of vertical motion is of a sense known as *indirect circulation*. Direct circulation decreases $\partial\theta/\partial n$, since isentropic surfaces are brought more horizontal with time. The nonconservative influences on horizontal distribution of potential temperature are a result of diabatic processes, all of which can be examined individually with respect to the lower row of drawings in Figure 10.391.

Because vertical moisture gradients are usually so many times larger than horizontal gradients, the effect of differential vertical motions on horizontal gradients of moisture is large in comparison with horizontal shrinking. The nonconservative influence on the moisture pattern is felt at the surface by differences in evaporation, such as across coastlines and across large surface temperature gradients, and by differences in condensation due primarily to differing surface temperatures. The nonconservative modification of moisture patterns aloft are brought about through removal of moisture in areas of condensation and addition of moisture to the air near clouds and in precipitation areas.

Redistribution of vorticity can be considered in this manner also. Potential absolute vorticity is conservative for most processes including horizontal divergence, and thus differential advection in the horizontal and vertical are principal factors changing its horizontal gradient. Horizontal convergence brings consecutive lines together without changing the value of potential vorticity. On the other hand, since absolute vorticity $f + \zeta$ is changed primarily by divergence, horizontal convergence and divergence modify not only the gradients by differential advection but also the values of absolute vorticity. Relative vorticity is changed in this way plus the additional modification by meridional displacements.

10.41. *The process of frontogenesis.*—We take as the definition of frontogenesis *the process by which horizontal temperature or potential temperature gradient is concentrated within a quasi-substantial zone of narrow but finite width continuous in the three space dimensions.* It is the process by which a frontal zone is formed or strengthened. The opposite is *frontolysis.* That definition of frontogenesis differs somewhat from the familiar Petterssen "frontogenetic factor," which we have termed "horizontal accumulation" and which does not consider adequately the vertical continuity of the frontal zone.

From our experience with fronts in daily situations, it appears that there are several frontal properties that must be accounted for in the eventual explanation of the process. Major features are:

1. Concentration of *baroclinity,* without which there can be no front. Its generation must require maximum horizontal accumulation of isentropic surfaces in the frontal zone.

2. Concentration of *static stability* in the frontal zone. This is not so fundamental a property as the first, and fronts are not necessarily the most stable regions of the atmosphere, although fronts do possess *local maxima* of stability.

3. Maximum of *cyclonic vorticity* (or minimum of anticyclonic vorticity) at the front in any level (Fig. 6.25). This alone suggests that horizontal convergence is an important part of the frontogenetic process.

4. Close relation between the *tilt* of the front and surfaces of potential temperature (and equivalent potential temperature). For given conditions conservative physical surfaces are accumulated most rapidly along a sheet parallel to them.

5. Distribution of *three-dimensional motions and weather* near the frontogenetic zone. In the succeeding discussion we refer primarily to the most common low-level frontogenesis, that in advance of moving cyclones, and the weather is identifiable with the ordinary warm front. But there are other types also.

Figure 10.411 depicts successive stages in the formation of a front in a horizontal plane. Coordinates are fixed relative to the isentrope θ about which the frontal zone forms. The pattern may be moving toward the cold air or toward the warm air, or it may be stationary, and the frontogenesis is of the warm, cold, or stationary type, respectively. Although we shall assume initially that isentrope θ preserves its position relative to the line or zone of frontogenesis,

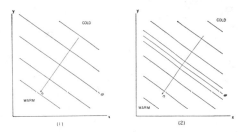

Fig. 10.411.—Frontogenesis viewed in a horizontal plane.

it is conceivable that the center of frontogenesis can and often does shift toward lower or higher temperatures during the process. The front is shown parallel to isentropes in the horizontal plane; this is not greatly different from usual situations. Direction n then can be assumed normal to both the isentropes and the front or frontogenetic zone. It is evident that frontogenesis in a plane involves (1) accumulation of isentropes (or isotherms) and (2) maximum rate of accumulation along the "line of frontogenesis." Similar provisions hold in three dimensions.

By the second provision, the location of the *line of frontogenesis* in a plane is the line along which the rate of accumulation of isentropes is greatest. The *surface of frontogenesis* is the extension of this line in three dimensions. It may or may not coincide with the present location of maximum $\partial\theta/\partial n$ or with most rapid velocity shrinking $(-\partial c_n/\partial n)$ normal to the isentropes, but it is where the product of the two is a maximum.[61]

It is convenient to examine the kinematic process of frontogenesis about a moving parcel by reference to a vertical plane (Fig. 10.412) chosen along the three-dimensional θ gradient through the point in question, P. Restriction to this plane is sufficient, since the character of motion normal to the plane does not contribute to accumulation of θ surfaces, and neither does the motion

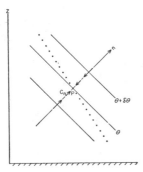

FIG. 10.412

parallel to them in the specified plane. P is located on the line of frontogenesis, shown by the series of dots in the diagram. This parcel may have vertical and horizontal velocity, but the velocity of the parcel itself is of no direct concern to frontogenesis about it. If we denote velocity in this plane along the ascendant of θ and *relative* to parcel P by C_n, accumulation of isentropes about P then is

$$A = -\frac{\partial C_n}{\partial n}\frac{\partial \theta}{\partial n} + \frac{\partial}{\partial n}\left(\frac{d\theta}{dt}\right). \quad (1)$$

Accumulation must be a maximum at P, since the parcel is on the line of frontogenesis.

Eq (1) describes frontogenesis in a minimum number of terms; it is just an adaptation of Eqs 10.39(1) and (2). The first term on the right incorporates all kinematic re-

61. Defining frontogenesis as a line or surface need not contradict our previous definition of a front as a zone.

quirements, and the second contains the nonconservative influences, both of which are usually operating simultaneously. The basic theory of frontogenesis is quite a simple one evident beforehand—it involves horizontal concentration of sloping isentropes in a vertical plane. Existing difficulties in complete explanations of frontogenesis must lie in the details and relative significance of each factor in Eq (1).

a) Kinematic considerations.—With reference only to $-(\partial C_n/\partial n)(\partial\theta/\partial n)$, for maximum accumulation at P in an initial field of uniform θ gradient, $-\partial C_n/\partial n$ must be a maximum at that point (or along the dotted line). Maximum shrinking along n requires a maximum either of convergence in this plane at P or of deformative contraction along direction n through P, or of their combination. Deformative contraction along n (or along its horizontal projection) will be attended by dilatation along θ surfaces in this plane, or normal to this plane, or both. Horizontal convergence, which we take as a requirement, has a conponent of contraction along n (n not vertical) and is compensated by vertical stretching. By the process illustrated, baroclinity is increased if θ surfaces are not horizontal initially, and static stability is increased if they are not vertical initially.

There are many conceivable combinations of horizontal and vertical motions in the vertical plane considered which satisfy the distribution of C_n in Figure 10.412; some of these appear in Figure 10.413. In scheme (a) the motion relative to P is horizontal only, indicating accumulation only by differential horizontal advection. Horizontal motion in (b) is translation only, and accumulation of isentropic surfaces results from indirect vertical circulation relative to P. In the next, motion relative to P is identical with C_n; accumulation results both from differential horizontal advection and from indirect rotation of θ surfaces.

Any of the schemes in Figure 10.413 can exist and be responsible for frontogenesis. However, from empirical evidence certain ones seem preferred. For example, if generation and concentration of cyclonic vorticity in the front are observed as part of frontogenesis, then horizontal convergence with or without deformation should be part of the process. This of itself would exclude indicates ascending motion in the warm air greater than in the cold air. Although other distributions surely exist, in this case relative motion in vertical plane across the line of frontogenesis resembles (*d*) and (*e*).

Viewed in a horizontal plane and excluding accumulation by indirect vertical circulation, frontogenesis is preferred parallel to the isotherms along an axis with a maximum

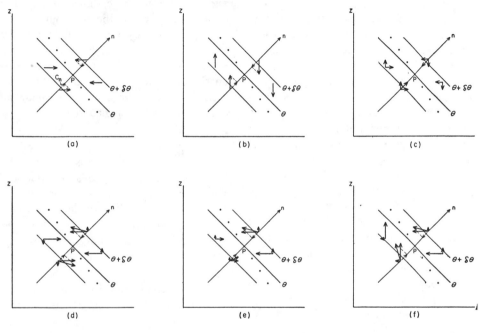

Fig. 10.413

scheme (*b*) in Figure 10.413. Pattern (*a*) should be dropped also, because it does not allow for the vertical expansion[62] associated with horizontal convergence and required through conservation of potential vorticity. We are thus left with the integrated schemes (*c*), (*d*), (*e*), and (*f*) or modifications thereof.

Distribution of motion in the vertical plane also has to satisfy preferred fields of vertical velocity in the frontogenetic region. In advance of traveling cyclones, where low-level frontogenesis is most prominent, the cloud and weather pattern usually

62. It does if θ surfaces are vertical.

of combined effects of horizontal convergence and deformation. The role of deformation in frontogenesis was first brought out by Bergeron (*Physikalische Hydrodynamik*) and later described in more detail by Petterssen, but the emphasis placed on *deformation fields*, characterized as neutral points and pressure cols on weather charts, has led to the general impression that such patterns are the most favorable for frontogenesis. This misconception is basically the important difference between deformation as a basic kinematic property and hyperbolic fields of flow—a difference analogous to horizontal divergence, on the one hand, and difluent streamlines, on the other. *Deformation does not require a flow pattern of hyperbolic streamlines;* larger rates of deformation are usually found in other types of flow.

In Figure 10.382c the preferred locations for frontogenesis by horizontal deformation are along two separate diagonal lines, one intersecting the trough to the right of the maximum wind and the other intersecting the ridge to the left of the maximum wind. These directions suggest a wave structure in the axis of frontogenesis which bears some agreement with the shape and location of the polar front relative to the upper flow pattern in the westerlies.

Necessity for horizontal convergence as part of frontogenesis brings up the importance of ageostrophic motion. Convergence by surface friction is of importance in frontogenesis near the ground, especially in cyclones and in pressure troughs aligned nearly along isotherms. Isallobaric wind is effective in frontogenesis near centers of pressure fall. Perhaps it is not just incidental that the most pronounced frontogenesis through the lower troposphere is found in areas of isallobaric minimum, as best illustrated ahead of mid-latitude cyclones. Horizontal convergence also may be partially due dynamically to horizontal motion into a pressure field of greater cyclonic vorticity.

Finally, vertical motions can be of help to frontogenesis if circulation is direct. Ascent of the warm air (or subsidence of the cold air) gives horizontal convergence in the intervening area, which can account at least partially for the concentration of cyclonic vorticity and for the required increase of baroclinity by horizontal accumulation of isentropic surfaces. However, there are two compensating factors in accumulating isentropes horizontally: direct vertical circulation decreases baroclinity by rotating the

isentropic surfaces toward horizontal, but it sets up ageostrophic motion with horizontal convergence. In order that baroclinity be increased in the process, a critical condition must be satisfied involving vertical and horizontal velocities, baroclinity, and stability.

We have defined frontogenesis as the line along which accumulation of isentropes is a maximum (or in three dimensions as the surface along which the accumulation of isentropic surfaces is a maximum); from Eq (1) it follows that $\partial A/\partial n = 0$ and $\partial^2 A/\partial n^2 < 0$ at this line or surface of frontogenesis. With reference to only the kinematic term in that equation, the following condition must be satisfied at this line:

$$\frac{\partial^2 C_n}{\partial n^2} \frac{\partial \theta}{\partial n} + \frac{\partial C_n}{\partial n} \frac{\partial^2 \theta}{\partial n^2} = 0.$$

By detailed analysis of this equation the following possibilities are seen with respect to the particular vertical plane:

1. If potential temperature is distributed linearly, the line of frontogenesis coincides with the axis of maximum velocity shrinking.

2. If velocity shrinking is uniform over the region, the line of frontogenesis coincides with the axis of maximum θ gradient.

3. If maxima of θ gradient and velocity shrinking both exist in the region, (a) the line of frontogenesis coincides with them only if they are coincident themselves (this is the ideal case in which the line of frontogenesis coincides with the largest horizontal temperature gradient) and (b) the line of frontogenesis is located between them if they are not coincident. Its relative distance from the axes of maximum velocity shrinking and of maximum temperature gradient depends on the relative magnitudes of the two terms in the last equation. This postulates that a front can displace itself across isentropic surfaces toward the axis of maxi-

mum velocity shrinking, which is also a matter of fact.

In summary, we find from kinematic considerations that frontogenesis is favored along axes of maximum horizontal convergence, along axes of greatest deformative dilatation, along zones of largest horizontal temperature gradient, and also along axes of

divergence). The upper portion of that front might be there by virtue of horizontal advection of a large temperature gradient produced within or to the rear of the upper-level pressure trough, where low-level conditions are frontolytic.

b) Nonconservative thermodynamic effects. —The diabatic term in Eq (1) alone could

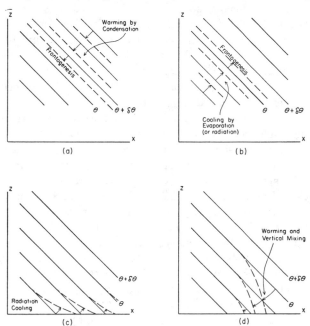

FIG. 10.414.—Frontogenesis through nonconservative influences

greatest indirect vertical circulation. The process is most rapid where the sum effect is greatest. Since there are several contributions to consider, in common cases of frontogenesis it is possible that one of the individual contributions might be negative. When viewing a front or frontogenesis in a vertical plane, it is well to bear in mind that processes in one part of the front may be opposite in another part—frontogenesis in one, frontolysis in the other. For example, there is some evidence that, with ordinary warm frontogenesis at the ground, attended by convergence, deformation, and ascent, processes above are frontolytic (horizontal

account for accumulation of isentropic surfaces within the frontal zone, and, except for the vorticity requirement, it apparently could be totally responsible for formation of a front. The line or surface of frontogenesis by this process is found where the gradient of diabatic change is the greatest—to the cold side of maximum warming or to the warm side of maximum cooling.

In the free atmosphere the most effective diabatic processes in frontogenesis are condensation of water vapor in the warm air mass (Fig. 10.414*a*) and evaporation within the cold air (*b*). As the principal cause of condensation is ascending motion, more rapid ascent by the warm air, as implied by Figures 10.413*de,* aids in frontogenesis through release of latent heat. Precipitation from the ascending warm air, falling through colder air beneath the generating front, may evaporate before reaching the ground and be responsible for further intensification of the front.

Near the ground frontogenesis, at least in the sense of accumulation, often results from favorable distributions of radiation cooling and warming and from effects of vertical mixing. Large gradients of radiation cooling (warming) are to be found along snow lines in higher latitudes, at large gradients of sea-surface temperature, along coasts, and at the boundary between clear and overcast skies over land; some of these radiation gradients may reverse in direction diurnally to give a phenomenon of the nature of land-and-sea-breeze "fronts." The frontogenetic process is acting on the same system of particles if the air is stationary or moving parallel to the radiation boundary. As the effect of cooling from below extends upward rather slowly, concentration of isentropic surfaces by this process alone usually would not penetrate high into the troposphere. Vertical mixing attends any pronounced warming from below, and the concentration of isentropic surfaces so formed may extend to greater heights with almost vertical orientation.

Although these nonconservative influences are a means of concentrating potential temperature gradient locally in the atmosphere and are frequently a part of the total process of frontogenesis (and frontolysis), it appears quite fruitless at present to attempt to explain the problem without introducing kinematics. Diabatic processes under favorable conditions serve for initiating frontogenesis or as a catalyst once the process has started. For example, with reference to the pattern of Figure 10.414*d,* it is possible that the troughlike pattern of

pressure in low levels, maintained hydrostatically by the concentration of isentropic surfaces, can result in horizontal convergence through ground friction; the frontogenesis so created then might extend itself by various means to greater depths in the troposphere.

10.42. *Notes on preferred frontogenetic areas.*—It is customary to say that frontogenesis is favored climatically over certain regions of large surface temperature gradient where opposing low-level currents meet. In winter the two frontogenetic zones most preferred climatically in the northern hemisphere are off the mid-latitude east coasts of the major land masses, and they extend along the large contrasts of sea-surface temperature, at least in partial agreement with what was said in the previous section.

In smaller scale, and on a daily basis, frontogenesis and fronts are found closely associated with the major currents through the troposphere, for quite obvious reasons. Frontogenesis at the surface normally is found to the warm side of the upper westerly current, and, if a surface front (cold front) separates from that current by some distance laterally, it is apt to frontolyze. These deductions follow from hydrostatics and need not infer cause and effect. Coastal temperature gradients and the boundary of a flat cold mass along a mountain barrier are excluded, since neither is strictly a front.

Fronts in pronounced pressure troughs at the surface are predominantly frontogenetic. Existence and maintenance of the trough is some indication of a steep frontal slope and that the slope is being conserved aloft. Further, a sharp trough is evidence of surface frictional convergence and deformation of favorable sense. A low-level front lying normal to anticyclonically curved flow in those levels is subject to frontolysis; both divergence and deformation fields are then probably frontolytic, and, as the front has little slope, it is rapidly destroyed near the ground. If the flow is parallel to the front, frontogenesis can proceed even with anticyclonic curvature; in this case there is either a real trough (in the form of a col) or an apparent trough (axis tt' in Fig. 4.104). These fronts are relatively stationary, and frontogenesis is especially marked when they coincide with large temperature gradients of the underlying surface. In general, a cold front must undergo low-level kinematic frontogenesis to maintain itself, since warming of the cold mass behind it is frontolytic.

PROBLEMS AND EXERCISES

1. Draw streamline and isotach patterns for the following conditions using Eqs 10.10(1). (The differential equation for any streamline is $v\,dx - u\,dy = 0$, and the value of an isotach is $\sqrt{[u^2 + v^2]}$.)

   a) $u_0 = 5 \text{ m sec}^{-1}$, $v_0 = -10 \text{ m sec}^{-1}$, $b_0 = -1 \times 10^{-5} \text{ sec}^{-1}$.

   b) $u_0 = -5 \text{ m sec}^{-1}$, $a_0 = 1 \times 10^{-5} \text{ sec}^{-1}$.

   c) $v_0 = 5 \text{ m sec}^{-1}$, $a_0 = b_0 = 1 \times 10^{-5} \text{ sec}^{-1}$.

   d) $b_0 = 2a_0 = 2 \times 10^{-5} \text{ sec}^{-1}$.

   e) $b_0 = c_0 = 2a_0 = 2 \times 10^{-5} \text{ sec}^{-1}$.

   f) $b_0 = -c_0 = 1 \times 10^{-5} \text{ sec}^{-1}$.

   g) $a'_0 = c_0 = 1 \times 10^{-5} \text{ sec}^{-1}$.

   h) $u_0 = v_0 = 5 \text{ m sec}^{-1}$, $a_0 = b_0 = 2c_0 = 2 \times 10^{-5} \text{ sec}^{-1}$.

2. What is the pattern of motion relative to an air parcel located at any point in 1(*b*) above?

3. Consider a small disk of area $A = \pi r^2$ centered at the origin in Figure 10.09c. Find by the increase in area its average divergence during period δt and how that result differs from Eq 10.11(2).

4. A vertical air column moves from 30° to 60° north latitude and doubles its pressure depth. If potential vorticity is conserved, what is the average horizontal divergence over the period?

5. Three stations, A, B, and C, outline an equilateral triangle of base 100 km. B and C are at the same latitude (B west), and A is farther north. Compute the average divergence, vorticity, and deformation if the winds (in knots) are (A) 240° 40, (B) 270° 30, and (C) 250° 35.

READING REFERENCES

BJERKNES, V., HESSELBERG, TH., and DEVIK, O. *Dynamic Meteorology and Hydrography*, Part II: *Kinematics*. Washington, D.C.: Carnegie Institution, 1911.

BYERS, H. R. *General Meteorology*, pp. 293–98. New York: McGraw-Hill Book Co., 1944.

PETTERSSEN, SVERRE. *Weather Analysis and Forecasting*, pp. 205–73. New York: McGraw-Hill Book Co., 1940.

CHAPTER 11

Certain Aspects of Broad-Scale Analysis

11.01. *Introduction*.—In preceding chapters emphasis was given to mid-latitude weather patterns part by part. Now we consider integrating the weather pattern hemispherically. Although this leads into climatic aspects of the general circulation, with which it is assumed the reader is sufficiently familiar, we will try to avoid that subject and adhere to analytical topics as far as possible. In what follows, reference is made to middle and high latitudes. In the final chapter we touch briefly on the tropics.

11.02. *Intensity of the westerlies*.—The intensity of the zonal westerlies is one important control in geographic locations and behavior of weather patterns over the entire hemisphere. It is a governing factor in the location, shape, and intensity of the semipermanent circulation cells;[1] meridional exchange of air masses; paths and speeds of pressure systems; and movement, area, and intensity of precipitation. The importance to forecasting, especially over periods of a day or more, is readily evident.

The simplest measure of westerly intensity is the *zonal index*, which is the difference in average pressure (or pressure height) along two latitude circles, or the corresponding geostrophic west-wind speed. This computation is made from most or all longitudes of the hemisphere. In the past two fixed latitudes—35° and 55°—were used. This band embraces the major portion of the average westerlies, at least during the colder part of the year (see Figs. 1.031*a* and 1.032*a*). Because the maximum westerlies may oscillate within this latitude band, and also out of it, especially in summer, computations of westerly speed are now being made over narrower bands (5° of latitude) for a wider range of latitude,[2] thus permitting a plot of the meridional profile of westerlies. However, neither method is completely reliable when the westerly current is greatly eccentric about the pole or when the current is extremely split.

A few elementary effects of zonal index on the flow pattern can be deduced from Figure 11.02, which indicates in an ideal manner the composition of any flow pattern in terms of (i) a certain basic zonal current and (ii) the "perturbation" pattern. (The diagrams also illustrate the variation of a cellular pressure pattern with height due to uniform meridional temperature gradient.) The figure demonstrates clearly that the amplitude of flow is inversely related to the zonal index. Another feature not indicated is the relation between zonal index and wave length. In the discussion of Figure 10.31 it was stated that, the stronger the winds, the greater is the wave length of a CAV trajectory and therefore the longer is the equilibrium wave length of the flow. In effect, both amplitude and wave length in the westerlies are dependent on the zonal wind.

At this point it is necessary to attempt a

1. C.-G. Rossby, "Relation between Variations in the Intensity of the Zonal Circulation of the Atmosphere and the Displacement of the Semipermanent Centers of Action," *Journal of Marine Research*, Vol. II, No. 1 (1939).

2. G. P. Cressman, "On the Forecasting of Long Waves in the Upper Westerlies," *Journal of Meteorology*, V (1948), 44–57.

distinction in scale of atmospheric waves. We consider theoretically at least two types: (i) the *planetary waves* (long waves or major waves), whose patterns are governed by the intensity of the zonal wind, by certain geographic controls, and by conservation of absolute vorticity, and which correspond to the semipermanent pressure patterns; and (ii) the traveling waves (minor waves, isallobaric waves, or frontal waves), of shorter length than, and which are "steered" nearly along, the planetary flow

of unusually strong westerly flow in middle latitudes ("high index") the following are *ideal* features:

(i) The westerly flow aloft is characterized by very long planetary waves (hemispheric wave number about 3) and by preferred locations of major waves aloft—troughs near east coasts of continents (and Russia) and ridges near west coasts (in northern hemisphere winter).

(ii) The sea-level pressure pattern, when averaged over a period of a few days, has

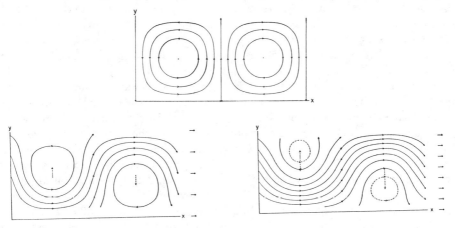

Fig. 11.02.—Effect of uniform zonal motion on the shape of the flow pattern; *upper*, idealized cellular pattern; *left*, same with moderate westerly flow; *right*, same with strong westerly flow.

pattern; these perturbations or secondary circulations superimposed on the large-scale waves have associated divergence and vertical motion patterns which produce the important daily weather phenomena. Proper analysis of both types of waves is obviously essential to forecasting; the longer the forecast period, the greater is the emphasis on planetary waves.

11.03. *Intensity of the westerlies and the hemispheric weather pattern.*—The essential features of the large-scale weather pattern in relation to west-wind speed can be deduced roughly from the foregoing and are discussed in more detail in the "Reading References." To summarize, during periods

well-developed centers near their average locations, and these systems are elongated west-east.

(iii) Cyclones along the poleward side of the westerlies move more rapidly eastward and reach maximum intensity in the Aleutian and Icelandic areas.

(iv) Precipitation is concentrated along the paths of the cyclones, with less than normal in the equatorward band of latitude.

(v) The tropical easterlies have fewer interruptions than usual.

(vi) There is subnormal meridional exchange of air masses; isotherms and frontal systems have abnormal west-east orientation.

Weak westerly flow in middle latitudes is

apparently associated with stronger than average westerlies over subtropical or subpolar latitudes, or both. It may be seen on a hemispheric upper-level chart as a greatly meandering westerly current or as westerly currents of little amplitude both north and south of the central latitudes. Because there is such a variety of flow patterns associated with "low index" in middle latitudes, its features can be outlined only in generality.

(i) There are more westerly waves around the hemisphere at least over middle and subtropical latitudes; they are distributed more randomly.

(ii) Cyclones are more frequent in the lower middle latitudes and anticyclones in the higher middle latitudes.

(iii) Systems in middle latitudes are displaced more meridionally; cyclones moving poleward have abundant precipitation.

(iv) The tropical easterlies have more frequent interruptions by mid-latitude perturbations.

(v) Meridional exchange of air masses can give large negative temperature departures in the lower middle latitudes and positive departures in higher latitudes.

The above outlines give simplified views of the relation of extremes in mid-latitude westerlies to the hemispheric weather pattern. Ideal cases are few, and seldom do the patterns approach uniformity around the hemisphere.

11.04. *The Rossby long-wave formula.*— For a sinusoidal wave pattern with uniform zonal west-wind velocity U and negligible horizontal divergence, Rossby[3] has shown that the following formula applies:

$$c = U - \beta L^2/4\pi^2 , \qquad (1)$$

where c is wave speed (positive eastward), β the poleward variation of Coriolis parameter, and L the wave length. In this wave pattern the air conserves its absolute

3. *Op. cit.*

vorticity, and the sinusoidal motion describes CAV trajectories.

In the baroclinic westerlies U increases upward, while c is practically invariant with height and, excluding perhaps the lowest third of the troposphere, L is nearly constant upward. Thus, the formula is applicable only at one intermediate level, which doubtless varies with the case and to some extent along the wave. The level of best fit should correspond with the level of least divergence, on the average near 600 mb. An increment equal to the zonal thermal wind should be added to U when applying the formula at levels below, and subtracted when applied at levels above. As shown by Rossby,[4] for $f + \zeta > 0$ horizontal divergence in such waves is directly proportional to $[(U - \beta L^2/4\pi^2) - c_a]v$, or to $(c_c - c_a)v$, where v is the poleward component of velocity, c_a the actual wave speed (or the computed wave speed at the level for which Eq [1] applies), and c_c the wave speed determined from Eq (1) for an arbitrary level. In the low levels c_c is usually smaller than c_a; convergence is prevalent east of the wave trough, and divergence to the west. At high levels the reverse is preferred, all of which seems verified empirically.

Eq (1) allows both progressive and retrogressive waves, as determined by the relative magnitudes of the two quantities on the right. For given zonal wind speed shorter waves move eastward, and longer waves can retrogress, and the speed is proportional to $U - \beta L^2/4\pi^2$.

From Eq (1) a *stationary wave length, L_s,* can be found for any latitude and observed value of U; it is a more useful parameter than computed wave speed for qualitative applications of the long-wave formula. The quantity L_s is defined by setting $c = 0$:

$$U = \beta L_s^2/4\pi^2 , \quad \text{or} \quad L_s = 2\pi\sqrt{(U/\beta)} . \quad (2)$$

4. See V. P. Starr, *Basic Principles of Weather Forecasting* (New York: Harper & Bros., 1942).

Thus, the stronger the zonal wind, the greater is the stationary ("equilibrium") wave length. Eq (2) also suggests that for uniform zonal wind the number of waves required increases equatorward.

Substitution of Eq (2) into Eq (1) gives

$$c = (\beta/4\pi^2)(L_s^2 - L^2), \qquad (3)$$

which shows that waves for which L < L_s *are progressive and that waves for which* L > L_s *are retrogressive*, in agreement with previous deductions.

Figures Q (Appendix) are nomograms for solving Eq (2) and also for applying Eq (1). Diagram (*a*) relates the three variables U, ϕ, and L_s, and (*b*) does the same with meridional pressure gradient replacing zonal velocity. In (*a*) the profile of U is also the profile of L_s with respect to latitude. If then the actual wave length L (usually the average for several waves) is entered at proper latitudes in the chart, the difference in U measured horizontally between the curves for L_s and L is the measure of wave speed, according to Eq (3).

When used with proper discretion, the simple wave formula is a valuable qualitative tool in forecasting the broad-scale flow pattern, as much in determining displacement as development of the upper flow pattern. It is subject to failure when there is no organized wave pattern, during periods of significant change of index or wave number, and when the westerlies are eccentric with respect to the pole. There are difficulties arising in analysis as well. Upper-level analysis is required over broad oceanic and continental regions with few direct data. In this case, accuracy of analysis remains as great a problem as whatever shortcomings might exist in forecasting formulas. A principal difficulty in applying and vertifying the long-wave formula is properly distinguishing between the two coexistent wave patterns. The effect of superposition of a series of traveling short waves along a longer standing wave can be seen graphically. The trough of the standing wave is sharply defined when it coincides with the trough in a short wave but is rounded or even split when it coincides with a short-wave ridge. The difficulty with smaller perburbations can be removed to some extent by reference to average pressure charts for periods of a few days. Also useful in this regard is a sort of

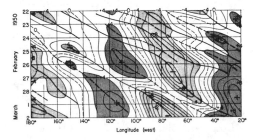

FIG. 11.04.—*x*, *t* analysis of average height of the 500-mb surface for latitudes 40°, 45°, and 50° N. The reference height is 17,800 feet, and departures are labeled in hundreds of feet. (All data from *Daily Series, Synoptic Weather Maps, Northern Hemisphere Sea Level and 500-Mb Charts.*)

x, *t* perturbation pressure analysis maintained daily (Fig. 11.04). A related variation is a diagram containing daily *x*, *Z* profiles of height of an isobaric surface along a latitude circle (or average height across a latitude band), with each day's profile arranged in vertical time scale.

In Figure 11.04 the region of positive departure along any horizontal line indicates a pressure ridge in the particular latitude band; the negative regions indicate troughs. These happen to be the components of the major waves. The minor waves along the axis of the current are in some places within this latitude band; in others, outside of it (see Fig. 7.02*b*, which represents the end of the period included in Fig. 11.04). During the period illustrated the average position of one ridge was about 115° W, the trough downstream near 80° W, and the next ridge near

25° W, representing major wave length 90° or hemispheric wave number of 4 in this latitude band. Even though there is some progression in the wave, it is quite persistent and well defined in the analysis. The pattern west of about 140° W is questionable, since it is based on no direct data within that region.

11.05. *Structural variations in the upper westerlies.*—With increasing amounts of data for the upper troposphere there has come a radical change in our impression of flow patterns in that region, and a similar turnabout seems necessary in methods of pressure and wind analysis. In the vicinity of 10,000 feet, once known as the "upper air," the westerlies appear quite uniform in a broad belt, but at higher levels the flow is more concentrated in one or several wavelike currents or jet streams.

a) Multiple westerly currents.—In Figure 7.02b the westerly current appears relatively singular across the Pacific Ocean and North America. It begins to split over the western Atlantic, with one branch following a northern course from north of Newfoundland, skirting Iceland, and passing over the Norwegian Sea, Lake Baikal, and northern Japan. The southern current takes a meandering course over western Europe, North Africa, and central Asia. Even over western North America the current has double (or even triple) maxima.[5] Notice that the split currents are found in the regions of most dense data. Certainly, the current at this level is considerably more streaky than is indicated over the oceans and central Asia. In those areas the partly necessary restriction to linear interpolation has resulted in broad currents of more uniform appearance. We might safely add that even in areas of dense data the analysis tends toward smoothed accounts of the actual wind pattern. Moreover, the westerly

5. See also Figure 6.13b.

current continues to get more streaky over the next 10,000–20,000 feet upward. A careful analysis by isotachs might reveal that a current center consists of several parallel wind maxima or that a number of "fingers" extend along the current from an area of maximum.

These features of the flow make proper analysis a difficult problem. We can no longer consider extreme winds (and wind shear) careless errors in observation or transmission and simply ignore them. Nowadays, aircraft operations demand analysis and forecasting of these phenomena, even though we know very little about their origin, their nature, and their influence on the weather observed at the surface.

In the case of well-defined multiple westerly currents, it would seem that each would be governed by its own long-wave relation. Considering that at some longitude the waves are in phase, or inseparable, they are apt to be out of phase downstream due to longer waves in the current nearer the pole or to different wind speeds. This structure permits cold and warm vortices to exist in the area where two currents are widely separated (Europe in Fig. 7.02b) and also presents areas of *difluence* and *confluence* of two currents, which are considered significant in forecasting.[6]

As these multiple westerly currents have associated baroclinic bands, the possibility of multiple frontal systems exists. Indeed, it is not uncommon to find separate frontal systems some 20° of latitude apart or more in the same region of longitude—the more tropical one usually the weaker—as was suggested by Sections 6.23 and 9.20.

b) Meridional drift of the westerlies.—

6. J. Namias, "Physical Nature of Some Fluctuations in the Speed of the Zonal Circulation," *Journal of Meteorology*, IV (1947), 125–33; J. Namias and P. F. Clapp, "Confluence Theory of the High Tropospheric Jet Stream," *Journal of Meteorology*, VI (1949), 330–36.

There is some evidence,[7] in both large and small scale, that over periods of several days the individual jet streams appear to drift equatorward and eventually dissipate, while new ones form in higher latitudes. The drift is slow and irregular and often is evident only in small ranges of longitude. When in the lower latitudes, the "jets" might be in evidence only in a shallow region near the tropopause. The net drift may not be apparent in the zonal wind profile, where instead the zonal wind maximum might appear to oscillate meridionally.

c) *"Kalt- und Warmluftinseln."*[8]—Cold cyclones are frequently found equatorward from the main belt of westerlies, and just as frequently warm anticyclones are found poleward from the main westerlies. Figure 7.02*b* illustrates both. At times they may have strong circulations around them, but most often the strongest flow is on the equatorward side of the cyclone and on the poleward side of the anticyclone. In general, they are slow movers and may take various paths. In the western sector of the hemisphere, regions of preferred stagnation of cold cyclones are the Pacific Ocean just off the Californias and the Atlantic east of the Azores. The warm polar anticyclones prefer the northern European and northern Alaskan regions. Note how these four areas are located relative to the semipermanent pressure centers at the surface, and they are areas of large-scale difluence in the mean upper flow.

The cyclone represents a dome of cold air cut off from the polar pool, with exception perhaps of the lowest layer, by subsidence on its poleward side. The anticyclone ap-

pears to form from part of the warm upper-level ridge, usually overlying the surface cold layer (shielding layer), and ordinarily undergoes radiation cooling during its stay in high latitudes. Formation of these vortices might appear as the cut-off meanders in a jet stream of large amplitude during change in hemispheric wave number or as the process of separation of a westerly current into two distinct ones of differing wave length and amplitude.[9]

The anticyclone is reflected prominently in the pressure pattern at the surface. Cyclone tracks trend north and some distance south of it, as they do near the Greenland and Himalayan barriers; hence, these anticlones may be known as "blocks" or "blocking HIGHs" in the large-scale flow. Frontal waves skirt the poleward periphery, but their warm sectors *at the surface* are void of tropical air. Unless they take paths far removed from the upper anticyclone, frontal wave cyclones taking the equatorward track are mostly innocuous.

The isolated cold cyclones often have only obscure counterparts at the surface, especially when over the cold continents in winter (see charts for March 1950 in Chaps. 7 and 9). Normally the cold air is spreading in the lower layers or being displaced eastward as a whole, so that something of the nature of a low-level cold front may be found east of the upper cyclone. Its frontogenetic field is often absent, and, with tendency for homogeneity when over ocean surfaces, a veritable surface front may be

9. For theory and more detailed discussion consult Department of Meteorology, University of Chicago, "On the General Circulation of the Atmosphere in Middle Latitudes," *Bulletin of the American Meteorological Society*, XXVIII (1947), 255–80; E. Palmén, "Origin and Structure of High-Level Cyclones South of the Maximum Westerlies," *Tellus*, I (1949), 22–31 (summarized in American Meteorological Society, *Compendium of Meteorology* [Boston, Mass., 1951]); H. Wexler, "Anticyclones," in American Meteorological Society, *Compendium of Meteorology*, pp. 621–29.

7. G. P. Cressman, "Variations in the Structure of the Upper Westerlies," *Journal of Meteorology*, VII (1950), 39–47; N. A. Phillips, "The Behavior of Jet Streams over Eastern North America during January and February 1948," *Tellus*, II (1950), 116–24.

8. W. König, "Über Kalt- und Warmluftinseln," *Zeitschrift für Meteorologie* (1947), pp. 128–30.

difficult to detect. Frequently the broad and weak frontal zone moves far out ahead of the LOW reflected at the surface, so that the surface cyclone is itself frontless, even though instability in the cold air might be some temptation for placing a "stationary" cold front or occlusion in the surface trough or a series of cold fronts pinwheeling about the surface LOW. An associated warm front ordinarily is even more difficult to find because of the usual shallowness of the surface cold air adjoining the cold dome.

d) Eddies and shear lines with the westerly currents.—Existence of multiple westerly waves of different length, amplitude, and position over middle and high latitudes in the upper troposphere suggests that a number of cyclonic and anticyclonic circulations are to be found in particular places within the intervening areas. They are frequently overlooked in pressure analysis, with the aim of simplifying the patterns. This is one of the matters often classified as trivial, but, when trivialities are vector-wind discrepancies as much as 50 knots or more, we should emphasize them regardless.

Although more extreme examples could be illustrated, we base the discussion solely on Figure 7.02*b*. In the European sector the interwesterly area contains several eddies. The HIGH is part of the polar westerly wave, and the Mediterranean LOW is part of the more tropical wave. Notice in this area that, with very little imagination, a trough line can be followed extending from the Siberian LOW, through the southern Baltic, and toward the LOW over Iceland. A similar structure might be found over the eastern Atlantic and Mediterranean. Some evidence for others is given over central and western Asia, over the southwestern United States, across west central Canada, across central Alaska, and across Labrador. Since these are found daily mostly in the regions of dense data, they cannot be discounted merely as figments of analysis.

The common thing observed is the presence across the extended ridges of (i) zones of cyclonic shear along curved shear lines,[10] connecting or partially connecting adjacent cyclones of a wave, or (ii) the trough line of a col along this shear line with a small anticyclonic eddy on the poleward side. In most cases the shear line is curved concave toward the pole. On some days these can be traced in series around a large part of the hemisphere.

The significance of these features is evident in high-level wind computations and forecasting. Where otherwise an extended uniform ridge might indicate moderate west winds on its axis, there can exist strong west winds with easterlies a short distance poleward. The semisaddles are important also hydrostatically and in development. A closed anticyclonic circulation might be found at a lower or higher level or at the same level before or after. Such arrangement might be considered part of the development of warm polar HIGHs.

This analysis points out the deception in considering that the upper tropospheric westerlies are a broad wavelike band with large subtropical HIGHs extending into the ridges and large polar cyclones extending into the troughs. The westerlies are more streaky, and similarly the closed pressure systems are of smaller dimensions. Closed circulations in the upper troposphere are rarely larger than 20° of latitude in diameter —cyclones smaller.

e) Blocking.—There are varied definitions, explanations, and viewpoints on the concept of blocking, but without too much contradiction in the existing state of knowledge we may say that blocking is a process in which there occurs *westward progression*

10. Further discussion of shear lines is given in Sec. 13.06*d*. See also Y.-P. Hsieh, "On the Formation of Shear Lines in the Upper Atmosphere," *Journal of Meteorology*, VII (1950), 382–87; C. W. Newton *et al.*, "Structure of Shear Lines near the Tropopause in Summer," *Tellus*, III (1951), 154-71.

in reduction of speed of the zonal westerlies. This may be seen as prominent difluence in the westerlies, sudden folding-up of the westerly waves as though impinging on a fixed barrier, significant retarding of eastward-moving systems at the surface, or even westward movement of a warm blocking anticyclone in or slightly poleward of the main westerlies. Blocking action has occupied considerable interest in extended-period forecasting.[11] An excellent illustration of blocking is given in a case study by Berggren, Bolin, and Rossby.[12]

11.06. *Remarks on traveling waves.*— Pressure systems at the surface are said to be steered in the direction of the deep current above them with "the speed of the 700-mb flow," "half the speed of the 500-mb flow," or other such estimates. The surface systems under or nearly under the main currents aloft travel rapidly, while the more barotropic tropical systems, the isolated cold-core cyclones and warm-core anticyclones, and generally the polar systems all move rather slowly. The more symmetric cold-core cyclones found some distance to either side of the westerlies appear quasi-stationary for days at times.

If the thermal flow is of the same sense as the motion at the surface, the circulation intensifies with height and is detectable in the upper flow patterns. In the westerlies, however, the low-level thermal patterns are usually somewhat out of phase with the surface pressure patterns. The amplitude of the flow damps out with height, and the corresponding wave forms may be difficult

to detect in the strong westerly stream aloft. Under such conditions the minor waves aloft might be followed more reliably as isallobaric waves rather than pressure waves.[13]

Minor waves are indicated in Figure 11.04 by the heavy curves, whose positions were obtained from the completely analyzed charts and from details in the pattern of the diagram. Troughs are shown by continuous curves, and ridges by dashed ones. Velocities of the waves are inversely proportional to the slopes of these curves (20° longitude per day is about 35 knots). The small waves can be detected more or less clearly while moving through the long-wave pattern. Minor troughs moving through the major ridge appear to be damped (and they decrease the ridge), but they intensify in the major trough. The opposite is the case with minor ridges.

Interesting facts revealed by this diagram are the slow and quite uniform motion of the major wave eastward while the process of retrogression is often discontinuous. The entire long-wave pattern over North America and the Atlantic jumps westward over the period 26–28 February by intensification of an initially minor ridge and a minor trough west of the corresponding components of the pre-existing major wave, and the components of the previous major wave move eastward with decreasing intensity. It is interesting to add that similar retrogression occurred over eastern Asia on 24–25 February.

Descriptions of polar-front cyclone families appearing in the literature[14] compare favorably with our discussion of planetary and minor waves aloft. Considering par-

11. E. B. Garriott, *Long Range Weather Forecasts* (U.S. Weather Bureau Bulletin, No. 34 [Washington, D.C., 1904]); R. D. Elliott and T. B. Smith, "A Study of the Effect of Large Blocking Highs on the General Circulation in the Northern Hemisphere," *Journal of Meteorology*, VI (1949), 67–85.

12. "An Aerological Study of Zonal Motion, Its Perturbations and Break-down," *Tellus*, I (1949), 14–37.

13. See Figure 12.06a.

14. Sverre Petterssen, *Weather Analysis and Forecasting* (New York: McGraw-Hill Book Co., 1940), pp. 348–50; Palmén, in American Meteorological Society, *Compendium of Meteorology*, pp. 608–12.

ticularly the normal positions of the major waves, and also the fact that the polar front at the surface is most clearly defined downstream from the upper trough, one can visualize ahead of the major upper trough a family of frontal waves corresponding to the isallobaric train aloft, both traveling eastward and poleward with the steering flow. The frontal waves develop in the lower latitudes of the front and are mostly occluded upon reaching the higher latitudes. Normally, the entire system is drifting eastward. Behind the final wave of the sequence the surface (continental) HIGH drifts into the subtropics with a surge of fresh polar air and is transformed into an oceanic subtropical HIGH by heating from below and by subsidence. There is some indication of such wave families in Figure 7.02a.

11.07. *Added remarks on oceanic analysis.* —Familiarity with the details of analysis over land is of great help in analysis over oceans. Primary differences are due to absence of immediate orographic controls, availability of moisture at the surface, and smaller time variations in temperature and temperature gradients of the sea surface, which result in greater low-cloud frequencies, weaker frontal gradients in surface-air temperature, more prolonged stability or instability conditions in certain types of air flow, and prominent weather contrasts in the regions of large sea-surface temperature gradient. Although the positions and intensities of sea-surface isotherms and temperature gradient do vary daily under influence of large- and small-scale air motion, average monthly sea-surface temperature charts are valuable references in surface analysis and also in analysis of upper-level charts by the controls on stability and convection.

As a rule, the air over oceans is least stable in early winter and most stable in early summer. Fogs are frequent over cold-water areas in summer, and during that season fronts are more difficult to detect from elements in surface reports.

The big problem is upper-level analysis. In the present aerological network over the Atlantic Ocean north of 30° N, a reasonable pressure and temperature analysis can be performed without much reference to the surface chart. By proper co-ordination with the surface map it is possible to bring out details in the upper flow pattern useful in maritime and marine weather forecasting. There is a different story over the North Pacific Ocean—aerological data are presently insufficient in most instances to obtain even the broad-scale pattern in middle latitudes. Commercial and reconnaissance aircraft reports are a valuable aid, but such data are given on few and fixed tracks. Analysis involves extrapolation of patterns from the peripheral data, extrapolation of tendency patterns, familiarity with the patterns, deductions from isolated aerological reports, and, most important, co-ordination with the more abundant data and more reliable analysis of the surface chart, including reports of cloud motion.

The beginner will welcome a few suggestions in co-ordinating surface and upper-level pressure analyses to avoid the clumsy analysis resulting otherwise. The graphical air-mass approach described below is objective and produces desired consistent patterns, even though some labor is involved. After practice, the amount of handwork can be reduced appreciably. The technique consists merely of adding a predetermined thickness pattern to the 1000-mb contour pattern. Determining this thickness pattern is the only difficulty of the method. We will refer to the layer up to 700 mb; similar methods hold for a deeper layer or for the layer between 700 mb and a desired higher level, but generally with less accuracy.

Experience with thickness analysis over central and eastern North America, and in the areas of more dense aerological data over the Atlantic Ocean, will indicate the close relations of thickness values and patterns to air-mass and weather patterns at the surface. Frontal waves are waves in thickness lines having particular distributions of gradients. Usually, little significant error results from assuming the warm sector almost homogeneous in this layer (not reliable in layers above). Certain values of thickness can be associated with surface frontal positions, with consideration for latitude and sea-surface temperature distribution. Over tropical oceans a round maximum thickness of 10,000 feet is useful as a guide; the upper limit is larger by roughly 50 feet over tropical islands during the day and still larger over low-latitude deserts in summer.

The pattern should be continuous with thickness values at coastal stations (allowing for the effect of coastal waters and coastal mountain barriers), and should fit the reports from island and ship stations. Furthermore, the values of thickness determined indirectly should obey stability limitations. A superadiabatic lapse rate for the layer must be avoided, and even those near dry adiabatic over the open sea are questioned. In the western sector of a cyclone a dry adiabatic layer perhaps 500 meters in depth with moist adiabatic rate above gives a good approximation to the thickness using the surface-air temperature or even the estimated sea-surface temperature. A value determined in this region is critical, since in that vicinity the thickness line reaches its lowest latitude, and all others in the cold mass can be arranged accordingly. Another useful guide in drawing thickness lines is the vertical wind shear vector for the layer at each station.

The use of patterns of thickness departure from normal (Fig. 9.07) serves the same general purpose, as so would thickness-tendency charts. Both should be subjected to checks against thermal wind vectors, lapse rates, and the values reported at oceanic and coastal stations.

After some feel for the relation between surface and upper flow patterns, it may be necessary to determine just a few extrapolated heights in this way to guide the drawing of contours. To do this properly, surface positions of fronts and pressure centers should be entered on the chart to be drawn, or the drawing can be performed on a light-table with the upper chart superimposed on the completed or partly completed surface chart. In those areas where the surface analysis is unreliable more emphasis must be placed on extrapolations from height-tendency charts and on broad-scale controls in the flow pattern.

11.08. *Differences from winter to summer.* —The increase in duration and angle of sunlight from winter to summer increases the temperatures, most rapidly over continents; and the greater duration of sunlight in high-latitude summer slackens the south-north temperature gradient. These, plus corresponding adjustments in pressure patterns, are the primary differences from winter to summer. Since most of the differences in details follow immediately, an outline will suffice. Reference should be made to the charts in Chapter 1.

In summer (*a*) the westerlies are weaker and are located nearer the pole; (*b*) the polar front changes accordingly (its mean surface position over North America is near the United States–Canadian border); (*c*) the oceanic subtropical HIGHs are located nearer the pole and are as a rule well developed; (*d*) tropical conditions extend farther poleward than in winter; (*e*) systems in general move and change more slowly; (*f*) there is greater persistence in the flow pattern, and interdiurnal changes at the surface are small (excluding polar margins of continents in the northern hemisphere);

(g) the lower atmosphere is more stable over oceans than continents (excluding diurnal variations); and (h) large temperature gradients and strong wind currents in the upper troposphere are frequently found above rather barotropic conditions in the tropical air below, in a manner more striking than in winter.

With reference to the United States in particular, during summer (a) the mean flow at the surface is determined on the west coast by the eastern end of the persistent Pacific anticyclone, in the southwest by the heat LOW, and over the eastern half of the country almost as persistent by the western end of the Atlantic anticyclone; (b) aloft there is a persistent pressure trough just off the west coast, a less persistent trough near the east coast, the average center of the westerly current is over southern Canada, and the flow over most of the country is weak and, in the mean, anticyclonic; (c) frontal cyclones have preferred tracks across southern and central Canada; (d) Pacific polar air masses affect the northwestern part of the country but are heated rapidly over the continent; (e) cP air masses reach-

ing the United States are coolest if they have a direct trajectory from Hudson Bay, and their surface fronts are destroyed usually before penetrating beyond the central latitudes; (f) there are occasional invasions of the northeast by mP air from the Atlantic, following shallow "back-door" cold fronts in the low levels; (g) the southeastern part of the country, and usually the entire region east of the Rockies, is covered by tropical air which is quite homogeneous in the surface layer except for the westward surface-dewpoint drop just east of the Rocky Mountains; (h) the horizontal moisture pattern aloft differs markedly from that at the surface (see Secs. 8.07, 13.05, and 13.06); (i) the usual weak pressure patterns are conducive to land- and sea-breeze effects, notably on the east coast and the Great Lakes; (j) the weather patterns in the tropical air are nonfrontal—they are really of tropical nature; and (k) there is a permanent temperature gradient across the west coast in the low levels, aided some by the coastal upwelling of ocean water.

READING REFERENCES

AMERICAN METEOROLOGICAL SOCIETY. *Compendium of Meteorology*, "The General Circulation," pp. 541–67; "Mechanics of Pressure Systems," pp. 577–652; and "Polar Meteorology," pp. 917–64. Boston: The Society, 1951.

RIEHL, H., et al. *Forecasting in Middle Latitudes*. ("American Meteorological Society Monographs," Vol. I, No. 5.) Boston: The Society, 1952.

STARR, V. P. *Basic Principles of Weather Forecasting*, Chaps. 1, 2, 4, 5, 6, and 7. New York: Harper & Bros., 1942.

WILLETT, H. C. *Descriptive Meteorology*, pp. 124–49. New York: Academic Press, 1944.

CHAPTER 12

Local Analysis

12.01. *Introduction.*—This chapter surveys local analysis in two separate objectives. One is the small-scale features playing a vital role in the local weather which are departures from the larger-scale patterns seen by the synoptic charts. As this subject is extensive, covering the field of micrometeorology (in particular microanalysis) and much of physical meteorology, the discussion here must be limited. Our other objective is based on the opposite approach: weather analysis in ordinary scale based on local indications.

12.02. *Local variations in the surface-weather elements.*—Figure 12.02 permits some insight into small-scale fluctuations, but the original records are far more impressive in detail. While examining these diagrams, reference should be made to Figures 9.02*ab*. As we can mention only a few of the significant items in the diagrams, the reader is asked to study them more carefully.

The indicated variability in wind direction is still pronounced in spite of the limitation imposed by the few direction contacts. Note the relation of unsteadiness to the turbulence indicated by snow showers at Pittsburgh and Chicago. At Bermuda, where the wind recorder gave continuous traces on a time scale 3 inches to the hour, very large fluctuations in direction and speed could be seen—fluctuations within the minute by 10–30 miles per hour and through an octant of direction (through a quadrant in the fresh cold air) were the rule, and variation in speed with periods of several minutes were also apparent. At all stations discrepancies are large between wind speed averaged over the hour (dots) and that determined over short periods (curve).

The semidiurnal wave in the barogram appears at each station but most prominently in low latitudes (San Antonio). In addition, there are numerous irregular variations of shorter period which give the trace its rough appearance. Important to note is that frontal passage is not always defined reliably in the barogram. There is also evidence for analyzing fronts as second-order discontinuities in pressure rather than as presently done.

The temperature curves give the advection of air masses only after correction for the diurnal wave, evaporation, turbulence, and local effects on temperature at each station. Again, frontal passage may be difficult to detect; examine Bermuda, for example.

The curve of cloudiness was drawn using reports an hour apart. It therefore does not show the shorter period fluctuations, and in places it is not consistent with reports of precipitation. At any rate, it does indicate that forecasting cloudiness even to the nearest hour is still a very difficult task.

Variability in ceiling and visibility is a function of precipitation. Examine Pittsburgh in particular, which gave some re-reports of "sky obscured" during the brief snow showers. Visual spot observations such as these are subject to some error and are lacking in detail of variation.

We should mention briefly a few related variations among the various traces. Notice that at Chicago[1] the weather was fine (but

1. Airport station.

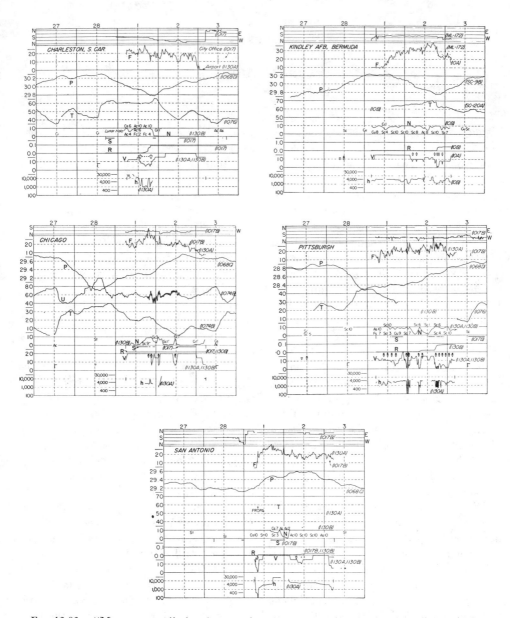

FIG. 12.02.—"Meteorograms" of surface weather elements for 27 February through 3 March 1950 (GCT). *Upper section:* wind direction; *F:* wind speed (mi hr⁻¹), dots are hourly average; *P:* pressure (inches); *U:* relative humidity (per cent); *T:* temperature (° F); *N:* total sky cover (tenths) with types and amount of cloud each 6 hours; *S:* sunshine; *R:* accumulated precipitation (inches); *V:* visibility (miles, 15+ entered as 20), with prevailing weather; *h:* ceiling (feet) in logarithmic scale. Numbers beside curves refer to the Weather Bureau and Signal Corps record forms from which the data were taken.

cold) on 1 March until near 1800 GCT, when cloudiness increased, snow showers began, there was a gentle trough in the barogram, and the wind was to veer into more northerly direction. In principle, this is a common sequence in a fresh cold-air mass for a station whose displacement is from the southern sector to the western sector relative to a deep cyclone. Observe also how the break in cloudiness and renewal of sunshine near 2200 GCT was related to the temperature and relative humidity variations, how the cloudiness and rain in the low pressure on 28 February affected the advective and diurnal course of temperature, and how the drop in visibility attended passage of the pressure ridge, decrease of wind, backing of wind into south, and warming aloft, besides diurnal cooling at the surface.

The dominating feature in the Pittsburgh diagram, in contrast with Chicago, is the effect of upslope terrain in an unstable cold-air mass blowing from the Great Lakes and perhaps also from the precipitation area on the northern side of the cyclone. This maintains persistent low cloudiness, snow flurries with variable visibility and ceiling, and small diurnal temperature variation. Clearing and stabilization finally occurred with proximity of the pressure ridge.

At the most southern station, San Antonio, quite a different sequence of events is observed with the frontal passage (FROPA) and within the cold air. Before the arrival of the front, wind was light and slowly veering from south through north into northeast, and pressure had begun to rise. With the frontal passage there was no sharp change in wind direction and pressure rise, but the wind increased sharply in speed; there were a splatter of rain and a sharp change in cloud conditions.[2] The latter part of the period was dominated by

2. In explaining the smoky conditions at the front, refer to Sec. 10.19a.

eastward motion of the California cold cyclone.

Charleston and Bermuda are affected by similar parts of the flow pattern, though at different times and with different effects by the surface. Throughout the 24 hours preceding the cold front, both stations experienced warming from mP conditions to mT. Rains preceded the front and ended abruptly with arrival of the cold-subsiding air mass. Note the difference in cloudiness at the two stations behind the front, the large prefrontal evaporation cooling at Bermuda, and the apparent separation of the rain area from the front at Charleston.

These diagrams serve to show the many additional considerations that have to be made in analyzing and forecasting the weather locally. We cannot depend only on model distributions of weather about fronts to answer all questions. For forecasting details in the weather, small-scale analysis is more necessary in some parts of the weather pattern than in others; but in all cases departures due to peculiarities of station location must be taken into account. There are those times also when "looking out the window," so to speak, is more helpful than staring at fronts and isobars on the weather map!

12.03. Local effects.—Influences on the weather locally by the underlying surface are known as the *local effects*. Some of these were mentioned in the preceding section and also in Section 4.23. It is not possible to discuss all the local effects on weather; there are unique effects for each point of observation.

Local effects may be grouped according to *orography* (mechanical), *nature of surface* (thermal and evaporative), and *sources of pollution*. In the first are the magnitude and direction of terrain slope, height and orientation of mountain ranges, rolling countryside, narrow river valleys, and even rivers with high bluffs. The second category involves differences between land and water, differences in soil cover and in sea-surface temperature gradient, and, as concerns fog, rivers, marshes, and other low places. Pollutive effects are notably those of smoke in visibility. These are very important in and about heavily industrialized areas and near larger cities during stable conditions in winter, but they are also present with a

outline of the coast, and orography.[3] At cities on the east side of the Rockies the slope of terrain is such as to produce abnormally good weather in westerly flow but especially poor conditions with easterly flow of moist stable air as exists in the north and northwest sectors of cyclones in winter. Chicago has the smoke problem, as do most industrial areas, which is a function of stability and wind direction; and its

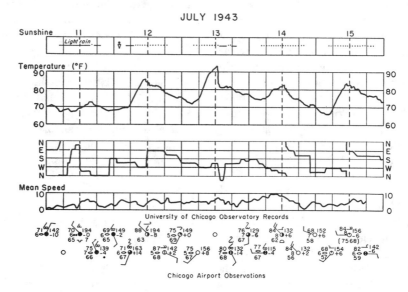

FIG. 12.03.—"Meteorogram" of 11–15 July 1943, 90th meridian time, for the University of Chicago (1 mile from lake) and surface observations at the airport on the west side of the city. Note onset of cool lake breeze near noon with weak pressure gradients prevailing.

coal-burning railroad near by. Each of these effects can produce large variations in the weather over just a small portion of the flow pattern.

It is necessary to know the local influences on low-level weather phenomena in order to analyze and forecast properly the weather for any locality. These influences occupy necessary emphasis on terminal forecasting for aircraft operations. Thus, at San Francisco conditions are very dependent on the direction of air flow and modifications due to the coastal water, the peculiar

weather is also influenced by the lake, which is a warm moist source in early winter and the cool source of breezes during weak pressure gradients in summer (Fig. 12.03). The Great Lakes area in general is featured by large variations in weather and climate over relatively short distances due to thermal and some orographic controls.

It is beyond the scope of any book to

3. Directorate of Weather, Headquarters Army Air Forces, *Terminal Weather on the San Francisco–New York Airway* (Washington, D.C.: Government Printing Office, 1942).

describe in detail the local modifications on weather at all the aircraft terminals, let alone all other places. Hundreds of useful local weather studies have been prepared for limited use, but most are not in generally available form.[4] Many more are made through experience and are never written. This is a commentary for reliance on length of experience in one certain area, with both its good and its bad features, and also for justified and unjustified criticism of emphasis on physical theory and generalization in meteorological education. Finally, we might say that accounts for local departures in the weather involve an understanding of (i) the details of geography—primarily physiography—and (ii) the physics of the lower atmosphere.

12.04. *Tracking of small-scale weather phenomena.*—The weather phenomena of small dimensions, including showers, thunderstorms, tornadoes, squall lines, and the like, require special modes of analysis, as they are difficult to follow adequately in the ordinary synoptic chart. Several rather simple methods have been in use, and others doubtless will be devised with improvements in observation.

There is provision in the hourly weather transmission for reporting these phenomena and their apparent direction of movement. Provided they persist, they can be followed from station to station in the network but more accurately if reports from aircraft are also available. Weather radar now provides for continuous tracking of rain areas. From a series of positions plotted on a large-scale map, it is possible to estimate velocity and perhaps acceleration. If the weather occurs along a line, successive positions (*isochrones*) will be curves on the map, from

4. Among these are numerous research papers prepared by airline, Weather Bureau, and Air Weather Service meteorologists. Only a few are ever published in the circulated journals.

which change of orientation and shape can be determined as well as movement. The principle is that of the continuity chart.

Though its use is restricted and it involves some effort, the horizontal time sec-

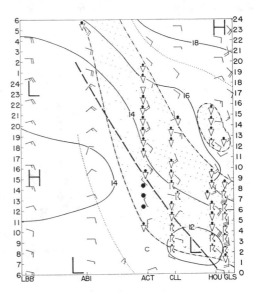

Fig. 12.04*a*.—Horizontal time section at the surface through an easterly wave trough, 29 June 1953, from Lubbock (LBB) to Abilene (ABI), Waco (ACT), College Station (CLL), Houston (HOU), and Galveston (GLS). (Time scale is GCT on left, and 90th meridian time on right; heavy dashed curve gives roughly the trough position, stippled areas are rain, dotted curves give the edges of overcast, and continuous curves are isobars of sea-level pressure. Winds are plotted in scale 10 knots to the full barb.) Notice the apparent narrowing of the rain area as it moved inland and also the effects of semidiurnal pressure variation.

tion (*x, t* diagram) has value in detailed analysis of movement and development of visible or nonvisible phenomena along a line of observation points. Figure 12.04*a* gives an example for a large perturbation in the weather pattern using quite a loose network of stations but nevertheless indicating some usefulness. In this diagram the slope and curvature of a characteristic line are proportional to its speed and acceleration, respectively. Depending on the layout and density of the observation network, this chart could be employed for analyzing movement and development of a number of weather elements, including fog and cloud decks.

pected that the reader will be exiled on a remote island with no external weather information, but we do believe the concept is a useful pedagogical tool (i) for exercise in integrating fact and physical theory, (ii) for analyzing the weather pattern daily with minimum expense and effort, and, most im-

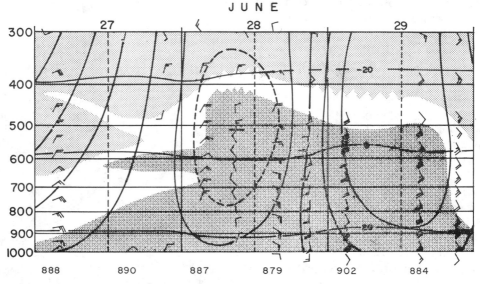

FIG. 12.04*b*.—Vertical time section for Lake Charles, 27–29 June 1953 (GCT) through the same easterly wave (see legend of Fig. 12.06*a*).

12.05. *Introduction to analysis from local indications.*—Analysis from single-station data is no doubt the oldest and most widely employed method of analysis. It is used by the mariner, the shepherd, the farmer, and the man on the street, often with results to the envy of the professional, who in some conditions is apt to be blinded in a mass of weather data. During the last war, when international exchange of weather information was restricted, and when weather analysis and forecasting over large areas had to be done from local observations, the concept of single-station analysis was expanded by the Chicago school[5] to incorporate the use of aerological data. Though the needs are different now, there is no hesitancy in giving life to the concept here. It is not ex-

portant, (iii) for developing attitude in ordinary analysis. In practice it is always of direct value in oceanic analysis particularly, and even in the denser data networks it is a method of deriving many clues from quick glances at data from a few stations.

Success with the technique depends on familiarity with atmospheric patterns—their structure, their movement, their development, their statistics—and not the least on keen observation by the individual. To the careful weather-watcher familiar

5. "Department of Meteorology, University of Chicago, Miscellaneous Reports," Nos. 1, 2, 6, 7, and 8; V. J. and M. B. Oliver, "Weather Analysis from Single-Station Data," in F. A. Berry, E. Bollay, and N. R. Beers (eds.), *Handbook of Meteorology* (New York: McGraw-Hill Book Co., 1945), pp. 858–79.

with these, there is valuable information in cloud developments, in local wind and pressure changes, in the temperature and humidity and their changes, in the vertical variation of wind and other elements, and so on. However reactionary this may seem, there are the occasions when crude estimates of patterns from local data alone will give equal or superior short-period forecasts than from the profusion of the weather map.

In the following paragraphs the objectives are to refer to existing literature on the subject, to expand on certain ideas—old and recent—found useful for instruction, and to mention a few points of value in the common methods of weather analysis. Although it is assumed that complete sequences of surface and aerological observations are available for the station, many reliable deductions could be obtained merely from amateur methods of observation.

12.06. *The vertical time section.*—The t, z diagram, or vertical time section, provides a means of co-ordinating all data in the vertical and in time sequence for a given station. Figures 12.06ab illustrate two series maintained on current basis but with most of the data omitted in reproduction. The first (a) was a period in which the center of the main westerly current aloft was generally near or south of the station; in the other the station was generally farther south of the westerly center, as the winds, temperatures, and season would indicate.

The surface observations were made at the airport on the west side of the city, and the aerological observations at Joliet, a short distance southwest. In several places the wind sounding is supplemented by the Rantoul (R) or Milwaukee (M) winds. The coarsely stippled areas represent temperature-dewpoint depression less than $5°$ C, and the more finely stippled areas depression of more than $10°$ C. As the moisture pattern is actually very irregular and diffi-

cult to define from 12-hourly observations, the analysis given must be considered crude.

The shaded moist areas are not to be taken wholly as cloud patterns. Besides the difficulties cited before in the analysis of moisture, cloud masses can depart considerably in shape from the line of $5°$ C temperature-dewpoint depression, as is clearly evident from examining the reports of cloud amount and cloud height in the surface observations. Clouds need not exist over all the shaded moist areas, and, owing to instrumental difficulties and method of humidity evaluation at cirrus levels particularly, clouds may exist even outside the indicated moist areas.

The types of analysis on the charts are largely a matter of choice. The 12-hour pressure tendency would be almost useless in low latitudes. The pressure analysis given in (b) is probably more easily drawn between soundings than are the patterns of tendency, but for the same reason it does not show the details of pressure variation as well. Isotherms of potential temperature would be desirable as a means of showing frontal zones, stable layers, the tropopause, and temperature change all at once. Indicating cloud layers would be useful if it were not for obscuration by lower clouds and lack of data concerning cloud tops.

The time coordinate is arranged to give proper space perspective when viewed from south (from the equator). In the westerlies, systems move from left to right. With movement dominantly from east, the reverse scale would apply (Fig. 12.04b).

The vertical time section can be maintained with all data and analysis in less than an hour a day. Depending on the purpose, it can be maintained in all detail or in simplified form approaching the surface time section in Figure 9.02d. They have proved value in analysis over oceans and remote land areas—in the tropics and elsewhere—since there the patterns for large

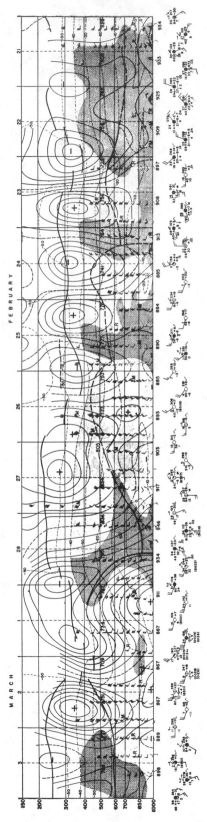

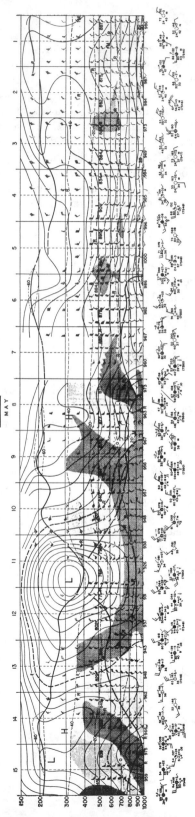

FIG. 12.06a.—Vertical time section for Chicago, 21 February–3 March 1950, with 12-hour pressure-height change patterns (100's of feet), tropopause, approximate frontal surface, isotherms, wind reports, humidity pattern, 500-mb height and (700/1000)-mb thickness (10's of feet), and 6-hourly surface observations. (Visibility plotted in eighths of miles up to 10 miles; surface winds in Beaufort scale; further description in text.) (Some of the fine stippling could not be reproduced.)

FIG. 12.06b.—Vertical time section for Chicago, 1–15 May 1952, with isopleths of pressure-height departure from the reference described in Section 6.13

areas must be determined from events at individual stations.

12.07. *The period 21 February–3 March 1950.*—To get a general setting for this period, we might first resort to normals in determining the large-scale flow pattern. For the normal height of 500 mb at this location and season we may use 17,900 feet. If actual height is much below this (by several hundred feet), the station is probably in a major trough of the upper flow pattern, and conversely for a ridge.[6] Thus, at the beginning of the period the station is in a ridge of little amplitude aloft (winds had recently backed from northwesterly). After passage of one smaller wave the deep trough passed by on the 22d, and the large ridge drifted by on the 27th and 28th only after an unusually regular series of minor waves. Immediately after this large ridge came a deep trough, to be followed by another progression of minor waves from northwest until about 6 March. Part of the period is reproduced in Figure 11.04, but neither in that chart nor from the synoptic pressure charts are the upper pressure waves so clearly indicated as by the tendencies in the vertical time section.

The appropriate normal value of (700/ 1000)-mb thickness is about 9200 feet. The time variation of this quantity is in good agreement with the flow pattern aloft— high values beneath and slightly to the rear of large ridges, low values located similarly with respect to large troughs, and superimposed on this are fluctuations corresponding to the minor (frontal) waves and also to diurnal temperature variation.

Now we should examine some of the interesting details given day by day in the time section. At the beginning there is

6. W. W. Jones *et al.*, *Weather-Map Construction and Forecasting in the Westerlies from Single-Station Aerological Data* ("Department of Meteorology, University of Chicago, Miscellaneous Reports," No. 7 [Chicago: University of Chicago Press, 1943]).

southeasterly flow at the surface with the usual veering through the friction layer. If we take 2000 feet as the gradient level and consider the rapid vertical change into west wind, a surface HIGH is centered a rather short distance east of the station. (Data the day before showed the surface center passed north of the station near 0630 GCT and its axis passed over the station near 4000 feet some 6 hours later. The horizontal temperature gradient, which can be estimated roughly from the time variation of temperature, will also aid in approximating the tilt of a pressure center with height). Because the wind soundings given in this portion of the chart are from two locations almost 100 miles apart, we must forego any detailed discussion using upper winds there.

The surface wind became more southerly and pressure continued to fall, until late on the 21st, when the wind shifted almost 180° and pressure began to rise, all suggesting passage of a cyclone south of the station with a sharp eastward-moving trough extending into the east winds north of the center. The winds from 2000 to 5000 feet at 0900 GCT on the 21st suggest the surface cyclone passed barely south of the station. Although this center was quite weak, it had a typical cloud sequence with some precipitation. The change from drizzle to freezing drizzle indicates cooling below freezing at the surface while temperatures above remain high.

The renewed pressure fall, the particular winds, and the persistent snowfall indicate either reversal of motion of the cyclone or its deepening. The former is the more unlikely in view of the horizontal patterns of temperature. The snowfall is no doubt aided by air motion over the lake from the northern sector of the cyclone. Finally, the cold air deepens, the station leaves the cyclonic circulation by northeastward motion of the center, and a weak pressure ridge goes by at the surface with center to the south and followed some time later by the ridge aloft.

The 23d gives a typical sequence in the retreating cold dome with approach of another cyclone—dry air aloft, beginning of warm advection, and then followed by general ascent and the upper cloud system near the cyclone. At 1500 GCT the winds at gradient level and 10,000 feet indicate for the (700/1000)-mb layer a mean temperature gradient almost 2° C per 1° latitude, with isotherms directed southeastward, and horizontal advection at the rate 4° C/12 hours, approximately. In the (500/700)-mb layer isotherms are directed ESE, and there is negligible advection; this is the ordinary variation of temperature and advective patterns with height for this location in the cold air at the surface. The next wind sounding gives consistent indications.

A check on the wind sounding at 0300 GCT shows it is unrepresentative or erroneous, most probably the latter; the next later one (0900 GCT) is also questionable in the upper part. Series of 500-mb charts indicate the wind backed slightly until about 0900 GCT on the 24th and then veered more rapidly into WNW. An inconsistent picture is obtained from these winds as reported and plotted.

The pressure trough goes by with occluded front at the surface (if any front at all) and LOW center passing north of the station. As the fresh air mass is the colder, the trough lags behind at upper levels. By 1500 GCT the cold advection has practically ceased—isotherms are generally parallel to the flow at all levels. Early on the 25th the temporary resumption of cold advection, combined with continued pressure rises at the surface and low temperatures, might suggest a renewed push of cold air behind a weak cyclone going by to the south.

Later on the 25th is a sequence of clouds and upper-level conditions characteristic of another frontal and pressure wave, but it is not so evident in the surface pressure, wind,

and temperature. In a 12-hour period the net temperature drop is 6° C at 500 mb, 2° C at 700 mb, and 0° C at 1000 mb. This is a clear-cut example of the utility of upper-level analysis in describing and forecasting weather at the surface. The surface synoptic pressure patterns for this period would hardly reveal this cloud system.

The pressure trough is followed by another southward push of cold air. At 1500 GCT on the 26th the cooling had ended, and isotherms throughout the troposphere were oriented NNW–SSE with appreciable gradient over the station, indicating the mass of cold air was deepest along a meridian to the east. Indications are that subsidence is or has been at work in this air.

Through the next 36 hours warmer and warmer air appears with approach of the large upper ridge and passage of the center of the surface anticyclone south of the station. The transition from light surface winds and clear conditions to strong flow from the south gives rapid warming at the surface. The warm front approaches from southwest and apparently comes within a few hundred miles of the station. (As the frontal slope near the ground is extremely variable, we do not attempt a guess at its surface location only from its position in the sounding but use in this connection the pressure, temperature, and isallobaric patterns.)

The occluded system passes with its rain area, wind shift, and attendant isallobaric field; these indicate the surface cyclone passed a short distance north of the station, say, over the northern Great Lakes. The wind soundings at 1500 and 2100 GCT suggest the surface front is oriented almost north-south at the latitude of the station and curves westward in lower latitudes. The persistent cold advection and very low temperatures to the rear indicate this cold mass extends far southward, as also do the depth of the attendant upper pressure trough and the strong westerly winds near the ground. These conditions also suggest that the surface cyclone to the northeast has deepened and perhaps retarded. The low-level moistening, stratocumulus clouds, and snow showers late on the 1st were mentioned previously in connection with Figures 9.15c and 12.02. The remainder of the sequence is a familiar repetition.

12.08. *The period 1–15 May 1952.*—Because of the weakening of the flow and poleward shift of the westerlies with approach of summer, this period does not have generally the large contrasts and rapid variations shown in the previous one. Most cold outbreaks are now shallow and do not extend as far southward. For describing this sequence a few helpful hints are in order. The normal height of the 500-mb surface is about 18,500 feet ($D \simeq 200$ feet) and the normal (700/1000)-mb thickness 9600 feet. Further, the local pressure-height tendency is the horizontal gradient of D in the chart.

At the beginning of the period we find a rather indifferent surface circulation pattern south of a small weak HIGH center; aloft the center is apparently west and south of the station. The surface air is of modified polar variety which has stagnated over this general region. From the circulation and temperatures aloft, indications are minimum activity of the polar front. The approaching trough tilts in abnormal sense in the vertical plane along its path, in agreement with the slightly higher temperature behind it at most levels. Just behind this trough, whose surface cyclone seems to have passed very near the station, came a shallow cold-air mass from Lake Michigan; its depth was about 2000 feet at 1500 GCT, 4000 feet 12 hours later, and difficult to determine thereafter. In spite of pressure rises at the surface attending the low-level cooling, there is moistening of the middle troposphere associated in the usual way with pretrough conditions in the westerlies.

The surface pressure ridge passes late on the 3d with an apparent center just south

of the station. The cool and smoky conditions persisting in the southerly winds indicate the surface air is arriving with a trajectory curved from east. Subsequently, the surface air arrives with higher temperature and moisture from a more southerly trajectory as the pressure ridge moves on— but the moderate dewpoint temperatures and the northwesterly flow aloft indicate that this air has not come from as far south as the Gulf of Mexico or environs. Notice from the vertical wind shear that the warmest air is west. As the next weak trough approaches, the moistening aloft combined with afternoon convection gives rise to high-level cumulus activity, but the extreme dryness above prohibits much development. The situation at this time lends some appearance of a *weak* warm front followed by air of cT characteristics, which was most likely formed by stagnation of polar air over the continent under anticyclonic conditions aloft.

A day later another trough beckons more strongly. This one gives a high-level thunderstorm and is followed by a shallow burst of fresh polar air arriving from northwest. The lowest surface temperatures occur in the air arriving later from the cold lake. The surface HIGH passes north of the station, giving air trajectories of longer travel over the lakes; and, even though the (700/1000)-mb layer as a whole begins to warm, surface-air temperatures remain low. During this period winds aloft from south of west appear and, along with general decrease of pressure aloft, indicate that the major upper ridge has drifted eastward. Gradual deterioration of weather is now expected.

This shallow cold dome deserves more careful analysis. At 0900 GCT on the 6th the wind sounding indicates cold advection and, from the (700/1000)-mb thermal vector, SW–NE orientation of the surface cold front east of the station. This thermal vector veers into WNW direction by 0300 GCT the following day. Meanwhile, it increases in magnitude to about 50 knots (thickness gradient about 100 feet/° ϕ), and cold advection continues a few hours longer. All things considered, this front remains relatively stationary south and west of the station while it is reinforced by low-level cold air from the northern Great Lakes and perhaps by south winds in the region south of it. The *exact* position of the front at the surface is of less consequence to the existing and future weather at Chicago than is the temperature pattern over the station.

Onset of warm advection, warm-front clouds and rain, and rapid pressure fall indicate approach of the next frontal cyclone. Since the easterly surface winds do not turn into south or southwest during the large pressure falls, one anticipates the cyclone will pass near or south of the station. Its axis passed about 3000 feet above the surface. The succeeding cold dome behaves like the previous one, although the air is somewhat cooler. Warm advection sets in by 0300 GCT on the 9th, at least above the friction layer, and other pronounced warm-front conditions have begun already.

Throughout the period 6–9 May, the general picture obtained is a quasi-stationary front lying at the surface generally west-east through the center of the country with small waves of short period moving along it. Aloft, the current consists of moderately strong west winds with similar short-wave pattern, but having as a whole some anticyclonic curvature, if pressure departures from normal are considered.

On the 9th another pressure trough goes by Chicago, apparently with a frontal wave crest to the south at the surface. The cooling behind is gradual and persistent. The small pressure and wind changes at the surface with unsettled weather suggest this surface cyclone takes a slow path north of east and deepens in accord with the south-westerly flow aloft and the slow motion of

the deep upper depression. Resumption of rain and upper clouds on the 12th and 13th attends a secondary pressure trough and renewed deepening of the cold air in the northwesterly flow. The surface front is now in low latitudes, as the deep cold air and westerly surface winds at Chicago on the 13th and 14th suggest, and the next frontal activity at the surface can be expected to arrive from north of west around the northern periphery of the major ridge we suspect over the western part of the country.

12.09. *Remarks on the relation of pressure and vertical motion patterns.*—The evidence shown in the two vertical time sections on the close relation of pressure and moisture patterns through the middle troposphere is worth emphasizing. The patterns of moisture and clouds in this region, and of attendant rainfall observed at the surface, are very sensitive to vertical motions. On the other hand, the moisture, clouds, and turbulent type of precipitation near the ground are a function of horizontal air trajectory, type of surface, and presence or absence of precipitation from above.

There is a remarkable association of fluctuations in depth of the moist layer with phase of the traveling upper waves. The more saturated conditions, with altostratus and rain, are found mostly in advance of the minor troughs, or, more precisely, near the pre-trough inflections of the wave. Dry subsidence areas are preferred near the pre-ridge inflection of the wave. This empiricism verifies Figures 10.20*bc* and also Section 10.33. Careful analysis of flow patterns in this region of the atmosphere brings out critical parts of the weather pattern not always so clearly associated with the surface pressure field.

12.10. *Use of the hodograph in local analysis.*—In Section 3.26 we mentioned briefly the wind variation in the friction layer and how it was affected by stability. Theoretically, the hodogram in this layer describes a logarithmic spiral about the vector wind at the gradient level. The spiral is clockwise upward in the northern hemisphere—veering of wind with height—and its amplitude and vertical depth are a function of stability, among other things. For stations of low elevation above sea level, we ordinarily assume the gradient-level wind is representative of the sea-level pressure or 1000-mb patterns. As it is difficult to determine the gradient level from wind reports as presently given (and doubtless also from more detailed methods of computing and reporting), use of a certain level (2000 feet) as the gradient level usually suffices with some allowance perhaps for stability.

Relations of wind to horizontal pressure pattern, and of thermal wind to thickness pattern, were discussed amply in Chapter 7. By using those principles, pressure and thermal patterns can be determined about the station from the hodogram. After a little practice the same can be done from plotted winds on the vertical time section, and even from the coded wind report. Rates of thermal advection may be computed or estimated qualitatively from the hodogram or visualized from the coded wind report.

Strong frontal zones should be indicated in the hodogram by large thermal wind directed nearly along contours of the front. The type of thermal advection distinguishes among cold, stationary, and warm fronts. Where the frontal zone fades into a horizontal stable layer, there is little indication in the wind variation with height.

As pointed out in several publications on single-station analysis, a rough idea of the lateral distribution of stability can be had from the vertical variation of thermal wind. In Figure 12.10 are shown the thermal wind and associated mean temperature patterns for two layers of about equal depth (bounded by levels *0*, *1*, and *2*). Along any vertical

the difference between the mean temperatures in the two layers may be considered a measure of the average lapse rate of temperature through both, and hence the dashed lines give roughly the pattern of stability. Notice how these conform with vector difference in thermal winds (dashed vector), stability increasing to the right of this vector. There are several places in the synoptic patterns where this indication is

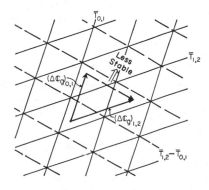

FIG. 12.10.—Showing graphically the pattern of mean stability (*dashed lines*) through two layers in relation to the thermal wind vectors in each.

useful; one is in the warm sector of a frontal wave at the approach of the cold front.

From the hodogram, as also in isobaric analysis, the "advective tendency of stability" between two levels can be determined from the vertical variation of temperature advection.

All uses of the hodogram in single-station analysis are based on the geostrophic assumption and on accuracy of wind observations. These methods are not wholly reliable in low latitudes, but otherwise their general merit would correspond to merits of synoptic pressure analysis.

12.11. *Outline for constructing synoptic patterns from local data.*—In preceding discussions we covered a number of considerations useful in constructing, at least mentally, the primary features of synoptic patterns at the surface and aloft in the vicinity of the station. Here we outline principles that can be used for this purpose primarily outside the tropics. Ideally, they all boil down to one principle: knowledge of structure and processes of the atmosphere.

(i) *Know the station and its environment.* Normal conditions and variability of the elements are useful references—freezing temperatures, northwest winds, or 500-mb change by 200 feet per day have plenty of meaning over the Florida peninsula but are more ordinary occurrences in the northeastern states. In using observations, allowance must be made for local influences.

(ii) *Determine the broad-scale steering pattern* from past events, from the circulation and from pressure departures aloft, and also from departures in temperature at most levels, the latter being effectively the air-mass parameter. If upper pressures are far above or below normal, large amplitude of the flow and "low-index" air-mass conditions are indicated. With winds aloft generally westerly and small departures from normal pressures, the high-index conditions with eastward propagation are in evidence. Positions of major wind currents aloft are given some indication by wind speeds, thermal wind, and temperatures.

(iii) *Determine the pressure pattern near the station* from wind directions and speeds (using the geostrophic approximation properly), from departures in pressure, from clouds and weather, from time variations of wind at various levels, and from shift of pressure systems with height. At the surface especially, location of fronts can be used in obtaining the detailed structure of the pressure pattern. Lower and upper pressure patterns must be consistent with the pattern of temperature. When drawing the analysis, all three patterns should be co-ordinated.

(iv) *Determine the location of fronts in the area.* Use of broad-scale and local flow patterns in detecting fronts were discussed in Chapters 6, 7, 9, and 11. The temperatures and pressure-layer thickness, which are air-mass indicators, will suggest the direction of fronts from the station. For example, if for the location and season and from past history the surface front is identified with thickness 9800 feet of the (700/1000)-mb layer, the thermal wind in the layer and thickness at the station will aid in orienting the front and in placing it the proper distance. With certain reservations, the front can be located in the vertical from wind shear and stability.

In a sequence through a frontal wave, approach of the warm front is detected from the pressure pattern, pressure change, weather, warm advection, and actual warming. The warm sector is evident from thermal indications and from approximate barotropy in the wind sounding. Approach of the cold front is indicated in the pressure and pressure-change patterns, frequently by instability weather in advance and also in the wind sounding. As indicated in Chapter 7 and as described earlier by Oliver and Oliver,[7] in the region up to about 200 miles in advance of the surface cold front there is usually negligible and indifferent advection in the lower few thousand feet, but cold advection above may appear as such from the wind sounding. Passage of the cold front is easily detected. Its orientation can be determined roughly from the thermal wind above a point in the fresh cold air at the surface, its motion from the normal gradient wind and from movement of pressure systems, and its later positions from thermal indications within the cold air.

(v) *Determine the pattern of temperature near the surface* from the thermal wind (or thickness pattern) with corrections for radiation, local influences, and variations in stability. Direction and strength of upper

7. *Op. cit.*

winds will give some immediate indication of the thermal pattern, and conversely.

(vi) *Determine the weather distribution* consistent with the flow pattern and history, with humidities in the air masses, and with effects by the surface.

READING REFERENCES

AMERICAN METEOROLOGICAL SOCIETY. "Local Circulations," *Compendium of Meteorology*, pp. 655–93. Boston: The Society, 1951.

OLIVER, V. J. and M. B. "Weather Analysis from Single-Station Data," in BERRY, F. A., BOLLAY, E., and BEERS, N. R. (eds.). *Handbook of Meteorology*, pp. 858–79. New York: McGraw-Hill Book Co., 1945.

STARR, V. P. *Basic Principles of Weather Forecasting*, Chap. 8. New York: Harper & Bros., 1942.

CHAPTER 13

Brief Survey of Analysis in the Tropics

13.01. *Introduction.*—From a survey of literature on the subject the meteorologist acquainted with analysis in middle and high latitudes might gain the impression that the tropics are the *other* half of the global area where the atmosphere is altogether different and where things do not apply as they should. Perhaps that is due to the great lead speculative theory has had in the past over concrete analysis in a large region with little data. In reality, the tropics afford the opportunity to employ basic concepts of analysis not complicated by frontal and geostrophic theory upon which we have come to rely elsewhere.

Though the geographer may not agree, here we define the tropics broadly as the area between the two subtropical axes of high pressure. This area shifts meridionally with the season and also slightly from day to day. Within the region are part of the horse latitudes, the easterly trades, the doldrums, and the tropical monsoon areas.

In this chapter we go into little descriptive detail that is properly in the field of climatology or that can be found among the "Reading References." In fact, no discussion is given on tropical storms, which would require many pages to handle adequately, and we have nothing to add on the Asiatic monsoon.

13.02. *The tropical pressure pattern.*—If average pressure patterns are indicative of daily patterns over any large area of the globe, that area is the tropics. Day-to-day changes of pressure are relatively small (excluding tropical storms), and they are usually overshadowed by the semidiurnal varia-

tion. The largest pressure changes occur along the poleward boundaries of the tropics, with moving disturbances in the westerlies. With these provisions, what was said in Section 1.03 and what can be seen by examining mean pressure charts of the tropics[1] should give fair indication of daily pressure patterns.

In general, pressure gradients decrease toward the equatorial trough of low pressure, where the pattern is usually ill defined. The trough shifts meridionally with the season, as indicated in Section 1.03, but by varying amounts in different longitudes. In the eastern Pacific and Atlantic oceans its mean position is north of the equator in both summer and winter. It is drawn poleward into or toward the continental heat LOWs of summer.

Pressure analysis is difficult near the equator, where gradients are weak and relation to winds is poor. With weak gradients, errors in observation of pressure and discrepancies by local dynamic effects (land and sea breezes and orography) appear magnified. From this standpoint, observations from ships at sea are better than from land stations. The pressure-wind relation can be used with some confidence down to roughly 10° or 15° latitude, although greater angles with the isobars can be expected there than in higher latitudes. Even within a few degrees from the equator the winds on the average should not intersect isobars at greater than right angles, so that *wind directions* can be of some use in pressure

1. See, e.g., B. Haurwitz and J. M. Austin, *Climatology* (New York: McGraw-Hill Book Co., 1944).

400

analysis there also. If the purpose is to determine the field of motion, which departs so greatly here from the pressure pattern, it appears that *detailed* pressure analysis near the equator is not so necessary as wind analysis. However, pressure analysis can have value there in broad scale or if the field of acceleration is desired.

In spite of weak pressure gradients and small interdiurnal changes, there are important weather patterns associated with isallobaric fields.[2] To separate these changes from the usually larger semidiurnal variation, use of the 24-hour local pressure change has been advocated and widely used.

13.03. *The mean surface flow patterns.*— Figures 13.03*abcd* are abstracts from *Atlas of Climatic Charts of the Oceans*[3] and give average wind data over the tropical oceans for January and July. For explanation of these, and for additional monthly charts, the reader is referred to the atlas.

Charts (*a*) and (*b*) give the resultant flow in January and July, months which in some areas represent the extreme seasons but in others do not. Resultant streamlines and features of the streamline pattern are determined readily by visual inspection. Subtropical anticyclonic centers can be found over each ocean in areas of weak resultant flow but perhaps not so clearly over the South Pacific Ocean. The trades are indicated by broad areas with winds between northerly and easterly (average east-north-easterly) in the northern hemisphere and between southerly and easterly (average east-southeasterly) in the southern hemisphere. The line along which the trades from both hemispheres converge, the *intertropical convergence line* (ICL),[4] can be seen rather clear-

ly around the globe except over the large land areas where data are not given.

In the January chart the ICL can be traced from the Liberian coast to the Amazon Delta, from the southern coast of Colombia to a northernmost location near 10° N 125° W and thence to the equator between 180° and 160° W. Here the line is either double or discontinuous. Another one can be detected beginning near 25° S 130° W (south of the "dry area of the Pacific"), extending south of the Solomon Islands, over or slightly off northern Australia, and then by Java, through Madagascar, and southern Africa. In July the ICL is roughly 10° farther north in the Atlantic, crosses the northern coast of South America and Panama, reaches near 15° N south of Lower California, dips near the equator in the longitudes of the Marshall, Gilbert, and Caroline island groups, and then trends northwestward to the China coast, from where it might be associated with the heat LOW of southern Asia and northern Africa. There is some indication of a secondary convergence line extending from the vicinity of 15° S 160° W and joining the other near 155° E.

The ICL is more evident in these charts and, somewhat correspondingly, more persistent in location from day to day in some areas than in others. It is most variable in the western Pacific, and there the position given by resultant winds is apt to differ most from its average daily position. The largest seasonal range in location occurs in the eastern hemisphere. From monthly charts it appears that the widest annual

2. G. E. Dunn, "Cyclogenesis in the Tropical Atlantic," *Bulletin of the American Meteorological Society*, XXI (1940), 215–29.

3. Washington, D.C.: U.S. Department of Agriculture, Weather Bureau, 1938.

4. The ICL, as used here, is the line along which streamlines converge and which separates the two primary wind systems of the tropics. This line may or may not coincide precisely with the zone of maximum horizontal velocity convergence or with the attendant trough line of low pressure. The three things can be distinctly different in certain areas, but the latter two and often the three are too often taken as synonymous.

Figs. 13.03*ab*.—Resultant surface wind velocities for each 5° unit area. Fine stippling indicates Beaufort wind forces 1 to 3, coarse stippling forces greater than 3. (*a*) January, (*b*) July. (From *Atlas of Climatic Charts of the Oceans* [U.S. Weather Bureau, 1938].)

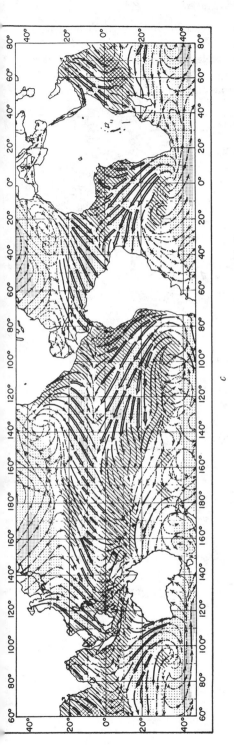

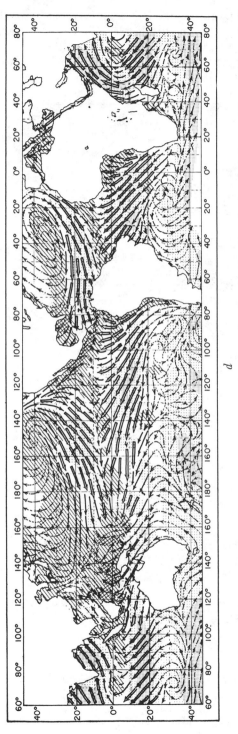

c

d

Figs. 13.03*cd*.—Predominant direction, constancy, and force of surface winds; direction lines based on dominant wind arrows computed for each 5° square, with relative constancy indicated as follows: *double-line shaft*, 81 per cent and over; *heavy broken arrow*, 61–80 per cent; *solid arrow*, 41–60 per cent; *fine arrow*, 25–40 per cent, of all winds from the quarter within which the line is a median. Hatched areas indicate predominance of Beaufort 0–3, 60–80 per cent of time (*line hatching*) and more than 80 per cent of time (*crosshatching*). Stippled areas indicate predominance of stronger winds, 60–80 per cent of time (*fine stippling*) and more than 80 per cent of time (*coarse stippling*). (*a*) January, (*b*) July. (Same source as Fig. 13.03*ab*.)

403

range in position of the ICL, and also of the pressure trough, occurs between February and August, in better agreement with the annual cycle of ocean-surface temperatures.

Charts (c) and (d) bring out additional prominent features of the tropical surface circulation. Among these we find: (i) light and variable average winds of the horse latitudes, notably in winter, indicating weak gradients and more or less zonal dis-

e

f

FIGS. 13.03*ef*.—Regional distribution of convergence and divergence (10^{-6} sec^{-1}) determined from resultant winds in (*e*) January, (*f*) July. (After H. Riehl, "On the Role of the Tropics in the General Circulation of the Atmosphere," *Tellus*, II [1950], 1–25.)

placements of the subtropical high-pressure cells; (ii) more persistence in location of the subtropical anticyclones in summer, especially in the northern hemisphere; (iii) the steadiness and strength of the northeast and southeast trades; (iv) somewhat wider trade-wind belts in the summer hemisphere; (v) regions of little daily variation and others of greater variation in location of the ICL; and (vi) the equatorial westerlies are unsteady, excluding the southwest monsoon of Asia. From both types of charts the broad areas of predominant velocity divergence and convergence can be determined visually. Large rates of divergence are found somewhat equatorward and eastward from the subtropical anticyclonic centers. Maximum convergence occurs in the general vicinity of the ICL; notice the net equatorward transport of air in charts (*a*) and (*b*).

In charts (*e*) and (*f*) are shown the patterns of divergence determined from (*a*) and (*b*). With exception of the confused pattern over the Philippines and East Indies, the fields of divergence are aligned fairly zonally with the major wind pattern, with average convergence in the vicinity of the ICL, and with divergence to both sides. Daily patterns of divergence would be more cellular with larger magnitudes, and, without stretching imagination, the ICL could be broken and have areas of positive divergence along some sections of its extent. We might add in passing that it is not incidental that the charts of cloud and rain frequencies published in the atlas agree well with charts (*e*) and (*f*) above.

13.04. *Indications of mean upper-level flow.*—From recently published charts of mean upper winds,[5] in spite of the sparse data over large areas, certain useful conclusions can be drawn regarding upper-level flow. For January at 700 mb a pressure ridge line extends zonally in the region 15°–

20° N and another about 20°–25° S, and a counterclockwise circulation elongated west-east is centered near 5° N 130° E. Flow over the equator is easterly except in the western Pacific and Indian Ocean areas. There is some indication of the ICL from the central Pacific westward to Africa, slightly north of its surface position at least over the East Indies.

For July at 700 mb the subtropical ridges are centered at 25°–30° N and 15°–20° S. Easterlies are found over the equatorial region except for the Indian Ocean and East Indies, where the flow is drawn from the South Indian Ocean first northward and then eastward over southeastern Asia around a clockwise center situated near southern Sumatra. The Indian westerlies meet the Pacific easterlies along a convergence line extending from New Guinea to Formosa, and thence both flow northwestward.

Still higher, at 300 mb, the mean patterns are more zonal, and easterlies are found over the equator in both seasons. The subtropical ridges are now near 10° N and 15° S in January and 20° N and 10° S in July. In the mean, the summer westerlies of southern Asia vanish near 600 mb.

Figure 13.04 shows, in an admittedly rough manner, the meridional variation of the pressure patterns with height. For a given longitude, Figure 13.04 would be least representative in the low levels. Notice that the tropical area decreases by about half through the troposphere. Displacement of the subtropical ridges toward the equator with height indicates poleward temperature gradient in their vicinity. Slight equatorward temperature drop might be expected in the eastern and equatorial sides of indi-

5. C. E. P. Brooks, *Upper Winds over the World* ("Geophysical Memoirs," No. 85); Y. Mintz and G. Dean, *The Observed Mean Field of Motion of the Atmosphere* ("Geophysical Research Papers," No. 17 [Cambridge, Mass.: Geophysics Research Directorate, AFCRC, 1952]).

vidual upper anticyclonic centers due to differential vertical motion. The great magnitude of the meridional displacement with height, however, is a reflection more of weak pressure gradients than of large temperature gradients. It is not certain that the ridges always tilt equatorward with height; there are surely reversed slopes in some layers in daily cases and perhaps in the mean.

The ridge lines in Figure 13.04 separate westerlies above from easterlies below; each

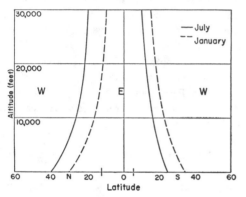

Fig. 13.04.—Schematic illustration of the average locations of the subtropical pressure ridges and the sea-level equatorial trough for the globe during January and July.

is therefore the *base of the westerlies*. In daily use the base of the westerlies is the tilted sheet separating winds with east and west components.

13.05. *The tropical air mass.*—Over the oceans influx of air into the tropics at the eastern ends of subtropical anticyclones, and also through their equatorward sides at the surface, maintains a well-mixed layer approaching equilibrium with the sea-surface temperature. Meantime, the strong subsidence, required in continuity with velocity divergence near the surface, carries down dry air from above and stabilizes the air column. A result is the dry *trade-wind inversion* above the surface mixed layer.[6]

The depth of the surface mixed layer, the depth of the inversion, and the temperature increase through the inversion all vary. For given surface and wind conditions it may be safe to assume that the inversion is lowest where divergence is greatest and generally that the inversion is larger where it is nearer the surface. The inversion may be a few hundred or a few thousand feet in depth. It rises and weakens downstream by further warming of the surface layer and slower rates of subsidence and, as recently demonstrated,[7] by cumulus convection and evaporation of cloud tops, so that air trajectories actually cross the inversion from the dry air above to the mixed layer below. At great distances from its source, the trade inversion may appear only as a prominent moisture discontinuity and small temperature lapse rate discontinuity (no inversion) in the vicinity of "6000–8000 feet," if found at all.

The trade inversion can be extremely low and pronounced over the cold-water coasts in subtropical latitudes, as near the California coast in summer. Above a shallow fog- or stratus-bound layer, the temperature may increase by more than 10° C and the relative humidity drop below limit of measurement in just a few hundred feet.

In general, the trade inversion is an almost barotropic layer sloping upward toward the equator and characterized by slight cold advection (backing of wind with height in the northern hemisphere). One might expect that the mixing below concentrates vertical shear within the inversion.

If there were no areas of organized ascent in the tropics, the tropical air mass could be visualized as a two-layer mass of almost barotropic air over oceans (as a

6. H. von Ficker, "Die Passat-Inversion," *Veröff. Met. Inst., Berlin*, Vol. I, No. 4 (1936).

7. H. Riehl *et al.*, "The North-east Trade of the Pacific Ocean," *QJRMS*, Vol. LXXVII, No. 334 (1951).

glance at sea-surface temperature charts will show), at least up through the middle troposphere, and whose moist layer deepens from subtropics to equator and generally from east to west due to accumulated effects of isolated cumulus convection in spite of the prevailing subsidence. Ascent and increase in depth of the moist layer occur in more or less organized manner in agreement with low-level convergence patterns attending both traveling and quasi-stationary dis-

land masses along with their seasonal and diurnal temperature variations. Australia, southwestern Asia, northern and southern Africa, and to some extent South America are capable of modifying polar air into cT air in summer and of harboring polar air in winter. By evaporation and transpiration at high temperatures, the tropical rain forests produce a hot, moist, and unstable air mass. Over continents and larger islands of the tropics, the diurnal variations of tempera-

TABLE 13.05*

MEAN MIDNIGHT HURRICANE-SEASON CARIBBEAN SOUNDING

Height (Km)	P (Mb)	T (° C)	U (%)	Height (Km)	P (Mb)	T (° C)
0	1013.6	25.3	85	10	286.3	−35.6
1	905.2	19.8	83	11	247.1	−43.6
2	805.1	14.3	73	12	212.7	−51.7
3	714.6	9.4	60	13	181.8	−59.9
4	633.0	4.0	53	14	154.4	−67.1
5	559.3	− 1.6	50	15	130.7	−73.1
6	492.6	− 7.5	48	16	110.3	−77.2
7	432.8	−13.8		17	92.6	−76.7
8	378.4	−20.6		18	77.8	−73.5
9	329.4	−27.8		19	65.9	−67.9

* E. J. Schacht, "A Mean Hurricane Sounding for the Caribbean Area," *Bulletin of the American Meteorological Society*, Vol. XXVII (1946).

turbances, and as already noted in connection with moist tongues extending from the tropics into middle latitudes. Fluctuations in depth of the moist layer are an important feature in analysis of tropical disturbances, much as we consider vertical oscillations of isentropic or frontal surfaces a part of mid-latitude disturbances.

Table 13.05 gives the average of the midnight soundings for Miami, San Juan, and Swan Island for the months August, September, and October, 1941 through 1944. In these average data the upper boundary of the moist layer is not clearly defined. This sounding shows potential instability and a temperature lapse rate greater than moist adiabatic up to 5 km. (The mean noon sounding is warmer by about 4° C at the surface and averages 2° C above to 10 km.) Until now we have ignored the larger

ture are many times larger than interdiurnal and seasonal variations, so that temperature gradients near coasts may be pronounced unless diurnal effects are subtracted.

13.06. *Disturbances in the tropical circulation over the oceans.*—One may have been led to believe that the tropical region consists of a narrow intertropical band of strong convection and rainfall paralleled on both sides by subtropical deserts with little or no rainfall. This is true to some extent, as there is such meridional variation in annual rainfall for the globe as a whole, but areas of little rainfall beneath the eastern ends of subtropical anticyclones (Lower California) alternate with areas of much rainfall at the western ends (Florida) along the same latitudes, and similar variations are found near the equator. Daily charts show that there are zonal and meridional shifts in the positions of the subtropical anticyclones and equatorial convergence zone and that there are traveling disturbances also in the easterlies. Seasonal rainfall charts are helpful in pointing out areas of preferred frequency and intensity of the disturbances.

In this section we outline, perhaps too briefly and ideally, several of the major features of the daily tropical flow pattern, leaving out those maintained locally by the continents and those of small scale such as land and sea breezes which require microanalysis. We include the waves in the easterlies, the equatorial convergence zone(s), other prominent convergence lines or shear lines in the field of motion, and the fresh invasions of polar air, but a most important phenomenon—the tropical storm—is excluded, at least in its mature stage.

a) Easterly waves.—Waves in the easterlies are indicated in the wind analysis given in Figures 13.06*ab* and also in the *t, z* sections in Figures 12.04*b* and 13.06*cd*. They are effectively a train of pressure (and wind) perturbations superimposed on the basic easterly current. Although they refer to the entire wave form in the motion, the name "easterly wave" has come to be applied to the pressure trough only.

Easterly troughs were described in detail by Riehl.[8] To summarize, they are normally characterized by (i) westward progression at speeds less than the basic current; (ii) greatest frequency and intensity in western parts of oceans; (iii) 24-hour pressure falls and rises of up to 3 mb; (iv) low-level divergence ahead and convergence in and behind the trough; (v) decrease in depth of the moist layer and fine weather in advance, general ascent and convective weather with

8. Herbert Riehl, *Waves in the Easterlies and the Polar Front in the Tropics* ("Department of Meteorology, University of Chicago, Miscellaneous Reports." No. 17 [Chicago: University of Chicago Press, 1943]).

and behind the trough (Figs. 12.04*ab*); (vi) the tilting of the trough to the east with height; and (vii) maximum intensity in the region between 5000 and 15,000 feet above the surface and weaker above (they are a feature of the lower troposphere). Of course, there are variations in speed, intensity, and structure.

The easterly trough which passed Kwajalein on September 8—rather weak at the time—appears with greater amplitude near 155° E in charts (*a*) and (*b*). Within 2 more days it gave birth to a typhoon. The trough in the eastern part of chart (*a*) passes Kwajalein on the 12th and is perhaps affected by the westward-moving cyclone shown in the high troposphere in chart (*e*). Notice the great difference in the tropical currents at 10,000 feet and 200 mb. The rather uniform and distinct easterly flow of the lower troposphere vanishes somewhere in the region between 15,000 and 25,000 feet (see the two *t, z* sections) and gives way to a more cellular pattern in the high troposphere.

b) The equatorial trough.—On 10 September 1945 the winds are rather uniformly easterly in the equatorial trough from 155° E to 180°, with no evidence of convergence and no significant wind shifts. Tarawa (1° N 173° E) reported fine weather over the period 8–12 September, and there was little unusual in the weather at Kwajalein some distance northwest except with the easterly wave troughs. Only in the extreme western Pacific is there anything resembling a convergence line (or zone) with wind shifts, but its exact location in the wind field is doubtful.

Palau is situated in the region of equatorial westerlies in the 1000-foot chart, and at 10,000 feet it is overlain by the easterlies. The analyses given seem to substantiate a convergence zone (with velocity convergence) in the vicinity of Palau. Chart (*c*) lends further evidence by the weather. The wind shift at Palau at 10,000 feet on the 9th and in lower levels on the 10th and 11th might indicate southwestward retreat of the convergence line.

We see, in effect, that, although the equatorial pressure trough is quite an extensive and real phenomenon, the wind field shows velocity convergence irregularly. The weather has similar variations—some places good and some places extremely convective, and often there are several more or less parallel bands of bad weather. The "Reading References" indicate our great lack of knowledge concerning the equatorial trough. Alpert gives a rather detailed discussion of just a portion of this trough.[9]

c) Interference in the subtropics.—In the vicinity of the subtropical high-pressure axes, the easterlies of the tropics are overlain by the mid-latitude westerlies in which features of the pressure and wind field move generally eastward. Thus, with reference to Figures 7.02*ab*, the wave pattern in the surface easterlies between 20° N and 30° N, and farther south in some areas, is expected to move in a manner similar to the features slightly north. The shearing of propagation between the subtropical easterlies and the deep easterlies nearer the equator produces important interactions between the two trains of systems. Some descriptions of these are given by the Civilian Staff of the Institute of Tropical Meteorology, by Cressman, and by Riehl in the "Reading References."

In the longitudes of major troughs in the upper westerlies, and slightly to the west, the surface polar front extends into the subtropics, and interaction is most evident. Notice in Figure 7.02*a* that a trough in the easterlies lies in the region 170° E–180°, coinciding in longitude with the split between subtropical HIGHs, nearly with the trough

9. L. Alpert, "The Intertropical Convergence Zone of the Eastern Pacific Ocean," *Bulletin of the American Meteorological Society*, XXVI (1945), 426–32, and XXVII (1946), 15–29, 62–66.

in middle latitudes, and very nearly with the "polar trough" in the upper westerlies. This trough in the easterlies is known by various names: "extended trough," "induced trough," "polar trough." Since these subtropical waves in the easterlies usually are quasi-stationary or move eastward, there is air motion westward relative to them in the low-level easterlies, with convergence and ascent to the east of the trough and divergence and descent to its west, all apart from the divergence pattern in the westerly wave aloft. It might be expected that the intensity of the weather pattern is proportional to the relative motion and to the intensity of the perturbation (Eq 10.33[3]).

Where the upper westerly current dips with strength into the subtropics (deep mass of polar air and major trough in the westerlies), frontal waves originating there may have properties like those in higher latitudes. Such is the case in the central North Pacific on 1 March 1950 and, with only slight reservation, in the California area also. Frequently the depressions formed in such areas are not identified as frontal waves, however.

Cold fronts, or most likely their remnants, may penetrate far into the tropics as a new *surge of the trades* after some weakening of the surface easterlies or even poleward-directed flow. In the vicinity of 500 mb the cold mass of air in winter seldom extends equatorward of 30° latitude, yet the surface front may be carried a great distance over the tropics with consequent spreading and sinking of the cold air. Through heating from below, the wedge of polar air is modified rapidly, and the low-level temperature contrast destroyed. This is occurring over the western Atlantic in Figure 7.02*a*, and a short time later the same occurs as the next front drifts southeastward.

d) Shear lines.—The name "shear line" has been used to identify a wide variety of

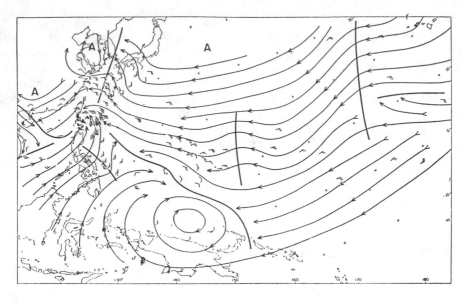

a

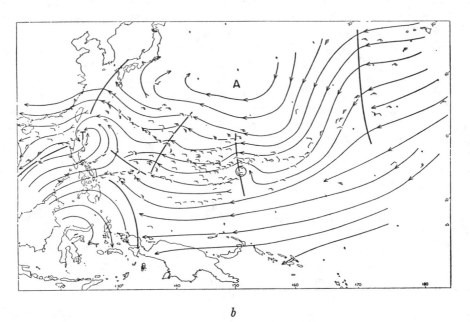

b

FIGS. 13.06*ab*.—Streamlines at 1000 feet (*a*) and 10,000 feet (*b*), 10 September 1945, 0000–0600 GCT. (After H. Riehl, "On the Formation of Typhoons," *Journal of Meteorology*, V [1948], 247–67.)

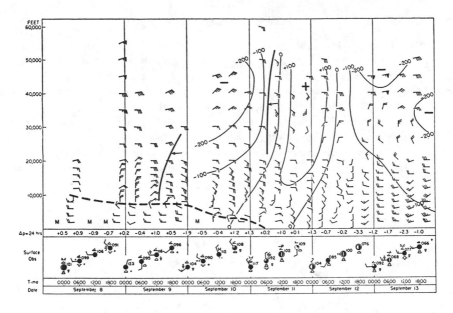

c

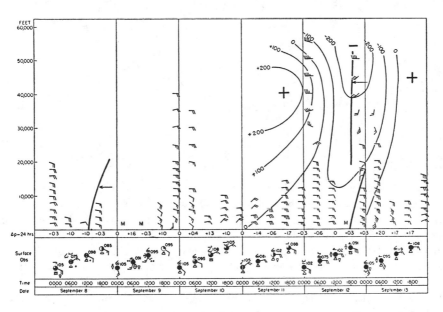

d

FIGS. 13.06cd.—t, z sections for Palau (7° N 135° E) (*c*) and Kwajalein (8° N 168° E) (*d*), for 8–13 September 1945. (After Riehl [1948].)

411

singularities in the wind and pressure fields. These can be convergence or divergence in the streamlines, the line lying parallel to one current at the meeting place of an adjacent current of different direction, the trough line through a col, or the separation of one current from a weaker parallel current. In most uses of the term, cyclonic shear of direction or speed is implied.

Shear lines or convergence lines in the

Subtropical shear lines aloft are frequent in summer in the western hemisphere, where in some longitudes this is the season of largest distance between the equatorial trough and the subtropical pressure ridge at the surface. In the place of only one distinct anticyclone at the surface, there may be two at the same longitude aloft, with the southernmost lying over surface easterlies; and the west-east trough line

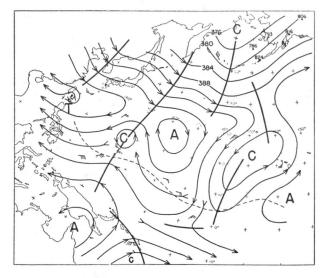

FIG. 13.06e.—Streamlines (and contours) at 200 mb, 10 September 1945, 0000 GCT. (After Riehl [1948].)

wind field, quite indefinite in the pressure field, are common in and near the equatorial trough in its regions of irregularity. The westerlies on the equatorward side have been explained as CAV trajectories of air crossing the equator. As with most such singularities in the wind field, it is not difficult to suppose that the equatorial shear line does not constitute a substantial airmass boundary; the line may be displaced relative to the air in response to large-scale controls and to hydrostatic adjustment. Examples of the equatorial shear line are seen in Figures 13.06ab. These may or may not coincide with maximum velocity convergence.

or shear line between them is seen at the surface at times only as shear in speed of the easterlies. The shear line is a common feature of the middle troposphere along the San Francisco–Honolulu route. The curved shear line aloft may extend from the semipermanent trough of the west coast into a trough over the central North Pacific, occasionally with closed cyclones at its intersection with the north-south troughs. Frequently, also, this situation extends zonally in series over several of the major westerly waves. These shear lines (or zones) may contain several small cyclonic vortices, and, if the picture is further confused as common by latitudinal split of the southern anticyclone, the intersection of the new north-south trough with the east-west shear line can result in a definite cyclone equatorward from the main anticyclone aloft. At the surface there may be only modest or indifferent indications in the pressure and wind fields, especially with few data, but some effect is expected in the weather pattern. Thunderstorm activity in the mT air mass of the southern United States is well related to these phenomena.[10]

There are many other singularities of this type in smaller scale in both summer and winter. Some in the low levels are attended by convection, others apparently not, at least in the immediate area. They are difficult to locate unless purposely sought, and, even so, their irregular formation and movement combined with inadequate networks of data are still a problem.

13.07. *Methods of tropical analysis.*— While analysis in low and in high latitudes has the same objective, the means to that end are quite different mainly because the structure of the atmosphere is different in degree. In high latitudes air-mass contrasts are large and pressure patterns well defined. In the tropics proper there is really only one air mass, and the usually weak pressure patterns do differ significantly from the field of motion. Analysis in the tropics must concentrate on the true motion and the related weather phenomena, for which the standard methods used in high latitudes have been found lacking. The difference even extends

10. Herbert Riehl, *Subtropical Flow Patterns in Summer* ("Department of Meteorology, University of Chicago, Miscellaneous Reports," No. 22 [Chicago: University of Chicago Press, 1947]).

to methods of reporting the weather. Elements of the present synoptic weather code such as uncorrected barometric tendency, cloud heights (uniform except in rain or orography), and cloud types do not apply as they do in other areas.

Adding to the problem is the paucity of data over oceans, which comprise about four-fifths the area of the tropics, and also over many continental interiors. With so few data and the small thermal and pressure gradients, diurnal variations over land are an important part of analysis.[11] In dealing with these in high latitudes, we usually subtract them from the large interdiurnal variations. In the tropics, however, the important elements are to be found rather in *the daily departures from the normal diurnal variations* for the station, including those of wind, clouds, and weather.

Under these conditions ordinary synoptic scalar analysis does not apply so readily. Tropical analysis must make use of single-station methods, analysis of departures, and more complete analysis of the weather, even after data networks become as dense as those existing in higher latitudes of the northern hemisphere. Desirable types of charts, representation, and analysis in the tropics are discussed in the "Reading References."[12]

11. For an excellent example see L. B. Leopold, "The Interaction of Trade Wind and Sea Breeze, Hawaii," *Journal of Meteorology*, VI (1949), 312–20.

12. See also in this connection Herbert Riehl and E. Schacht, "Methods of Analysis for the Caribbean Region," *Bulletin of the American Meteorological Society*, XXVII (1946), 569–75; M. E. Lopez, "A Technique for Detailed Radiosonde Analysis in the Tropics," *Bulletin of the American Meteorological Society*, XXIX (1948), 227–36.

READING REFERENCES*

AMERICAN METEOROLOGICAL SOCIETY. *Compendium of Meteorology*, pp. 859–913 (articles by C. E. PALMER, A. GRIMES, G. E. DUNN, and HERBERT RIEHL). Boston: The Society, 1951.

* This list does not give a complete survey of the literature on tropical meteorology, nor does it give all different ideas on the subject. The list contains only the more readily available survey material in this country.

CIVILIAN STAFF, INSTITUTE OF TROPICAL METEOROLOGY, RIO PIEDRAS, PUERTO RICO. "Tropical Synoptic Meteorology," in BERRY, F. A., BOLLAY, E., and BEERS, N. R. (eds.), *Handbook of Meteorology*, pp. 763–803. New York: McGraw-Hill Book Co., 1945. (See also section on "Southern Hemisphere Synoptic Meteorology," pp. 804–12.)

CRESSMAN, G. P. *Studies of Upper-Air Conditions in Low Latitudes*, Part II: *Relations between High- and Low-Latitude Circulations.* "Department of Meteorology, University of Chicago, Miscellaneous Reports," No. 24. Chicago: University of Chicago Press, 1948.

GARBELL, M. A. *Tropical and Equatorial Meteorology.* New York: Pitman Publishing Corp., 1947.

JENKINS, G. R. "Diurnal Variation of the Meteorological Elements," in BERRY, BOLLAY, and BEERS (eds.), *Handbook of Meteorology*, pp. 746–53.

RIEHL, HERBERT. *Waves in the Easterlies and the Polar Front in the Tropics.* "Department of Meteorology, University of Chicago, Miscellaneous Reports," No. 17. Chicago: University of Chicago Press, 1943.

———. *Tropical Meteorology.* New York: McGraw-Hill Book Co., 1954.

TANNEHILL, I. R. *Hurricanes.* Princeton: Princeton University Press, 1943.

Appendix

CONVERSION FACTORS

1 km = 10^3 m = 10^5 cm   1 ft = 30.48 cm

1 m = 3.2808399 ft = 39.370079 in   1 U.S. naut mi = 1.151555 stat mi
                                                                          = 1.853248 km

1 deg lat = 111.137 km = 69.057 stat mi = 59.969 naut mi

Angular velocity of earth = 1 mean solar (MS) day = 86,400 MS sec

1 MS sec = 1.002738 sidereal sec

1 m sec^{-1} = 3.6 km hr^{-1} = 1.94254 knots = 2.23694 mi hr^{-1} = 0.77742° ϕ per day

1 knot = 1 naut mi hr^{-1} = 1.151555 mi hr^{-1} = 0.514791 m sec^{-1} = 0.40021° ϕ per day

1 mi hr^{-1} = 0.868391 knot = 0.44704 m sec^{-1} = 0.34754° ϕ per day

1° ϕ per day = 1.2863 m sec^{-1} = 2.8774 mi hr^{-1} = 2.4987 knots

1 mb = 10^{-3} bar = 10^3 dynes cm^{-2} = 0.750099 mm Hg (45°) = 0.0295315 in Hg (45°)

1 in Hg (45°) = 33.8622 mb = 25.4 mm Hg (45°)

1 erg = 1 dyne-cm = 10^{-7} abs. joule = 2.38844 × 10^{-8} IT cal = 2.3892 × 10^{-8} cal (15° C)

1 abs. joule = 10^7 ergs = 0.238844 IT cal = 0.23892 cal (15° C)

TABLE B

BEAUFORT WIND SCALE
(Limits of Wind Speed at 10 m)*

Force	Knots	Mi Hr^{-1}	M Sec^{-1}
0........	<1	<1	0– 0.2
1........	1– 3	1– 3	0.3– 1.5
2........	4– 6	4– 7	1.6– 3.3
3........	7– 10	8– 12	3.4– 5.4
4........	11– 16	13– 18	5.5– 7.9
5........	17– 21	19– 24	8.0–10.7
6........	22– 27	25– 31	10.8–13.8
7........	28– 33	32– 38	13.9–17.1
8........	34– 40	39– 46	17.2–20.7
9........	41– 47	47– 54	20.8–24.4
10........	48– 55	55– 63	24.5–28.4
11........	56– 63	64– 72	28.5–32.6
12........	64– 71	73– 82	32.7–36.9
13........	72– 80	83– 92	37.0–41.4
14........	81– 89	93–103	41.5–46.1
15........	90– 99	104–114	46.2–50.9
16........	100–108	115–125	51.0–56.0
17........	109–118	126–136	56.1–61.2

* Resolution 9, International Meteorological Committee, Paris, 1946.

TABLE C

SCALE VARIATION FOR STANDARD MAP PROJECTIONS
(After Gregg and Tannehill)

LATITUDE	MERCATOR PROJECTION STANDARD PARALLEL 22½°		LAMBERT CONFORMAL CONIC PROJECTION STANDARD PARALLELS 30° AND 60°		POLAR STEREOGRAPHIC PROJECTION STANDARD PARALLEL 60°	
	Sphere	Ellipsoid*	Sphere	Ellipsoid*	Sphere	Ellipsoid*
0°.........	0.924	0.924	1.283	1.281	1.932	1.860
5..........	0.927	0.928	1.210	1.208	1.777	1.712
10.........	0.938	0.938	1.149	1.148	1.590	1.586
15.........	0.956	0.957	1.099	1.098	1.482	1.480
20.........	0.983	0.983	1.058	1.058	1.390	1.388
25.........	1.019	1.019	1.025	1.025	1.312	1.310
30.........	1.067	1.066	1.000	1.000	1.244	1.243
35.........	1.128	1.127	0.982	0.982	1.186	1.185
40.........	1.206	1.205	0.970	0.970	1.136	1.136
45.........	1.307	1.305	0.966	0.966	1.093	1.093
50.........	1.437	1.435	0.968	0.969	1.057	1.057
55.........	1.611	1.608	0.979	0.979	1.026	1.026
60.........	1.848	1.844	1.000	1.000	1.000	1.000
65.........	2.186	2.181	1.033	1.033	0.979	0.979
70.........	2.701	2.694	1.084	1.083	0.962	0.962
75.........	3.570	3.560	1.162	1.162	0.949	0.949
80....... .	5.320	5.306	1.293	1.292	0.940	0.940
85.........	10.600	10.570	1.566	1.564	0.934	0.936

* *The International Ellipsoid of Reference* (U.S. Coast and Geodetic Survey, Special Pub. No. 200 [1935]) has the following dimensions:
Semimajor axis: $a = 6378388$ m   Semiminor axis: $b = 6356911.946$ m
Mean radius: $(2a + b)/3 = 6371229.315$ m   Flattening: $f = 1/297$
Radius of sphere of same area: 6371227.709 m
Radius of sphere of same volume: 6371221.266 m
Length of meridian quadrant: 10002288.299 m
Length of equatorial quadrant: 10019148.4 m

TABLE D

VIRTUAL TEMPERATURE INCREMENT, $(T^* - T) = 0.61rT$

				T (° C)					r (‰)
− 30	− 20	− 10	0	10	20	30	40		
0.1	0.2	0.2	0.2	0.2	0.2	0.2	0.2	1
0.3	0.3	0.3	0.3	0.3	0.4	0.4	0.4	2
	0.5	0.5	0.5	0.5	0.5	0.6	0.6	3
	0.6	0.6	0.7	0.7	0.7	0.7	0.8	4
	0.8	0.8	0.8	0.9	0.9	0.9	1.0	5
		1.0	1.0	1.0	1.1	1.1	1.1	6
		1.1	1.2	1.2	1.3	1.3	1.3	7
		1.3	1.3	1.4	1.4	1.5	1.5	8
		1.4	1.5	1.6	1.6	1.7	1.7	9
		1.6	1.7	1.7	1.8	1.8	1.9	10
		1.8	1.8	1.9	2.0	2.0	2.1	11
		1.9	2.0	2.1	2.1	2.2	2.3	12
			2.2	2.2	2.3	2.4	2.5	13
			2.3	2.4	2.5	2.6	2.7	14
			2.5	2.6	2.7	2.8	2.9	15
			2.7	2.8	2.9	3.0	3.1	16
			2.8	2.9	3.0	3.1	3.2	17
			3.0	3.1	3.2	3.3	3.4	18
				3.3	3.4	3.5	3.6	19
				3.5	3.6	3.7	3.8	20
				3.6	3.8	3.9	4.0	21
				3.8	3.9	4.1	4.2	22
				4.0	4.1	4.3	4.4	23
				4.1	4.3	4.4	4.6	24
				4.3	4.5	4.6	4.8	25
					4.7	4.8	5.0	26
					4.8	5.0	5.2	27
					5.0	5.2	5.3	28
					5.2	5.4	5.5	29
					5.4	5.5	5.7	30

417

TABLE E

MEAN VIRTUAL TEMPERATURE IN °C FOR GIVEN VALUES OF THICKNESS (ΔZ) IN GEOPOTENTIAL FEET (GPFT) IN VARIOUS PRESSURE LAYERS

ΔZ	1	2	3	4	5	6	7	8	9	10	11	12	13
850/1000	3700 / −36.2	3800 / −29.8	3900 / −23.4	4000 / −17.0	4100 / −10.6	4200 / −4.2	4300 / 2.2	4400 / 8.6	4500 / 15.0	4600 / 21.4	4700 / 27.8		
700/850	4300 / −42.7	4400 / −37.3	4500 / −32.0	4600 / −26.6	4700 / −21.2	4800 / −15.9	4900 / −10.5	5000 / −5.2	5100 / 0.2	5200 / 5.6	5300 / 10.9	5400 / 16.3	5500 / 21.6
600/700	3400 / −43.6	3500 / −36.9	3600 / −30.1	3700 / −23.4	3800 / −16.6	3900 / −9.9	4000 / −3.1	4100 / 3.6	4200 / 10.4	4300 / 17.1			
500/600	3900 / −50.6	4000 / −44.8	4100 / −39.1	4200 / −33.4	4300 / −27.7	4400 / −22.0	4500 / −16.3	4600 / −10.6	4700 / −4.9	4800 / 0.8	4900 / 6.5	5000 / 12.2	
400/500	4700 / −54.0	4800 / −49.3	4900 / −44.6	5000 / −40.0	5100 / −35.3	5200 / −30.7	5300 / −26.0	5400 / −21.3	5500 / −16.7	5600 / −12.0	5700 / −7.3	5800 / −2.7	
350/400	2700 / −62.7	2800 / −54.9	2900 / −47.1	3000 / −39.4	3100 / −31.6	3200 / −23.8	3300 / −16.0	3400 / −8.2					
300/350	3100 / −63.9	3200 / −57.1	3300 / −50.4	3400 / −43.6	3500 / −36.9	3600 / −30.1	3700 / −23.4	3800 / −16.6					
250/300	3600 / −67.7	3700 / −62.0	3800 / −56.3	3900 / −50.6	4000 / −44.8	4100 / −39.1	4200 / −33.4	4300 / −27.7					
200/250	4300 / −72.6	4400 / −68.0	4500 / −63.3	4600 / −58.6	4700 / −54.0	4800 / −49.3	4900 / −44.6	5000 / −40.0	5100 / −35.3				
150/200	5300 / −81.4	5400 / −77.8	5500 / −74.2	5600 / −70.6	5700 / −67.0	5800 / −63.4	5900 / −59.7	6000 / −56.1	6100 / −52.5	6200 / −49.0	6300 / −45.3	6400 / −41.7	6500 / −38.0
125/150	3200 / −90.5	3300 / −84.8	3400 / −79.1	3500 / −73.4	3600 / −67.7	3700 / −62.0	3800 / −56.3	3900 / −50.6	4000 / −44.8	4100 / −39.1	4200 / −33.4	4300 / −27.7	
100/125	3900 / −91.3	4000 / −86.6	4100 / −82.0	4200 / −77.3	4300 / −72.6	4400 / −68.0	4500 / −63.3	4600 / −58.6	4700 / −54.0	4800 / −49.3	4900 / −44.6	5000 / −40.0	

TABLE F

PRESSURES ABOVE WHICH THE 1000-MB SURFACE IS LOCATED THE DISTANCE ΔZ (GPFT) FOR GIVEN VALUES OF MEAN VIRTUAL TEMPERATURE IN THE LAYER
(Thousands Digit Omitted from the Pressure)

ΔZ (GPFT)	$\overline{T} * (°C)$								
	−40	−35	−30	−25	−20	−15	−10	−5	0
1500...	069.3	067.8	066.3	065.0	063.6	062.4	061.1	060.0	058.8
1400...	064.5	063.1	061.8	060.5	059.3	058.1	057.0	055.9	054.8
1300...	059.8	058.5	057.3	056.1	054.9	053.8	052.8	051.8	050.8
1200...	055.1	053.9	052.7	051.6	050.6	049.6	048.6	047.7	046.8
1100...	050.4	049.3	048.2	047.2	046.3	045.4	044.5	043.6	042.8
1000...	045.7	044.7	043.8	042.9	042.0	041.2	040.4	039.6	038.9
900...	041.0	040.1	039.3	038.5	037.7	037.0	036.3	035.6	034.9
800...	036.4	035.6	034.9	034.1	033.5	032.8	032.2	031.6	031.0
700...	031.8	031.1	030.4	029.8	029.2	028.6	028.1	027.6	027.0
600...	027.2	026.6	026.0	025.5	025.0	024.5	024.0	023.6	023.1
500...	022.6	022.1	021.6	021.2	020.8	020.4	020.0	019.6	019.2
400...	018.0	017.6	017.3	016.9	016.6	016.3	016.0	015.7	015.4
300...	013.5	013.2	012.9	012.7	012.4	012.2	011.9	011.7	011.5
200...	009.0	008.8	008.6	008.4	008.3	008.1	007.9	007.8	007.7
100...	004.5	004.4	004.3	004.2	004.1	004.0	004.0	003.9	003.8
0...	000.0	000.0	000.0	000.0	000.0	000.0	000.0	000.0	000.0
− 100...	995.6	995.6	995.7	995.8	995.9	996.0	996.1	996.1	996.2
− 200...	991.1	991.3	991.5	991.6	991.8	992.0	992.1	992.3	992.4
− 300...	986.7	987.0	987.3	987.5	987.7	988.0	988.2	988.4	988.6
− 400...	982.3	982.7	983.0	983.4	983.7	984.0	984.3	984.6	984.9
− 500...	977.9	978.4	978.8	979.2	979.6	980.0	980.4	980.8	981.1
− 600...	973.6	974.1	974.6	975.1	975.6	976.1	976.5	977.0	977.4
− 700...	969.2	969.8	970.5	971.1	971.6	972.1	972.7	973.2	973.7
− 800...	964.9	965.6	966.3	967.0	967.6	968.2	968.8	969.4	970.1
− 900...	960.6	961.4	962.2	962.9	963.7	964.3	965.0	965.7	966.3
−1000...	956.3	957.2	958.1	958.9	959.7	960.5	961.2	961.9	962.6
−1100...	952.1	953.0	954.0	954.9	955.8	956.6	957.4	958.2	958.9
−1200...	947.8	948.9	949.9	950.9	951.9	952.7	953.6	954.5	955.3
−1300...	943.6	944.7	945.8	946.9	947.9	948.9	949.9	950.8	951.6
−1400...	939.4	940.6	941.8	942.9	944.0	945.1	946.1	947.1	948.0
−1500...	935.2	936.5	937.8	939.0	940.2	941.3	942.4	943.4	944.4

ΔZ (GPFT)	$\overline{T}$* (° C)								
	0	5	10	15	20	25	30	35	40
1500...	058.8	057.8	056.7	055.7	054.7	053.8	052.9	052.0	051.1
1400...	054.8	053.8	052.8	051.9	051.0	050.1	049.3	048.4	047.6
1300...	050.8	049.9	049.0	048.1	047.3	046.4	045.7	044.9	044.2
1200...	046.8	045.9	045.1	044.3	043.5	042.8	042.1	041.4	040.7
1100...	042.8	042.0	041.3	040.5	039.8	039.2	038.5	037.9	037.2
1000...	038.9	038.1	037.5	036.8	036.2	035.5	034.9	034.4	033.8
900...	034.9	034.3	033.6	033.1	032.5	031.9	031.4	030.9	030.3
800...	031.0	030.4	029.9	029.3	028.8	028.3	027.9	027.4	026.9
700...	027.0	026.5	026.1	025.6	025.2	024.7	024.3	023.9	023.5
600...	023.1	022.7	022.3	021.9	021.5	021.1	020.8	020.5	020.1
500...	019.2	018.9	018.5	018.2	017.9	017.6	017.3	017.0	016.8
400...	015.4	015.1	014.8	014.6	014.3	014.1	013.8	013.6	013.3
300...	011.5	011.3	011.1	010.9	010.7	010.5	010.4	010.2	010.0
200...	007.7	007.5	007.4	007.3	007.1	007.0	006.9	006.8	006.7
100...	003.8	003.8	003.7	003.6	003.6	003.5	003.4	003.4	003.3
0...	000.0	000.0	000.0	000.0	000.0	000.0	000.0	000.0	000.0
− 100...	996.2	996.3	996.3	996.4	996.4	996.5	996.6	996.6	996.7
− 200...	992.4	992.5	992.7	992.8	992.9	993.0	993.1	993.3	993.4
− 300...	988.6	988.8	989.0	989.2	989.4	989.6	989.7	989.9	990.1
− 400...	984.9	985.1	985.4	985.7	985.9	986.1	986.4	986.6	986.8
− 500...	981.1	981.5	981.8	982.1	982.4	982.7	983.0	983.2	983.5
− 600...	977.4	977.8	978.2	978.5	978.9	979.3	979.6	979.9	980.2
− 700...	973.7	974.1	974.6	975.0	975.4	975.8	976.2	976.6	977.0
− 800...	970.1	970.5	971.0	971.5	972.0	972.4	972.9	973.3	973.7
− 900...	966.3	966.9	967.4	968.0	968.6	969.1	969.6	970.0	970.5
−1000...	962.6	963.3	963.9	964.5	965.1	965.7	966.2	966.8	967.3
−1100...	958.9	959.6	960.4	961.0	961.7	962.3	962.9	963.5	964.1
−1200...	955.3	956.1	956.8	957.6	958.3	959.0	959.6	960.3	960.9
−1300...	951.6	952.5	953.3	954.1	954.9	955.6	956.3	957.0	957.7
−1400...	948.0	948.9	949.8	950.7	951.5	952.3	953.0	953.8	954.5
−1500...	944.4	945.4	946.3	947.2	948.1	949.0	949.8	950.6	951.4

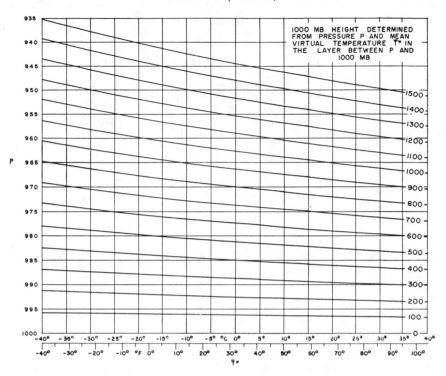

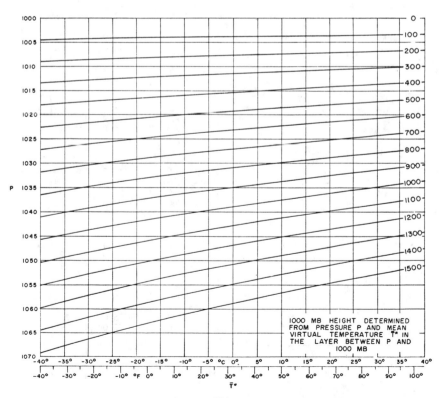

FIGURE G

THICKNESS OF THE (700/1000)-MB LAYER FROM THE 1000-MB VIRTUAL TEMPERATURE AND THE AVERAGE TEMPERATURE LAPSE RATE IN THE LAYER

(*D:* Dry Adiabatic; ½ *D:* Half the Dry Adiabatic; *I:* Isothermal)

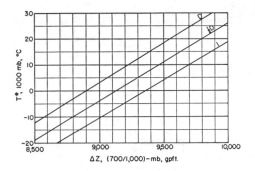

TABLE H

U.S. (NACA) STANDARD ATMOSPHERE

Pressure (Mb)	Temperature (° C)	Altitude (Feet)	Altitude (Feet)	Temperature (° C)	Pressure (Mb)
1050	17.0	− 990	− 1000	17.0	1050.5
1013.25	15.0	0	0	15.0	1013.25
1000	14.3	370	1000	13.0	977.3
950	11.5	1780	2000	11.0	942.1
900	8.6	3240	3000	9.1	907.9
850	5.5	4780	4000	7.1	875.0
			5000	5.1	842.9
800	2.3	6390			
750	− 1.0	8080	6000	3.1	812.1
700	− 4.6	9880	7000	1.1	781.9
650	− 8.3	11770	8000	− 0.8	752.5
600	−12.3	13790	9000	− 2.8	724.0
			10000	− 4.8	696.9
550	−16.6	15950			
500	−21.2	18280	12000	− 8.8	644.4
450	−26.2	20800	14000	−12.7	595.0
400	−31.7	23560	16000	−16.7	548.9
350	−37.7	26610	18000	−20.7	505.9
			20000	−24.6	465.6
300	−44.5	30050			
250	−52.3	33980	25000	−34.5	375.9
200	−55.0	38660	30000	−44.4	300.7
150	−55.0	44680	35000	−54.3	238.3
100	−55.0	53170	35332	−55.0	234.5
			40000	−55.0	187.6
			45000	−55.0	147.8
			50000	−55.0	116.4
			55000	−55.0	91.67
			60000	−55.0	72.20
			65000	−55.0	56.89

TABLE I

Atmosphere with the Same Properties as the NACA except Height Is in Geopotential Feet and the Lapse Rate of Temperature Is Constant at 6.5° C per Geopotential Kilometer

(See Sec. 6.13)

Pressure (Mb)	Temperature (° C)	Altitude (Gpft)
1013.25	15.0	0
1000	14.3	370
850	5.5	4780
700	− 4.6	9880
600	−12.4	13800
500	−21.2	18290
400	−31.7	23580
300	−44.6	30070
250	−52.4	34000
200	−61.5	38630
150	−72.8	44320
125	−79.6	47770
100	−87.7	51830

423

TABLE J

RELATIVE HUMIDITY (PER CENT) FROM TEMPERATURE AND TEMPERATURE-DEWPOINT DEPRESSION*

DEWPOINT DEPRESSION (° C)	40	35	30	25	20	15	10	5	0	−5	−10	−15	−20	−25	−30	−35	−40
1	95	95	94	94	94	94	94	93	93	93	92	92	92	91	91	91	90
2	90	89	89	89	88	88	87	87	86	86	85	85	84	83	83	82	81
3	85	85	84	83	83	82	82	81	80	79	79	78	77	76	75	74	73
4	81	80	79	78	78	77	76	75	74	73	73	71	70	69	68	67	66
5	76	75	75	74	73	72	71	70	69	68	67	66	64	63	62	60	59
6	72	71	70	69	68	67	66	65	64	63	61	60	59	57	56	54	53
7	68	67	66	65	64	63	62	61	59	58	57	55	54	52	51	49	47
8	64	63	62	61	60	59	57	56	55	53	52	50	49	47	46	44	42
9	61	60	59	57	56	55	54	52	51	49	48	46	45	43	41	39	38
10	58	56	55	54	53	51	50	48	47	45	44	42	41	39	37	35	34
11	54	53	52	50	49	48	46	45	43	42	40	39	37	35	33	32	27
12	51	50	49	47	46	45	43	41	40	38	37	35	34	32	30	28	24
13	48	47	46	44	43	42	40	38	37	35	34	32	30	29	27	25	21
14	46	44	43	41	40	39	37	36	34	32	31	29	28	26	24	23	19
15	43	42	40	39	37	36	34	33	31	30	28	27	25	23	22	20	16
16	40	39	38	36	35	33	32	30	29	27	26	24	23	21	20	16	
17	38	37	35	34	32	31	29	28	27	25	23	22	21	19	18	14	
18	36	34	33	32	30	29	27	26	24	23	21	20	19	17	16	12	
19	34	32	31	30	28	27	25	24	22	21	20	18	17	15	14	11	
20	32	30	29	28	26	25	23	22	21	19	18	16	15	14	12	09	
21	30	28	27	26	24	23	22	20	19	17	16	15	14	12	09		
22	28	27	25	24	23	21	20	19	17	16	15	13	12	11	08		
23	26	25	24	22	21	19	18	17	16	15	15	12	11	10	07		
24	25	23	22	21	19	18	17	16	14	13	12	11	10	09	06		
25	23	22	21	19	18	17	16	14	13	12	11	10	09	08	06		
26	22	20	19	18	17	16	14	13	12	11	10	09	08	05			
27	20	19	18	17	16	14	13	12	11	10	09	08	07	05			
28	19	18	17	15	14	13	12	11	10	09	08	07	06	04			
29	18	17	15	14	13	12	11	10	09	08	07	06	06	04			
30	17	16	14	13	12	11	10	09	08	07	07	06	05	03			

*Computed from saturation vapor pressures (with respect to water surface) given in *Smithsonian Meteorological Tables* (6th rev. ed.; Washington, D.C., 1951).
Refer to Tables 1.05ab for differences between saturation over water and saturation over ice.

TABLE K

SATURATION MIXING RATIO OVER WATER (‰)*

TEMPERATURE (° C)	PRESSURE (MB)							
	1000	850	700	600	500	400	300	200
40........	49.81	59.41						
35........	37.25	44.27						
30........	27.69	32.82	40.29					
25........	20.44	24.17	29.59					
20........	14.95	17.66	21.56	25.29				
15........	10.83	12.78	15.57	18.24				
10........	7.762	9.146	11.13	13.02	15.69			
5........	5.495	6.471	7.870	9.198	11.07			
0........	3.839	4.519	5.492	6.415	7.710			
− 5........	2.644	3.112	3.780	4.412	5.300	6.637		
−10........	1.794	2.110	2.562	2.990	3.590	4.492		
−15........	1.197	1.408	1.709	1.994	2.393	2.993	3.995	
−20........	0.785	0.923	1.120	1.306	1.568	1.960	2.615	
−25........	0.505	0.594	0.720	0.840	1.008	1.260	1.680	2.523
−30........	0.318	0.374	0.454	0.530	0.635	0.794	1.058	1.588
−35........	0.196	0.231	0.280	0.327	0.392	0.490	0.653	0.979
−40........	0.118	0.139	0.169	0.197	0.236	0.295	0.393	0.589
−45........	0.070	0.082	0.099	0.116	0.139	0.173	0.231	0.346
−50........	0.040	0.047	0.057	0.066	0.079	0.099	0.132	0.198

* Extracted from *Smithsonian Meteorological Tables* (Washington, D.C., 1951).

TABLE L

POTENTIAL TEMPERATURE (° K) AS FUNCTION OF TEMPERATURE AND PRESSURE*

$$(R_d/c_p = 2/7)$$

TEMPERA- TURE (° C)	PRESSURE (MB)									
	1000	850	700	600	500	400	300	200	150	100
40....	313.16	328.0								
35....	308.16	322.8								
30....	303.16	317.6	335.7							
25....	298.16	312.3	330.2							
20....	293.16	307.1	324.6	339.2						
15....	288.16	301.8	319.1	333.4						
10....	283.16	296.6	313.5	327.6	345.2					
5....	278.16	291.4	308.0	321.9	339.1					
0....	273.16	286.1	302.5	316.1	333.0					
− 5....	268.16	280.9	296.9	310.3	326.9	348.4				
−10....	263.16	275.7	291.4	304.5	320.8	341.9				
−15....	258.16	270.4	285.9	298.7	314.7	335.4	364.2			
−20....	253.16	265.2	280.3	292.9	308.6	328.9	357.1			
−25....	248.16	259.9	274.8	287.1	302.5	322.4	350.1	393.0		
−30....	243.16	254.7	269.3	281.4	296.4	315.9	343.0	385.1	418.1	469.5
−35....	238.16	249.5	263.7	275.6	290.3	309.4	335.9	377.2	409.5	459.8
−40....	233.16	244.2	258.2	269.8	284.2	302.9	328.9	369.3	400.9	450.2
−45....	228.16	239.0	252.6	264.0	278.1	296.4	321.8	361.4	392.3	440.5
−50....	223.16	233.8	247.1	258.2	272.0	290.0	314.8	353.4	383.7	430.9
−55....	218.16	228.5	241.6	252.4	265.9	283.5	307.7	345.5	375.1	421.2
−60....	213.16	223.3	236.0	246.6	259.8	277.0	300.7	337.6	366.5	411.5
−65....	208.16	218.0			253.7	270.5	293.6	329.7	357.9	401.9
−70....	203.16	212.8				264.0	286.6	321.8	349.3	392.2
−75....	198.16						279.5	313.8	340.7	382.6
−80....	193.16							305.9	332.1	372.9

* Extracted from *Smithsonian Meteorological Tables* (Washington, D.C., 1951).

426

TABLE M

PRESSURE (MB) AT GIVEN VALUES OF POTENTIAL TEMPERATURE (° K) AND TEMPERATURE (° C)

θ° K	T° C											
	30	25	20	15	10	5	0	−5	−10	−15	−20	−25
250.........												
255.........										1044	975	909
260.........									1043	976	911	849
265.........								1042	976	913	852	795
270.........							1042	976	914	855	798	744
275.........						1040	977	916	857	802	749	698
280.........					1040	977	917	860	805	753	703	655
285.........				1039	978	918	862	808	757	707	661	616
290.........			1039	978	920	864	811	760	712	666	622	580
295.........		1038	978	922	866	814	764	716	671	627	585	546
300.........	1037	979	923	869	817	767	720	675	632	591	552	515
305.........	979	924	871	820	771	724	680	637	597	558	521	486
310.........	925	873	822	774	728	684	642	602	564	527	492	459
315.........	874	825	778	732	689	647	607	569	533	498	465	434
320.........	828	781	736	693	652	612	575	539	504	472	440	411
325.........	784	740	697	656	617	580	544	510	478	447	417	389
330.........	743	701	661	622	585	550	516	484	453	424	395	369
335.........	705	665	627	590	555	522	490	459	430	402	375	350
340.........	669	632	595	560	527	495	465	436	408	381	356	335
345.........	636	600	566	533	501	471	442	414	388	362	338	316

$\theta°$ K	$T°$ C												
	-25	-30	-35	-40	-45	-50	-55	-60	-65	-70	-75	-80	-85
300......	515	479	446	414	384	355	328	302	278	256	234	214	195
305......	486	452	421	391	362	335	309	285	263	241	221	202	184
310......	459	427	397	369	342	317	292	270	248	228	209	191	174
315......	434	404	376	349	323	299	276	255	235	215	198	181	165
320......	411	382	356	330	306	283	262	241	222	204	187	171	156
325......	389	362	337	313	290	268	248	228	210	194	177	162	148
330......	369	343	319	296	275	254	235	217	199	183	168	153	140
335......	350	326	303	281	261	241	223	206	189	174	159	146	133
340......	335	309	288	267	247	229	212	195	180	165	151	138	126
345......	316	294	273	254	235	218	201	185	171	156	144	131	120
350......	300	279	260	241	224	207	191	176	162	149	137	125	114
355......	286	266	247	230	213	197	182	168	154	142	130	119	108
360......	272	253	236	219	203	188	173	160	147	135	124	113	103
365......		241	224	208	193	179	165	152	140	129	118	108	98
370......		230	214	199	184	170	157	145	134	123	112	103	94
375......			204	190	176	163	150	138	127	117	107	98	89
380......			195	181	168	155	143	132	122	112	102	94	85
385......			186	173	160	148	137	126	116	107	98	89	82
390......			178	165	153	142	131	121	111	102	94	86	78
395......			170	158	146	136	125	115	106	98	89	82	75
400......			163	151	140	130	120	110	102	93	86	78	71
405......			156	145	134	124	115	106	97	89	82	75	68
410......			149	139	129	119	110	101	93	86	78	72	65
415......			143	133	123	114	105	97	89	82	75	69	63
420......			137	127	118	109	101	93	86	79	72	66	60
425......			132	122	113	105	97	89	82	76	69	63	58
430......			127	117	109	101	93	86	79	73	66	61	55
435......			121	113	105	97	89	82	76	70	64	58	53
440......			117	108	100	93	86	79	73	67	61	56	51
445......			112	104	96	89	83	76	70	64	59	54	49

FIGURE N
Geostrophic Wind–Isobaric Contour Scale
(After J. C. Bellamy)

NOTE: The dashed latitude lines are to be used with ten times the labeled value of either the spacing or the velocity

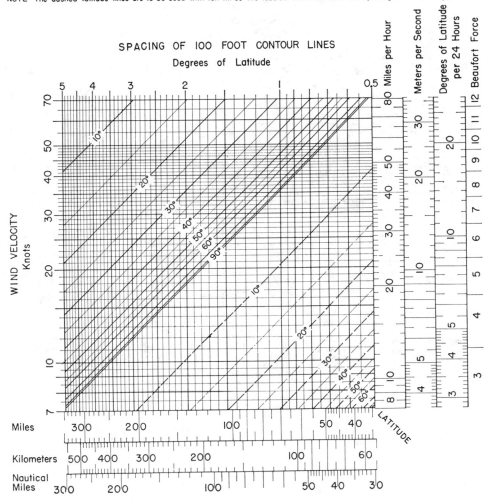

429

FIGURE O

GRADIENT WIND SCALE

(After J. C. Bellamy)

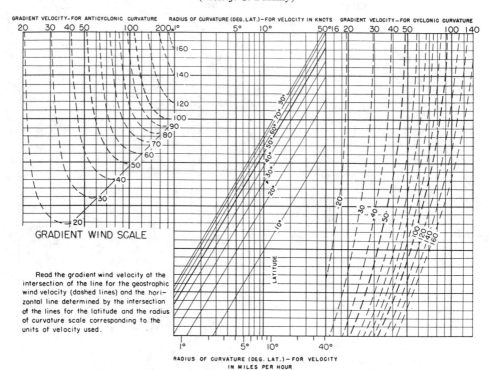

GRADIENT VELOCITY-FOR ANTICYCLONIC CURVATURE   RADIUS OF CURVATURE (DEG. LAT.)-FOR VELOCITY IN KNOTS   GRADIENT VELOCITY-FOR CYCLONIC CURVATURE

GRADIENT WIND SCALE

Read the gradient wind velocity at the intersection of the line for the geostrophic wind velocity (dashed lines) and the horizontal line determined by the intersection of the lines for the latitude and the radius of curvature scale corresponding to the units of velocity used.

RADIUS OF CURVATURE (DEG. LAT.)—FOR VELOCITY
IN MILES PER HOUR

FIGURE P

GEOSTROPHIC WIND SCALE—CONSTANT LEVEL

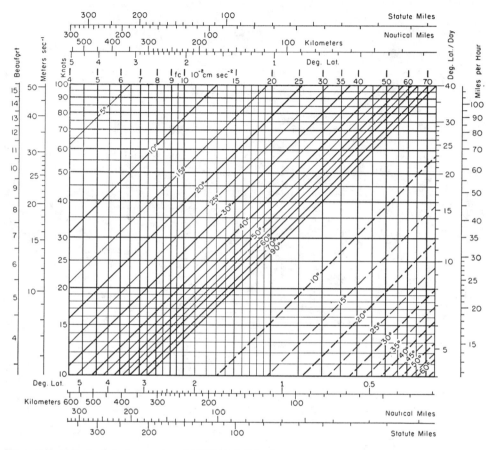

Upper scale is for 2-mb isobar spacing at 10,000 feet (3 km).

Lower scale is for 3-mb isobar spacing at sea level.

(Scales prepared assuming density in U.S. Standard Atmosphere.)

$c = (1/\rho f)(\Delta p/\Delta n);$ $f = 2\omega \sin \phi.$

For finding the geostrophic wind c: Locate the distance between isobars Δn in the appropriate distance scale. Follow the vertical line for fc to the intersection with the slanting line for latitude ϕ. Read the value of the horizontal line through this intersection.

For finding the distance between isobars Δn: Follow the horizontal line for speed to the intersection with the line for latitude, and read the value vertically opposite this point on the distance scale.

(The dashed lines for latitude are to be used after the speed c is divided by 10.)

431

FIGURE Q

LONG-WAVE DIAGRAMS

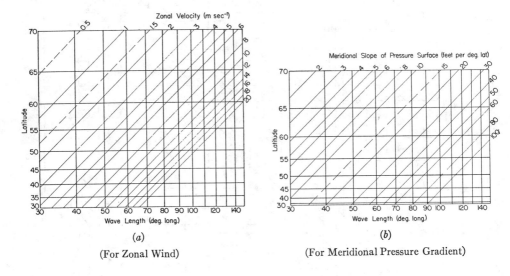

(a)

(For Zonal Wind)

(b)

(For Meridional Pressure Gradient)

Index

A CATALOG OF SELECTED
DOVER BOOKS
IN ALL FIELDS OF INTEREST

A CATALOG OF SELECTED DOVER BOOKS IN ALL FIELDS OF INTEREST

DRAWINGS OF REMBRANDT, edited by Seymour Slive. Updated Lippmann, Hofstede de Groot edition, with definitive scholarly apparatus. All portraits, biblical sketches, landscapes, nudes. Oriental figures, classical studies, together with selection of work by followers. 550 illustrations. Total of 630pp. 9⅛ × 12¼.
21485-0, 21486-9 Pa., Two-vol. set $25.00

GHOST AND HORROR STORIES OF AMBROSE BIERCE, Ambrose Bierce. 24 tales vividly imagined, strangely prophetic, and decades ahead of their time in technical skill: "The Damned Thing," "An Inhabitant of Carcosa," "The Eyes of the Panther," "Moxon's Master," and 20 more. 199pp. 5⅜ × 8½. 20767-6 Pa. $3.95

ETHICAL WRITINGS OF MAIMONIDES, Maimonides. Most significant ethical works of great medieval sage, newly translated for utmost precision, readability. Laws Concerning Character Traits, Eight Chapters, more. 192pp. 5⅜ × 8½.
24522-5 Pa. $4.50

THE EXPLORATION OF THE COLORADO RIVER AND ITS CANYONS, J. W. Powell. Full text of Powell's 1,000-mile expedition down the fabled Colorado in 1869. Superb account of terrain, geology, vegetation, Indians, famine, mutiny, treacherous rapids, mighty canyons, during exploration of last unknown part of continental U.S. 400pp. 5⅜ × 8½. 20094-9 Pa. $6.95

HISTORY OF PHILOSOPHY, Julián Marías. Clearest one-volume history on the market. Every major philosopher and dozens of others, to Existentialism and later. 505pp. 5⅜ × 8½. 21739-6 Pa. $8.50

ALL ABOUT LIGHTNING, Martin A. Uman. Highly readable non-technical survey of nature and causes of lightning, thunderstorms, ball lightning, St. Elmo's Fire, much more. Illustrated. 192pp. 5⅜ × 8½. 25237-X Pa. $5.95

SAILING ALONE AROUND THE WORLD, Captain Joshua Slocum. First man to sail around the world, alone, in small boat. One of great feats of seamanship told in delightful manner. 67 illustrations. 294pp. 5⅜ × 8½. 20326-3 Pa. $4.95

LETTERS AND NOTES ON THE MANNERS, CUSTOMS AND CONDITIONS OF THE NORTH AMERICAN INDIANS, George Catlin. Classic account of life among Plains Indians: ceremonies, hunt, warfare, etc. 312 plates. 572pp. of text. 6⅛ × 9¼. 22118-0, 22119-9 Pa. Two-vol. set $15.90

ALASKA: The Harriman Expedition, 1899, John Burroughs, John Muir, et al. Informative, engrossing accounts of two-month, 9,000-mile expedition. Native peoples, wildlife, forests, geography, salmon industry, glaciers, more. Profusely illustrated. 240 black-and-white line drawings. 124 black-and-white photographs. 3 maps. Index. 576pp. 5⅜ × 8½. 25109-8 Pa. $11.95

ILLUSTRATED DICTIONARY OF HISTORIC ARCHITECTURE, edited by Cyril M. Harris. Extraordinary compendium of clear, concise definitions for over 5,000 important architectural terms complemented by over 2,000 line drawings. Covers full spectrum of architecture from ancient ruins to 20th-century Modernism. Preface. 592pp. 7½ × 9⅜. 24444-X Pa. $14.95

THE NIGHT BEFORE CHRISTMAS, Clement Moore. Full text, and woodcuts from original 1848 book. Also critical, historical material. 19 illustrations. 40pp. 4⅝ × 6. 22797-9 Pa. $2.50

THE LESSON OF JAPANESE ARCHITECTURE: 165 Photographs, Jiro Harada. Memorable gallery of 165 photographs taken in the 1930's of exquisite Japanese homes of the well-to-do and historic buildings. 13 line diagrams. 192pp. 8⅜ × 11¼. 24778-3 Pa. $8.95

THE AUTOBIOGRAPHY OF CHARLES DARWIN AND SELECTED LETTERS, edited by Francis Darwin. The fascinating life of eccentric genius composed of an intimate memoir by Darwin (intended for his children); commentary by his son, Francis; hundreds of fragments from notebooks, journals, papers; and letters to and from Lyell, Hooker, Huxley, Wallace and Henslow. xi + 365pp. 5⅜ × 8. 20479-0 Pa. $5.95

WONDERS OF THE SKY: Observing Rainbows, Comets, Eclipses, the Stars and Other Phenomena, Fred Schaaf. Charming, easy-to-read poetic guide to all manner of celestial events visible to the naked eye. Mock suns, glories, Belt of Venus, more. Illustrated. 299pp. 5¼ × 8¼. 24402-4 Pa. $7.95

BURNHAM'S CELESTIAL HANDBOOK, Robert Burnham, Jr. Thorough guide to the stars beyond our solar system. Exhaustive treatment. Alphabetical by constellation: Andromeda to Cetus in Vol. 1; Chamaeleon to Orion in Vol. 2; and Pavo to Vulpecula in Vol. 3. Hundreds of illustrations. Index in Vol. 3. 2,000pp. 6⅛ × 9¼. 23567-X, 23568-8, 23673-0 Pa., Three-vol. set $37.85

STAR NAMES: Their Lore and Meaning, Richard Hinckley Allen. Fascinating history of names various cultures have given to constellations and literary and folkloristic uses that have been made of stars. Indexes to subjects. Arabic and Greek names. Biblical references. Bibliography. 563pp. 5⅜ × 8½. 21079-0 Pa. $7.95

THIRTY YEARS THAT SHOOK PHYSICS: The Story of Quantum Theory, George Gamow. Lucid, accessible introduction to influential theory of energy and matter. Careful explanations of Dirac's anti-particles, Bohr's model of the atom, much more. 12 plates. Numerous drawings. 240pp. 5⅜ × 8½. 24895-X Pa. $4.95

CHINESE DOMESTIC FURNITURE IN PHOTOGRAPHS AND MEASURED DRAWINGS, Gustav Ecke. A rare volume, now affordably priced for antique collectors, furniture buffs and art historians. Detailed review of styles ranging from early Shang to late Ming. Unabridged republication. 161 black-and-white drawings, photos. Total of 224pp. 8⅜ × 11¼. (Available in U.S. only) 25171-3 Pa. $12.95

VINCENT VAN GOGH: A Biography, Julius Meier-Graefe. Dynamic, penetrating study of artist's life, relationship with brother, Theo, painting techniques, travels, more. Readable, engrossing. 160pp. 5⅜ × 8½. (Available in U.S. only) 25253-1 Pa. $3.95

ILLUSTRATED GUIDE TO SHAKER FURNITURE, Robert Meader. All furniture and appurtenances, with much on unknown local styles. 235 photos. 146pp. 9 × 12. 22819-3 Pa. $7.95

WHALE SHIPS AND WHALING: A Pictorial Survey, George Francis Dow. Over 200 vintage engravings, drawings, photographs of barks, brigs, cutters, other vessels. Also harpoons, lances, whaling guns, many other artifacts. Comprehensive text by foremost authority. 207 black-and-white illustrations. 288pp. 6 × 9. 24808-9 Pa. $8.95

THE BERTRAMS, Anthony Trollope. Powerful portrayal of blind self-will and thwarted ambition includes one of Trollope's most heartrending love stories. 497pp. 5⅜ × 8½. 25119-5 Pa. $8.95

ADVENTURES WITH A HAND LENS, Richard Headstrom. Clearly written guide to observing and studying flowers and grasses, fish scales, moth and insect wings, egg cases, buds, feathers, seeds, leaf scars, moss, molds, ferns, common crystals, etc.—all with an ordinary, inexpensive magnifying glass. 209 exact line drawings aid in your discoveries. 220pp. 5⅜ × 8½. 23330-8 Pa. $4.50

RODIN ON ART AND ARTISTS, Auguste Rodin. Great sculptor's candid, wide-ranging comments on meaning of art; great artists; relation of sculpture to poetry, painting, music; philosophy of life, more. 76 superb black-and-white illustrations of Rodin's sculpture, drawings and prints. 119pp. 8⅜ × 11¼. 24487-3 Pa. $6.95

FIFTY CLASSIC FRENCH FILMS, 1912–1982: A Pictorial Record, Anthony Slide. Memorable stills from Grand Illusion, Beauty and the Beast, Hiroshima, Mon Amour, many more. Credits, plot synopses, reviews, etc. 160pp. 8¼ × 11. 25256-6 Pa. $11.95

THE PRINCIPLES OF PSYCHOLOGY, William James. Famous long course complete, unabridged. Stream of thought, time perception, memory, experimental methods; great work decades ahead of its time. 94 figures. 1,391pp. 5⅜ × 8½. 20381-6, 20382-4 Pa., Two-vol. set $19.90

BODIES IN A BOOKSHOP, R. T. Campbell. Challenging mystery of blackmail and murder with ingenious plot and superbly drawn characters. In the best tradition of British suspense fiction. 192pp. 5⅜ × 8½. 24720-1 Pa. $3.95

CALLAS: PORTRAIT OF A PRIMA DONNA, George Jellinek. Renowned commentator on the musical scene chronicles incredible career and life of the most controversial, fascinating, influential operatic personality of our time. 64 black-and-white photographs. 416pp. 5⅜ × 8¼. 25047-4 Pa. $7.95

GEOMETRY, RELATIVITY AND THE FOURTH DIMENSION, Rudolph Rucker. Exposition of fourth dimension, concepts of relativity as Flatland characters continue adventures. Popular, easily followed yet accurate, profound. 141 illustrations. 133pp. 5⅜ × 8½. 23400-2 Pa. $3.50

HOUSEHOLD STORIES BY THE BROTHERS GRIMM, with pictures by Walter Crane. 53 classic stories—Rumpelstiltskin, Rapunzel, Hansel and Gretel, the Fisherman and his Wife, Snow White, Tom Thumb, Sleeping Beauty, Cinderella, and so much more—lavishly illustrated with original 19th century drawings. 114 illustrations. x + 269pp. 5⅜ × 8½. 21080-4 Pa. $4.50

THE BLUE FAIRY BOOK, Andrew Lang. The first, most famous collection, with many familiar tales: Little Red Riding Hood, Aladdin and the Wonderful Lamp, Puss in Boots, Sleeping Beauty, Hansel and Gretel, Rumpelstiltskin; 37 in all. 138 illustrations. 390pp. 5⅜ × 8½. 21437-0 Pa. $5.95

THE STORY OF THE CHAMPIONS OF THE ROUND TABLE, Howard Pyle. Sir Launcelot, Sir Tristram and Sir Percival in spirited adventures of love and triumph retold in Pyle's inimitable style. 50 drawings, 31 full-page. xviii + 329pp. 6½ × 9¼. 21883-X Pa. $6.95

AUDUBON AND HIS JOURNALS, Maria Audubon. Unmatched two-volume portrait of the great artist, naturalist and author contains his journals, an excellent biography by his granddaughter, expert annotations by the noted ornithologist, Dr. Elliott Coues, and 37 superb illustrations. Total of 1,200pp. 5⅜ × 8.
Vol. I 25143-8 Pa. $8.95
Vol. II 25144-6 Pa. $8.95

GREAT DINOSAUR HUNTERS AND THEIR DISCOVERIES, Edwin H. Colbert. Fascinating, lavishly illustrated chronicle of dinosaur research, 1820's to 1960. Achievements of Cope, Marsh, Brown, Buckland, Mantell, Huxley, many others. 384pp. 5¼ × 8¼. 24701-5 Pa. $6.95

THE TASTEMAKERS, Russell Lynes. Informal, illustrated social history of American taste 1850's–1950's. First popularized categories Highbrow, Lowbrow, Middlebrow. 129 illustrations. New (1979) afterword. 384pp. 6 × 9.
23993-4 Pa. $6.95

DOUBLE CROSS PURPOSES, Ronald A. Knox. A treasure hunt in the Scottish Highlands, an old map, unidentified corpse, surprise discoveries keep reader guessing in this cleverly intricate tale of financial skullduggery. 2 black-and-white maps. 320pp. 5⅜ × 8½. (Available in U.S. only) 25032-6 Pa. $5.95

AUTHENTIC VICTORIAN DECORATION AND ORNAMENTATION IN FULL COLOR: 46 Plates from "Studies in Design," Christopher Dresser. Superb full-color lithographs reproduced from rare original portfolio of a major Victorian designer. 48pp. 9¼ × 12¼. 25083-0 Pa. $7.95

PRIMITIVE ART, Franz Boas. Remains the best text ever prepared on subject, thoroughly discussing Indian, African, Asian, Australian, and, especially, Northern American primitive art. Over 950 illustrations show ceramics, masks, totem poles, weapons, textiles, paintings, much more. 376pp. 5⅜ × 8. 20025-6 Pa. $6.95

SIDELIGHTS ON RELATIVITY, Albert Einstein. Unabridged republication of two lectures delivered by the great physicist in 1920–21. *Ether and Relativity* and *Geometry and Experience.* Elegant ideas in non-mathematical form, accessible to intelligent layman. vi + 56pp. 5⅜ × 8½. 24511-X Pa. $2.95

THE WIT AND HUMOR OF OSCAR WILDE, edited by Alvin Redman. More than 1,000 ripostes, paradoxes, wisecracks: Work is the curse of the drinking classes, I can resist everything except temptation, etc. 258pp. 5⅜ × 8½. 20602-5 Pa. $4.50

ADVENTURES WITH A MICROSCOPE, Richard Headstrom. 59 adventures with clothing fibers, protozoa, ferns and lichens, roots and leaves, much more. 142 illustrations. 232pp. 5⅜ × 8½. 23471-1 Pa. $3.95

THE ART NOUVEAU STYLE BOOK OF ALPHONSE MUCHA: All 72 Plates from "Documents Decoratifs" in Original Color, Alphonse Mucha. Rare copyright-free design portfolio by high priest of Art Nouveau. Jewelry, wallpaper, stained glass, furniture, figure studies, plant and animal motifs, etc. Only complete one-volume edition. 80pp. 9⅜ × 12¼. 24044-4 Pa. $8.95

ANIMALS: 1,419 COPYRIGHT-FREE ILLUSTRATIONS OF MAMMALS, BIRDS, FISH, INSECTS, ETC., edited by Jim Harter. Clear wood engravings present, in extremely lifelike poses, over 1,000 species of animals. One of the most extensive pictorial sourcebooks of its kind. Captions. Index. 284pp. 9 × 12. 23766-4 Pa. $9.95

OBELISTS FLY HIGH, C. Daly King. Masterpiece of American detective fiction, long out of print, involves murder on a 1935 transcontinental flight—"a very thrilling story"—NY Times. Unabridged and unaltered republication of the edition published by William Collins Sons & Co. Ltd., London, 1935. 288pp. 5⅜ × 8½. (Available in U.S. only) 25036-9 Pa. $4.95

VICTORIAN AND EDWARDIAN FASHION: A Photographic Survey, Alison Gernsheim. First fashion history completely illustrated by contemporary photographs. Full text plus 235 photos, 1840–1914, in which many celebrities appear. 240pp. 6½ × 9¼. 24205-6 Pa. $6.00

THE ART OF THE FRENCH ILLUSTRATED BOOK, 1700–1914, Gordon N. Ray. Over 630 superb book illustrations by Fragonard, Delacroix, Daumier, Doré, Grandville, Manet, Mucha, Steinlen, Toulouse-Lautrec and many others. Preface. Introduction. 633 halftones. Indices of artists, authors & titles, binders and provenances. Appendices. Bibliography. 608pp. 8⅜ × 11¼. 25086-5 Pa. $24.95

THE WONDERFUL WIZARD OF OZ, L. Frank Baum. Facsimile in full color of America's finest children's classic. 143 illustrations by W. W. Denslow. 267pp. 5⅜ × 8½. 20691-2 Pa. $5.95

FRONTIERS OF MODERN PHYSICS: New Perspectives on Cosmology, Relativity, Black Holes and Extraterrestrial Intelligence, Tony Rothman, et al. For the intelligent layman. Subjects include: cosmological models of the universe; black holes; the neutrino; the search for extraterrestrial intelligence. Introduction. 46 black-and-white illustrations. 192pp. 5⅜ × 8½. 24587-X Pa. $6.95

THE FRIENDLY STARS, Martha Evans Martin & Donald Howard Menzel. Classic text marshalls the stars together in an engaging, non-technical survey, presenting them as sources of beauty in night sky. 23 illustrations. Foreword. 2 star charts. Index. 147pp. 5⅜ × 8½. 21099-5 Pa. $3.50

FADS AND FALLACIES IN THE NAME OF SCIENCE, Martin Gardner. Fair, witty appraisal of cranks, quacks, and quackeries of science and pseudoscience: hollow earth, Velikovsky, orgone energy, Dianetics, flying saucers, Bridey Murphy, food and medical fads, etc. Revised, expanded In the Name of Science. "A very able and even-tempered presentation."—The New Yorker. 363pp. 5⅜ × 8. 20394-8 Pa. $6.50

ANCIENT EGYPT: ITS CULTURE AND HISTORY, J. E Manchip White. From pre-dynastics through Ptolemies: society, history, political structure, religion, daily life, literature, cultural heritage. 48 plates. 217pp. 5⅜ × 8½. 22548-8 Pa. $4.95

AMERICAN CLIPPER SHIPS: 1833–1858, Octavius T. Howe & Frederick C. Matthews. Fully-illustrated, encyclopedic review of 352 clipper ships from the period of America's greatest maritime supremacy. Introduction. 109 halftones. 5 black-and-white line illustrations. Index. Total of 928pp. 5⅜ × 8½.
25115-2, 25116-0 Pa., Two-vol. set $17.90

TOWARDS A NEW ARCHITECTURE, Le Corbusier. Pioneering manifesto by great architect, near legendary founder of "International School." Technical and aesthetic theories, views on industry, economics, relation of form to function, "mass-production spirit," much more. Profusely illustrated. Unabridged translation of 13th French edition. Introduction by Frederick Etchells. 320pp. 6⅛ × 9¼. (Available in U.S. only)
25023-7 Pa. $8.95

THE BOOK OF KELLS, edited by Blanche Cirker. Inexpensive collection of 32 full-color, full-page plates from the greatest illuminated manuscript of the Middle Ages, painstakingly reproduced from rare facsimile edition. Publisher's Note. Captions. 32pp. 9⅜ × 12¼.
24345-1 Pa. $4.95

BEST SCIENCE FICTION STORIES OF H. G. WELLS, H. G. Wells. Full novel *The Invisible Man*, plus 17 short stories: "The Crystal Egg," "Aepyornis Island," "The Strange Orchid," etc. 303pp. 5⅜ × 8½. (Available in U.S. only)
21531-8 Pa. $4.95

AMERICAN SAILING SHIPS: Their Plans and History, Charles G. Davis. Photos, construction details of schooners, frigates, clippers, other sailcraft of 18th to early 20th centuries—plus entertaining discourse on design, rigging, nautical lore, much more. 137 black-and-white illustrations. 240pp. 6⅛ × 9¼.
24658-2 Pa. $5.95

ENTERTAINING MATHEMATICAL PUZZLES, Martin Gardner. Selection of author's favorite conundrums involving arithmetic, money, speed, etc., with lively commentary. Complete solutions. 112pp. 5⅜ × 8½. 25211-6 Pa. $2.95

THE WILL TO BELIEVE, HUMAN IMMORTALITY, William James. Two books bound together. Effect of irrational on logical, and arguments for human immortality. 402pp. 5⅜ × 8½. 20291-7 Pa. $7.50

THE HAUNTED MONASTERY and THE CHINESE MAZE MURDERS, Robert Van Gulik. 2 full novels by Van Gulik continue adventures of Judge Dee and his companions. An evil Taoist monastery, seemingly supernatural events; overgrown topiary maze that hides strange crimes. Set in 7th-century China. 27 illustrations. 328pp. 5⅜ × 8½. 23502-5 Pa. $5.95

CELEBRATED CASES OF JUDGE DEE (DEE GOONG AN), translated by Robert Van Gulik. Authentic 18th-century Chinese detective novel; Dee and associates solve three interlocked cases. Led to Van Gulik's own stories with same characters. Extensive introduction. 9 illustrations. 237pp. 5⅜ × 8½.
23337-5 Pa. $4.95

Prices subject to change without notice.
Available at your book dealer or write for free catalog to Dept. GI, Dover Publications, Inc., 31 East 2nd St., Mineola, N.Y. 11501. Dover publishes more than 175 books each year on science, elementary and advanced mathematics, biology, music, art, literary history, social sciences and other areas.